GAME THEORY AND RELATED TOPICS

GAME THEORY AND RELATED TOPICS

Proceedings of the Seminar on
Game Theory and Related Topics,
Bonn / Hagen, 26-29 September, 1978

managing editors:

O. MOESCHLIN
Department of Mathematics
University of Hagen

D. PALLASCHKE
Insitute for Applied Mathematics
University of Bonn

1979

NORTH-HOLLAND PUBLISHING COMPANY—AMSTERDAM · NEW YORK · OXFORD

ISBN: 0 444 85342 1

Publishers:

NORTH-HOLLAND PUBLISHING COMPANY
AMSTERDAM • NEW YORK • OXFORD

Sole Distributors for the U.S.A. and Canada:

ELSEVIER NORTH-HOLLAND INC.
52 VANDERBILT AVENUE, NEW YORK, N.Y. 10017

Library of Congress Cataloging in Publication Data

Seminar on Game Theory and Related Topics, Bonn and
Hagen, 1978.
Game theory and related topics.

1. Game theory--Congresses. I. Moeschlin, Otto.
II. Pallaschke, Diethard. III. Title.
QA269.S4 1978 519.3 79-15339
ISBN 0-444-85342-1

PRINTED IN THE NETHERLANDS

PREFACE

The Theory of Games plays an important rôle not only because of its applications in the fields of Statistics, Economics and Political Sciences etc. but also as an independent mathematical discipline. In this connection we emphasize that not only results and methods of other mathematical disciplines have been introduced into the Theory of Games but that it also has influenced the developement of other mathematical disciplines by its problems. Although a considerable number of mathematicians have already found interest in the field of Game Theory it still remains an important objective to make the Theory of Games known and attractive more widely among mathematicians. Besides, there is a constant need to improve the interactions between those who apply Game Theory in their own fields and those who are engaged in the Mathematical Theory of Games.

An International Seminar on Game Theory and Related Topics took place as an activity of the University of Hagen and the Gesellschaft für Mathematik und Datenverarbeitung (GMD), Schloß Birlinghoven, near Bonn, from the 26th - 29th September 1978 at Hagen (26 and 27 September) and at Bonn (28 and 29 September). The GMD is a large-scale research institution of the Federal Republic of Germany and the Land of North Rhine-Westphalia. This research institution has close personal ties with the University of Bonn.

Apart from the one main topic "Game Theory" there were others such as "Fixed-Point- and Optimization Theory", "Measure Theoretic Concepts" and "Mathematical Economics" whereby, of course, the second and the third topic have to be considered as methods of working within "Game Theory" and the related field "Mathematical Economics".

The seminar was held under the honorary chairmanship of Prof.E. Sperner. He also delivered the Inaugural address in which he pointed out in a humurous way and with the acclamation of the audience that it was now fifty years since the "Lemma" was found, the name of which he, Prof. Sperner is bearing. In most of

the lectures new results were presented. Furthermore lectures were given in which distinguished scholars discussed the current state-of-the-art.
The book contains almost all the papers delivered during the seminar. In addition to them a "Problem" is included which was formulated by Prof.K. Jacobs and discussed among the participants of the seminar too.

We wish to express our gratitude to the Minister für Wissenschaft und Forschung of the Land of North Rhine-Westphalia in Düsseldorf, the Gesellschaft für Mathematik und Datenverarbeitung, Schloß Birlinghoven and the Deutscher Akademischer Austauschdienst in Bonn-Bad Godesberg for their financial support, without which the seminar could not have taken place.
We are also indebted to Mr. H. Engel, Ministerialrat im Ministerium für Wissenschaft und Forschung of the Land of North Rhine-Westphalia, to Prof.Dr. Dr.h.c. O. Peters, Rector of the University of Hagen as well as to Mr. R. Bartz, Chancellor of the University of Hagen and his staff.
In the same way we are indebted to Prof.Dr. F. Krückeberg and Dr. M. Flitner, member resp. former member of the Board of Directors of the GMD, the latter is now Chancellor of the University of Hamburg, as well as to the Ministerium für Wissenschaft und Forschung of the Federal Government in Bonn, which is responsible for the GMD.

Special thanks are due to Prof.Dr. Dr.h.c. E. Sperner for his engagement, to the participants of the seminar, to the contributors to this volume and to the members of the Editorial Board for their advice. We appreciate the excellent cooperation of the North-Holland Publishing Company. Last but not least we wish to express our appreciation to the Department of Mathematics of the University of Hagen, the Institute of Mathematics of the Gesellschaft für Mathematik und Datenverarbeitung; especially we are indebted to Dipl.-Math. G. Brentzel, Dr. Chr. Klein, Dr. R. Lorentz, Dipl.-Math. H. Meister, Dipl.-Math. W. Wübbels as well as Mrs. E. Harf and Mrs. G. Soentgen for the trouble they have taken in helping to organize the seminar.

Otto Moeschlin, Diethard Pallaschke

TABLE OF CONTENTS

PART I : GAME THEORY

PART I
GAME THEORY

GAME THEORY AND RELATED TOPICS,
O. Moeschlin, D. Pallaschke (eds.)

CORE-AND KERNEL-VARIANTS BASED ON IMPUTATIONS AND DEMAND PROFILES

W. Albers
Institute of Mathematical Economics
University of Bielefeld

Some solution concepts for real valued characteristic function games are considered, based on imputations, preimputations or demand profiles. (Preimputations are characterized by the condition $x(N)=v(N)$, demand profiles by $x(S)\geq v(S)$ for all $S\subset N$.) For imputations and preimputations the concepts are modifications of the least core (MASCHLER, PELEG, SHAPLEY [1977]), the kernel (DAVIS, MASCHLER [1965]) and the superkernel (a generalization of the kernel), which are obtained by the excess functions $\delta(x,S):=(v(S)-x(S))/|S|$ and $\lambda(x,S):=(v(S)-x(S)/x(S)$. Reasonable concepts for demand profiles (ALBERS [1974], [1979]) correspond to the least core and the superkernel. For the concepts based on preimputations this correspondence is given by projections on the hyperplane of preimputations.

0. Basic Notations.

We consider real valued characteristic function games (N,v), where $N=\{1,2,...,n\}$ is a finite set of players and $v: P(N)\to \mathbb{R}_+$ is a function which assigns a value $v(S)\in \mathbb{R}$, $v(S)\geq 0$ to any coalition $S\subset N$. For mathematical convenience we assume that $v(\emptyset)=0$. For paragraphs 4 - 6 we presume that v is 0-normalized (i. e., $v(\{i\})=0$ for all $i\in N$ and that $v(N)\geq 0$. - We shortly write $x(S)$ for $\Sigma_{i\in S}\ x_i$.

1. Imputations and Demands.

The solution concepts are based on the following types of n-vectors:

1.1 Definition: $x\in \mathbb{R}^N$ is a preimputation, if $x(N)=v(N)$, x is an imputation, if $x(N)=v(N)$ and $x(\{i\})\geq v(\{i\})$ for all $i\in N$, x is a d-profile, if $x(S)\geq v(S)$ for all $S\subset N$.

Preimputations and imputations can be realized as payoffs, but only imputations meet the individual rationality condition $x(\{i\})\geq v(\{i\})$ for all $i\in N$. Preimputations can result in solution concepts based on equal concessions below a reference point $x\in \mathbb{R}^N$ with $x(N)>v(N)$.

If for example in a two-person-game with $v(N)=4$, $v(\{1\})=v(\{2\})=0$ the reference point is (7,1), then the solution with equal concession amounts below this point is (5,-1). If, however, the concessions below the reference point are requested to be proportionally equal to $x_i - v(\{i\})$, then the compromise will in any case be an imputation, here $(3.5,0.5)$.

d-profiles will usually not be realized as outcomes. They result from a fully different approach with the basic assumption that during the bargaining procedure every player obtains an idea of the payoff he should get if he enters a coalition. This payoff should be independent from the coalition and may be interpreted as a demand $d_i \in \mathbb{R}$. The demands of all players give a demand profile $d=(d_1,\ldots d_n)$.

We assume that a coalition $S \subset N$ will be entered only if the demands of all players in S can be satisfied, i. e., if $d(S) \leq v(S)$. The set of all these feasible coalitions corresponding to $d=(d_1,\ldots,d_n)$ is

<u>1.2 Definition:</u> $\mathrm{coal}(v,d) := \{S \subset N \mid d(S) \leq v(S)\}$.

For a player $i \in N$ let

<u>1.3 Definition:</u> $\mathrm{coal}_i(v,d) := \{S \subset N \mid d(S) \leq v(S) \text{ and } i \in S\}$.

From experimental results (SELTEN, SCHUSTER [1968] or [1970], ALBERS [1978]) it is known, that players enter a coalition S only, if their demands can be satisfied and additionally the sum of demands is not essentially less than $v(S)$. A corresponding idea is that players with unequatly low demand levels preferably enter coalitions where high additional payoffs $v(S)-d(S)$ can be realized. Our approach excludes such speculative payoff improvements and assumes that the players in a coalition $S \subset N$ with $d(S)<v(S)$ will raise their demands until $d(S)=v(S)$. d-profiles are the possible results of such bargaining procedures.

Of course, generally d-profiles cannot be realized as payoffs for all players. If for a given demand profile $d=(d_1,\ldots,d_n)$ a feasible coalition $S \in \mathrm{coal}(v,d)$ is entered, then the corresponding payoffs are $x(d,S)$ with the components

$$x_i(d,S) := \begin{cases} d_i & \text{if } i \in S \\ v(\{i\}) & \text{otherwise} \end{cases}.$$

The set of all possible outcomes corresponding to d are

$x(d) := \{x(d,S) \mid S \in coal(v,d)\}$.

Those d-profiles, which can be realized as payoffs define the core.

<u>1.4 Definition:</u> $core(v) := \{x \in \mathbb{R}^N \mid x(N)=v(N), x(S) \geq v(S)$ for all $S \subset N\}$.

2. The d-Core.

A reasonable solution concept for d-profiles is the d-core.

<u>2.1 Definition:</u> $d\text{-}core(v) := \{d \in \mathbb{R}^N \mid d(N)$ minimal s.t. $d(S) \geq v(S)$ for all $S \subset N\}$

This solution concept has been introduced by CROSS [1964] who argued that demand vectors in the d-core require minimal additional payoffs if they shall be realized as payoffs for all players.

From the definition follows immediately that the d-core can be computed by a linear programm, that it is convex and not empty for all v and that $d\text{-}core(v) = core(v)$ if $core(v) \neq \emptyset$. The latter result characterizes the d-core as a generalization of the core.

Another characterization of the d-core is based on the following balance condition for a set $\mathfrak{S}$ of coalitions:

<u>2.2 Definition:</u> $\mathfrak{S} \subset P(N)$ is <u>weakly balanced</u>, if there are weights $\lambda_S \geq 0$ (for all $S \subset N$) such that $\Sigma_{S \in \mathfrak{S}} \lambda_S 1_S = 1_N$. [1][2]

This condition may be explained in the following way: If one interprets the terms $\lambda_S / \Sigma_{T \in \mathfrak{S}} \lambda_T$ as the probability that the coalition S is formed, then any player $i \in N$ has the same chance $1/\Sigma_{S \in \mathfrak{S}} \lambda_S$ to enter a coalition. (This interpretation should be seen as a theoretical explanation. Experimental results show that the probabilities that coalitions $S \in \mathfrak{S} = coal(v,d)$ are formed are usually not propor-

---

[1]) $1_S \in \mathbb{R}^N$ is the indicator function of S, i. e. the vector with the components $(1_S)_i = 1$ if $i \in S$ and $(1_S)_i = 0$ if $i \notin S$.

[2]) A similar balance condition has been used by SHAPLEY [1968] to characterize games with nonempty cores.

tionally to the weights λ_S and depend essentially on the bargaining rules - see ALBERS [1978].)

Applying this balance condition to coal(v,d) we get:

2.3 Theorem: d-core(v)= $\{x$ d-profile | coal(v,d) weakly balanced$\}$[1])

It should be remarked, that by this characterization the d-core concept can be transferred to set-valued characteristic function games and location games (compare ALBERS [1979]).

There are relations to the main simple solution (VON NEUMANN, MORGENSTERN [1944]), since for quota games the quota vector is contained in the d-core.

3. The d-Superkernel.

A different balance condition for $\mathcal{S}$=coal(v,d) which was first formulated by MASCHLER, PELEG [1966] or [1967] in papers on the kernel is

3.1 Definition: $\mathcal{S}\subset P(N)$ is unseparated, if for all $i,j\in N$:
$\mathcal{S}_i\subset\mathcal{S}_j \Rightarrow \mathcal{S}_i=\mathcal{S}_j$ (where $\mathcal{S}_i := \{S\in\mathcal{S} \mid i\in S\}$).

Applying this definition on d-profiles one gets

3.2 Definition: d-superkernel(v): = $\{x\in \mathbb{R}^N \mid x(S)\geq v(S)$ for all $S\subset N$, coal(v,d) nonempty and unseparated$\}$.

In order to interprete the unseparatedness condition we consider a d-profile d and two players $i\neq j\in N$ such that $\text{coal}_i(v,d)\subset\text{coal}_j(v,d)$. Here player i is in so far dependent from j as he can only enter coalitions containing j. If now player j has a feasible coalition $S\in\text{coal}_j(v,d)$ that excludes i , then j can threat player i to enter S unless i reduces his demand in favor of j. Of course i must follow this proposal. This shows that the situation will only be stable, if there is no such coalition S, i. e., if coal(v,d) is unseparated. -

We now give some properties of the d-superkernel.

---

[1]) This characterization was given by TURBAY [1978]. It can also be found in ALBERS [1974].

<u>3.3 Lemma:</u> rint d-core(v) $\subset$ d-superkernel(v) [1])

<u>Proof:</u> If d$\in$rint d-core(v), then there exist weights $\lambda_S>0$ ($S\in$coal(v,d)) such that $\Sigma_{S\in \text{coal}(v,d)}\lambda_S 1_S = 1_N$, from this follows that coal(v,d)$\neq\emptyset$ and that $\text{coal}_i(v,d)\subsetneqq\text{coal}_j(v,d)$ for no pair of players i,j$\in$N, and vice versa.

Since the d-core is nonempy for all v we get from this lemma

<u>3.4 Theorem:</u> d-superkernel(v)$\neq\emptyset$ for all v .

We remark that the d-superkernel need not be convex (see example 7.1. in ALBERS [1974]). However we suggest that the d-superkernel is connected for all games v.

It should be mentioned that the d-core and the d-superkernel will not solve any game in a one-step procedure:

<u>3.5 Example:</u> n=3, v(N)=1.3, v(S)=1 if |S|=2, v({i})=v($\emptyset$)=0 for all i$\in$N.

For this game d-core(v)=d-superkernel(v)={d},where d=(1/2,1/2,1/2). The corresponding feasible coalitions are

coal(v,d)={{1,2},{1,3},{2,3}} and

the corresponding payoffs

x(d)={(1/2,1/2,0),(1/2,0,1/2),(0,1/2,1/2)}.

In any case the sum of payoffs reached after this first step is 1. But if in a second step the players extend their coalition to N, they get the additional payoff .3. - in ALBERS [1979] a solution concept for such multistep procedures is described which may be formulated for the d-core or the d-superkernel. In this article however we only consider the d-core and the d-superkernel corresponding to a single bargaining step.

---

1) rint M is the relative interior of M, i. e. the largest open set contained in M, open with respect to the topology induced by the Euclidean topology to the smallest affine subspace of $\mathbb{R}^N$ that contains M.

4. ε-, δ- and λ-Excess.

A solution concept based on preimputations or imputations must permit $v(S)>x(S)$ for some coalitions $S\subset N$, if it shall define nonempty solution sets for any game v. A solution x with this property will only be stable, if the players agree not to realize possible payoff improvements. The possible improvements can be evaluated by the following excess functions:

4.1 Definitions:

$$\varepsilon(x,S) := v(S)-x(S)$$

$$\delta(x,S) := (v(S)-x(S))/|S| \quad ^{1)}$$

$$\lambda(x,S) := v(S)/x(S) \quad ^{2)}\ ^{3)}$$

The excesses may be seen as measures for the degree of temptation to get $v(S)>x(S)$, where $\varepsilon(x,S)$ is the total improvement, $\delta(x,S)$ gives the possible improvement of one player if the additional payoff is distributed equally, $\lambda(x,S)$ is the relative improvement if the outcome of S is $v(S)$ instead of $x(S)$, of course this improvement can be realized for any player in S, if $v(S)-x(S)$ is distributed proportionally to the payoffs x_i.

5. The ε-Cores.

Within a normative model it seems reasonable that those payoff distributions have the highest stability for which the maximal temptation (evaluated by the excess) is minimal. This consideration gives the following solution concepts depending on the excess function:

5.1 Definition: [pre-]ε-core(v): = $\{x\in \mathbb{R}^N \mid x$ [pre-]imputation, $\max_{S\subset N}\varepsilon(x,S)=\text{minimal}\}$ [4)]

---

1) $|S|$ denotes the number of players in S.

2) If one defines $\lambda(x,S) := v(S)-x(S)/x(S)$ all results hold as well.

3) If not $v(\{i\})=0$ for all $i\in N$, then one must consider the corresponding 0-normalized game or equivalently replace $v(S)$ by $v(S)-\Sigma_{i\in S}v(\{i\})$ and $x(S)$ by $x(S)-\Sigma_{i\in S}v(\{i\})$.

4) The [pre-] may be omitted.

We shall only consider the δ- and λ-versions of this concept which are obtained if one replaces ε by δ or λ.

<u>5.2 Theorem:</u> [pre-]δ-core(v)=[pre-]imputation(v)$\cap\{x-1_N\cdot\Delta \mid$ x d-profile$\}$
where $\Delta := \min_{x[\text{pre-}]\text{imputation}} \max_{S\subset N} \delta(x,S)$.

This means that one obtains the δ-core by shifting the set of d-profiles down in the direction -1_N until the set of [pre-]imputations is touched. The corresponding intersection is the [pre-]δ-core(v). - Before we prove the theorem we first remark what happens with the excess if sets are shifted in this way:

<u>5.3 Remark:</u> If $x\in \mathbb{R}^N$, $\Delta\in\mathbb{R}$ and $\bar{x} := x+1_N\cdot\Delta$,
then $\delta(\bar{x},S)=\delta(x,S)-\Delta$. [1])

<u>Proof</u> (of theorem 5.2):

Let $M :=$ [pre-]imputations(v) $\cap\ \{x-1_N\cdot\Delta \mid x\in\text{d-profile(v)}\}$,
x a preimputation and $\bar{x} := x+1_N\cdot\Delta$.
If $x\in M$, then $\bar{x}$ is a d-profile, which for one $S\subset N$ meets $\bar{x}(S)=v(S)$, hence $\max_{S\subset N}\delta(\bar{x},S)=0$. Applying remark 5.3 we get $\max_{S\subset N}\delta(x,S)=\Delta$.
If however $x\notin M$, then $\bar{x}$ is no d-profile, $v(S)>\bar{x}(S)$ for one $S\subset N$, $\max_{S\subset N}\delta(\bar{x},S)>0$ and $\max_{S\subset N}\delta(x,S)>\Delta$. Thus x is in the [pre-]δ-core, iff $x\in M$.

From this theorem follows that the [pre-]δ-core can also be characterized as the intersection of all nonempty sets of the type $\{x-1_N\cdot\Delta \mid$ x d-profile$\}$ $\cap$ [pre-]imputations(v), $\Delta\in\mathbb{R}$.
These are the cores of the games (N,v_Δ) with

$$v_\Delta(S) := \begin{cases} v(N) & \text{if } S=N \\ v(S)-|S|\cdot\Delta & \text{if } S\neq N \end{cases} \quad , \ \Delta\in\mathbb{R}.$$

The function v_Δ is obtained if one imposes taxes $|S|\cdot\Delta$ in all cases where a coalition $S\neq N$ is formed. The minimal taxe $\tilde{\Delta}$ for

---

1) The corresponding property of λ-excesses is $\lambda(\bar{\bar{x}},S)=\lambda(x,S)/\Delta$, if $\bar{\bar{x}}=x\cdot\Delta$.

which core($v_{\tilde{\Delta}}$) is nonempty defines the δ-core(v)=core($v_{\tilde{\Delta}}$). (Note that $\tilde{\Delta}$ is equal to the value Δ of theorem 5.2.) [1])

The following theorem characterizes the pre-δ-core as a projection of the d-core on the set of preimputations:

<u>5.4 Theorem:</u> pre-δ-core(v)=$\{x-1_N\cdot(x(N)-v(N))/|N| \mid x\in$d-core(v)$\}$

<u>Proof:</u> From the characterization of the d-core as d-profiles with minimal $x(N)$ follows that by shifting the set of d-profiles in the direction -1_N until the set of preimputations is touched, this intersection is given by the shifted d-core. (Note that $\Delta(x)\colon = (x(N)-v(N))/|N|$ is equal for all $x \in$ d-core(v) and gives the (minimal) Euclidian distance between the d-core and the set of preimputations.)

From these theorems we get immediately:

<u>5.5 Corollary:</u>
(1) δ-core(v) = pre-δ-core(v), iff pre-δ-core(v) $\subset$ imputations(v)
(2) δ-core(v) $\subset$ pre-δ-core(v), iff pre-δ-core(v) $\cap$ imputations(v)$\neq\emptyset$
(3) δ-core(v) $\cap$ pre-δ-core(v)=$\emptyset$, iff pre-δ-core(v)$\cap$imputations(v)=$\emptyset$

The following example shows that in fact the δ-core and the pre-δ-core may have a nonempty intersection:

<u>5.6 Example:</u>
Let $n=5$ $(q_1,q_2,q_3,q_4,q_5) = (10,10,10,9,1)$, and v the lowest superadditive function with $v(\{i\})=0$ for all $i\in N$ and $v(\{i,j,k\})=q_i + q_j + q_k$ if i,j,k are pairwise different.

For this game d-core(v)=$\{(10,10,10,9,1)\}$,
pre-δ-core(v)=$\{(8,8,8,7,-1)\}$ and
δ-core(v)=$\{(7.75,7.75,7.75,6.75,0)\}$

---

[1]) A corresponding characterization of the ε-core is given in MASCHLER, PELEG, SHAPLEY [1977]. They use the notation "least core" instead of "ε-core".

Similar results to 5.2, 5.3 and 5.4 are obtained for the [pre-]λ-core. In this case however the projection of the d-core ist a central projection with the center $(v(\{i\}) \mid i\in N)$ instead of an orthogonal projection. [1]) Since such a projection keeps the condition $x(\{i\})\geqq v(\{i\})$, we get that the pre-λ-core and the λ-core are equal:

<u>5.6 Theorem:</u> pre-λ-core(v)=λ-core(v)=$\{x\cdot v(N)/x(N) \mid x\in$ d-core(v)$\}$.

<u>Proof:</u> Similar to that of theorems 5.2 and 5.4.

6. Kernels and Superkernels.

Similar relations as those between the pre-δ-core, pre-λ-core and the d-core are obtained by corresponding kernel-variants.

<u>6.1 Definition:</u> The pre-ε-kernel(v) is the set of all preimputations for which for all $i,j\in N$:

$$\max_{S\subset N,i\in S,j\notin S}\varepsilon(x,S) = \max_{S\subset N,j\in S,i\notin S}\varepsilon(x,S).$$

If one considers imputations instead of preimputations one has to exclude the players with $x_i=v(\{i\})$ from the max-condition:

<u>6.2 Definition:</u> The ε-kernel(v) is the set of all imputations for which for all $i,j\in N$:

$$\max_{S\subset N,i\in S,j\notin S}\varepsilon(x,S) = \max_{S\subset N,j\in S,i\notin S}\varepsilon(x,S)$$

or $x_i= v(\{i\})$ or $x_j= v(\{j\})$

One immediately gets

<u>6.3 Lemma:</u> x pre-ε-kernel(v) ⇒ for all $i,j\in N$

$$\max_{S\subset N,i\in S}\varepsilon(x,S) = \max_{S\subset N,j\in S}\varepsilon(x,S) .$$

(For the ε-kernel the max-equation only holds for players with $x_i>v(\{i\})$.)

Let for any $x\in \mathbb{R}^N$, $i\in N$:

<u>6.4 Definition:</u> $\text{coal}^\varepsilon(v,x) := \{S\subset N \mid \varepsilon(x,S)=\max_{S\subset N}\varepsilon(x,S)\}$

and $\text{coal}_i^\varepsilon(v,x) := \{S\subset N \mid i\in S, \varepsilon(x,S)=\max_{i\in S\subset N}\varepsilon(x,S)\}$.

---

[1]) Compare the footnote to remark 5.3.

Then we get from Lemma 6.3

6.5 Remark: For any $x \in$ pre-ε-kernel(v) and for all $i \in N$
$coal_i^{\varepsilon}(v,x) = \{S \in coal^{\varepsilon}(v,x) \mid i \in S\}$.

Moreover

6.6 Lemma: For any $x \in$ pre-ε-kernel(v) the set $coal^{\varepsilon}(v,x)$ is unseparated.

Thus the pre-ε-kernel is included in the following pre-ε-superkernel:

6.7 Definition:
pre-ε-superkernel(v) = $\{x \in \mathbb{R}^N \mid x$ preimputation and $coal^{\varepsilon}(v,x)$ unseparated$\}$.

The difference between the pre-ε-superkernel and the pre-ε-kernel is that the latter has an additional condition for players $i,j \in N$ with

$$coal_i^{\varepsilon}(v,x) = coal_j^{\varepsilon}(v,x).$$

We call such two players unseparated by x.

6.8 Lemma: If $x \in \mathbb{R}^N$ and $coal^{\varepsilon}(v,x)$ is unseparated, then there is a unique partition $P(x) := \{S_1,\ldots,S_r\}$ of N such that

(1) $i,j \in S_k \in P(x) \Rightarrow i,j$ unseparated by x

(2) $i \in S_k \in P(x)$, $j \in S_l \in P(x)$, $S_k \neq S_l \Rightarrow i$ and j are "separated by x" (i. e., $coal_i^{\varepsilon}(v,x) \neq coal_j^{\varepsilon}(v,x)$).

While the [pre-]ε-superkernel only gives a common payoff to the players within a set $S_k \in P(x)$, which may be distributed within a certain range, the [pre-]ε-kernel also considers the excesses $\max_{S \subset N, i \in S, j \notin S} \varepsilon(x,S)$ and $\max_{S \subset N, j \in S, i \notin S} \varepsilon(x,S)$, which must be equal for all $i,j \in S_k$. -

The following example shows that by this condition for a given partition P of N and payoffs $\bar{x}(S_i) \in \mathbb{R}$, $(S_i \in P)$, there need not be singled out at most one solution x in the [pre-]ε-kernel with $P(x)=P$ and $x(S_i)=\bar{x}(S_i)$ for all $S_i \in P$:

6.9 Example: $n=5$

$$v(S) = \begin{cases} 40 & \text{if } S \supset \{1,2,3,4\} \text{ or } S \supset \{1,2,3\} \text{ or } S \supset \{3,4,5\} \\ 15 & \text{if } S \supset \{1,3\} \text{ or } S \supset \{2,4\} \\ 0 & \text{otherwise} \end{cases}$$

For this game the [pre-]ε-kernel(v) = conv{(5,15,15,5,20), (15,5,5,15,20)} . [1])

The same results (except for the solution of example 6.9) hold if one replaces every ε by δ or λ respectively. For δ and λ we give some additional results:

<u>6.10 Theorem:</u> pre-δ-superkernel(v)=

$$\{x-1_N\cdot(x(N)-v(N))/|N| \mid x\in\text{d-superkernel}(v)\}.$$

<u>Proof:</u> The proof is obvious, if one remembers from remark 5.3 that by a projection $x \to \bar{x}=x-1_N\cdot\Delta$ all δ-excesses are changed equally to $\delta(\bar{x},S)=\delta(x,S)+\Delta$, such that the sets with maximal excesses are unchanged.

This theorem says that the pre-δ-superkernel can be obtained from the d-superkernel as the pre-δ-core from the d-core by orthogonal projection of the d-superkernel.
Similarly we get the pre-λ-superkernel from the d-superkernel by central projection to the point $(v(\{i\}) \mid i\in N)$, which is the origin if $v(\{i\})=0$ for all $i\in N$. Since the condition $x_i \geq v(\{i\})$ is kept by this projection and is met by all elements of the d-superkernel, the pre-λ-superkernel is a subset of the imputations and therefore equal to the λ-superkernel:

<u>6.11 Theorem:</u> λ-superkernel(v) = pre-λ-superkernel(v) =

$$\{x\cdot v(N)/x(N) \mid x\in\text{d-superkernel}(v)\} .$$

<u>Proof:</u> Similar to that of theorem 6.10.

We do not consider the shape of the δ-superkernel in cases where it is different from the pre-δ-superkernel. That this may happen can again be seen from the game given in example 5.6. There the d-superkernel ist equal to the d-core and thus the pre-δ-superkernel is equal to the pre-δ-core which is not contained in the set of imputations and thereby disjoint to the δ-superkernel.

<u>7. On the Attractivity of the Different Solution Concepts.</u>

It is hard to say which of the solution concepts given above should generally be preferred to another. May be for any concept there is a

---

1) A unique solution can be obtained if one considers not only the second largest excess but all excesses in a lexicographic order. But such an approach involves normative problems.

special situation in which it is useful. - Nevertheless the reader should see the essential difference between d-profile concepts which try to analyse bargaining situations with different possible outcomes and the corresponding imputation or preimputation concepts which - except the case that the outcomes are in the core - are compromises in one or another way.

Let us first consider the concepts based on d-profiles. Here it seems reasonable to restrict ones considerations to profiles in the d-superkernel. An example shows that by a further restriction to the d-core essential alternatives may be omitted:

<u>7.1 Example:</u> $n=7$,

$$v(K) = \begin{cases} 1 & \text{if } |K\cap\{1,2,3\}|\geq 2 \text{ and } |K\cap\{4,5,6,7\}|\geq 3 \\ 0 & \text{otherwise} \end{cases}$$

For this game the d-superkernel is the convex closure of $\{x,y\}$ and d-core$(v)=\{x\}$, where $x=(0,0,0,1/3,1/3,1/3,1/3)$ and $y=(1/2,1/2,1/2,0,0,0,0)$. Since any winning coalition requires a simple majority of both sets $\{1,2,3\}$ and $\{4,5,6,7\}$ it seems reasonable that both of these groups should be treated somehow similarly. This is the case in the d-superkernel, but not in the d-core. [1])
On the other hand one may argue that, if a state would want to modify the game v by imposing a tax on the formation of any coalition different from N (compare the considerations following remark 5.3), then the lowest tax to get a nonempty core for the new game v_Δ would be $\Delta = 1/4$, and the corresponding core-solution would be core$(v_\Delta) = \{(0,0,0,1/4,1/4,1/4,1/4)\}$ which corresponds to the d-core. It should however be seen that this explanation of the d-core essentially changes the given game v. (Note however, that the problem of unequately low payoffs to players 1,2,3 can be excluded if the game is analysed via the power instead of the original characteristic function (see MASCHLER [1964] , which gives the value of 1/2 or 3/7 to the coalition $\{1,2,3\}$ and 1/2 or 4/7 to $\{4,5,6,7\}$.)
A second question is which of the concepts based on imputations should be preferred. From a mathematical point of view one could prefer the pre-δ-versions to the δ-versions because they are more con-

[1]) Note that this disadvantage of the d-core similarly holds for the core, since, if in example 7.1 one changes $v(N)$ to 4/3, one gets core$(v) = \{(0,0,0,1/3,1/3,1/3,1/3)\}$.

venient. They however give results where players have to agree to outcomes $x_i<v(\{i\})$. This seems unreasonable, but again the story from above makes such a result reasonable as long as x_i is not less than the imposed tax.

On the other hand the δ-core and the δ-superkernel may lead to solutions $x_i = v(\{i\})$ for some players and for other players not. This seems in some sense irrational, if the solution concept is based on the idea that the adequate demands of players are given by the d-concept and one obtains corresponding payoff solutions by numerically equal concessions of all players. It does not make sense that within such an agreement of all players one player gets $x_i = v(\{i\}) = 0$, while others get $x_j >> v(\{j\})$. From this point of view the λ-versions should be preferred.

These arguments lead us to a slight preference of the superkernel-versions to the core-versions and of the λ-versions to the δ- and ε-versions.

LITERATURE

ALBERS, W.: Zwei Lösungskonzepte für kooperative Mehrpersonenspiele, die auf Anspruchsniveaus der Spieler basieren. OR-Verfahren/Methods of Operations Research XVIII, p. 1-8, Meisenheim/Glan 1974.

ALBERS, W.: Bloc Forming Tendencies as Characteristics of the Bargaining Behavior in Different Versions of Apex-Games. In: H. Sauermann (ed.): Coalition Forming Behavior, p. 172-203, Tübingen 1978.

ALBERS, W.: Grundzüge einiger Lösungskonzepte, die auf Forderungsniveaus der Spieler basieren. In: W. Albers, G. Bamberg, R. Selten (ed.): Entscheidungen in kleinen Gruppen, p. 9-37, Meisenheim/Glan 1979.

DAVIS, M. and MASCHLER, M.: The Kernel of a Cooperative Game. Naval Research Logistics Quart. 12, 223-259, 1965.

MASCHLER, M.: The Power of a Coalition. Management Science, Vol. 10, 8-29, 1963.

MASCHLER, M. and PELEG, B.: A Characterization, Existence Proof and Dimension Bounds for the Kernel of a Game. Pacific J. of Math. 18, 289-328, 1966.

MASCHLER, M., PELEG, B. and SHAPLEY, L.S.: Geometric Properties of the Kernel, Nucleolus and Related Solution Concepts. P-6027, The Rand Corporation, Santa Monica/Cal. 1977.

von NEUMANN, J. and MORGENSTERN, O.: Theory of Games and Economic Behavior, Princeton 1944.

SELTEN, R. and SCHUSTER, K.G.: Psychological Variables and Coalition Forming Behavior. In: K.Borch and J. Mossin (eds.): Risk and Uncertainty, p. 221-240, London 1968.

SELTEN, R. and SCHUSTER, K.G.: Psychologische Faktoren bei Koalitionsverhandlungen. In: H. Sauermann (ed.): Beiträge zur experimentellen Wirtschaftsforschung, Vol. II, Tübingen 1970.

SHAPLEY, L.S.: On Balanced Sets and Cores. Naval Research Logistics Quart. 14, 453-460, 1967.

SHAPLEY, L.S. and SHUBIK, M.: The Core of an Economy with Nonconvex Preferences. RM-3518, The Rand Corporation, Santa Monica/Cal. 1963.

SHAPLEY, L.S. and SHUBIK, M.: Quasi-Cores in a Monetary Economy with Nonconvex Preferences. Econometrica 34, p. 805-827, 1966.

GAME THEORY AND RELATED TOPICS
O. Moeschlin, D. Pallaschke (eds.)

BAYESIAN GAME THEORY

W. Armbruster, W. Böge
Institute of Applied Mathematics, University of Heidelberg

The definition of solution concepts for n-person games with unknown utility functions involves well-known technical and conceptual difficulties if the set of utility functions is non-finite, or if the players' marginal probability distributions over opponent utilities do not satisfy Harsanyi's consistency assumption. These difficulties, which reflect inadequacies of classical game theory, are avoided by means of a "Bayesian" game theory, in which suitable canonical representations for the players' subjective probability measures are constructed. The set of n-tuples of subjective probability measures which are possible in the game, if each player optimizes his expected utility and knows the probability measures of his opponents, turns out to be a suitable generalization of equilibrium points.

1. Motivation

I am interested in three applications of game theory: 1) new democratic voting procedures, 2) economic models, 3) the synthesis of both. The question with 1) is to find a procedure which, chance steps being allowed, leads in every case at least approximately to pareto-optimal voting results and is invariant with respect to permutations of the committee members. No such procedure is known so far. The diploma theses of some students of mine were concerned with this question, and recently we found such a procedure that works when the utility function of each player lies outside a low dimensional exception space and is known to all players. To find a procedure that works without this latter assumption, i.e. when the utility functions are unknown to the other players -as is usually the case in politics- seems to be a very difficult problem. The idea is to look for procedures that give the players the possibility of learning enough about the utility functions of the others. But first one has to give a foundation of game theory with unknown utility functions, which I did several years ago, from a Bayesian point of view. Several students working on voting procedures have used this foundation; others have further developed the theory. The learning behavior of the players has also been examined for special repeated games, particularly in the Ph. D. thesis of W. Armbruster.

The Bayesian assumption means, that someone for whom it is essential to know which of several alternatives is true, will have at least a subjective probability distribution on these alternatives. It seems to me that this is the only way of getting a consistent theory. The objection, that appropriate subjective a-priori distributions are difficult to determine, is not relevant in many of our applications. For example, we look for a single voting procedure that works for all subjective a-priori distributions that the players may have. In other cases, subjective probabilities are just results of our applications, as in example 7.1 of this paper.

For 2) some students are working on differential games with known utility functions. However, even in very simple situations we are faced with unsolved analytical problems -equivalent to problems in partial differential equations- when attempting to prove the existence and uniqueness of equilibrium points. When these problems are solved, we intend to study differential games with unknown utility functions as well, using the Bayesian foundation.

So far I have hesitated to publish anything about Bayesian game theory, since a simplification of the theory seemed desirable, and since its relationship to classical game theory was not apparent. Fortunately, through his thesis problem, Mr. Armbruster was led to such a relationship and to a convenient exposition of the theory, both of which he will now present:

The basic result of Bayesian decision theory, that the personal preference pattern of the decision maker can be parametrized by a subjective utility function and a subjective probability measure, will be accepted as an axiom for the purpose of this paper. For details, the work of W. Böge [3],[7], or L.J. Savage [8] should be consulted. Bayesian game theory, which is an application of this result to n-person games, will be briefly outlined following the approach of W. Böge with some minor modifications. An equilibrium concept will be defined in terms of this new theory and compared with classical equilibrium concepts. The proofs of most theorems must be omitted for lack of space, but can be found in a manuscript [1] of mine, which I hope to publish in the near future.

2. Notation

If X is a compact space (Hausdorff) $C(X)$ denotes the space of continous real valued functions on X with the topology of uniform convergence. $C'(X)$ denotes its dual space with the weak* topology. $W(X)$, the set of regular Borel probability measures on X, is a compact subset of $C'(X)$. If $x \in X$, let $\hat{x} \in W(X)$ denote the probability measure supported by $\{x\}$, that is $\hat{x}(f) = f(x)$ for $f \in C(X)$.
If Y is compact, and $\varphi: X \to Y$ is continuous, define the continuous linear map $o\varphi: C(Y) \to C(X)$ as $o\varphi(g) = go\varphi$. Defining $oo\varphi$ as the restriction of $o(o\varphi)$ to $W(X)$, we have $oo\varphi(P)(g) = P(go\varphi)$, for all $g \in C(Y)$ and $P \in W(X)$. $oo\varphi(P)$, which is also written $Poo\varphi$, is just the image measure of P with respect to φ.
For $1 \le p \le n$, let $\hat{p}$ and $\hat{\not p}$ denote the projection maps from some Cartesion product $A_1 \times .. \times A_n$ onto A_p and onto $\prod_{q \neq p} A_q$ respectively. The actual domain of these two maps will not always be specified, since this will be evident from the context.Finally, for arbitrary sets and mappings $f_i: A \to B_i$ and $g_i: A_i \to B_i$ $(i=1,..,n)$,let $(f_1,..,f_n)$ denote the map $a \mapsto (f_1(a),..,f_n(a))$, and define $\overset{n}{\underset{i=1}{\times}} g_i(a_1,..,a_n) := (g_1(a_1),..,g_n(a_n))$.

3. Bayesian principle

The consequences which result when a person makes a decision or follows some course of action h, usually depend not only on h, but also on unknown and uncontrollable external influences. Suppose that the set of possible actions H, the set of consequences C, and the set of external influences or "states of the world" S, are compact, and that the consequence function, $c: H \times S \to C$, describing how the consequences depend on the action chosen and on the actual state of the world, is continuous. We shall say that a person p, described by H,S,C,c, acts according to the *Bayesian principle of choice*, if there exists a continuous, real valued function $u: C \to \mathbb{R}$, called his (subjective) utility function, and a measure $P \in W(S)$, called his (subjective) probability measure, s.t. p chooses an act h only if $N(h) = \max_{h' \in H} N(h')$, where $N(h) = \int_S u(c(h,s))\, P(ds)$

4. Games with unknown utility functions

If, for a given n-person game we wish to say that each player acts according to the Bayesian principle, we must first define compact sets and continuous maps H_p, S_p, C_p, and c_p for $p=1,..,n$. Actions $h_p \in H_p$ shall be defined as pure strategies in the normal form of the game. Thus each player chooses his strategy before observing any of the actual moves occurring in the game; in particular, his strategy choice is independent of the chance moves $z \in Z$. Consequences are defined as outcomes of the game (e.g. endpoints of the "game tree") and are the same for all players. The dependence of consequences on the strategies and chance moves will be denoted by $\lambda: Z \times H_1 \times .. \times H_n \to C$. It will always be assumed that $Z, H_1, .., H_n, C$ are compact and that λ is some given continuous function. In order to define the consequence function $c_p: H_p \times S_p \to C$, a compact set S_p and continuous functions ζ_p, χ_p will be constructed below, satisfying $\zeta_p: S_p \to Z$ and $\chi_p: S_p \to \prod_{q \neq p} H_q$. Hence c_p may be suitably defined as $\lambda \circ (id/H_p \times (\zeta_p, \chi_p))$, where id/H_p denotes the identity map on H_p. If $u_p: C \to \mathbb{R}$ is player p's utility function and $P_p \in W(S_p)$ his probability measure, his expected utility of h_p is

$$N(h_p) = \int_{S_p} u_p(\lambda(h_p, (\zeta_p, \chi_p)(s_p)))\, P_p(ds_p)$$

$$= \int_{Z \times \prod_{q \neq p} H_q} u_p(\lambda(z, h_1, .., h_n))\, P_p \circ\circ (\zeta_p, \chi_p)(d(z, h_1, .. \not{p} .., h_n))$$

(1) $$=: N(h_p, u_p \circ \lambda, P_p \circ\circ (\zeta_p, \chi_p)).$$

We shall not assume that u_p is given, but that it is contained in some given compact set U_p of continuous, real-valued functions. Each player p regards the probability measure and the utility function of player q $(q \neq p)$, as random variables, taking values in $W(S_q)$ and U_q respectively. What remains to be shown is the existence of compact sets S_p and continuous functions ζ_p, $\chi_p (p=1,..,n)$ mentioned above, as well as the existence of random variables υ_p, π_p, such that $\upsilon_p: S_p \to \prod_{q \neq p} U_q$ and $\pi_p: S_p \to \prod_{q \neq p} W(S_q)$.

The following definition and subsequent construction originate from a lecture on game theory given by W. Böge in 1970.

4.1 Definition: Let $S_1^o, .., S_n^o$ be compact sets. An n-tuple of compact sets and continuous maps $(S_1, .., S_n, \rho_1, .., \rho_n)$ is called an *oracle system* for $S_1^o, .., S_n^o$ if for all p $\rho_p: S_p \to S_p^o \times \prod_{q \neq p} W(S_q)$.

4.2 Theorem: Such oracle systems exist

Proof: For i=1,2,... define $S_p^i := S_p^o \times \prod_{q \neq p} W(S_q^{i-1})$. Let ${}^1\varphi_p^o$ be the projection map from S_p^1 onto S_p^o and define ${}^{i+1}\varphi_p^i := id/S_p^o \times \mathop{\times}_{q \neq p} \circ\circ\, {}^i\varphi_q^{i-1}$ for i=1,2,... (id/S_p^o = identity map on S_p^o). Let ${}^i\varphi_p^i := id/S_p^i$, and ${}^j\varphi_p^i := {}^{i+1}\varphi_p^i \circ \ldots \circ {}^{j-1}\varphi_p^{j-2} \circ {}^j\varphi_p^{j-1}$ for $j > i$. Each S_p^i is compact and ${}^j\varphi_p^i : S_p^j \to S_p^i$ is continuous for all $j \geq i \geq 0$ and $p=1,..,n$. $(S_p^i, {}^j\varphi_p^i)$ is a projective system in the category of compact sets and continuous maps, and its projective limit, (S_p, φ_p^i) exists. We also write $S_p = \lim S_p^i$. In the same category, $(W(S_p^i), \circ\circ\, {}^j\varphi_p^i)$ is a projective system and the spaces $\lim W(S_p^i)$ and $W(\lim S_p^i)$ are homeomorph. Since $\lim(S_p^o \times \prod_{q \neq p} W(S_q^i))$ and $S_p^o \times \prod_{q \neq p} \lim W(S_q^i)$ are also homeomorph, there exist homeomorphisms $\rho_1, .., \rho_n$, such that $\rho_p: S_p \to S_p^o \times \prod_{q \neq p} W(S_q)$ for all p.

$(S_1, .., S_n, \rho_1, .., \rho_n)$ as constructed above will be called the *canonical oracle system* for $S_1^o, .., S_n^o$. It has the following important property: given any other oracle system $(S'_1, .., S'_n, \rho'_1, .., \rho'_n)$ for $S_1^o, .., S_n^o$, there exists a unique n-tuple $(\Psi'_1, .., \Psi'_n)$ of continuous maps $S'_p \to S_p$ such that $\rho_p \circ \Psi'_p = (id/S_p^o \times \mathop{\times}_{q \neq p} \circ\circ\, \Psi'_q) \circ \rho'_p$ for all p. In other words, the canonical oracle system is a terminal object in the category of oracle systems and n-tuples of commuting maps. As a result many theorems, proved only for the canonical oracle system, can immediately be generalized to arbitrary oracle systems.

Returning now to the n-person game, let $S_p^o = Z \times \prod_{q \neq p}(H_q \times U_q)$ and let $(S_1,..,S_n,\rho_1,..,\rho_n)$ be an oracle system for $S_1^o,..,S_n^o$. The set of external influences for player p can now be defined as S_p, and the components of $\rho_p : S_p \to Z \times \prod_{q \neq p} H_q \times \prod_{q \neq p} U_q \times \prod_{q \neq p} W(S_q)$ are the required functions $\zeta_p, \chi_p, \upsilon_p, \pi_p$.

Although the oracle system for $S_1^o,..,S_n^o$ defined above allows us to represent the subjective probability measures of the players, and hence to state the Bayesian principle for n-person games, it is sometimes more convenient to use somewhat different constructions, which also serve this objective. Two such possibilities will now be described.

(i) Since the "payoff function" of a classical game in normal form does not depend on the parameter z, we define the continuous function $\alpha_p : U_p \times W(Z) \to C(H_1 \times \ldots \times H_n)$ as follows

$$\alpha_p(u_p,Q)(h_1,..,h_n) = \int_Z u_p(\lambda(z,h_1,..,h_n))\, Q(dz).$$

The set of "utility tables" for player p, $T_p := \alpha_p(U_p \times W(Z))$ is compact. Define $S_p^o := \prod_{q \neq p}(H_q \times T_q)$, and let $(S_1,..,S_n,\rho_1,..,\rho_n)$ denote the canonical oracle system for $S_1^o,..,S_n^o$. Write $\rho_p =: (\chi_p, \tau_p, \pi_p)$. Let $(S_1',..,S_n',\rho_1',..,\rho_n')$, $\rho_p' =: (\zeta_p', \chi_p', \upsilon_p', \pi_p')$, be the canonical oracle system for $S_1^{o\prime},..,S_n^{o\prime}$, where $S_p^{o\prime} := Z \times \prod_{q \neq p}(H_q \times U_q)$.

4.3 Lemma: There exists a continuous map $\beta_p : S_p' \to S_p$, such that $\chi_p \circ \beta_p = \chi_p'$.

The proof is straightforward. Let us now assume, each player believes that the chance moves of the game are independent of the strategy choices of his opponents. Hence, if $P_p' \in W(S_p')$ is player p's probability measure, $P_p' \circ\circ (\zeta_p', \chi_p') = P_p' \circ\circ \zeta_p' \otimes P_p' \circ\circ \chi_p'$.

4.4 Theorem: a) Player p acts according to the Bayesian principle iff b) there exists some $t_p \in T_p$ and $P_p \in W(S_p)$, such that p chooses h_p only if this strategy maximizes

$$N(h_p,t_p,P_p \circ\circ \chi_p) := \int_{\prod_{q \neq p} H_q} t_p(h_1,..,h_n)\, P_p \circ\circ \chi_p(d(h_1,..\not{p}..,h_n)).$$

Proof: 'a)' means that some $u_p \in U_p$ and $P_p' \in W(S_p')$ exist, such that p chooses h_p only if it maximizes $N(h_p, u_p \circ \lambda, P_p' \circ\circ(\zeta_p', \chi_p'))$ as in (1).

Define β_p as in 4.3. Let $P_p := P_p' \circ\circ \beta_p$, and $t_p := \alpha_p(u_p, P_p' \circ\circ \zeta_p')$. Since $P_p \circ\circ \chi_p = P_p' \circ\circ \chi_p'$, $N(h_p, u_p \circ \lambda, P_p' \circ\circ(\zeta_p', \chi_p')) = N(h_p, u_p \circ \lambda, P_p' \circ\circ \zeta_p' \otimes P_p \circ\circ \chi_p) = N(h_p, t_p, P_p \circ\circ \chi_p)$, proving b). Conversely, suppose for some t_p, P_p p maximizes $N(h_p,t_p,P_p \circ\circ \chi_p)$. There exist $u_p \in U_p$, $Q \in W(Z)$, such that $\alpha_p(u_p,Q) = t_p$. Since $(\zeta_p', \chi_p') : S_p' \to Z \times \prod_{q \neq p} H_q$ is surjective, some $P_p' \in W(S_p')$ exists, such that $P_p' \circ\circ(\zeta_p', \chi_p') = Q \otimes P_p \circ\circ \chi_p$. Again $N(h_p, u_p \circ \lambda, P_p' \circ\circ(\zeta_p', \chi_p')) = N(h_p, t_p, P_p \circ\circ \chi_p)$, proving a).

Because of this result, each $P_p \in W(S_p)$ will also be called a (subjective) probability measure and $t_p \in T_p$ a (subjective) utility function of player p.

(ii) Since each player knows his own strategy choice, utility function and probability measure, it is not necessary to consider these as random variables from the player's own point of view. Nevertheless, it is sometimes convenient to do so. In this case, state spaces $S_1',\ldots,S_n'$, and random variables $\rho_1',\ldots,\rho_n'$, must be defined satisfying $\rho_p'\colon S_p' \to \prod_{p=1}^{n}(H_p\times T_p\times W(S_p'))$.

4.5 Definition: Let S^o be a compact set. A compact set S together with a continuous map $\rho\colon S \to S^o\times W(S)^n$ is called a multioracle for S^o.

The *canonical multioracle* for S^o can be defined using the sets $S^{i+1} = S^o\times W(S^i)^n$, and the mappings ${}^1\varphi^o\colon S^1 \to S^o$ (projection map) and ${}^{i+1}\varphi^i := \mathrm{id}/S^o\times(\mathrm{oo}\,{}^i\varphi^{i-1})^n$. Defining $(S,\varphi^i) = \lim(S^i,{}^j\varphi^i)$, a homeomorphism $\rho\colon S \to S^o\times W(S)^n$ can be shown to exist, as in 4.2. Letting (S,ρ) be some multioracle for $S^o := \prod_{p=1}^{n}(H_p\times T_p)$, the spaces and maps S_p',ρ_p' above may be defined as $S_p'=S$, and $\rho_p'=\rho$, for all p. The relationship between this multioracle and the canonical oracle system for $S_p^o = \prod_{q\neq p}(H_q\times T_q)$ is given by

4.6 Theorem: There exists a unique n-tuple $(\Psi_1,\ldots,\Psi_n)$ of continuous maps, such that for all p the following diagramm is commutative

$$\begin{array}{ccccc}
S & & \xrightarrow{\quad\quad\quad\Psi_p\quad\quad\quad} & & S_p \\
\downarrow\rho & & & & \downarrow\rho_p \\
S^o\times W(S)^n & \xrightarrow{\mathrm{id}/S^o\times\overset{n}{\underset{p=1}{\times}}\mathrm{oo}\Psi_p} & S^o\times\prod_{p=1}^{n}W(S_p) & \xrightarrow{\not{p}} & S_p^o\times\prod_{q\neq p}W(S_q)
\end{array}$$

4.7 Theorem: Player p acts according to the Bayesian principle iff there exist $t_p\in T_p$, $P_p\in W(S)$, such that p chooses h_p only if this strategy maximizes

$$N(h_p,t_p,P_p\mathrm{oo}\Psi_p\mathrm{oo}\chi_p) := \int_{\prod_{q\neq p}H_q} t_p(h_1,\ldots,h_n)\,P_p\mathrm{oo}\Psi_p\mathrm{oo}\chi_p(d(h_1,\ldots\not{p}\ldots,h_n)).$$

For the remainder of this paper, H_p and $T_p\subset C(H_1\times\ldots\times H_n)$ will denote given compact sets. S_p^o will mean $\prod_{q\neq p}(H_q\times T_q)$, and $S^o=\prod_{p=1}^{n}(H_p\times T_p)$. (S,ρ) will refer to the canonical multioracle for S^o, $\rho=(\chi,\tau,\pi)$, and $(S_1,\ldots,S_n,\rho_1,\ldots,\rho_n)$ to the canonical oracle system for $S_1^o,\ldots,S_n^o$. Ψ_p will denote the mapping $\Psi_p\colon S \to S_p$, defined as in 4.6.

5. Restricted games

The canonical multioracle, (S,ρ), describes a game in which each player knows merely the sets of possible strategies and utility functions $H_1,\ldots,H_n,T_1,\ldots,T_n$. We now intend to analyze games, in which each player has some additional knowledge about the random variable ρ, that is, about his own strategy, utility function and probability measure, and those of his opponents. As Harsanyi [6] has pointed out, the strategy choice of the players generally depends not only on this "first order" information about ρ, but also on the players' "second order" information, which is defined as their information about the first order information. Furthermore, it depends on their "third order" information, that is, their information about the second order information, and so on. The complete and precise definition of a game hence requires the description of countably many levels of information. We proceed to define a class of games, for which this description is relatively straightforward.

5.1 Definition: A compact subset $R\subset S^o\times W(S)^n$ is called a *restriction* (wrt. S^o). The *game restricted by* $R\subset S^o\times W(S)^n$ is defined by the following statements ${}^oR$, ${}^1R$, ...

${}^oR$: The strategies, utility functions and probability measures of the n players are elements of R.

${}^1R$: Each player knows, that ${}^oR$. ...

... ${}^iR$: Each player knows, that ${}^{i-1}R$. ...

If we define ${}^1S := \rho^{-1}(R)$, ${}^{i+1}S := \rho^{-1}(S^o \times W({}^iS)^n)$, then ${}^iR$ (for $i \geq 1$) simply says, that each player's subjective probability measure is supported by ${}^iS$. If we define $S|R := \bigcap_{i\geq 1} {}^iS$, then $R \cap S^o \times W(S|R)^n = \rho(S|R)$, and ${}^oR \wedge {}^1R \wedge {}^2R \wedge \ldots$ is equivalent to the statement: The strategies, utility functions and probability measures of the n players are elements of $\rho(S|R)$. Hence, the latter statement is a somewhat more precise definition of the game restricted by R.
$S|R$ can also be characterized as follows: Call (S,ρ) an R-restricted multioracle, if it is a multioracle and if $\rho(S) \subset R$. The category of multioracles and commuting maps can be defined in analogy to 4.2, and it can be shown, that $S|R$ is a terminal object in the subcategory of R-restricted multioracles.

5.2 Definition: The set of possible strategies and utility functions, $(\chi,\tau)(S|R)$, is called the *solution set* of the game restricted by R.

5.3 Example: In classical games with known utility functions, i.e. $T_p=\{t_p\}$, the solution set $\mathcal{M}$ is defined as some prespecified subset of $H_1 \times \ldots \times H_n$, which does not depend on the players' probability measures. (e.g. the set of equilibrium points, the set of pareto-optimal points, etc. Often H_p is convexified, i.e. $W(H_p)$ is used in place of H_p.) Whether or not the players know that $\mathcal{M}$ is the solution set, i.e. whether they know, that they will choose some strategy n-tuple from $\mathcal{M}$, is not usually explicitly stated in classical game theory, but let us assume they do. Furthermore, assume that each player knows, that each player knows that they will choose such strategies, etc. We are thus considering the game restricted by $R := \mathcal{M} \times \prod_p(\{t_p\} \times W(S))$. We shall show, that the solution set of this game is just the classical solution set, more precisely, that $(\chi,\tau)(S|R) = \mathcal{M} \times \{(t_1,\ldots,t_n)\}$. Hence, the only difference between the classical game and the game restricted by R, is that the description of the latter is more detailed.
Since $W(S|R) \subset W(S)$, we have $\rho(S|R) = \mathcal{M} \times \prod_p(\{t_p\} \times W(S|R))$. Let $\sigma : S^o \times W(S)^n \to S^o$ denote the projection map. Since $(\chi,\tau) = \sigma\circ\rho$, it suffices to show, that $W(S|R) \neq \emptyset$ if $\mathcal{M} \neq \emptyset$. If $(h_1,\ldots,h_n) \in \mathcal{M}$, define $P^o := \bigotimes_{p=1}^n(\hat{h}_p \otimes \hat{t}_p) \in W(S^o)$, $P^{i+1} := P^o \otimes \bigotimes_{p=1}^n \hat{P}^i \in W(S^{i+1})$, for $i=0,1,\ldots$ The projective limit $P := \lim P^i \in W(S)$ satisfies $P\circ o\circ\chi = \bigotimes_{p=1}^n \hat{h}_p$ and hence $P \in W({}^1S)$. Using $P\circ o\circ\pi = \bigotimes_{p=1}^n \hat{P}$, $P \in W({}^iS)$ for $i \geq 2$ by induction. Thus $P \in \bigcap_{i\geq 1} W({}^iS) = W(\bigcap_{i\geq 1} {}^iS) = W(S|R)$.

5.4 Example: The above result, that $(\chi,\tau)(S|R) = \sigma(R)$, can be generalized to games with unknown utility functions: Suppose each player knows his own utility function, and has a probability distribution $w_p \in W(\prod_{q\neq p} T_q)$ on the utility functions of his opponents. In other words, if t_p is player p's utility function and P_p his probability measure on S, then $P_p\circ o\circ\tau = \hat{t}_p \otimes w_p$. Let X_p be the set of all Borel-measurable mappings $x_p : T_p \to H_p$, and let $\mathcal{M}$ be a given subset of $X_1 \times \ldots \times X_n$. We interpret $\mathcal{M}$ as follows: if the players have utility functions $t_1,\ldots,t_n$, they will choose strategies $x_1(t_1),\ldots,x_n(t_n)$ only if $(x_1,\ldots,x_n) \in \mathcal{M}$. Altogether, we are assuming, that the players' strategies, utility functions, and probability measures are elements of $L \cap M$, where

$$L = \prod_{p=1}^n (H_p \times \{(t_p,P_p) \in T_p \times W(S) : P_p\circ o\circ\tau = \hat{t}_p \otimes w_p\}), \text{ and}$$

$$M = \{(h_1,\ldots,h_n,t_1,\ldots,t_n) \in S^o : \exists\, (x_1,\ldots,x_n) \in \mathcal{M} \;\; \forall p \;\; x_p(t_p)=h_p\} \times W(S)^n.$$

The solution set of the game restricted by $R = L \cap M$ can be determined as above: $(\chi,\tau)(S|R) = \sigma(R) = \sigma(M)$, if M is compact.

6. Bayesian games

Although the game restricted by $L \cap M$ above has a particularly simple solution set, which can be determined without taking higher levels of information into consideration, it is not easy to decide how M should be defined. Therefore, instead of confining ourselves to this special type of restriction, we begin with more elementary ones. The assumption, that the players act according to the Bayesian principle of choice, will briefly be called the *Bayesian hypothesis*. In the notation of 4.7, this means that the strategies, utility functions and probability measures of the players are elements of

$$B = \prod_{p=1}^{n} \{(h_p, t_p, P_p) \in H_p \times T_p \times W(S): N(h_p, t_p, P_p \circ\circ \Psi_p \circ\circ \chi_p) = \max_{h'_p \in H_p} N(h'_p, t_p, P_p \circ\circ \Psi_p \circ\circ \chi_p)\}.$$

Related to the Bayesian hypothesis is the assumption that each player knows his own utility function and probability measure as well as the strategy he is choosing. We call this assumption the *self-knowledge hypothesis*. It says that if h_p is player p's strategy choice, t_p his utility function and P_p his probability measure, then his probability distribution $P_p \circ\circ \rho \circ\circ \hat{p}$ on $H_p \times T_p \times W(S)$ is supported by $\{(h_p, t_p, P_p)\}$. In other words, the strategies, utility functions and probability measures of the n players are elements of

$$C = \prod_{p=1}^{n} \{(h_p, t_p, P_p) \in H_p \times T_p \times W(S): \quad P_p \circ\circ \rho \circ\circ \hat{p} = \hat{h}_p \otimes \hat{t}_p \otimes \hat{P}_p\}.$$

Games restricted by some compact subset of $B \cap C$ are called *Bayesian games*. Since B and C are compact, we can of course study games restricted merely by $B \cap C$. However, the solution set of such games often seems unrealistically large, as may be illustrated by the following example.

6.1 Example: If $n = 2$, $T_p = \{t_p\}$, where $t_1 = -t_2 =$

	H_2	
H_1	2	2
	1	3

then $\chi(S|B \cap C) = H_1 \times H_2$, even though this zero-sum 2-person game has a unique saddle point.

6.2 Example: On the other hand, if we replace $t_1 = -t_2$ by

	h_2	h'_2
h_1	2	3
h'_1	1	3

then $\chi(S|B \cap C) = \{(h_1, h_2)\}$. (To obtain these solutions, we used the fact that $\chi(S|B \cap C) = \prod_{p=1}^{2} (\bigcap_{i \geq 0} H_p^i)$, where $H_p^0 = H_p$, and for $q \neq p$ and $i \geq 0$

$$H_p^{i+1} = \{h_p \in H_p: \exists P_p \in W(H_q^i) \quad N(h_p, t_p, P_p) = \max_{H_p} N(h'_p, t_p, P_p)\} \;)$$

Before attempting to reduce the size of the solution sets by introducing restrictions which are proper subsets of $B \cap C$, the following question must be clarified: How small can and should a solution set be, ideally? That the solution set should be non-empty is obvious, since some strategy is always chosen in a real game. We define: If $S|R \neq \emptyset$, the restriction R is said to be *consistent* (the axioms ${}^0R$, ${}^1R$, ${}^2R$,.. defining the game being logically consistent in this case). On the other hand, to demand that the solution set should consist of a single point in the case of known utility functions, or more generally, that for each $(t_1,..,t_n) \in \prod_p T_p$ there should exist a unique $(h_1,..,h_n) \in \prod_p H_p$, such that $(h_1,...,h_n,t_1,..,t_n)$ is an element of the solution set, also seems unrealistic: for example, if (h_1,h_2,t_1,t_2) is an element of the solution set, and $t_1(h_1,h'_2) = t_1(h'_1,h'_2)$ for all $h'_2 \in H_2$, then (h'_1,h_2,t_1,t_2) must also be considered an element of the solution set. In order to define a more convenient criterion for the "smallness" of the solution set,

let $\tilde{W}(R)$ denote the set of probability distributions $\tilde{P}_p \in W(\prod_{q \neq p} H_q)$, which can possibly occur in the game restricted by R, that is,
$\tilde{W}(R) = \{(P_1 \circ\circ \Psi_1 \circ\circ \chi_1, \ldots, P_n \circ\circ \Psi_n \circ\circ \chi_n) : (P_1, \ldots, P_n) \in \pi(S|R)\}$

$= \pi(S|R) \overset{\times}{_p} \circ\circ \Psi_p \overset{\times}{_p} \circ\circ \chi_p$

We define: A restriction $R \subset B \cap C$ is said to be *complete*, if $\tilde{W}(R)$ has exactly one element. Note that, if $R \subset B \cap C$, then
$(\chi,\tau)(S|R) \subset \{(h_1, \ldots, h_n, t_1, \ldots, t_n) : \exists\ (\tilde{P}_1, \ldots, \tilde{P}_n) \in \tilde{W}(R)$ such that

$$\forall\ p \quad N(h_p, t_p, \tilde{P}_p) = \max_{H_p} N(h_p', t_p, \tilde{P}_p)\},$$

and hence, if $|\tilde{W}(R)| = 1$, the strategy choice of the players is uniquely determined for almost all utility functions. More precisely, suppose $|\prod_p H_p| = k$, and let m denote Lebesgue measure on $\mathbb{R}^k$. If $\tilde{W}(R) = \{(\tilde{P}_1, \ldots, \tilde{P}_n)\}$, the set of utility functions $(t_1, \ldots, t_n)$ such that for some $h_p' \neq h_p$, both $(h_1, \ldots, h_p, \ldots, h_n, t_1, \ldots, t_n)$ and $(h_1, \ldots, h_p', \ldots, h_n, t_1, \ldots, t_n)$ are elements of $(\chi,\tau)(S|R)$, is a set of m^n-measure zero, since the linear subspace $\{t_p : N(h_p, t_p, \tilde{P}_p) = N(h_p', t_p, \tilde{P}_p)\}$ of $\mathbb{R}^k$ has m-measure zero.

7. Games with known utility distributions

As a generalization of games with known utility functions, we now consider games in which the utility functions are not necessarily known, but where each player p's probability distribution $P_p \circ\circ \tau \circ\circ \hat{\mathcal{P}}$ on the utility functions of his opponents is some given probability distribution $w_p \in W(\prod_{q \neq p} T_q)$. More precisely, we consider games restricted by
$F = F(w_1, \ldots, w_n) = S^o \times \prod_{p=1}^{n} \{P_p \in W(S) : P_p \circ\circ \tau \circ\circ \hat{\mathcal{P}} = w_p\}$.
If $w_p = \otimes_{q \neq p} \hat{t}_q$ for some $(t_1, \ldots, t_n) \in \prod_{q \neq p} T_q$, we have the special case of known utility functions.

As in 6.2, the solution set of a game restricted by $R = B \cap C \cap F$ can be found without calculating $S|R$ explicitly. Define the projection maps $\chi_p^o : S_p^o \to \prod_{q \neq p} H_q$, and $\tau_p^o : S_p^o \to \prod_{q \neq p} T_q$, and construct the compact sets $A_p^0, A_p^1, \ldots$ as follows: $A_p^0 = H_p \times T_p$, $A_p^{i+1} = \{(h_p, t_p) : \exists\ P_p^o \in W(\prod_{q \neq p} A_q^i)$ such that $P_p^o \circ\circ \tau_p^o = w_p$, $N(h_p, t_p, P_p^o \circ\circ \chi_p^o) = \max_{H_p} N(h_p', t_p, P_p^o \circ\circ \chi_p^o)\}$.
Then $(\chi,\tau)(S|R) = \prod_{p=1}^{n} (\bigcap_{i \geq 0} A_p^i)$.

7.1 Example: Let n=2, $H_p = \{h_p^1, h_p^2\}$, $T_p = \{t_p \in \mathbb{R}^{H_1 \times H_2} : \sum_{H_1 \times H_2} t_p^2(h_p, h_q) = 1\}$, and let w_p be the uniform distribution on T_q, that is, on the surface of the unit sphere in $\mathbb{R}^4$ $(q \neq p)$. Using the iterative procedure described above, the game restricted by $B \cap C \cap F(w_1, w_2)$ can be solved analytically ([5],[9]), and has the solution set $A_1 \times A_2$, where
$A_p = \bigcup_{\substack{i=1 \\ j \neq i}}^{2} (\{h_p^i\} \times \{t_p \in T_p : \sum_{H_q} t_p(h_p^i, h_q) \geq \sum_{H_q} t_p(h_p^j, h_q)\})$.
Note that the strategies of player p are everywhere uniquely determined as a function of t_p, except on the w_q -zero set $\{t_p \in T_p : \sum t_p(h_p^1, h_q) = \sum t_p(h_p^2, h_q)\}$. Furthermore, $R = B \cap C \cap F(w_1, w_2)$ is complete: $\tilde{W}(R) = \{(P_1, P_2)\}$, where each P_p is the uniform distribution on H_q. Finally, unlike classical games, the solution of the Bayesian game above required the use of countably many levels of information: A_p^{i+1} is a proper subset of A_p^i, for all $i \geq 0$.

8. Equilibrium points

Coming back to example 6.1, we shall now propose an additional hypothesis, which is consistent with the Bayesian hypothesis and the hypothesis of self-knowledge and which, in the special case of two person games with known utility functions and a unique equilibrium point in pure strategies, will yield the equilibrium point as the solution set.

The *equilibrium hypothesis* states that everything, which player q knows about his opponents, is also known to player p; more precisely, for all p and all $q \neq p$ player p knows player q's probability distribution $P_q \circ\circ \Psi_q \in W(S_q)$. If $P_p \in W(S)$ denotes player p's probability measure, his knowledge concerning $W(S_q)$ can be represented by the distribution $P_p \circ\circ \pi \circ\circ \hat{q} \circ\circ (\circ\circ \Psi_q) = P_p \circ\circ \Psi_p \circ\circ \pi_p \circ\circ \hat{q} \in W(W(S_q))$. To say that p knows $P_q \circ\circ \Psi_q$ means $P_p \circ\circ \Psi_p \circ\circ \pi_p \circ\circ \hat{q} = \widehat{P_q \circ\circ \Psi_q}$. Hence the equilibrium hypothesis corresponds to the restriction

$$E = S^o \times \{(P_1,\dots,P_n) \in W(S)^n : \forall\, p \;\; P_p \circ\circ \Psi_p \circ\circ \pi_p = \bigotimes_{q \neq p} \widehat{P_q \circ\circ \Psi_q}\}.$$

If the restrictions B,C,E and F above are defined with respect to arbitrary compact sets H_p, $T_p \subset C(H_1 \times \dots \times H_n)$ and arbitrary probability measures $w_p \in W(\prod_{q \neq p} T_q)$, the following theorem holds (see [1])

8.1 Theorem: $S \mid B \cap C \cap E \cap F \neq \emptyset$

In order to compare the solution set of games restricted by $B \cap C \cap E \cap F$ to classical equilibrium point solution sets, we use the following

8.2 Definition: If R is an arbitrary restriction, the elements of $\pi(S \mid R \cap E)$ are called *equilibrium points* of the game restricted by R.

Note that $\pi(S \mid R \cap E)$ is just the set of n-tuples of subjective probability measures, which are possible in the game restricted by $R \cap E$. The above theorem implies, that equilibrium points exist for all Bayesian games with known utility distributions. The following abbreviations are convenient

$$\mathcal{E} = \mathcal{E}(w_1,\dots,w_n) = \pi(S \mid B \cap C \cap E \cap F(w_1,\dots,w_n)) \subset W(S)^n$$

$$\mathcal{E}^o = \mathcal{E}^o(w_1,\dots,w_n) = \mathcal{E}(w_1,\dots,w_n) \underset{p}{\times} \circ\circ \Psi_p \underset{p}{\times} \circ\circ (\chi_p, \tau_p) \subset \prod_{p=1}^{n} W(S_p^o)$$

$$\tilde{\mathcal{E}} = \tilde{\mathcal{E}}(w_1,\dots,w_n) = \mathcal{E}(w_1,\dots,w_n) \underset{p}{\times} \circ\circ \Psi_p \underset{p}{\times} \circ\circ \chi_p \subset \prod_{p=1}^{n} W(\prod_{q \neq p} H_q).$$

The elements of $\mathcal{E}^o$ and $\tilde{\mathcal{E}}$ (as well as those of $\mathcal{E}$) will also be called equilibrium points of the game restricted by $B \cap C \cap F(w_1,\dots,w_n)$.

$\tilde{\mathcal{E}}$ can be determined directly, without first calculating $\mathcal{E}$, as will now be shown. Let $\mathcal{P}(X)$ denote the set of all subsets of a set X, and let χ_p^o, τ_p^o be the projection maps defined in 7. Define $\gamma_p : W(\prod_{q \neq p} H_q) \to \mathcal{P}(H_p \times T_p)$ and $\Phi_p : \prod_{q \neq p} W(\prod_{r \neq q} H_r) \to \mathcal{P}(W(\prod_{q \neq p} H_q))$ as

$$\gamma_p(\tilde{P}_p) = \{(h_p, t_p) : N(h_p, t_p, \tilde{P}_p) = \max_{H_p} N(h_p', t_p, \tilde{P}_p)\}$$

$$\Phi_p(\tilde{P}_1,\dots \not{p} \dots,\tilde{P}_n) = \{P_p^o \in W(\prod_{q \neq p} \gamma_q(\tilde{P}_q)) : P_p^o \circ\circ \tau_p^o = w_p\} \circ\circ \chi_p^o$$

8.3 Theorem: $\tilde{\mathcal{E}} = \{(\tilde{P}_1,\dots,\tilde{P}_n) \in \prod_{p=1}^{n} W(\prod_{q \neq p} H_q) : \forall\, p \;\; \tilde{P}_p \in \Phi_p(\tilde{P}_1,\dots \not{p} \dots,\tilde{P}_n)\}$

Once $\tilde{\mathcal{E}}$ is found, the solution of the game restricted by $B \cap C \cap E \cap F$ is easily determined, using

8.4 Theorem: $(\chi,\tau)(S \mid B \cap C \cap E \cap F) = \bigcup_{(\tilde{P}_1,\dots,\tilde{P}_n) \in \tilde{\mathcal{E}}} \prod_{p=1}^{n} \gamma_p(\tilde{P}_p)$

Particularly interesting from the technical point of view is the case where for all p and all $\tilde{P}_p \in W(\prod_{q \neq p} H_q)$, the set $\prod_{q \neq p} \gamma_q(\tilde{P}_q)$ is w_p-almost surely a graph, that is, for w_p-almost all $(t_1,\dots \not{p} \dots,t_n)$ there exists a unique $(h_1,\dots \not{p} \dots,h_n)$ such that for all $q \neq p$ $(h_q, t_q) \in \gamma_q(\tilde{P}_q)$. In this case there exists a unique $P_p^o \in W(\prod_{q \neq p} \gamma_q(\tilde{P}_q))$ such that

$P^o_p \circ \tau^o_p = w_p$. Hence the set $\Phi_p(\tilde{P}_1,..\not{p}..,\tilde{P}_n)$ consists of a single point which we denote $\phi_p(\tilde{P}_1,..\not{p}..,\tilde{P}_n)$. If we define

$\phi(\tilde{P}_1,...,\tilde{P}_n) = (\phi_1(/_1,...,\tilde{P}_n),...,\phi_p(\tilde{P}_1,..\not{p}..,\tilde{P}_n),...,\phi_n(\tilde{P}_1,...,/_n))$,

then $\tilde{E}$ is the set of fixed points of the point-to-point mapping ϕ. If each H_p is finite, we have

8.5 $\phi_p(\tilde{P}_1,..\not{p}..,\tilde{P}_n)(h_1,..\not{p}..,h_n)$
$= w_p(\prod_{q \neq p}\{t_q\colon N(h_q,t_q,\tilde{P}_q) = \max_H N(h_q',t_q,\tilde{P}_q)\})$,

and hence $\tilde{E}$ is the set of solutions of a finite system of (nonlinear) equations, which can be found by the usual numerical means.

8.6 Example: Define H_p,T_p,w_p as in example 7.1. Knowing the solution of this example, it is of course unnecessary to calculate $\tilde{E}$, since $\tilde{E} = \tilde{W}(B\cap C\cap E\cap F) \subset \tilde{W}(B\cap C\cap F)$, $\tilde{E} \neq \emptyset$, $|\tilde{W}(B\cap C\cap F)| = 1$, and hence $\tilde{E} = \tilde{W}(B\cap C\cap F)$. However, whereas the determination of $\tilde{W}(B\cap C\cap F)$ in 7.1 required the recursive calculation of A^1_p, $\tilde{E}$ can be found directly with much less effort: Since the hyperplane $\{t_p \in R^4\colon N(h^1_p,t_p,\tilde{P}_p) = N(h^2_p,t_p,\tilde{P}_p)\}$ has w_p-measure zero and devides T_p into two equal parts, we have $\phi_q(P_p)(h^1_p) = w_q\{t_p\colon N(h^1_p,t_p,\tilde{P}_p) \geq N(h^2_p,t_p,\tilde{P}_p)\} = \frac{1}{2}$. Hence the unique solution of $\phi_q(\tilde{P}_p)(h_p) = \tilde{P}_q(h_p)$ is $\tilde{P}_p(h_q) = \frac{1}{2}$ for all p, $q \neq p$ and $h_q \in H_q$.

9. Comparison with classical equilibrium concepts

The assertion at the beginning of 8., relating classical equilibrium points in pure strategies to solutions of games restricted by $B\cap C\cap E\cap F$ remains to be shown. In general, the equilibrium points of games restricted by $B\cap C\cap F(w_1,..,w_n)$ are closely related to classical equilibrium concepts as defined for games with unknown utility functions and given distributions $w_1,..,w_n$. Two such classical equilibrium concepts will now be defined, existence theorems will be given, and their relation to $E^o(w_1,..,w_n)$ will be examined.

(i) If we wish to bring a game with unknown utility functions and given utility distributions $w_1,..,w_n$ into the classical normal form, we must interpret an n-tuple of utility functions as a preliminary chance move governed by some probability measure $w \in W(T_1\times...\times T_n)$. Player p may "observe" his own utility function, t_p, and hence has pure normalized strategies $x_p\colon T_p \to H_p$ at his disposal. Since w_p is player p's probability distribution on $\prod_{q\neq p} T_q$ at a stage where he knows his own utility function, t_p, w_p must be interpreted as the conditional distribution $w(\cdot|t_p)$ with respect to w, given t_p. However, since in the particular game situation we are considering, w_p is fixed and hence independent of t_p, w must be a product measure of the form $w = w_p \otimes (w\circ\hat{p})$. Hence, a game with given utility distributions $w_1,..,w_n$ can be brought into normal form iff a probability measure $w \in W(T_1\times..\times T_n)$ exists such that $w = w_p \otimes (w\circ\hat{p})$ for all p, that is, iff $w_1,..,w_n$ are "consistent" in the sense of Harsanyi [6]. An equivalent condition is that there exist $\bar{w}_1,..,\bar{w}_n$, $\bar{w}_p \in W(T_p)$, such that $w_p = \bigotimes_{q\neq p} \bar{w}_q$ for all p. Since equilibrium points in pure normalized strategies need not exist, and since mixed normalized strategies cannot be defined properly for non-finite T_p (cf.[2]), we shall work with suitably generalyzed behavioral strategies.
We need the following notation: If X,Y are compact sets, let $K(Y,X)$ denote the set of Markov kernels from Y to X, i.e., the set of all mappings $K\colon Y\times C(X) \to \mathbb{R}$ such that $K(y,\cdot) \in W(X)$ for all $y \in Y$, and $K(\cdot,f)$ is Borel-measurable for all $f \in C(X)$. If $P \in W(Y)$, we can define a probability measure $K\times P \in W(X\times Y)$ as follows: for $f \in C(X)$,

$g\in C(Y)$, let $f\otimes g(x,y) := f(x)g(y)$, and $K\times P(f\otimes g) := \int K(\cdot,f)g\,dP$. Note that $(K\times P)\circ\hat{2} = P$, where $\hat{2}: X\times Y \to Y$ is the projection map. For a given $P \in W(Y)$ define the equivalence relation $\sim$ on $K(Y,X)$ as follows: $K\sim K'$ iff there exists a Borel set $M \subset Y$, $P(M) = 0$, such that $K(y,f) = K'(y,f)$ for all $y \in M^c$ and all $f \in C(X)$. If $K\sim K'$, then $K\times P = K'\times P$. Conversely, if X is metrizable, then to each $\bar{P} \in W(X\times Y)$ satisfying $\bar{P}\circ\hat{2} = P$, there exists a unique equivalence class $\tilde{K}$ of kernels from Y to X such that $K\times P = \bar{P}$ for all $K \in \tilde{K}$. In other words, if X is metrizable, $K(Y,X)$ can be identified (modulo $\sim$) with the set $\{\bar{P} \in W(X\times Y) : \bar{P}\circ\hat{2} = P\}$.

A kernel $K_p\in K(T_p,H_p)$, being a mapping $K_p: T_p \to W(H_p)$, is a behavioral strategy (in the sense of Kuhn) for the game defined above. In order to avoid metrizability assumptions in theorem 9.1, and in view of the 1-1 correspondence between kernels and measures explained above, we shall however define the set of behavioral strategies of player p as $X_p:=\{\bar{P}_p\in W(H_p\times T_p) : \bar{P}_p\circ\hat{2} = \bar{w}_p\}$. If the players use behavioral strategies $\bar{P}_1,..,\bar{P}_n$, the expected "payoff" for player p is

$$V_p(\bar{P}_1,..,\bar{P}_n) := \int_{S^o} t_p(h_1,..,h_n)d\bar{P}_1(h_1,t_1)\ldots d\bar{P}_n(h_n,t_n),\quad (S^o=\prod_{p=1}^{n}(H_p\times T_p))$$

and the set of equilibrium points in behavioral strategies is

$$M = \{(\bar{P}_1,..,\bar{P}_n)\in \prod_p X_p: \forall p\ \ V_p(\bar{P}_1,..,\bar{P}_n) = \max_{\bar{P}'_p\in X_p} V_p(\bar{P}_1,..,\bar{P}'_p,..,\bar{P}_n)\}$$

$= M(\bar{w}_1,..,\bar{w}_n)$. If H_p and $T_p\subset C(H_1\times..\times H_n)$ are compact, the generalized Kakutani fixpoint theorem [4] can be applied to prove

9.1 Theorem: $M(\bar{w}_1,..,\bar{w}_n) \neq \emptyset$ for all $\bar{w}_p\in W(T_p)$, $p=1,..,n$.

The following theorem states the relationship between equilibrium points in behavioral strategies and equilibrium points of Bayesian games with known utility distributions $w_1,..,w_n$.

9.2 Theorem: Let $\bar{E}(w_1,..,w_n) =$
$= \{(\bar{P}_1,..,\bar{P}_n)\in \prod_p W(H_p\times T_p): (\bigotimes_{q\neq 1}\bar{P}_q,..,\bigotimes_{q\neq n}\bar{P}_q) \in E^o(w_1,..,w_n)\}$.
If $\bar{w}_1,..,\bar{w}_n$, $\bar{w}_p\in W(T_p)$ exist such that $w_p= \bigotimes_{q\neq p}\bar{w}_q$ for all p, then $\bar{E}(w_1,..,w_n) = M(\bar{w}_1,..,\bar{w}_n)$, otherwise $\bar{E}(w_1,..,w_n) = \emptyset$.

In case of known utility functions, where $\bar{w}_p= \hat{t}_p$ and $w_p= \bigotimes_{q\neq p}\hat{t}_q$, X_p can be identified with $W(H_p)$, M is simply the set of equilibrium points in mixed strategies, and $E^o(w_1,..,w_n) = \tilde{E}(w_1,..,w_n)$. In the case of 2-person games $w_p= \bar{w}_q$, and $E^o(w_1,w_2) = \bar{E}(w_1,w_2) = M(\bar{w}_1,\bar{w}_2)$. Hence for 2-person games with known utility functions

9.3 $\tilde{E}(\hat{t}_2,\hat{t}_1) = M(\hat{t}_1,\hat{t}_2)$,

that is, $(\tilde{P}_1,\tilde{P}_2) \in W(H_2) \times W(H_1)$ is an equilibrium point of the game restricted by $B\cap C\cap F(\hat{t}_2,\hat{t}_1)$ iff it is an equilibrium point in mixed strategies. However the interpretations of $\tilde{P}_p$ differ: whereas Bayesian game theory interprets $\tilde{P}_1$ as player 1's probability distribution on the strategies of player 2, classical game theory interprets this measure as a randomization procedure used by player 2 to determine his pure strategy choice. Suppose the equilibrium point $(\tilde{P}_1,\tilde{P}_2)$ is unique. Then classical game theory asserts that player p will choose strategy h_p with probability $\tilde{P}_q(h_p)$, whereas theorem 8.4 merely states, that he chooses some strategy, which maximizes $N(\cdot,t_p,\tilde{P}_p)$. However, if $(\tilde{P}_1,\tilde{P}_2)$ is an equilibrium point in pure strategies, e.g. $\tilde{P}_p= \hat{h}_q$ both theories conclude, that p chooses h_p.

(ii) Whereas equilibrium points as defined above exist for arbitrary compact $T_p \subset C(H_1 \times .. \times H_n)$ but only for "consistent" $w_1,..,w_n$, the following alternatively defined equilibrium points exist for arbitrary $w_1,..,w_n$, if each T_p is finite: Define the set of behavioral strategies for player p as $\mathcal{K}_p := \mathcal{K}(T_p, H_p)$. For $(K_1,..,K_n) \in \prod_p \mathcal{K}_p$, $t_p \in T_p$, let

$$\int_{\prod_{q \neq p} T_q} \int_{\prod_{p=1}^{n} H_p} t_p(h_1,..,h_n) K_1(t_1, dh_1) \ldots K_n(t_n, dh_n) \, dw_p(t_1,..\not{p}..,t_n) =: t_p^*(K_1,..,K_n),$$

and define the set of equilibrium points (cf.[6])

$$\{(K_1,..,K_n) \in_p \mathcal{K}_p : \forall p \ \forall t_p \in T_p \ t_p^*(K_1,..,K_n) = \max_{K_p' \in \mathcal{K}_p} t_p^*(K_1,..,K_p',..,K_n)\}$$

$=: \mathcal{M}'(w_1,..,w_n)$. If H_p is compact and T_p finite, then $\mathcal{K}_p = W(H_p)^{T_p}$, and the usual fixed point theorem can be applied to show that

9.4 Theorem: $\mathcal{M}'(w_1,..,w_n) \neq \emptyset$ for all $(w_1,..,w_n) \in \prod_{p=1}^{n} W(\prod_{q \neq p} T_q)$

However, if T_p is an arbitrary compact subset of $C(H_1 \times .. \times H_n)$, I do not know, whether such equilibrium points exist. A relationship to equilibrium points of Bayesian games with known utility distributions can be established as follows: For $K_q \in \mathcal{K}_q$, the product kernel $\bigotimes_{q \neq p} K_q \in \mathcal{K}(\prod_{q \neq p} T_q, \prod_{q \neq p} H_q)$ is defined by

$$\bigotimes_{q \neq p} K_q(t_1,..\not{p}..,t_n, f_1 \otimes .. \not{p} .. \otimes f_n) = K_1(t_1, f_1) K_2(t_2, f_2) .. \not{p} .. K_n(t_n, f_n),$$

for $f_q \in C(H_q)$. Hence $(\bigotimes_{q \neq p} K_q) \times w_p \in W(S_p^o)$ and we can prove

9.5 Theorem:

$$\{((\bigotimes_{q \neq 1} K_q) \times w_1,..,(\bigotimes_{q \neq n} K_q) \times w_n) : (K_1,..,K_n) \in \mathcal{M}'(w_1,..,w_n)\} \subset \mathcal{E}^o(w_1,..,w_n)$$

Various other equilibrium concepts, which generalize the mixed strategy equilibrium points of games with known utility functions can be defined; the above comparison between classical and Bayesian equilibrium concepts is not meant to be exhaustive. It does, however illustrate some difficulties faced by classical game theory, which can be overcome using the Bayesian approach.

1. Existence of equilibrium points: Whereas all Bayesian games with known utility distributions have equilibrium points, to my knowledge no classical equilibrium points have been defined and shown to exist for arbitrary compact $H_p, T_p \subset C(H_1 \times .. \times H_n)$ and arbitrary $w_p \in W(\prod_{q \neq p} T_q)$.
2. Theoretical foundation: The assumption, that the players act according to one of the classical equilibrium concepts above, is not only stronger than the equilibrium hypothesis, but also less general, being defined only for games with given distributions $w_1,..,w_n$.
3. Practical calculation: If T_p is large, the sets $\mathcal{M}$ or $\mathcal{M}'$ are difficult to calculate. However it is not necessary to do so, since the behavior of the players can be determined using $\mathcal{E}$, as demonstrated by 8.4. This set may readily be found even for non-finite T_p if the conditions of 8.5 apply.

References

[1] W. Armbruster: Spiele mit unbekannten Nutzenfunktionen und ihre Gleichgewichtspunkte, Heidelberg 1978

[2] R.J. Aumann: Borel structures for function spaces, Illinois J. of Math. 5 (1961), 614-630

[3] W. Böge: Gedanken über die Angewandte Mathematik, in: Mathematiker über die Mathematik, M. Otte(ed.), Springer Heidelberg 1974

[4] I. Glicksberg: A further generalization of the Kakutani fixed point theorem..., Proc.Amer.Math.Soc. 3 (1952), 170-174

[5] H. Dieses: Spiele mit unbekannten Nutzenfunktionen, Heidelbg.1971

[6] J.C. Harsanyi: Games with incomplete information played by Bayesian players, Parts I-III, Management Science 14 (1967-68) 159-82

[7] R. Rabusseau, W. Reich: Entscheidung einer Person unter Unsicherheit, Heidelberg, 1972

[8] L.J. Savage: The Foundations of Statistics,Wiley&So. New York 195

[9] Th. Eisele: On solutions of Bayesian games, to be published

GAME THEORY AND RELATED TOPICS
O. Moeschlin, D. Pallaschke (eds.)

ON THE SUM OF THE WEIGHTS OF MINIMAL BALANCED SETS

G. Bruyneel
Seminarie voor Wiskundige Beleidstechnieken,
Rijksuniversiteit te Gent, Gent, Belgium

We prove that for each element of a particular set V_n of finitely continued fractions there exists a minimal balanced set of $N=\{1,\ldots,n\}$ for which the sum of the weights equals the rational value of that element. The elements of V_n can easily be listed, either in the form of continued fractions or in the form of rational numbers. A method is given to construct a minimal balanced set for which the sum of the weights equals any given rational number.

1. Introduction

Let $N=\{1,\ldots,n\}$ and $\mathcal{S}=\{S_1,\ldots,S_p\}$ be finite sets, where all S_j $(j=1,\ldots,p)$ are different proper subsets of N. $\mathcal{S}$ is a balanced set of N (Shapley [5]) if there exist positive numbers γ_j $(j=1,\ldots,p)$, called weights for $\mathcal{S}$, such that $\sum_{S_j\ni i} \gamma_j = 1$ $(i=1,\ldots,n)$. The vector $\gamma=[\gamma_1,\ldots,\gamma_p]^T$ is a weight vector for $\mathcal{S}$.
A balanced set is minimal if it contains no proper subset which is also balanced.
A minimal balanced set is proper if $S_j \cap S_k \neq \phi$ $(j,k=1,\ldots,p)$.
It is well known that a balanced set is minimal if and only if there is a unique weight vector, and that $p \leqslant n$ for a minimal balanced set. Shapley [5] characterizes the cooperative n-person games v with nonempty core C(v) by means of a system of inequalities involving the weights of all minimal balanced sets of N:

$C(v)\neq\phi \Leftrightarrow \sum_{j=1}^{p} \gamma_j v(S_j) \leqslant v(N)$ for all minimal balanced sets $\mathcal{S}=\{S_1,\ldots,S_p\}$ (weight vector γ) of N.

We put $\Gamma = \sum_{j=1}^{p} \gamma_j$ for a minimal balanced set, without denoting explicitly the dependency of Γ on $\mathcal{S}$. By means of Γ, Shapley's result can be extended to the ε-core $C_\varepsilon(v) = \{x\in E^n \| e(N,x)=0 \text{ and } e(S,x)\leqslant\varepsilon, \forall S\in 2^N\setminus\{\phi,N\}\}$ for any real ε, where $e(S,x) = v(S)-\sum_{i\in S} x_i$ is the excess of coalition S for payoff vector x. Indeed, it is easy to prove that

$C_\varepsilon(v)\neq\phi \Leftrightarrow \sum_{j=1}^{p} \gamma_j v(S_j) \leqslant v(N)+\varepsilon\Gamma$ for all minimal balanced sets $\mathcal{S}=\{S_1,\ldots,S_p\}$ (weight vector γ) of N.

The purpose of this paper is to investigate the possible values of Γ to throw more light on the structure of minimal balanced sets. Peleg [4] gives an inductive method to construct all minimal balanced sets of N. To obtain the values of Γ by means of Peleg's method requires the construction of the minimal balanced sets and the computation of their weight vectors. Our results however do not require knowledge of the minimal balanced sets themselves and can be used for any n without an induction argument. Some theory on continued fractions is used to obtain the results.

Denote a continued fraction $a_1+\frac{1}{a_2+}\dots\frac{1}{+a_q}$ ($a_1\geq 0$ integer, $a_2,\dots,a_q>0$ integer) by $(a_1,\dots,a_q)$, its rational value by T/Q, and its rth approximation $(a_1,\dots,a_r)$ by T_r/Q_r ($r=1,\dots,q$). The following properties are well known (see e.g. Hancock [3]); they are also valid if $q=\infty$.

To compute the rational value $T/Q\equiv T_q/Q_q$ of $(a_1,\dots,a_q)$, we can use the recursive formula $\frac{T_r}{Q_r}=\frac{a_rT_{r-1}+T_{r-2}}{a_rQ_{r-1}+Q_{r-2}}$ ($r=3,\dots,q$), where T_{r-1}/Q_{r-1} and T_{r-2}/Q_{r-2} are irreducible quotients; the quotient T_r/Q_r will also be irreducible.

$$T_{r-1}/Q_{r-1} - T_r/Q_r = \begin{cases} 1/Q_{r-1}Q_r & \text{if r is odd} \\ -1/Q_{r-1}Q_r & \text{if r is even} \end{cases} \quad (r\geq 2)$$

$\{T_r/Q_r\ (r=1,3,5,\dots)\}$ is an increasing sequence.

$\{T_r/Q_r\ (r=2,4,6,\dots)\}$ is a decreasing sequence.

$T_r/Q_r > T_s/Q_s$ for any even r and any odd s.

$\{|T_r/Q_r - T/Q|\ (r=1,\dots,q)\}$ is a decreasing sequence, converging to 0.

T/Q is contained in the open interval defined by any two consecutive approximations T_{r-1}/Q_{r-1} and T_r/Q_r ($r=2,\dots,q-1$) if $q>2$.

Where needed we will suppose $a_q>1$ to suppress the ambiguity $(a_1,\dots,a_{q-1},1)=(a_1,\dots,a_{q-2},a_{q-1}+1)$. Then each rational number T/Q can be written in just one way as a continued fraction: $a_1,\dots,a_q$ is the sequence of quotients when we apply Euclid's algorithm to compute the greatest common divisor of T and Q.

We define for $n\geq 2$ the sets of continued fractions

$V_n = \{(a_1,\dots,a_q) \| a_1>0, a_q>1, 1\leq q\leq n-1, \sum_{i=1}^{q} a_i\leq n\}$ and $W_n = \{(a_1,\dots,a_q)\in V_n \| \sum_{i=1}^{q} a_i=n\}$. It is obvious that V_n is the disjoint union of V_{n-1} and W_n.

We denote $N\setminus\{n\}$ by N^-.

2. Some Basic Properties on Minimal Balanced Sets and their Weights

The proof of the following propositions is easy. Moreover, propositions 4 and 5 are part of Peleg's inductive construction [4].

Proposition 1

$\Gamma>1$ for every minimal balanced set of N.

Proposition 2

If $\mathcal{S}=\{S_1,\dots,S_p\}$ is a minimal balanced set of N with weight vector γ, then

$\mathcal{S}^c=\{N\backslash S_1,\dots,N\backslash S_p\}$ is a minimal balanced set of N with weight vector $\delta=\gamma/(\Gamma-1)$. (We call $\mathcal{S}^c$ the complement of $\mathcal{S}$.)

Proposition 3

$n/(n-1)\leq\Gamma\leq n$ for every minimal balanced set of N.

Proposition 4

Let $\mathcal{S}=\{S_1,\dots,S_p\}$ be a minimal balanced set of N^- with weight vector γ. Put $T_j=S_j$ $(j=1,\dots,p)$, $T_{p+1}=\{n\}$. Then $\mathcal{T}=\{T_1,\dots,T_{p+1}\}$ is a minimal balanced set of N with weight vector $\delta=[\gamma^T\ 1]^T$.

Proposition 5

Let $\mathcal{S}=\{S_1,\dots,S_p\}$ be a minimal balanced set of N^- with weight vector γ, and $P\subset\{1,\dots,p\}$ an index set such that $\sum_{j\in P}\gamma_j=1$. Put $T_j=S_j\cup\{n\}$ if $j\in P$, $T_j=S_j$ if $j\notin P$. Then $\mathcal{T}=\{T_1,\dots,T_p\}$ is a minimal balanced set of N with weight vector γ.

3. The Sum of the Weights

Theorem 1

For each $(a_1,\dots,a_q)\in V_n$ there exists a minimal balanced set of N for which $\Gamma=(a_1,\dots,a_q)$.

Proof: The theorem holds for n=2: $V_2=\{(2)\}$, and the partition $\{\{1\},\{2\}\}$, with $\gamma_1=\gamma_2=1$, $\Gamma=2$, is a (and the only) minimal balanced set of $\{1,2\}$.
We now prove the theorem by induction; $M=\{1,\dots,m\}$.
Assume that for each $(b_1,\dots,b_q)\in V_{m-1}$ $(3\leq m\leq n)$ there exists a minimal balanced set of M^- for which the weight sum equals $(b_1,\dots,b_q)$. Let $(a_1,\dots,a_q)\in V_m$, then either $(a_1,\dots,a_q)\in V_{m-1}$ or $(a_1,\dots,a_q)\in W_m$.
If $(a_1,\dots,a_q)\in V_{m-1}$, there exists a minimal balanced set of M^- with weight sum $(a_1,\dots,a_q)$. By proposition 5 we obtain a minimal balanced set of M with weight sum $(a_1,\dots,a_q)$.
If $(a_1,\dots,a_q)\in W_m$, we distinguish two cases.
Case 1. If $a_1=1$ (thus $q\geq2$), put $b_i=a_{i+1}$ $(i=1,\dots,q-1)$. Then $(b_1,\dots,b_{q-1})\in V_{m-1}$, and there exists a minimal balanced set of M^- with weight sum $(b_1,\dots,b_{q-1})$. By proposition 4 we obtain a minimal balanced set of M with weight sum $(b_1,\dots,b_{q-1})+1$. Its complement (proposition 2) is a minimal balanced set of M with weight sum $((b_1,\dots,b_{q-1})+1)/(b_1,\dots,b_{q-1})=(1,b_1,\dots,b_{q-1})=(a_1,\dots,a_q)$.
Case 2. If $a_1>1$ (thus $q<n-1$), put $b_1=a_1-1$, $b_i=a_i$ $(i=2,\dots,q)$. Then $(b_1,\dots,b_q)\in V_{m-1}$, and there exists a minimal balanced set of M^- with weight sum $(b_1,\dots,b_q)$. By proposition 4 we obtain a minimal balanced set of M with weight sum $(b_1,\dots,b_q)+1=(a_1,\dots,a_q)$. ∎

Theorem 2

$|W_n|=2^{n-2}$ and $|V_n|=2^{n-1}-1$.

Proof: Denote by A_n^q the number of different q-tuples $(a_1,\dots,a_q)$ for which $\sum_{i=1}^{q} a_i=n$ and $a_i>0$ $(i=1,\dots,q)$. $A_n^1=1$ for all n. By the formula $A_n^r = \sum_{p=1}^{n-r+1} A_{n-p}^{r-1}$ $(r=2,\dots,n)$ we obtain $A_n^2 = \sum_{p=1}^{n-1} A_{n-p}^1 = \binom{n-1}{1}$,
$A_n^3 = \sum_{p=1}^{n-2} A_{n-p}^2 = \sum_{p=1}^{n-2} \binom{n-1-p}{1} = \binom{n-1}{2}$, $A_n^4 = \sum_{p=1}^{n-3} A_{n-p}^3 = \sum_{p=1}^{n-3} \binom{n-1-p}{2} = \binom{n-1}{3}$, ...,
$A_n^q = \binom{n-1}{q-1}$. By excluding the A_{n-1}^{q-1} q-tuples for which $a_q=1$, we obtain
$|W_n| = \sum_{q=1}^{n-1} A_n^q - \sum_{q=2}^{n-1} A_{n-1}^{q-1} = 2^{n-2}$. Now $|V_n| = |V_{n-1}|+|W_n| = \dots = |V_2|+\sum_{p=3}^{n} |W_p| =$
$= 2^{n-1}-1$. ∎

Theorem 3

If $T/Q=(a_1,\dots,a_q)\in W_n$ and $q>1$, then
1° $T_{q-1}/Q_{q-1}\in V_{n-2}$ (except if $q=2$ and $a_1=1$),
2° $T'/Q'\equiv(a_1,\dots,a_{q-1},a_q-1)\in W_{n-1}$ (also for $q=1$, except if $a_1=2$),
3° $\frac{T}{Q} = \frac{T'+T_{q-1}}{Q'+Q_{q-1}}$.

Proof: The exceptions exclude the number 1. For 1° we remark that $q\leqslant n-2$ if $a_{q-1}>1$ and $q>2$ if $a_{q-1}=1$. For 2° we remark that $q\geqslant 2$ if $a_q=2$ and $q\leqslant n-2$ if $a_q>2$. Now the proof of 1° and 2° is easy; 3° follows from the formula given in section 1 if we remark that $(a_1,\dots,a_q-1,1)=(a_1,\dots,a_q)$. ∎

4. Listing of the Elements of V_n

We will list the elements of V_n in increasing order, firstly as continued fractions, afterwards as rational numbers.
From section 1 it follows that
1° $(a_1,\dots,a_{q-1})$ <(>) $(a_1,\dots,a_q)$ if q is even (odd),
2° if $a_q<b_q$ and $r\geqslant q$, then $(a_1,\dots,a_q)$ >(<) $(a_1,\dots,a_{q-1},b_q,\dots,b_r)$ if q is even (odd).
The second inequality includes the first if we extend $(a_1,\dots,a_{q-1})$ formally to $(a_1,\dots,a_{q-1},\infty)$. Then we can proceed as follows, in hierarchical order of operations. We order the elements of V_n by increasing a_1, for equal a_1 by decreasing a_2, for equal a_2 by increasing a_3, for equal a_3 by decreasing a_4, etc.
The first element of the list is $(1,n-1)=n/(n-1)$, the last element is $(n)=n$ (compare with proposition 3); both belong to W_n.
The $2^{n-1}-1$ elements of V_n are obtained from the $2^{n-2}-1$ elements of V_{n-1} by adding the 2^{n-2} elements of W_n (theorem 2). By theorem 3, each element of W_n, except (n) (if $q=1$) and $(1,n-1)$ (if $q=2$ and $a_1=1$), is obtained from 2 elements of V_{n-1} (more

precisely, one of V_{n-2} and one of W_{n-1}) by summing their numerators and their denominators. Moreover, the listing as continued fractions shows that these elements of V_{n-1} are consecutive. Thus to obtain V_n we place a new element between every two elements of V_{n-1}. This operation is also valid at the beginning and the end of the list, if we formally consider 1/1 and 1/0 resp. as least and greatest weight sum for any n; indeed, $\frac{(n-1)+1}{(n-2)+1} = \frac{n}{n-1}$ and $\frac{(n-1)+1}{1\ \ +0} = \frac{n}{1}$.
Table 1 contains the listing in both forms for n=2,…,6.

continued fractions					rational numbers				
n=2	n=3	n=4	n=5	n=6	n=2	n=3	n=4	n=5	n=6
					[1/1]	[1/1]	[1/1]	[1/1]	[1/1]
				(1,5)					6/5
			(1,4)	(1,4)				5/4	5/4
				(1,3,2)					9/7
		(1,3)	(1,3)	(1,3)			4/3	4/3	4/3
				(1,2,1,2)					11/8
			(1,2,2)	(1,2,2)				7/5	7/5
				(1,2,3)					10/7
	(1,2)	(1,2)	(1,2)	(1,2)		3/2	3/2	3/2	3/2
				(1,1,1,3)					11/7
			(1,1,1,2)	(1,1,1,2)				8/5	8/5
				(1,1,1,1,2)					13/8
		(1,1,2)	(1,1,2)	(1,1,2)			5/3	5/3	5/3
				(1,1,2,2)					12/7
			(1,1,3)	(1,1,3)				7/4	7/4
				(1,1,4)					9/5
(2)	(2)	(2)	(2)	(2)	2/1	2/1	2/1	2/1	2/1
				(2,4)					9/4
			(2,3)	(2,3)				7/3	7/3
				(2,2,2)					12/5
		(2,2)	(2,2)	(2,2)			5/2	5/2	5/2
				(2,1,1,2)					13/5
			(2,1,2)	(2,1,2)				8/3	8/3
				(2,1,3)					11/4
	(3)	(3)	(3)	(3)		3/1	3/1	3/1	3/1
				(3,3)					10/3
			(3,2)	(3,2)				7/2	7/2
				(3,1,2)					11/3
		(4)	(4)	(4)			4/1	4/1	4/1
				(4,2)					9/2
			(5)	(5)				5/1	5/1
				(6)					6/1
					[1/0]	[1/0]	[1/0]	[1/0]	[1/0]

Table 1: V_n

Remark

There are some remarkable relations between V_n and the sequence of Fibonacci numbers $F_1=1$, $F_2=1$, $F_i=F_{i-1}+F_{i-2}$ (i=3,4,…).
For any n⩾2, V_n contains in particular the elements $\frac{2}{1} = \frac{1+1}{1+0}$, $\frac{3}{2} = \frac{2+1}{1+1}$, $\frac{5}{3} = \frac{3+2}{2+1}$, $\frac{8}{5} = \frac{5+3}{3+2}$, $\frac{13}{8} = \frac{8+5}{5+3}$, etc., i.e. $F_3/F_2,\ldots,F_{n+1}/F_n$. The corresponding continued fractions are (2),(1,2),(1,1,2),(1,1,1,2),…, or (1,1),(1,1,1),(1,1,1,1),…. Thus

$F_{n+1}/F_n=(a_1,\dots,a_n)$ with $a_1=\dots=a_n=1$, and $\lim_{n\to\infty} F_{n+1}/F_n=(1,1,1,\dots)=(1+\sqrt{5})/2$.
For any $n\geq 3$, V_n contains in particular the elements $\frac{3}{1}=\frac{2+1}{1+0}$, $\frac{5}{2}=\frac{3+2}{1+1}$, $\frac{8}{3}=\frac{5+3}{2+1}$, $\frac{13}{5}=\frac{8+5}{3+2}$, etc., i.e. $F_4/F_2,\dots,F_{n+1}/F_{n-1}$. The corresponding continued fractions are (3),(2,2),(2,1,2),(2,1,1,2),..., or (2,1),(2,1,1),(2,1,1,1),.... Thus $F_{n+2}/F_n=(a_1,\dots,a_n)$ with $a_1=2$, $a_2=\dots=a_n=1$, and $\lim_{n\to\infty} F_{n+2}/F_n=(2,1,1,1,\dots)=(3+\sqrt{5})/2$.
The properties of V_n imply that all F_{n+1}/F_n and F_{n+2}/F_n $(n=1,2,\dots)$ are irreducible.
Remark that $\lim_{n\to\infty} F_{n+2}/F_n=(\lim_{n\to\infty} F_{n+1}/F_n)^2$.
Other similar and general properties of the Fibonacci numbers can be deduced, but they are well known: see e.g. Dickson [2], pp. 393-396, 441.

5. Weight Sums not in V_n

The converse of theorem 1 holds only for n=2,3,4. For $n\geq 5$ there are weight sums not in V_n. Indeed, $\{\{1,2,3\},\{1,2,4\},\{3,4,5\},\{1,5\},\{2,5\}\}$ and $\{\{1,2,3\},\{1,2,4\},\{1,5\},\{2,5\},\{3,4\}\}$ are minimal balanced sets of $\{1,\dots,5\}$; the weight vectors are resp. $[2/5,2/5,3/5,1/5,1/5]^T$ and $[1/4,1/4,1/2,1/2,3/4]^T$, and the weight sums resp. 9/5=(1,1,4) and 9/4=(2,4), not belonging to V_5.
If a weight sum $\Gamma\notin V_{n-1}$ exists for N^-, then also $\Gamma/(\Gamma-1)\notin V_{n-1}$ is a weight sum for N^- (proposition 2). By propositions 2, 4 and 5, it is easy to prove that Γ, $\Gamma/(\Gamma-1)$, $\Gamma+1$, $(2\Gamma-1)/(\Gamma-1)$, $1+1/\Gamma$ and $2-1/\Gamma$ are weight sums for N. The last four (and possibly all) of them are not in V_n. Thus the number of weight sums not in V_n is increasing with n.

6. Construction of a Minimal Balanced Set with Given Weight Sum

Theorem 1 assures the existence of a minimal balanced set for which $\Gamma=(a_1,\dots,a_q)\in V_n$ if the element of V_n is given. Now we construct such a minimal balanced set.
More generally, for any given rational number $T/Q>1$, it is obvious that $T/Q=(a_1,\dots,a_q)\in W_n$ for $n=\sum_{i=1}^{q} a_i$.
Consider a partition $\{K_1,\dots,K_q\}$ of N, where $K_1=\{k_{11},\dots,k_{1,a_1-1}\}$ ($K_1=\phi$ if $a_1=1$), $K_i=\{k_{i1},\dots,k_{ia_i}\}$ $(i=2,\dots,q-1)$, $K_q=\{k_{q1},\dots,k_{q,a_q+1}\}$. Define

$L_{1j}=\{k_{1j}\}$ $(j=1,\dots,a_1-1)$ if $a_1>1$, $L_{ij}=\bigcup_{m=1}^{i/2-1} K_{2m}\cup(K_i\setminus\{k_{ij}\})\cup(\bigcup_{m=i+1}^{q} K_m)$ (i even, $i\leq q$; $j=1,\dots,|K_i|$), $L_{ij}=\bigcup_{m=1}^{(i-1)/2} K_{2m}\cup\{k_{ij}\}$ (i odd, $3\leq i\leq q$; $j=1,\dots,|K_i|$).

Theorem 4

$\{L_{ij}$ $(i=1,\dots,q;\ j=1,\dots,|K_i|)\}$ if $a_1>1$ or $\{L_{ij}$ $(i=2,\dots,q;\ j=1,\dots,|K_i|)\}$ if $a_1=1$ is a minimal balanced set of N. The weights of L_{1j} $(j=1,\dots,a_1-1)$ are 1 (if $a_1>1$), those of L_{ij} ($i=2,\dots,q$; $j=1,\dots,|K_i|$) are $1/\prod_{h=2}^{i}(a_h,\dots,a_q)$. The sum of the weights is $T/Q=(a_1,\dots,a_q)$.

Proof: We abbreviate "minimal balanced set" to "MBS" in this proof.
The partition $\{\{k_{q1}\},\dots,\{k_{q,a_q+1}\}\}$ of K_q is a MBS of K_q, with weights $1,\dots,1$ and weight sum a_q+1.
By proposition 2, $\{K_q\setminus\{k_{q1}\},\dots,K_q\setminus\{k_{q,a_q+1}\}\}$ is a MBS of K_q, with weights $1/a_q,\dots,1/a_q$, and weight sum $1+1/a_q=(1,a_q)$.
By proposition 4, $\{\{k_{q-1,1}\},\dots,\{k_{q-1,a_{q-1}}\},K_q\setminus\{k_{q1}\},\dots,K_q\setminus\{k_{q,a_q+1}\}\}$ is a MBS of $K_{q-1}\cup K_q$, with weights $1,\dots,1,1/a_q,\dots,1/a_q$ and weight sum $a_{q-1}+(1,a_q)=1+(a_{q-1},a_q)$.
By proposition 2,
$\{K_{q-1}\cup K_q\setminus\{k_{q-1,1}\},\dots,K_{q-1}\cup K_q\setminus\{k_{q-1,a_{q-1}}\},K_{q-1}\cup\{k_{q1}\},\dots,K_{q-1}\cup\{k_{q,a_q+1}\}\}$ is a MBS of $K_{q-1}\cup K_q$, with weights $1/(a_{q-1},a_q),\dots,1/(a_{q-1},a_q),1/a_q(a_{q-1},a_q),\dots,1/a_q(a_{q-1},a_q)$ and weight sum $(1,a_{q-1},a_q)$.
By proposition 4, $\{\{k_{q-2,1}\},\dots,\{k_{q-2,a_{q-2}}\},K_{q-1}\cup K_q\setminus\{k_{q-1,1}\},\dots,K_{q-1}\cup K_q\setminus\{k_{q-1,a_{q-1}}\},$ $K_{q-1}\cup\{k_{q1}\},\dots,K_{q-1}\cup\{k_{q,a_q+1}\}\}$ is a MBS of $K_{q-2}\cup K_{q-1}\cup K_q$, with weights $1,\dots,1,$ $1/(a_{q-1},a_q),\dots,1/(a_{q-1},a_q),1/a_q(a_{q-1},a_q),\dots,1/a_q(a_{q-1},a_q)$ and weight sum $1+(a_{q-2},a_{q-1},a_q)$.
Applying alternatively propositions 2 and 4 we obtain a MBS of $\bigcup_{i=2}^{q}K_i$, with weight sum $(1,a_2,\dots,a_q)$. If $a_1=1$, this is also a MBS of N, with weight sum $(a_1,\dots,a_q)$. If $a_1>1$, we add K_1 and obtain a MBS of N, with weight sum $(a_1-1)+(1,a_2,\dots,a_q)=$ $=(a_1,\dots,a_q)$. The elements of the MBS are the sets L_{ij}, and their weights are as stated in the theorem. ∎

Remark

If we denote the equal weights of L_{ij} ($j=1,\dots,|K_i|$) by γ_{L_i} ($i=1,\dots,q$), then $\gamma_{L_{i+1}}=\gamma_{L_i}/(a_{i+1},\dots,a_q)$ ($i=1,\dots,q-1$), thus $1=\gamma_{L_1}>\dots>\gamma_{L_q}>0$.[1] Putting $T/Q=\gamma_{L_0}$, we have $\gamma_{L_2}=\gamma_{L_0}-a_1\gamma_{L_1}$, $\gamma_{L_3}=\gamma_{L_2}/(a_3,\dots,a_q)=\gamma_{L_2}((a_2,\dots,a_q)-a_2)=\gamma_{L_1}-a_2\gamma_{L_2}$, and generally $\gamma_{L_0}=T/Q$, $\gamma_{L_1}=1$, $\gamma_{L_i}=\gamma_{L_{i-2}}-a_{i-1}\gamma_{L_{i-1}}$ ($i=2,\dots,q$).
It can also be proved that $\gamma_{L_{i-1}}/(1+a_i)<\gamma_{L_i}<\gamma_{L_{i-1}}/a_i$ ($i=2,\dots,q-1$).

7. Weight Sum of a Proper Minimal Balanced Set

The minimal value $n/(n-1)$ in proposition 3 is attained for $\{N\setminus\{1\},\dots,N\setminus\{n\}\}$, being a proper minimal balanced set for all $n>2$. The maximal value n is attained for the partition $\{\{1\},\dots,\{n\}\}$, which is never proper. In the next theorem we give a maximal value for proper minimal balanced sets in function of p, viz. p/2. Of course, this implies $\Gamma\leqslant n/2$, but the latter value is not attained for all n.

Theorem 5

$\Gamma \leqslant p/2$ for every proper minimal balanced set.

Proof: Let $\mathcal{S}=\{S_1,\dots,S_p\}$ be a proper minimal balanced set. $|S_j \cap S_k| \geqslant 1$ $(j,k=1,\dots,p)$ implies $\gamma_j+\gamma_k \leqslant 1$ $(j,k=1,\dots,p;\ j\neq k)$. Thus

$$\frac{1}{2}\sum_{\substack{j,k=1\\ j\neq k}}^{p}(\gamma_j+\gamma_k) \leqslant \binom{p}{2}\text{ , i.e. } (p-1)\sum_{j=1}^{p}\gamma_j \leqslant \binom{p}{2} \text{ or } \Gamma\leqslant p/2.$$

Now we construct for each $p \geqslant 3$ a proper minimal balanced set for which $\Gamma=p/2$ and $n=\binom{p}{2}$.

$S_1=\{1,2,3,4,\dots,p-1\}$, $S_2=\{1,p,p+1,p+2,\dots,2p-3\}$, $S_3=\{2,p,2p-2,2p-1,\dots,3p-6\}$,
$S_4=\{3,p+1,2p-2,3p-5,\dots,4p-10\}$, ..., $S_{p-1}=\{p-2,2p-4,3p-7,4p-11,\dots,n\}$,
$S_p=\{p-1,2p-3,3p-6,4p-10,\dots,n\}$

$\mathcal{S}=\{S_1,\dots,S_p\}$ is a minimal balanced set of N for which $\gamma_j=1/2$ $(j=1,\dots,p)$, $\Gamma=p/2$. $\mathcal{S}$ is proper, because $|S_j\cap S_k|=1$ $(j,k=1,\dots,p;\ j\neq k)$, viz. $S_1\cap S_2=\{1\}$, $S_1\cap S_3=\{2\}$, ..., $S_1\cap S_p=\{p-1\}$, $S_2\cap S_3=\{p\}$, $S_2\cap S_4=\{p+1\}$, ..., $S_2\cap S_p=\{2p-3\}$, ..., $S_{p-1}\cap S_p=\{n\}$. ∎

Remark

Let $\mathcal{S}$ be as constructed in the proof of theorem 5. If we consider the elements of N as nodes and the elements of $\mathcal{S}$ as edges of a hypergraph $H=(N,\mathcal{S})$, then its dual H^* is the complete graph K_p of order p. In [1], it is proved more generally that the dual hypergraph H^* of any connected linear graph H is minimal balanced if and only if H is not bichromatic.

References

[1] Bruyneel G.: Some Applications of Minimal Balanced Sets in Graph Theory, paper presented at the III. Symposium über Operations Research, Mannheim, September 6-8, 1978

[2] Dickson L.E.: History of the Theory of Numbers, Vol. I: Divisibility and Primality (1971), Chelsea Publ. Co., New York

[3] Hancock H.: Development of the Minkowski Geometry of Numbers (1939), The Macmillan Co., New York

[4] Peleg B.: An Inductive Method for Constructing Minimal Balanced Collections of Finite Sets, Naval Res. Log. Q. 12 (1965), 155-162

[5] Shapley L.S.: On Balanced Sets and Cores, Naval Res. Log. Q. 14 (1967), 453-460

GAME THEORY AND RELATED TOPICS
O. Moeschlin, D. Pallaschke (eds.)

REGULAR GAMES

M. EGEA

Departement de Mathématiques
Université de Lyon 1
VILLEURBANNE - FRANCE

Preliminaries.

The assumption of a finite number of players is incompatible with the concept of perfect competition. [cf.(5)]

The case of a continuum of players has been studied by R.J. Aumann and L.S. Shapley who take in consideration the Borel σ-algebra of the segment $[0, 1]$.

Since the family of coalitions likely to be set up owns a structure of Boolean algebra, we have especially studied the algebraic aspect of coalition games.

To any Boolean algebra $\mathscr{B}$, it is possible to associate, by the mean of Stone representation, a Boolean space Ω such that the set clopen sets ["clopen" means both open and closed] of Ω is isomorphic to algebra $\mathscr{B}$.[cf.(7)]

Thus, we can assume that the coalition game $\mathscr{B}$ likely to be set up, is in coincidence with the Boolean algebra of clopen sets of a topological Boolean space , called fundamental space.

Lemma. Let Ω be a fundamental space. Ω is finite if and only if the coalition family is in coincidence with the set $\mathscr{P}(\Omega)$ of subsets of Ω .

Proof. [cf.(7)] * Let us suppose that $\Omega = \{1, \ldots, n\}$ is a compact totally disconnected space. If s is an element of Ω, then $\{s\}$ is a clopen set of Boolean space Ω. Let $S \in \mathscr{P}(\{1, \ldots, n\})$: $S = \bigcup_{s \in S}\{s\}$ is a clopen set.

* Let us suppose that the coalition family $\mathscr{B} = \mathscr{P}(\Omega)$.
$\Omega = \bigcup_{i \in \Omega} \{i\}$ and Ω is compact space, then there are elements $i_1, \ldots, i_n$ in Ω, such that $\Omega = \{i_1, \ldots, i_n\}$.

Let Ω be a Boolean space ; a player x is identified to an isolated point of topological space Ω, i.e., $\{x\}$ is by definition an atom of Boolean algebra $\mathscr{B}$ of coalitions.

So, it is possible to give a mathematical expression of the idea that any coalition cannot be reduced to a unique player. [We consider, for instance, the coalition algebra is an infinite free Boolean algebra (cf. (4))].

It is also possible to express that in an infinite coalition set, only an individual x_0 is not a player by considering the lemma:

Lemma. Let E be an infinite set such that $x_0 \notin E$. The set $E \cup \{x_0\}$ is a fundamental space such that the coalitions are finite or cofinite subsets of E.

Proof. [cf. (4)]. We denote by $\mathscr{P}_0(E)$ the set of finite or cofinite subsets of E. Then $\mathscr{P}_0(E)$ is a Boolean algebra ; a clopen in $E \cup \{x_0\}$, i.e. a coalition, is an element of $\mathscr{P}_0(E)$; x is a player iff $\{x\} \in \mathscr{P}_0(E)$, i.e. iff $x \neq x_0$.

In a Boolean coalition algebra $\mathscr{B}$, an allocation will be a finitely additive mapping μ; with real values, i.e., such that for all disjoint coalitions S and T, $\mu(S \cup T) = \mu(S) + \mu(T)$.

If Ω is the fundamental space, $\mathscr{A}(\Omega)$ will denote the allocation set in $\mathscr{B}$. We say that a Boolean algebra is finely measured if it exists $\mu \in \mathscr{A}(\Omega)$, such that $\mu \geqslant 0$, and such that $\mu(S) = 0 \iff S = \emptyset$.

Proposition. When $\Omega = \{1,\dots,n\}$, $\mathscr{A}(\Omega)$ is isomorphic to the vectorial $\mathbb{R}$-space $\mathbb{R}^n$.

Proof. [c.f.(2)] . Let $\mu \in \mathscr{A}(\Omega)$. We put $\varphi(\mu) = (\mu(\{i\}))_{i=1,\dots,n}$. $\mu \longmapsto \varphi(\mu)$ is a one-to-one linear function φ from $\mathscr{A}(\Omega)$ into $\mathbb{R}^n$.

Proposition. Let $\mathscr{B}$ be a Boolean algebra freely generated by an infinite coalition family. Then $\mathscr{B}$ is finely measured.

Proof. [cf.(4) p. 271, or c.f. Kelley J.L. : Measures in Boolean algebras. Pacific. J. Math.]

Similary, Boolean algebra of sub-intervals of the real line $\mathbb{R}$ is finely measured by Lebesgue measure. We can also notice that Ω is finite, then $\mathscr{P}(\Omega)$ is finely measured by $S = \{1, \dots, p\} \longrightarrow \text{Card}(S) = p$.

Regular games.

A game, in its characteristic function form, consists in determining pair (Ω, v) such that Ω is a Boolean space and where v is a real-valued function, defined on the coalition family $\mathscr{B}$, such that $v(\emptyset) = 0$ and such that if S and T are two disjoint coalitions : $v(S \cup T) \geqslant v(S) + v(T)$.

When Ω is finite, it is possible to put :

$\forall\, S \in \mathscr{P}(\Omega)$, $v°(S) = \sum_{i \in S} v(\{i\})$; thus by ordering the set of real-valued functions of $\mathscr{P}(\Omega)$, by $f \leqslant g$ iff for any $S \in \mathscr{P}(\Omega)$, $f(S) \leqslant g(S)$; the imputation concept as being an allocation $\mu \in \mathscr{A}(\Omega)$, such that $\mu(\Omega) = v(\Omega)$, and such that $v° \leqslant \mu$.

<u>Lemma.</u> Let (Ω, v) be a game such that Ω is finite. We have :
$v^\circ = \text{Max}\ \{\mu \in \mathscr{A}(\Omega)/\mu \leqslant v\}$.

<u>Proof.</u> [cf.(2)]. When $\Omega = \{1,\ldots,n\}$, $\mathscr{A}(\Omega) \cong \mathbb{R}^n$, and $\hat{v}_\circ = (v\{i\})_{i=1,\ldots n} \in \mathbb{R}^n$ is such that $\varphi(\hat{v}_\circ) = v_\circ \in \mathscr{A}(\Omega)$, and $\dot{v}_\circ \lneq v$. If $\mu \in \mathscr{A}(\Omega)$, and if $\mu \leqslant v$, we have for any coalition $S \subseteq \{1,\ldots,n\}$: $\mu(S) = \sum_{i \in S} \mu(\{i\}) \leqslant v^\circ(S)$, i.e., $\mu \leqslant v^\circ$.

<u>Definition.</u> We call regular game any game (Ω, v) such that the allocation $v^\circ = \text{Max}\ \{\mu \in \mathscr{A}(\Omega)/\mu \leqslant v\}$ exists.

The conception of generalization of games (Ω, v) to an infinity of coalitions without imposing to some of them the existence of the allocation v°, would not let us to express imputation concept.

<u>Definition.</u> The imputation of a regular game (Ω, v) consists in determining an allocation μ in the coalition algebra, such that $v^\circ \leqslant \mu$ and such that $v(\Omega) = \mu(\Omega)$.

<u>Theorem.</u> Any game whose values are non negative is a regular game.

<u>Proof.</u> [cf.(2)] . Let S be a coalition. We denote by A(S) the set of finite family $\{S_{i_1}, \ldots, S_{i_n}\}$ of coalitions such that : $S = S_{i_1} \cup \ldots \cup S_{i_n}$ and $\forall\ i_p \neq i_q,\ S_{i_p} \cap S_{i_q} = \emptyset$. We put : $A_v(S) = \{\sum_{p=1}^{n} v(S_{i_p}) \ /\ \{S_{i_1},\ldots,S_{i_n}\} \in A(S)\}$

We assume $v \geqslant 0$, and we put :

$$v^\circ(S) = \text{Inf}\ A_v(S).$$

The function $v^\circ : \mathscr{B} \longrightarrow \mathbb{R},\ S \longmapsto v^\circ(S)$, verifies :

$$v^\circ = \text{Max}\ \{\mu \in \mathscr{A}(\Omega)\ /\mu\ \leqslant\ v\}.$$

We prove the result in [2].

<u>Remark.</u> Let $\mathscr{P}_\circ(E)$ be the set of the finite and cofinite subsets of an infinite set E of players. The function which attributes the gain $-n$ to a coalition with finite cardinal n and the gain 2^{-n} to a coalition S such that card $(E-S) = n$ defines the characteristic function v: $\mathscr{P}_\circ(E) \longrightarrow \mathbb{R}$ of an infinite game. Let $(x_n)_{n \geqslant 1}$ be infinite players, and we put :

$S_n = \{x_p\ /\ 1 \leqslant p \leqslant n\}$. S_n is a coalition such that :

$v(S_n) = -n$ and, $v\ (E-S_n) = 2^{-n}$. Let us suppose μ is an allocation in $\mathscr{P}_\circ(E)$ such that $\mu \leqslant v$. We have, for any n : $E = S_n \cup (E-S_n) \Longrightarrow \mu(E) \leqslant -n + 2^{-n}$: we obtain a contraction. So $v : \mathscr{P}_o(E) \longrightarrow \mathbb{R}$ is not a regular game.

Let us consider a regular game (Ω, v), and let us put : $e_v(\Omega) = v(\Omega) - v^\circ(\Omega) \geqslant 0$. For any coalition S, we put : $I_S(x) = \begin{cases} 0 & \text{if } x \notin S \\ 1 & \text{if } x \in S \end{cases}$
At last, let $k_v : \Omega \longrightarrow \mathscr{A}(\Omega)$ be the function defined by : $\forall\, x \in \Omega$, $k_v(x) : \mathscr{B} \longrightarrow \mathbb{R}$, where : $\forall S \in \mathscr{B}$, $k_v(x)(S) = v^\circ(S) + I_S(x)e_v(\Omega)$

We get the following proposition:

<u>Proposition.</u> k_v takes its values in the set of imputations of (Ω, v).

<u>Proof</u>. [cf.(2)] . (1) When $e_v(\Omega) = 0$, i.e., $v(\Omega) = v^\circ(\Omega)$, we have $v = v^\circ$, and for any $x \in \Omega$, $k_v(x) = v^\circ$. In this case v° is an imputation, and we say that (Ω, v) is a <u>non essential</u> game.

(2) Let us suppose that $e_v(\Omega) \neq 0$. For any $x \in \Omega$, the set $\mathscr{M}_x = \{S \in \mathscr{B} / x \notin S\} = \mathscr{P}(\Omega - \{x\}) \cap \mathscr{B}$ is an <u>maximal ideal</u> of Boolean algebra $\mathscr{B}$ of coalitions ; then the function : ν_x : $\nu_x(S) = 0$ if $S \in \mathscr{M}_x$ and $\nu_x(S) = e_v(\Omega)$ if $S \notin \mathscr{M}_x$, verifies $\nu_x \in \mathscr{A}(\Omega)$. We obtain, for any $x \in \Omega$:

$$k_v(x) = \nu_x + v^\circ \in \mathscr{A}(\Omega), \text{ with } k_v(x)(\Omega) = v(\Omega) \text{ and } v^\circ \leqslant k_v(x), \text{ i.e.,}$$

$k_v(x)$ is an imputation.

Assuming $k_v(x) = k_v(y)$, i.e., $\nu_x = \nu_y \Longrightarrow x = y$.
The function k_v is an injection of Ω in the set of imputations of (Ω, v) ; then, it is said that these games are <u>essentials</u>.

Let μ be an imputation of a regular game (Ω, v). We put : $\mathscr{E}(\mu, v) = \{S \in \mathscr{B} / \mu(S) = v^\circ(S)\}$. Then $\mathscr{E}(\mu, v)$ is an ideal of the algebra $\mathscr{B}$. We say that μ is an <u>isolated imputation</u> if $\mathscr{E}(\mu, v)$ is a maximal ideal of $\mathscr{B}$.

<u>Proposition.</u> Let μ be an imputation ; μ is an imputation isolated iff there exists $x \in \Omega$ such that $\mu = k_v(x)$.

<u>Proof.</u> [cf.(2)]. We can remark that :

$$\forall\, x \in \Omega, \quad \mu = k_v(x) \Longleftrightarrow \mathscr{E}(\mu, v) = \mathscr{M}_x$$

and for any maximal ideal $\mathscr{M}$ of $\mathscr{B}$, there is $x \in \Omega$, such that $\mathscr{M} = \mathscr{M}_x$.

<u>Cores of an essential regular game</u>.

<u>Definition.</u> Let (Ω, v) be an essential regular game.
We say that an allocation μ in the coalition algebra $\mathscr{B}$ is <u>strong</u> solution of the game (Ω, v) if $v \leqslant \mu$ and $\mu(\Omega) = v(\Omega)$.

Such allocations are imputations of the game (Ω, v), and form a set $\mathscr{C}_v^s(\Omega)$, which is called the <u>strong core</u> of the game.

<u>Remark</u>. It exists infinite <u>essential</u> regular game with empty strong core, for instance, all essential regular game whose sum is constant [i.e., for any coalition S, $v(\Omega) = v(S)+v(\Omega-S)$] : let $\mu \in \mathscr{C}_v^s(\Omega)$. We have $\mu(\Omega) = v(\Omega)$, with $v(S) \leqslant \mu(S)$ for any coalition S ; $v(S) + v(\Omega-S) = \mu(S) + \mu(\Omega-S) \Longrightarrow v(S) = \mu(S)$ i.e., $v = \mu$, and then $v = v^\circ$: we obtain a contradiction.

On the other hand, allocation μ in an infinite coalition algebra $\mathscr{B}$ such that for any coalition, $0 \leqslant \mu(S) \leqslant 1$, and $\mu(\Omega) = 1$, generate regular game whose characteristic function is $v(S) = [\mu(S)]^2$, and such that : $\mu(\Omega) = v(\Omega)$, and $v \leqslant \mu$, i.e., $\mu \in \mathscr{C}_v^s(\Omega)$.

Let $\mathscr{P}_\circ(E)$ be the coalition algebra composed by finite and cofinite subsets of an infinite set E. Let us denote by $\mathscr{F}$ the set of cofinite subsets of E. We put : $I_{\mathscr{F}} : \mathscr{P}_\circ(E) \longrightarrow \{0, 1\}$ such that : $I_{\mathscr{F}}(S) = 1 \Longleftrightarrow S \in \mathscr{F}$.

$I_{\mathscr{F}}$ is an allocation in $\mathscr{P}_\circ(E)$. Let us denote by $v : \mathscr{P}_\circ(E) \longrightarrow \mathbb{R}$ the function defined by : for any $S \in \mathscr{P}_\circ(E)$:

$$v(S) = \begin{cases} 0, & \text{if } S \text{ is finite} \\ 2^{-n}, & \text{if } S \text{ is cofinite such that } \\ & \text{card}(E-S) = n. \end{cases}$$

v is a function characteristic of an essential regular game, and we have $I_{\mathscr{F}}(\Omega) = 1 = v(\Omega) = 2^{-0}$, $v(S) \leqslant I_{\mathscr{F}}(S)$ for any $S \in \mathscr{P}_\circ(E)$, i.e., $I_{\mathscr{F}}$ is a strong solution of v.

For finite games, the core of these games is generally introduced by the domination concept. The definition of domination is generalized in following way :

<u>Definition</u>. Let (Ω, v) be an essential regular game. One says that an imputation μ of (Ω, v) dominates an imputation ν of (Ω, v), if there exists a non-empty coalition S such that $\mu(S) \leqslant v(S)$ and such that for any coalition $T \subsetneq S$, $T \neq \emptyset$, we have $\nu(T) < \mu(T)$.

Let us denote by $\mathscr{C}_v^W(\Omega)$ the set of all undominated imputations of (Ω, v). The elements of $\mathscr{C}_v^W(\Omega)$ are called weak solutions of (Ω, v), and $\mathscr{C}_v^W(\Omega)$ is called weak core of (Ω, v).

Lemma. | Let (Ω, v) be an essential regular game. Then, $\mathscr{C}_v^s(\Omega) \subseteq \mathscr{C}_v^W(\Omega)$

Proof. Let us suppose $\mathscr{C}_v^s(\Omega) \nsubseteq \mathscr{C}_v^W(\Omega)$; there exists $\mu \in \mathscr{C}_v^s(\Omega)$ such that $\mu \notin \mathscr{C}_v^W(\Omega)$, i.e., there exists an imputation ν , and a coalition $S_\circ$ such that : $\mu(S_\circ) < \nu(S_\circ) \leqslant v(S_\circ)$; $\mu \in \mathscr{C}_v^s(\Omega) \Longrightarrow (\forall T \in \mathscr{B})(v(T) \leqslant \mu(T)) \Longrightarrow v(S_\circ) \leqslant \mu(S_\circ)$
We obtain a contradiction.

Definition. | We say that a game (Ω, v) is a perfect game if (Ω, v) is an essential regular game such that $\mathscr{C}_v^S(\Omega) = \mathscr{C}_v^W(\Omega)$.

The existence of an essential infinite regular essential game, and non perfect game is a problem which has not been solved up to know.

In the following proposition, we give examples of infinite perfect games.

Proposition. | A sufficient condition for an essential regular game to be a perfect game is that its coalitions algebra are finely measured.

Proof. Let us suppose $\mathscr{C}_v^s(\Omega) \subsetneqq \mathscr{C}_v^W(\Omega)$. There exists an imputation μ in $\mathscr{C}_v^W(\Omega)$ such that $\mu \in \mathscr{C}_v^s(\Omega)$, i.e., there exists a coalition S such that $\mu(S) < v(S)$. We assume $\mathscr{B}$ is finely measured : let $\Psi : \mathscr{B} \longrightarrow \mathbb{R}^+$ be a finely measure .

We put : $a = v(S) - \mu(S) > 0$ and $b = v(\Omega) - v(S) - v^\circ(\Omega-S) \geqslant 0$

For any $K \in \mathscr{B}$, we put :

$$\theta(K) = \mu(S \cap K) + \frac{a}{\Psi(S)} \cdot \Psi(S \cap K) + v^\circ(K \cap (\Omega-S)) + \frac{b}{\Psi(\Omega-S)} \cdot \Psi(K \cap (\Omega-S))$$

$\theta : \mathscr{B} \longrightarrow \mathbb{R}, K \longmapsto \theta(K)$, is an imputation of (Ω, v), such that : $\theta(S) = v(S)$ and $\forall T \neq 0, T \subseteq S \Longrightarrow \mu(T) < \theta(T)$, i.e., θ dominates μ : we obtain a contradiction.

Examples. 1) When Ω is a finite set of players to any essential games is a perfect game.

2) To any essential regular game on the Boolean algebra of sub-intervals of $\mathbb{R}$, is a perfect game.

3) Similarly, if a coalition algebra is free generated by an infinite coalition family, any essential regular game on this algebra is a perfect game.

Over-coalitions and over-game of an infinite non negative valued game.

When the fundamental space Ω is infinite, the coalition family $\mathscr{B}$ is strictly included in the set $\mathscr{P}(\Omega)$ of subsets of the topological space Ω : $\mathscr{B} \subsetneqq \mathscr{P}(\Omega)$. For $H \in \mathscr{P}(\Omega) - \mathscr{B}$, let us define $\alpha(H)$ as the coalition family

$S \in \mathcal{B}$, such that $S \subsetneq H$. Then, $\alpha(H)$ is an ideal of the algebra $\mathcal{B}$, and $\bigcup_{S \in \alpha(H)} S$ is an open in Ω . [cf.(7)]

Let us consider (Ω, v) a game such that $v \geqslant 0$. We define a function $\tilde{v}$ from $\mathcal{P}(\Omega)$ into $\mathbb{R}^+$, by :

$$\text{for any } H \in \mathcal{P}(\Omega) : \tilde{v}(H) = \sup_{S \in \alpha(H)} \{v(S)\}.$$

Proposition. The function $\tilde{v}$ is the smallest continuation of v such that $\tilde{v}$ would be a super-additive function on $\mathcal{P}(\Omega)$.

Proof. [cf. (2)] . For any $S \in \mathcal{B}$, we have : $\tilde{v}(S) = v(S)$, and if $S \in \alpha(H)$, $v(S) \leqslant \tilde{v}(H)$. We can easily see that for any $(H, P) \in \mathcal{P}(\Omega)^2$, such that $H \cap P = \emptyset$, we have $\tilde{v}(H \cup P) \geqslant \tilde{v}(H) + \tilde{v}(P)$. So, $\tilde{v}$ is a super-additive continuation of v. Let f be a super-additive continuation of v : $f : \mathcal{P}(\Omega) \longrightarrow \mathbb{R}^+$. Then f is increasing : $\forall S \in \alpha(H) : S \subsetneq H \Longrightarrow f(S) \leqslant f(H)$, with $f(S) = v(S)$, so, we have : $\sup_{S \in \alpha(H)} f(S) = \tilde{v}(H) \leqslant f(H)$, i.e., $\tilde{v} \leqslant f$.

Proposition. Let (Ω, v) be a game such that $v \geqslant 0$. Let μ be an imputation of (Ω, v). For any family $(H_i)_{i \in I}$ of open sets in Ω, such that $\forall (i, j) \in I^2$, $i \neq j$, $H_i \cap H_j = \emptyset$, we have the following properties :

1) $\tilde{v}(\bigcup_{i \in I} H_i) \geqslant \sum_{i \in I} \tilde{v}(H_i)$

2) $\tilde{\mu}(\bigcup_{i \in I} H_i) = \sum_{i \in I} \tilde{\mu}(H_i)$

Proof. [cf.(2)] . We put $\bigoplus_{i \in I} \alpha(H_i) = \{S_{i_1} \cup \ldots \cup S_{i_n} / S_{i_k} \in \alpha(H_i)\}$

We remark that Ω is a compact space, and any clopen S is compact space. Let S be an element of $\alpha(\bigcup_{i \in I} H_i)$. There exists $\{H_{i_1}, \ldots, H_{i_n}\}$, $i_p \in I$, such that $S \in \oplus\, \alpha(\bigcup_{i \in I} H_i)$, i.e., $\alpha(\bigcup_{i \in I} H_i) \subsetneq \bigoplus_{i \in I} \alpha(H_i)$. [For details, see [2]]

Obviously $\bigoplus_{i \in I} \alpha(H_i)$ is included in $\alpha(\bigcup_{i \in I} H_i)$, so: $\oplus\, \alpha(\bigcup_{i \in I} H_i) = \alpha(\bigcup_{i \in I} H_i)$

1) Then : $\tilde{v}(\bigcup_{i \in I} H_i) = \sup_{S \in \bigoplus_{i \in I} \alpha(H_i)} v(S) \geqslant \sup_{J_\circ \subset I} \{\sum_{j \in J_\circ} \tilde{v}(H_j)\}$

for any finite subset $J_\circ$ of set I, i.e., $\tilde{v}(\bigcup_{i \in I} H_i) \geqslant \sum_{i \in I} \tilde{v}(H_i)$.

2) Let μ be an imputation of (Ω, v). Let $S \in \bigoplus_{i \in I} \alpha(H_i)$:

$S = S_{i_1} \cup \ldots \cup S_{i_n}$, $S_{i_p} \in \alpha(H_{i_p})$, where : $H_{i_p} \cap H_{i_q} = \emptyset$, so $S_{i_p} \cap S_{i_q} = \emptyset$,

and we have : $\sum_{i \in I} \tilde{\mu}(H_i) \geqslant \mu(S_{i_1}) + \ldots + \mu(S_{i_p}) = \mu(S)$.

Then : $\sum_{i \in I} \tilde{\mu}(H_i) \geqslant \sup_{S \in \underset{i \in I}{\oplus} \alpha(H_i)} \mu(S) = \tilde{\mu}(\bigcup_{i \in I} H_i) \Longrightarrow \sum_{i \in I} \tilde{\mu}(H_i) = \tilde{\mu}(\bigcup_{i \in I} H_i)$

Definition. | We say that $\mathscr{P}(\Omega)$ is the set of over-coalitions, and that $(\mathscr{P}(\Omega), \tilde{v})$ is the over-game of the game (Ω, v).

We characterize an injective hull of a coalition algebra $\mathscr{B}$, as being the smallest over-algebra $\mathscr{B}_r$ of $\mathscr{B}$ such that $\mathscr{B}_r$ is complete and such that $\mathscr{B}_r$ is included in $\mathscr{P}(\Omega)$ [cf.(7)] .

Then, $\mathscr{B}_r$ is the set of regular opens in Ω, i.e., the set of opens in Ω such that $\omega = \overset{\circ}{\overline{\omega}}$. Let us suppose $\omega_i \in \mathscr{B}_r$, $i \in I$:

We have : $\sup_{\mathscr{B}_r} \{\omega_i\}_{i \in I} = \underset{i \in I}{\nabla} \omega_i = \overset{\circ}{\overline{\bigcup_{i \in I} \omega_i}} \supseteq \bigcup_{i \in I} \omega_i$

$$\inf_{\mathscr{B}_r} \{\omega_i\}_{i \in I} = \underset{i \in I}{\Delta} \omega_i = \overset{\circ}{\overline{\bigcap_{i \in I}}} \omega_i \subseteq \bigcap_{i \in I} \omega_i$$

$$\omega' = \overset{\circ}{\overline{\Omega - \omega}}, \text{ for any } \omega \in \mathscr{B}_r.$$

We denote by $\tilde{v}_r$ the restriction of $\tilde{v}$ to $\mathscr{B}_r$.

Proposition. | $\tilde{v}_r$ is the characteristic function of an over-game of (Ω,v).

Proof. [cf.(2)]. Let us suppose $(\omega_i)_{i \in I}$ is a disjoint $\cap$-family of $\mathscr{B}_r$. Then we have :

$$\tilde{v}_r(\underset{i \in I}{\nabla} \omega_i) = \tilde{v}(\overline{\bigcup_{i \in I} \omega_i}) \geqslant \tilde{v}(\bigcup_{i \in I} \omega_i) \geqslant \sum_{i \in I} \tilde{v}_r(\omega_i).$$

[For any $\omega_i \in \mathscr{B}_r$, we have : $\alpha(\underset{i \in I}{\nabla} \omega_i) = \alpha(\overline{\bigcup_{i \in I} \omega_i})$.]

Definition. | The pair $(\mathscr{B}_r, \tilde{v}_r)$ is called the injective hull of the game (Ω, v).

Definition. | We say that over-game $f : \mathscr{B}_r \longrightarrow \mathbb{R}^+$ of (Ω, v) (i.e. f is superadditive in $\mathscr{B}_r$, and for any $S \in \mathscr{B}_r$, $f(S) = v(S)$), is sup-α-continuous if for any $\omega \in \mathscr{B}_r$, $f(\omega) = \sup_{S \in \alpha(\omega)} f(S)$.

Proposition. | The injective hull $(\mathscr{B}_r, \tilde{v}_r)$ of (Ω, v) is the unique over-game of (Ω, v), with coalition algebra $\mathscr{B}_r$, which is sup- -continuous.

Proof. Let $f : \mathscr{B}_r \longrightarrow \mathbb{R}^+$ be over-game sup-α-continuous of (Ω,v). For any

$S \in \mathscr{B}$, we have $f(S) = v(S)$, then for any $\omega \in \mathscr{B}_r$, $\sup_{S \in \alpha(\omega)} f(S) = \sup_{S \in \alpha(\omega)} v(S)$, i.e., $f(\omega) = v_r(\omega)$.

Definition. A strong $\cup$-solution of $(\mathscr{B}_r, \tilde{v}_r)$ is a function $\mu : \mathscr{B}_r \longrightarrow \mathbb{R}^+$, such that $\mu \geqslant \tilde{v}_r$, $\mu(\Omega) = \tilde{v}_r(\Omega)$, and if for any disjoint $\cap$-family $(\omega_i)_{i \in I}$ of $\mathscr{B}_r$, we have :

$$\sup_{S \in \alpha(\bigcup_{i \in I} \omega_i)} \mu(S) = \sum_{i \in I} \mu(\omega_i).$$

The strong $\cup$-core of $(\mathscr{B}_r, \tilde{v}_r)$ is the set of strong $\cup$-solution of $(\mathscr{B}_r, \tilde{v}_r)$.

Proposition. Let (Ω, v) be a game such that $v \geqslant 0$. The strong core of (Ω, v) is non-empty iff the strong $\cup$-core of $(\mathscr{B}_r, \tilde{v}_r)$ is non-empty.

Proof. We assume $\mathscr{C}^s_v(\Omega)$ is non-empty. There exists $\mu \in \mathscr{A}(\Omega)$ such that $\mu \geqslant v$. Then $\nu = \tilde{\mu}_r \geqslant \tilde{v}_r$, verifies $\nu(\Omega) = \mu(\Omega)$, i.e., $\nu(\Omega) = v(\Omega)$, i.e., $\nu(\Omega) = \tilde{v}_r(\Omega)$, and for any disjoint $\cap$-family $(\omega_i)_{i \in I}$ of $\mathscr{B}_r$:

$$\sup_{S \in \alpha(\bigcup_{i \in I} \omega_i)} \nu(S) = \sup_{S \in \alpha(\bigcup \omega_i)} \mu(S) = \sum_{i \in I} \tilde{\mu}_r(\omega_i) = \sum_{i \in I} \nu(\omega_i)$$

So, ν is a strong $\cup$-solution of $(\mathscr{B}_r, \tilde{v}_r)$.

Let us consider θ a strong $\cup$-solution of $(\mathscr{B}_r, \tilde{v}_r)$ and let us put : $\mu = \theta\,|_{\mathscr{B}}$; we can easily see that μ is a strong solution of (Ω, v).

Remark. In a similar way, we can define the strong ∇-core of $(\mathscr{B}_r, \tilde{v}_r)$. An element μ of strong ∇-core is function $\mu : \mathscr{B}_r \longrightarrow \mathbb{R}^+$, such that for any disjoint $\cap$-family $(\omega_i)_{i \in I}$ of $\mathscr{B}_r$, $\mu(\nabla_{i \in I} \omega_i) = \sum_{i \in I} \mu(\omega_i)$, and such that $\tilde{v}_r \leqslant \mu$, with $\tilde{v}_r(\Omega) = \mu(\Omega)$. We can easily see that any strong ∇-solution is a strong $\cup$-solution of $(\mathscr{B}_r, \tilde{v}_r)$.

Some other aspects of regular games are developed in [(2)] , and especially we introduce and study the class of "well-framed games", and the concept of "ψ-imputations". Under a new lightening, regular games allow an approach of the value concept (in the Shapley sense) and of the bargaining set concept for an infinite number of players.

REFERENCES

(1) R.J. AUMANN and L.S. SHAPLEY ; "Values of non atomic games" - Princeton University Press - New-Jersey 1974.

(2) M. EGEA : "Jeux réguliers sur un anneau Booléen de Coalitions" Thèse de Doctorat de Spécialité - Université Claude Bernard Lyon 1.

(3) M. EGEA : "Jeux réguliers à un nombre infini de joueurs" Publication Econométrique.

(4) R. FAURE et E. HEURGONS ; "Structures ordonnées et Algèbres de Boole" Collection "PROGRAMMATION", Editions Gautiers-Villars.

(5) C. MOUTON : "L'hypothèse d'un ensemble continu d'agents dans l'économie générale". Thèse de Doctorat d'Etat - Université de Rennes.

(6) G. OWEN : "Game Theory" - Sauders 1968.

(7) D. PONASSE et J.C. CARREGA ; "Algèbre et Topologie Booléenne" Editions Masson.

GAME THEORY AND RELATED TOPICS
O. Moeschlin, D. Pallaschke (eds.)

ON EQUILIBRIUM STRATEGIES IN NONCOOPERATIVE DYNAMIC GAMES

Luuk P.J. Groenewegen and Jaap Wessels

Data Processing Division / Department of Mathematics
Eindhoven University of Technology
Eindhoven, The Netherlands

SUMMARY. In this paper a characterization is given for equilibrium strategies in noncooperative dynamic games. These dynamic games are formulated in a very general way without any topological conditions. For a Nash-equilibrium concept, it is shown that equilibrium strategies are conserving and equalizing. Moreover, it is shown that a set of strategies with these properties satisfies the equilibrium conditions.

With these characterization earlier characterizations for one-person decision processes, gambling houses and dynamic games have been generalized. Especially, this paper shows that such a characterization is basic for a very general class of dynamic games and does not depend on special structure. Of course, in dynamic games with more structure a more refined formulation of the characterization is possible.

1. INTRODUCTION

In the analysis of any type of decision processes (with one or more decision makers) one may distinguish three essentially different kinds of activities:

1. The construction of a decent mathematical model based on the decision structure and the propulsion mechanism, usually both being formulated only conditionally in local time.
   In discrete time, this activity is usually not very difficult, since it can make use of the well-known Ionescu Tulcea-construction for handling random elements. In continuous time this activity presents essential difficulties, but there are techniques for handling these difficulties. These are based for instance on the so-called Girsanov measure transformations.
2. The proof of the existence of "good" strategies of a nice type.
   "Good" can mean here: optimal or nearly optimal, Nash-equilibrium or nearly Nash-equilibrium, Pareto-optimal etc. A nice type of strategies can mean: memoryless or even stationary, pure, monotone etc.
3. The search for necessary (and preferably sufficient) conditions for "good" strategies.
   Such conditions can have the form of a set of optimality equations (resulting from Bellman's optimality principle), a maximum principle, a set of Hamilton-

Jacobi equations etc. As is well-known all these types of conditions are strongly related.

In the literature, the activities of the second and third type are often intensively interwoven, since results on conditions are often obtained in order to use them for obtaining existence results. In this paper however, we will concentrate on the search for necessary and sufficient conditions for good strategies. Namely, it appears that the formulation of such conditions is possible in a very general way without using the specific structure of the actual problem. Of course, specification of the generally formulated conditions for actual problems can give extra insight. However, it is equally important to see the general principles which produce the results.

Bellman's *optimality principle* [1, chapter 3, section 3], which runs as follows: *an optimal policy has the property that, whatever the initial state and initial decision are, the remaining decisions must constitute an optimal policy with regard to the state resulting from the first decision*, is a good starting point for an analysis of a very general decision process. Although it is rather vague in its initial formulation, the optimality principle has proved to constitute a very stimulating guide for the analysis. In the theory of one-person decision processes this has led to characterizations for optimal strategies in different situations to begin with Bellman's optimality equations in [1]. Dubins and Savage [4] and Sudderth [17] gave precise and elegant formulations for the situation of gambling houses. They were the first who made it explicit that a condition in the form of optimality equations is only sufficient for optimality of a strategy if there is some strong form of fading in the process. If not, a supplementary condition should be added. Essentially, the optimality equations-type-of-condition says that the strategy is such that the player does not give any potential reward away. Therefore, this type of condition has been called the *conservingness* property. If the process or reward does not fade away in time,one should add a condition which requires a strategy to cash its potential rewards. This condition has been called the *equalizingness* property. Together these two conditions have appeared to be necessary and sufficient for optimality in several types of one-person decision processes. As mentioned already, this has been proved for gambling houses. Later, Hordijk [9] gave a similar formulation for a certain class of Markov decision processes. More recently, this has been generalized to discrete time decision processes with a much more general propulsion mechanism and reward structure by Kertz and Nachman in [11]. In another line of development, typical control theoretical structures (with continuous time) have been treated by several authors. Relatively general formulations for the fading case have been given by Striebel in [16] and by Boel and Varaiya in [2]. These lines of thought have been combined and further generalized by Groenewegen in his monograph [7]. However, the most important fea-

ture of Groenewegen's approach is that it is very straightforward and relying on intuitively clear notions. In doing so, it is proved on one hand that conservingness and equalizingness are very central and essential notions, on the other hand it is proved that equalizing and conserving provide necessary and sufficient conditions for optimality of a strategy in many types of situations nearly without any specific side conditions e.g. of a topological nature.
So far for one-person decision processes.

For multi-person decision processes or dynamic games an analogous development can be traced. The two basic papers for this topic are Kuhn [12] and Shapley [15]. Kuhn formulates the optimality principle for multi-stage games and Shapley formulates the optimality equations for discounted (i.e. strongly fading) stochastic games. The most striking feature of these papers is, that they have been written before their one-person counterparts. For later developments the one-person theory took the lead. For stochastic games with some strong fading mechanism the conservingness property has been proved to be necessary and sufficient for optimality by many authors (for an overview of such conditions, we refer to Parthasarathy and Stern [13] and to van der Wal and Wessels [18]). Analogous results may be found in the literature about differential games (both deterministic and stochastic), see e.g. the book by Isaacs [10] and the paper by Elliott [5]. The first direct attempts for the establishment of a characterization of optimality in dynamic games can be found in Groenewegen and Wessels [8], Groenewegen [6], and Couwenbergh [3]. However, in his monograph [7], Groenewegen seems to be presenting the most elegant and intuitively appealing approach for the characterization problem in noncooperative dynamic games. Like for one-person decision processes, this approach does not require any extra conditions (e.g. of a topological nature) and it also allows for a very general propulsion mechanism and reward structure. The analysis in this paper will be based essentially on the approach of that monograph.

In section 2 the set-up of our general dynamic (noncooperative) game will be given. Section 3 gives some examples. Section 4 contains our main results and in section 5 some ramifications are indicated.

2. THE SET-UP

In this section we will present a (nonconstructive) set-up for a rather general class of noncooperative stochastic dynamic games for an arbitrary number of players.

L = set of players;

X = state space; X is endowed with a σ-field $\mathcal{X}$;

T = time space; for simplicity we will take T as $\{0,1,2,\ldots\}$ or $[0,\infty)$;

$A^{(\ell)}$ = action space for player ℓ ($\ell \in L$); $A^{(\ell)}$ is endowed with a σ-field $\mathcal{A}^{(\ell)}$;
ν = starting distribution, so ν is a probability measure on $(X,\mathcal{X})$.

The idea is that a starting state in X is determined by random selection from X with probability distribution ν. At any time $t \in T$, the state of the system (an element of X) is observed by the players and they all may choose an action from their action space. These actions have some influence on the behaviour of the system. In order to define this behaviour, we have first to introduce strategies for the players. Since these strategies may depend on the history of the process, we start by introducing such histories.

$A = \times_{\ell\in L} A^{(\ell)}$, A is Cartesian product of the individual action spaces; A is endowed with the appropriate product-σ-field; an element of A denotes a compound action of all the players;

$H^{(t)} = [\times_{\tau\in T, \tau<t} (X \times A)] \times X$, is the space of state-action paths ending with the state of the system at time t; $H^{(t)}$ is endowed with the appropriate product-σ-field $\mathcal{H}^{(t)}$. Similarly, $H = \times_{\tau\in T} (X \times A)$, endowed with the appropriate product-σ-field $\mathcal{H}$. H is the set of all allowed realizations with respect to state and compound actions.

$U^{(\ell)}$ = space of allowed strategies (or controls) for player ℓ; an arbitrary element $u^{(\ell)}$ of $U^{(\ell)}$ gives for any $t \in T$ and any history $h_t \in H^{(t)}$ a probability distribution $u^{(\ell)}(t,h_t,\cdot)$ on $A^{(\ell)}$, it is required that $u^{(\ell)}(t,\cdot,\cdot)$ is a transition probability from $H^{(t)}$ to $A^{(\ell)}$; moreover, it is required that $U^{(\ell)}$ is closed with respect to tail exchanges of individual controls, this means: if $u_1^{(\ell)}, u_2^{(\ell)} \in U^{(\ell)}$, $t \in T$, $B \in \mathcal{H}^{(t)}$, then $u^{(\ell)} \in U^{(\ell)}$ with $u^{(\ell)}(\tau,h_\tau,\cdot) = u_2^{(\ell)}(\tau,h_\tau,\cdot)$ for $\tau \geq t$, h_t (the restriction of h_τ until time t) $\in B$ and $u^{(\ell)}(\tau,h_\tau,\cdot) = u_1^{(\ell)}(\tau,h_\tau,\cdot)$ elsewhere.

$U = \times_{\ell\in L} U^{(\ell)}$, $u \in U$ is a compound strategy.

Now we are able to formulate the main assumption, viz.

> for any (starting) state $x \in X$ and any compound strategy $u \in U$ a probability measure $\mathbb{P}_{x,u}$ on $(H,\mathcal{H})$ is given; moreover the probabilities are measurable as a function of x.

By this assumption we circumvent the obligation to construct a probabilistic structure from more elementary data. With these probabilities we can easily construct the probability measures for the decision process for the given starting distribution ν by

$$\mathbb{P}_u(H') := \int_X \mathbb{P}_{x,u}(H')\nu(dx) \quad \text{for any } H' \in \mathcal{H} .$$

In fact we need a slight extension of the assumption, namely we need the probabilities for the remainder of the process for any given path until time t. Conditioning of $\mathbb{P}_u$ only gives them almost surely on $H^{(t)}$; especially in the case of continuous time the exceptionset might grow out of hands. Therefore we prefer to assume the existence of all these conditioned probabilities (note that these assumptions cover exactly the activity that is described in the introduction as model construction): For any $h_t \in H^{(t)}$ ($t \in T$), and any compound strategy $u \in U$ there exists a probability measure $\mathbb{P}_{h_t,u}$ on $(H,\mathcal{H})$ such that

≡ $\mathbb{P}_{h_t,u}$ is concentrated on $\{h_t\} \times A \times \underset{\substack{\tau\in T\\ \tau>t}}{\times} (X \times A)$

≡ $\mathbb{P}_{h_t,u}(H')$ is $\mathcal{H}^{(t)}$-measurable as a function of h_t for any $H' \in \mathcal{H}$

≡ $\mathbb{P}_{h_t,u}(H^{(t)}, A', \underset{\tau>t}{\times} (X \times A)) = u(t,h_t,A')$ for all $t \in T$, $u \in U$, $A' \in \mathcal{A}$

≡ (nonanticipativity) $\mathbb{P}_{h_t,u_1}(H') = \mathbb{P}_{h_t,u_2}(H')$,

if $H' = B \times \underset{\tau>s}{\times} (X \times A)$ with B a measurable subset of $\underset{\tau\leq s}{\times} (X \times A)$ for some $s > t$ and $u_1(\sigma,\cdot,\cdot) = u_2(\sigma,\cdot,\cdot)$ for all σ with $t \leq \sigma < s$.

≡ (conditioning properties)

a) $$\int_H f(h)\mathbb{P}_{h_t',u}(dh) = \int_H \int_H f(h)\mathbb{P}_{h_s'',u}(dh)\mathbb{P}_{h_t',u}(dh'')$$

for any $t \leq s$ and any nonnegative $\mathcal{H}$-measurable function f.

b) $$\int_H f(h)g(h)\mathbb{P}_{h_t',u}(dh) = f(h_t') \int_H g(h)\mathbb{P}_{h_t',u}(dh)$$

for any $t \in T$ and any nonnegative $\mathcal{H}$-measurable f and g with f only depending on h_t.

Now the process-part of the dynamic game has been defined appropriately. Only the criterion is still to be defined. This will also be done in a very general way: $r^{(\ell)}$ is a real valued measurable function on H for any $\ell \in L$ and denotes the reward of player ℓ as a function of the realization of the game.
$r^{(\ell)}$ is supposed to be quasi-integrable with respect to $\mathbb{P}_u$ for all $u \in U$.
For given $h_t \in H^{(t)}$, $u \in U$ the expected reward for player ℓ is defined by

$$v_t^{(\ell)}(h_t,u) := \begin{cases} \mathbb{E}_{h_t,u} r^{(\ell)}, & \text{if } r^{(\ell)} \text{ is quasi-integrable} \\ -\infty, & \text{otherwise} . \end{cases}$$

Now we are able to introduce our equilibrium concept:

$u_* \in U$ is a *compound equilibrium strategy* iff

$$v_t^{(\ell)}(h_t,u_*) \geq v_t^{(\ell)}(h_t,u_*/u^{(\ell)}) \quad \mathbb{P}_{u*}\text{-a.s.} \quad \text{for all } \ell, u^{(\ell)} ,$$

where $u_*/u^{(\ell)}$ denotes the compound strategy u_* with $u_*^{(\ell)}$ replaced by $u^{(\ell)}$.

3. SOME EXAMPLES

a) *Markov or stochastic games* (compare references [13,15,18]). $T = \{0,1,...\}$, $X = \{1,2,...\}$, $L = \{1,2,...\}$.
The probability measures $\mathbb{P}_{\nu,u}$ are generated by the transition probabilities $p(i,j;a_1,a_2,...)$, denoting the probability for finding state j at time t + 1, if at time t the system is in state i and the players choose actions $a_1,a_2,...$ respectively.
In this type of problem the utility is usually based on the local income function $r_\ell(i;a_1,a_2,...)$ denoting the actual reward for player ℓ at time t if the system is in state i and the players choose actions $a_1,a_2,...$ respectively.
Standard forms for the utility then become

$$r^{(\ell)}(h) := \sum_{t=0}^{\infty} r_\ell(x_t;a^{(t)}) \quad \text{or} \quad \sum_{t=0}^{\infty} \beta^t r_\ell(x_t;a^{(t)})$$

and

$$r^{(\ell)}(h) := \liminf_{t\to\infty} \frac{1}{t} \sum_{\tau=0}^{t-1} r_\ell(x_t;a^{(t)}) ,$$

where $h = (x_0,a^{(0)},x_1,a^{(1)},...)$, $x_t \in X$, $a^{(t)} \in A$.
In the literature many variants of such models can be found.

b) *Differential* (and difference) *games* (references [5,10]). $T = [0,\tau]$ or $[0,\infty)$, $X = \mathbb{R}^m$.
The propulsion mechanism of the process is

$$\dot{x}(t) = f(x(t),a) ,$$

which generates the path of the process and its probability distribution if the instantaneous compound action a is chosen according to some (mixed) strategy. The utility function for a given state realization over time x(t) and a given compound action as a function of time a(t) can be

$$r^{(\ell)}(x(\cdot),a(\cdot)) = \int_0^\tau f_\ell(x(t),a(t))dt .$$

With the same utility function the propulsion mechanism may also be stochastic e.g.

$$dx(t) = f(x(t),a)dt + A(x(t))dB(t) ,$$

where B(t) is Brownian motion.
In the literature many variants of such models are studied.

4. THE BASIC CONCEPTS

Here we will introduce the basic concepts of our characterization. These concepts are formulated in terms of the value functions:

$$\psi_t^{(\ell)}(h_t,u_0) := \sup_{u^{(\ell)}} v_t^{(\ell)}(h_t,u_0/u^{(\ell)}) \; . \tag{4.1}$$

The optimality concept can now be rewritten as:
u is an equilibrium strategy iff

$$\psi_t^{(\ell)}(h_t,u) = v_t^{(\ell)}(h_t,u) \quad \text{for all } \ell,t \quad \mathbb{P}_u\text{-a.s.}$$

Now we obtain for an arbitrary strategy u for $\tau \geq t$ and all ℓ

$$v_t^{(\ell)}(h_t,u) = \mathbb{E}_{h_t,u} r^{(\ell)} = \mathbb{E}_{h_t,u}\mathbb{E}_{h_\tau,u} r^{(\ell)} = \mathbb{E}_{h_t,u} v_\tau^{(\ell)}(h_\tau,u) \quad \mathbb{P}_u\text{-a.s.}$$

Using this relation, we obtain for an equilibrium strategy that the functions $\psi_t^{(\ell)}$ have the martingale property:

Lemma. If u is an equilibrium strategy, then for all $\ell, \tau \geq t$

$$\psi_t^{(\ell)}(h_t,u) = \mathbb{E}_{h_t,u}\psi_\tau^{(\ell)}(h_\tau,u) \quad \mathbb{P}_u\text{-a.s.} \tag{4.2}$$

The question now arises whether any strategy which satisfies (4.2) is an equilibrium strategy. An extremely simple example shows that this is not true.

Counterexample. (one-person game)

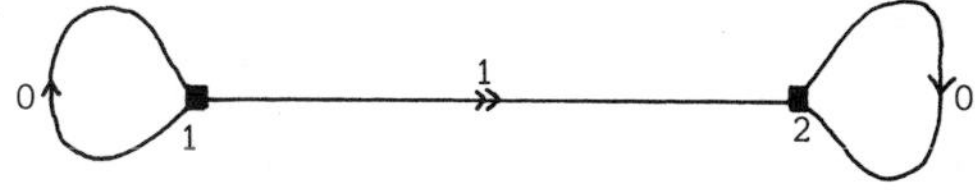

This example has 2 states. State 2 is absorbing and brings no rewards.
In state 1 the only player of the game has two options: staying another period ($T = \{0,1,2,\ldots\}$) without reward or jumping to state 2 with reward 1.
Apparently, the strategy "stay in 1" satisfies (4.2) and is not optimal.

So (4.2) is a necessary condition for equilibriumness but not a sufficient one. (4.2) only requires from a strategy that the players don't loose their prospective rewards. However, it does not guarantee that the players really cash their prospective rewards (see the example above). For this reason we call a strategy that satisfies (4.2) a *conserving* strategy.

For finding an additional condition which might ensure optimality if it is combined with conservingness, we turn to the definition of equilibrium strategy. Suppose u is an equilibrium strategy, then we have by definition

$$\mathbb{E}_u[\psi_t^{(\ell)}(h_t,u) - v_t^{(\ell)}(h_t,u)] = 0 .$$

So we obtain the following trivial statement. If u is optimal, then

$$(4.3) \qquad \lim_{t\to\infty} \mathbb{E}_u[\psi_t^{(\ell)}(h_t,u) - v_t^{(\ell)}(h_t,u)] = 0 \quad \text{for all } \ell .$$

This statement says that in the end the prospective reward is really cashed. A strategy which satisfies (4.3) is said to be *equalizing*. So, we have two properties for equilibrium strategies, namely conservingness saying that prospective reward should be maintained - and equalizingness - saying that prospective reward should be cashed in the end. Now we can hope that these two conditions are (also) sufficient for a strategy to be an equilibrium strategy.

Theorem. A necessary and sufficient condition for a strategy to be an equilibrium strategy is that it is conserving (4.2) and equalizing (4.3).

Proof. The necessity has already been proved. So suppose that u satisfies (4.2) and (4.3). For any t and τ ($\tau \geq t$) we have from (4.2):

$$(4.4) \qquad \mathbb{E}_u\psi_t^{(\ell)}(h_t,u) = \mathbb{E}_u\mathbb{E}_{h_t,u}\psi_\tau^{(\ell)}(h_\tau,u) = \mathbb{E}_u\psi_\tau^{(\ell)}(h_\tau,u) .$$

With (4.3) this implies (remember (4.4) holds for any $\tau \geq t$)

$$\mathbb{E}_u\psi_\tau^{(\ell)}(h_t,u) = \lim_{\tau\to\infty} \mathbb{E}_u v_\tau^{(\ell)}(h_\tau,u) = \mathbb{E}_u r^{(\ell)} = \mathbb{E}_u v_t^{(\ell)}(h_t,u) .$$

Since

$$v_t^{(\ell)}(h_t,u) \leq \psi_t^{(\ell)}(h_t,u)$$

and since both functions have equal expectations, we may conclude that they are equal $\mathbb{P}_u$-a.s.

5. SOME RAMIFICATIONS

A compound equilibrium strategy is not necessarily a sensible strategy. To illustrate one weakness of the concept we give an example of a deterministic 2-person 0-sum game.

Example.

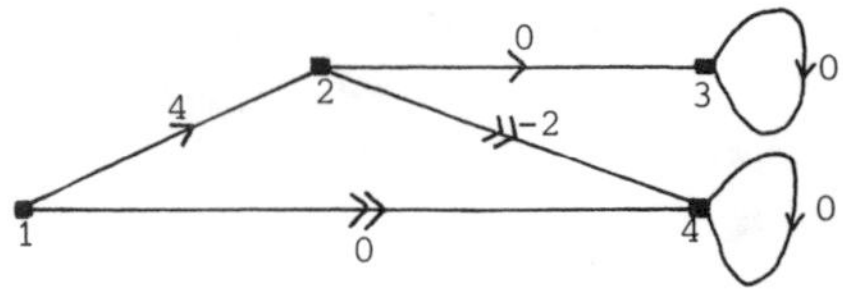

States 3 and 4 are absorbing states without reward. In state 1 the second player can choose between going to 2 (which costs him 4) and going to 4 (which costs him nothing). In state 2 the first player may choose between going to state 3 (without reward) and going to 4 (which costs him 2). Now the strategy for player 1 "go to 3 if the state is 2 at t = 0, otherwise go to 4" is part of an equilibrium strategy when it is combined with "go to 4" for the second player. However, in this way the first player takes unnecessary risks. Namely, if the second player would play stupid, the first player might win 4 units and now only wins 2.

From the example we see that the equilibrium concept might be improved. In fact some improvements have been suggested in the literature (see e.g. Selten [14]).

Below we present 3 types of equilibrium concepts, which are largely based on concepts from the literature. The third implies the second, the second implies the first and the first implies the concept from section 4. The notations are largely self-evident, but will be explained after the definitions.

A strategy u_* is *semi-subgame perfect* iff for all t,ℓ,u

$$\psi_t^{(\ell)}(h_t,u_*) = v_t^{(\ell)}(h_t,u_*/_t u^{(\ell)}) \quad \mathbb{P}_{u_*/u^{(\ell)}}\text{-a.s.}$$

A strategy u_* is *tail optimal* iff for all t,k,ℓ,u

$$\psi_t^{(\ell)}(h_t,u_*/_t u^{(k)}) = v_t^{(\ell)}(h_t,u_*/u^{(k)}) \quad \mathbb{P}_{u_*/u^{(k)}}\text{-a.s.}$$

A strategy u_* is *subgame perfect* iff for all t,ℓ,u

$$\psi_t^{(\ell)}(h_t,{}_t uu_{*t}) = v_t^{(\ell)}(h_t,{}_t uu_{*t}) \quad \mathbb{P}_u\text{-a.s.}$$

Here $u_*/_t u^{(\ell)}$ means the strategy u_* with $u_*^{(\ell)}$ before time t replaced by $u^{(\ell)}$. ${}_t uu_{*t}$ means the strategy which combines the strategies u (before time t) and u_* from time t on.

The difference between these equilibrium concepts can best be seen from examples as the following of a deterministic 2-person 0-sum game in discrete time:

<u>Example</u>.

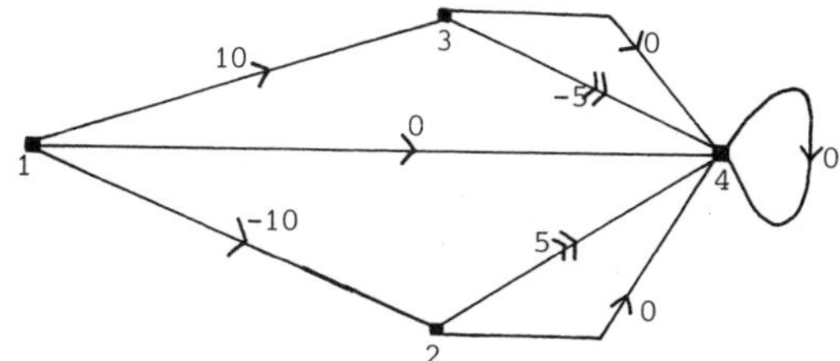

State 4 is absorbing without any further reward. In state 3 the first player can choose between rewards 0 and -5 (without influence for the other player). In state 2 the second player can choose between losses 0 and 5. In state 1 both players have two actions. The reward is 0 if both choose the same action and the first earns 10 for the combination (1,2) and looses 10 for the combination (2,1).
Consider the strategy for player 1 which chooses always action 1 in state 1 and in state 3 uses action 1 if the game starts in 3 and action 2 otherwise.
For player 2 we consider the analogous strategy.
This pair of strategies is semi-subgame perfect but not tail optimal.

Completely analogous to the situation in section 4 for the standard equilibrium concept, one can define conservingness and equalizingness related to these stronger equilibrium concepts. Equally similar one proves that the appropriate conservingness and equalizingness are necessary and sufficient for a compound strategy to be an equilibrium strategy in the related sense (for details see [7, Ch. 6].

Another extension of our theory may found by putting somewhat more structure on the dynamic game. An important example of such a structure is recursiveness, which requires basically that the process (allowed action set and propulsion mechanism) from time t on do not depend on the history before t and it requires that the future rewards except for additive and multiplicative factors only depend on the future of the process. In such a structure it is possible to reformulate the characterization of optimality in local quantities instead of the global quantities of section 4. For one-person decision processes such a reformulation can be found in [7, Ch. 3,4]. For dynamic games it will be worked out in a forthcoming paper. In such a reformulation the characterization is more akin to the usual optimality conditions.

REFERENCES

[1] R. Bellman, Dynamic programming. Princeton, Princeton University Press, 1957.

[2] R. Boel, P. Varaiya, Optimal control of jump processes. SIAM J. Control Optimization 15 (1977), 92-119.

[3] H.A.M. Couwenbergh, Characterization of strong (Nash) equilibrium points in Markov games. Memorandum COSOR 77-09 (April 1977) Eindhoven University of Technology (Dept. of Math.).

[4] L.E. Dubins, L.J. Savage, How to gamble if you must. New York, McGraw-Hill, 1965.

[5] R.J. Elliott, The existence of optimal strategies and saddle in stochastic differential games. p. 123-135 in P. Hagedorn, H.W. Knobloch, G.J. Olsder (eds.), Differential games and Applications. Springer (Lecture Notes in Control and Information Sciences no. 3), Berlin 1977.

[6] L.P.J. Groenewegen, Markov games; properties of and conditions for optimal strategies. Memorandum COSOR 76-24 (November 1976) Eindhoven University of Technology (Dept. of Math.).

[7] L.P.J. Groenewegen, Characterization of optimal strategies in dynamic games. MC-tracts no. 90, Mathematical Centre, Amsterdam 1979 (to appear).

[8] L.P.J. Groenewegen, J. Wessels, On the relation between optimality and saddle-conservation in Markov games. pp. 183-211 in Dynamische Optimierung, Bonn, Math. Institut der Universität Bonn (Bonner Mathematische Schriften 98) 1977.

[9] A. Hordijk, Dynamic programming and Markov potential theory. MC-tract no. 51, Mathematical Centre, Amsterdam 1974.

[10] R. Isaacs, Differential games. J. Wiley, New York 1975 (2nd ed.).

[11] R.P. Kertz, D.C. Nachman, Persistently optimal plans for non-stationary dynamic programming: the topology of weak convergence case. Annals of probability (to appear).

[12] H.W. Kuhn, Extensive games and the problem of information. Annals of Mathematics Study 28 (1953), p. 193-216.

[13] T. Parthasarathy, M. Stern, Markov games - a survey. Technical Report of University of Illinois at Chicage Circle, Chicago 1976.

[14] R. Selten, Reexamination of the perfectness concept for equilibrium points in extensive games. Internat. J. Game Th. 4 (1975), 25-55.

[15] L.S. Shapley, Stochastic games. Proc. Nat. Acad. Sci. USA 39 (1953), 1095-1100.

[16] C. Striebel, Optimal control of discrete time stochastic systems. Springer (Lecture Notes in Econ. and Math. Systems no. 110), Berlin 1975.

[17] W.D. Sudderth, On the Dubins and Savage characterization of optimal strategies. Ann. Math. Statist. 43 (1972), 498-507.

[18] J. van der Wal, J. Wessels, Successive approximation methods for Markov games. p. 39-55 in H.C. Tijms, J. Wessels (eds.), Markov decision theory. MC-tract 93, Mathematical Centre, Amsterdam 1977.

GAME THEORY AND RELATED TOPICS
O. Moeschlin, D. Pallaschke (eds.)

A DYNAMIC MODEL FOR SETTING RAILWAY NOISE STANDARDS

E. Höpfinger
Institute for Economics and Operations Research,
University of Karlsruhe,
Germany

D. von Winterfeldt,
Institute for Applied Systems Analysis,
Laxenburg,
Austria

This paper describes the application of a multistage game theoretical model to setting noise standards which is illustrated by the case of trains. The problem was structured to match the decision problem which the Environment Agency faced when setting standards for Shinkansen trains. The model considers three players: the regulator (environment agency), the producer (railway corporation), and the impactees (residents along the railway line who suffer from noise). The game has seven stages characterized by the actions of the impactees ranging from petitions to legal litigation. The final stages are the outcomes of a possible lawsuit. The case is either won by the producer or the impactees, or a compromise is reached. Transition probabilities between stages are considered parameters of the game. They depend mainly on the noise level the impactees consider acceptable, the standard set by the regulator, and the actual level of noise emitted. Only the regulator and the producer are active players in the sense that they have a set of choices characterized as standard levels (regulator) and noise protection measures (producer). The impactees are modeled as a response function. Several solutions according to a hierarchical solution concept of the game are derived. In particular, conditions are given under which the regulator or the producer would prefer a compromise solution to awaiting the outcome of the court case. These conditions can be expressed directly as functions of noise levels and transition probabilities, given some simple assumptions about the shape of the utility functions of the regulator and the producer.

1. INTRODUCTION

Since the superrapid "bullet train", the Shinkansen, began operations in Japan in 1964, complaints about train noise have never ceased. Peak noise levels can reach over 100 dB leading to substantial disturbances of residential living. Since the responses of the government and the railway corporation to these complaints have been slow, citizens began to go through various forms of protest, including petitions, organizations, and legal litigation. In 1972 the government asked the railway corporation to take urgent steps against Shinkansen noise. But it was not until 1975 that noise standards (70-75 dB) were issued to force the railway corporation to respond to the citizens' need for quietness. Residents, however, were not content with these standards and the railway corporation's subsequent attempts at improving sound protection measures. A legal battle between residents and the railway corporation is still going on in which residents ask to reduce Shinkansen noise to a "nondisturbing" level.

In a recent paper (see [1]) the decision process of the Environment Agency and the railway corporation was described and analyzed. In this analysis the need was recognized for more formal methodologies to study decision making involving the conflict between environmental and developmental interests. The present paper is a first attempt at developing such a methodology based on dynamic game theoretic models. The purpose of such models is to explore alternative strategies of the conflicting actors in environmental standard setting decisions, and to derive essential properties of "optimal" strategies depending on the parameters of the game and alternative solution concepts. In the present absence of a well founded quantitive theory on the consequences of citizens' initiatives such a model has to be rather coarse-structured.

Essentially three groups are involved in typical environment-development conflicts the regulator, the producer (developer), and the impactee (sufferer of pollution). In the case of train noise these groups are an environmental agency (regulator), a railway corporation (producer), and the residents along the line (impactees). Neglecting institutional arrangements, the regulator and the producer are considered single rational players for the purposes of the model. The decisions of the residents are considered (possibly probabilistic) reactions to the decision of the regulator and the producer. Thus the impactee is not modeled as a rational player but rather as a response function: this reflects the experience that citizens' initiatives tend to react only on potential outcomes of developments. The conflict situation between regulator, producer, and residents is formalized as a multistage two-person game, where a state is characterized by the action of the residents or a judgement by a court.

2. THE MODEL

Two-person dynamic or multistage games in extensive form (see [2] or [3]) are regarded that are similar to stochastic games. At each stage a component game of perfect information is played that is completely specified by a state. The players' choices do not control only the payoffs but also the transition probabilities governing the component game to be played at the next stage. It is assumed that the regulator and the producer have the same estimates of the transition probabilities.

A state of the game is characterized by the present upper bound for an admitted noise level, the last action of the residents or a judgement by the court. Seven states are considered. State 1 denotes the state after construction of the railway line. State 2 indicates that a petition has taken place. State 3 records that the population affected by noise has built up an organization for negotiation with government in order to arrive at a low noise standard. If the negotiations fail the residents can start a lawsuit. This is indicated by state 4. The upper bound for the admitted noise level equals the maximum value $\bar{n}$ of noise produced by the train without special sound protection measures for the first four states.

A substantial property of the model is the assumption of a threshold $n_I \leqq \bar{n}$, so that a noise level below n_I is not considered a relevant disturbance of the residents. This seems to be a tolerable simplification. A permanent compromise between all parties, denoted by state 5 will therefore prescribe an upper bound for the admitted noise which is not greater than n_I. State 6 indicates that the lawsuit was decided in a neutral or positive way for the railway corporation and the government which have to take into account an upper bound n_R which might be close to $\bar{n}$ such that $n_I < n_R \leqq \bar{n}$. State 7 indicates that the lawsuit was decided in favor of the residents. Hence the upper bound for the admitted noise may be assumed to be n 5,6, and 7 are final or absorbing states. See also Figure 1. For each state the component game and the transition probability are specified after the introduction of the basic utility functions.

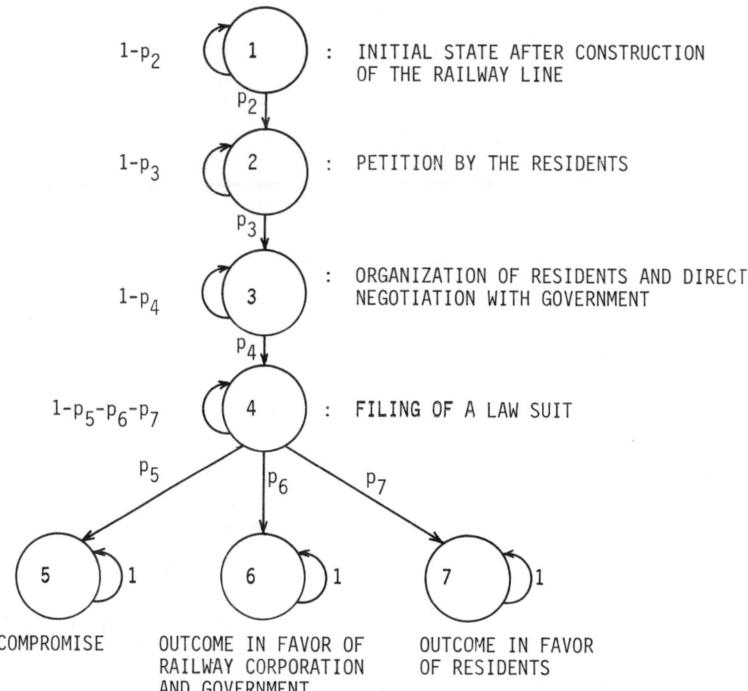

Figure 1. States of the game and transition probabilities (p_i).

The model assumes that the costs and benefits of restricting or increasing noise levels from the train can be expressed as utility functions on noise levels. The utility funstion of the railway corporation is given as

$$u_P : [\underline{n},\bar{n}] \rightarrow \mathbb{R} ,$$

as long as there is no effective action by the resicents. In general, this function will be strictly increasing. $\underline{n}>0$ denotes the minimum value of noise under which the train can be run under economic considerations. In fact, there exists evidence that within reasonable values of $\underline{n}$ and $\bar{n}$ (e.g. 60 and 100 dB, respectively) this function may be linear (see [1]). Thus in some cases it may be possible to express u_P as

$$u_P(n) = n + e \quad ,$$

neglecting a scaling factor.

The utility function of the regulator is also assumed to be defined directly on noise levels:

$$u_R : [\underline{n},\bar{n}] \rightarrow \mathbb{R} .$$

u_R is to reflect a compromise between the economic importance of the train and the noise pollution effects on residents along the line. In the model u_R is assumed to be unimodal with a peak at $\underline{n} \leq L^+ \leq \bar{n}$. The following argument supports the assumption that u_R is unimodal. Assuming that u_R balances environmental and developmental interests, a crude approximation of u_R could be given by

$$u_R = Wu_P + u_I ,$$

where $W>0$ is an importance weight factor which indicates the relative weight of economic considerations, and u_I is the impactee's utility function. From survey data [4,5] one can infer that the strength of complaints to noise (an indicator of u_I) is approximately quadratically related to noise level. Thus neglecting scaling factors

$$u_I = -(n-\underline{n})^2 + f$$

Substituting u_I and u_P in u_R gives

$$u_R = W(n+e) - (n-\underline{n})^2 + f ,$$

which is unimodal with a maximum at $L^+ = \frac{W}{2} + \underline{n}$.

Nevertheless the subsequent analysis only requires a strictly increasing u_P and a unimodal u_R which has a peak at L^+.

In case of the first states i=1,2,3 the component game is specified as follows. First the regulator chooses his measure $m_R \varepsilon M_R(i)$, where $M_R(i)$ denotes the set of measures available to him. Knowing m_R the producer chooses $m_P \varepsilon M_P(i,m_R)$ where $M_P(i,m_R)$ is the set of measures available to him . M_R and M_P are specified by

$$M_R(i) = \{l \mid \underline{n} \leq l \leq \bar{n}\} ,$$

$$M_P(i,l) = \{n \mid \underline{n} \leq n \leq \bar{n}\},$$

where l denotes the highest level of noise the regulator allows, and n the value of noise generated by operating the railway. The residents' choices are not specified because they are formalized by a response function resulting in special transition probabilities.

Given state $i \varepsilon \{1,2\}$ only states i and i+1 can succeed. Regulator and producer believe the transition probabilities to be

$$P(i|i,l,n) = \begin{cases} 1 & \text{if } n \leq n_I \\ 1-p_{i+1} & \text{if } n > n_I \end{cases}$$

and

$$P(i+1|i,l,n) = 1 - P(i|i,l,n)$$

where $p_i>0$ (i=2,3) represent the experts' subjective probabilities that residents will choose a petition, or form an organization, respectively if $n>n_I$.

In case of a formation of an organization , i.e. state 3, residents will begin negotiations aimed at forcing the regulator to give in and set an acceptable standard. Then

$$P(3|3,l,n) = \begin{cases} 1 & \text{if } l \leq n_I \\ 1-p_4 & \text{if } l > n_I \end{cases}$$

$$P(4|3,l,n) = 1 - P(3|3,l,n)$$

where $p_4>0$ and state 4 denotes the start of a lawsuit. The payoff of the component game are specified by

$$U_j(i,l,n) := u_j(n) \quad (j=R,P;\ i=1,2,3)$$

Three outcomes of a lawsuit are considered. There is a compromise (state 5) suspending the lawsuit, or a judgement in favor of regulator and producer (state 6), or a judgement in favor of the residents (state 7).

After the filing of a lawsuit the regulator's measure may fix a bound l for the noise at the current stage and at the same time may make a permanent commitment for a fixed bound Λ at the later stages. Λ could be interpreted as a quality standard to be effective permanently after a fixed period of time has passed. To keep the model small we assume that Λ becomes effective immediately. Since by assumption a threshold n_I exists Λ will only be accepted if $\Lambda \leqq n_I$. Hence the regulator's measure set can be defined by

$$M_R(4) := \{l \mid \underline{n} \leqq l \leqq \bar{n}\}$$

where $l \leqq n_I$ is interpreted as a permanent upper bound Λ. Analogously a measure of the producer might be fixed by a pair (n,N) of numbers where n denotes the actual noise level at the current stage, while N denotes a commitment made by the producer to regard this limit from now on. By the same arguments the producer's measure set can be defined by

$$M_P(4,l) = \{n \mid \underline{n} \leqq n \leqq l\}$$

where $n \leqq n_I$ may be interpreted as a permanent upper bound for noise.

Let $$M_C := \{(l,n) \mid \min(l,n) \leqq n_I,\ l \varepsilon M_R(4),\ n \varepsilon M_P(4,l)\}$$

be called the set of compromise pairs of choices. M_C contains just the pairs (l,n) of measures guaranteeing to the residents that from now on no noise level greater than n_I will occur. Let

$$P(5|4,l,n) = \begin{cases} 1 \text{ if } (l,n) \notin M_C \\ 0 \text{ else} \end{cases}$$

$$P(6|4,l,n) = \begin{cases} p_6 & \text{if } (l,n) \notin M_C \\ 0 & \text{else} \end{cases}$$

$$P(7|4,l,n) = \begin{cases} p_7 & \text{if } (l,n) \notin M_C \\ 0 & \text{else} \end{cases}$$

where $\underline{n} \leqq n_I \leqq n_R \leqq \bar{n}$ holds for the maximum noise level n_R decreed by a court judgement in favor of the producer, and p_6+p_7 need not equal 1. Hence

$$P(4|4,l,n) = \begin{cases} 0 & \text{if } (l,n) \varepsilon M_C \\ 1-p_6-p_7 & \text{else} \end{cases}$$

The payoffs are specified by

$$U_j(4,l,n) = u_j(n) \quad (j=R,P)$$

State 5 means that either the regulator has agreed to take n_I as the maximum level of noise, or that the producer has bound himself to noise levels not larger than n_I. Let the sets of measures be given by

$$M_R(5) := \{l \mid \underline{n} \le l \le n_I\}$$

$$M_P(5,l) := \{n \mid \underline{n} \le n \le l\}.$$

Then

$$P(5 \mid 5,l,n) = 1$$

The payoffs are specified by

$$U_j(5,l,n) = u_j(n) \qquad (j=R,P;\ \ \underline{n} \le n \le l)$$

State 6 indicates a sentence unfavorable to the residents. Let

$$M_R(6) = \{l \mid \underline{n} \le l \le n_R\} \ ,$$

$$M_P(6,l) = \{n \mid \underline{n} \le n \le l\} \ .$$

Then $P(6 \mid 6,l,n) = 1$ and $U_j(6,l,n) = u_j(n) \quad (j=R,P)$

State 7 denotes a sentence unfavorable to regulator and producer. Let

$$M_R(7) = \{l \mid \underline{n} \le l \le n_I\},$$

$$M_P(7,l) = \{n \mid \underline{n} \le n \le l\}.$$

Then $P(7 \mid 7,l,n) = 1$

In the case of a lost lawsuit the producer's and the regulator's utilities change. This is because such a sentence would have much wider reaching consequences than a voluntary agreement to a standard. First of all, implementation time, rules of operation, etc. prescribed in a sentence would mean substantial restriction of freedom to the railway corporation. Secondly, the sentence would most likely be applied throughout the railway network. Thus the model assumes that

$$U_P(7,l,n) = u_P(n) + c_P \ ,$$

where $c_P<0$ is a fixed penalty as a result of the sentence. Also, the regulator stands to lose both in prestige and in lost flexibility if the court should decide in favor of the impactees. Again this loss is expressed in his utility function.

$$U_R(7,l,n) = u_R(n) + c_R \ .$$

In the case of $L^+>n_I$ it appears not unreasonable to assume that c_j $(j=P,R)$ is a negative multiple m_j of $u_j(L^+) - u_j(n_I)$, i.e.

$$c_j = -m_j[u_j(L^+) - u_j(n_I)] \quad (j=P,R)$$

A play π of the game is given by an infinite sequence $(s^1,m_R^1,m_P^1;\ s^2,m_R^2,m_P^2;\ \ldots)$ of states and measures. We define the utility of a play π by the discounted infinite sum of the stage utilities

$$\underline{U}_j(\pi) := \sum_{i=1}^{\infty} \rho^{i-1} U_j(s^i,m_R^i,m_P^i) \qquad (j=R,P)$$

where $0<\rho<1$ is a discount factor.

The game is now completely described except for the definition of strategies and the solution concept. To keep the analytical part of the paper small we admit only stationary strategies where the choices depend only on the last state and the last measure of the other player. This is consistent with the definition of the response function of the impactee. However it seems that for the values of the parameters for which the solutions were derived non-stationary strategies might be proven to be irrelevant. This need not be the case for values of the parameters for which a solution was not derived.

Definition: A strategy σ_R of the regulator is a map

$$\sigma_R : S \to [\underline{n},\bar{n}] ,$$

such that

$$\sigma_R(s) \in M_R(s) \quad (s \in S) ,$$

where S denotes the set of states.

A strategy σ_P of the producer is a map $\sigma_P : \{(s,l) | s \in S, l \in M_R(s)\} \to [\underline{n},\bar{n}]$, such that

$$\sigma_P(s,l) \in M_P(s,l)$$

The sets of strategies are denoted by Σ_R and Σ_P.

For each strategy pair (σ_R,σ_P) a play $\pi = (s^1,l^1,n^1; s^2,l^2,n^2; \ldots)$ is realized. Since the strategies are stationary, two components (s^i,l^i,n^i) and (s^r,l^r,n^r) are equal as soon as $s^i=s^r$. By the definition of the transition probabilities at most seven states can occur with probability greater than zero and only one will be repeated infinitely often. From this it follows that the set $\Pi(\sigma_R,\sigma_P)$ of possibly realized plays π is finite or denumerable. The probability $P(\pi|\sigma_R,\sigma_P)$ for $\pi \in \Pi(\sigma_R,\sigma_P)$ is given as an infinite product of the terms $P(s^{i+1}|s^i,l^i,n^i)$ defined above. The payoff of player $j \in \{R,P\}$ is supposed to be his expected utility of the plays:

$$V_j(\sigma_R,\sigma_P) = \sum_{\pi \in \Pi(\sigma_R,\sigma_P)} \underline{U}_j(\pi) P(\pi|\sigma_R,\sigma_P) .$$

The strategies are to be determined according to the following solution concept.

Definition: A hierarchical solution is a pair (τ_R,τ_P) of a strategy $\tau_R \in \Sigma_R$ and a map $\tau_P : \Sigma_R \to \Sigma_P$ such that

$$V_P(\sigma_R,\tau_P(\sigma_R)) = \max_{\sigma_P \in \Sigma_P} V_P(\sigma_R,\sigma_P) ,$$

$$V_R(\tau_R,\tau_P(\tau_R)) = \max_{\sigma_R \in \Sigma_R} V_R(\sigma_R,\tau_P(\sigma_R)) .$$

3. THE GAME-THEORETIC SOLUTION

In order to keep the analytical part as small as possible we shall only discuss heuristic equations which, however, can be justified sa soon as one establishes the analytical framework in full detail. At least part of it can be found in [7].

For the rest of the paper let $\Gamma_b(b=1,\ldots,7)$ denote subgames of the original game such that b is the first state. Hence Γ_1 is the original game. $\Gamma_b(b=5,6,7)$ has only the state b. $\Gamma_b(b=1,2,3,4)$ covers states b, b+1,...,7. Though in principle one has to distinguish the strategies for different Γ_b we denote by abuse of notation the reduction of $\sigma_j \varepsilon \Sigma_j$ to Γ_b by σ_j. Let $V_{j,b}$ denote the payoff function for player j in game Γ_b. Without the simple proof we state that

$$V_{j,7}(\sigma_R,\sigma_P) = \frac{1}{1-\rho}(u_j(\sigma_P(7,\sigma_R(7)) + c_j)$$

$$V_{j,b}(\sigma_R,\sigma_P) = \frac{1}{1-\rho} u_j(\sigma_P(b,\sigma_R(b))) \qquad (b = 5,6)$$

for j=R,P. Now let $(\sigma_R,\sigma_P)(i) := (\sigma_R(i),\sigma_P(i,\sigma_R(i)) \quad (i=1,\ldots,7)$. Then

$$\begin{aligned} V_{j,4}(\sigma_R,\sigma_P) = {} & u_j(\sigma_P'(4,\sigma_R(4))) \\ & + \rho\, P(4|4,(\sigma_R,\sigma_P)(4))V_{j,4}(\sigma_R,\sigma_P) \\ & + \rho\, P(5|4,(\sigma_R,\sigma_P)(4))V_{j,5}(\sigma_R,\sigma_P) \\ & + \rho\, P(6|4,(\sigma_R,\sigma_P)(4))V_{j,6}(\sigma_R,\sigma_P) \\ & + \rho\, P(7|4,(\sigma_R,\sigma_P)(4))V_{j,7}(\sigma_R,\sigma_P) \end{aligned}$$

By backward iteration

$$\begin{aligned} V_{j,b}(\sigma_R,\sigma_P) = {} & u_j(\sigma_P(b,\sigma_R(b))) \\ & + \rho\, P(b|b,(\sigma_R,\sigma_P)(b))V_{j,b}(\sigma_R,\sigma_P) \\ & + \rho\, P(b+1|b,(\sigma_R,\sigma_P)(b))V_{j,b+1}(\sigma_R,\sigma_P) \qquad (b=1,2,3;\ \ j=R,P) \end{aligned}$$

Three situations are conceivable:

(1) The regulator can enforce his maximum utility;
(2) If regulator and producer have won the lawsuit, the regulator has to offer $l > L^+$ in order to keep the producer from compromising;
(3) If regulator and producer have won the lawsuit, not even the offer $l = n_R$ can keep the producer from compromising.

Though the calculation of the hierarchical solution for three situations is not difficult for any given set of values of the parameters, the derivation of the hierarchical solution as a function of the parameters would require a lot of space. Therefore we consider only the first and the third situation. The two classes of parameters given, however, do not in general exhaust the set of all the parameter values possible.

At first we establish a pair of strategies yielding the maximum utility to the regulator.

Definition: Let $L^+ > n_I$. The vector of real numbers $(L^+,n_I,n_R, ,p_6,p_7)$ satisfies the compromise condition of player j (C,j) if

$$u_j(n_I) > \frac{1}{1-\rho(1-p_6-p_7)}\{(1-\rho)u_j(L^+) + p_6\rho u_j(\min(L^+,n_R)) + p_7\rho[u_j(n_I)+c_j]\}$$

holds.

As can be seen by the formulae above, (C,j) indicates that a compromise is more advantageous to player j.

Theorem: Let $\phi\varepsilon\Sigma_R$ and $\Psi\varepsilon\Sigma_P$ be defined by

$$\phi(i) := L^+(i=1,2,3), \quad \phi(5) := \phi(7) := n_I$$

$$\phi(6) := \min(L^+,n_R)$$

$$\phi(4) := \begin{cases} n_I & \text{if } L^+>n_I \text{ and } (C,R) \text{ holds} \\ L^+ & \text{if } L^+\leqq n_I \text{ or } (C,R) \text{ is violated} \end{cases}$$

$$\Psi(i,1) := 1(i=1,2,\ldots\ldots,7)$$

Then (ϕ,Ψ) yields the maximal utility to the regulator:

$$V_R(\phi,\Psi) = \sup_{\Sigma_R\times\Sigma_P} V_R(\sigma_R,\sigma_P)$$

In order to avoid a lengthy and not instructive proof we only give the idea of the proof. First let $L^+\leqq n_I$. Because of the definition of $U_R(i,m_R,m_P)$ the inequality holds for all possible states i and measures $m_R$und m_P. Hence $V_R(\sigma_R,\sigma_P)\leqq\frac{1}{1-\rho}u_R(L^+)$ for each $(\sigma_R,\sigma_P)\,\varepsilon\,\Sigma_R\times\Sigma_P$. But $V_R(\phi,4) = \frac{1}{1-\rho}u_R(L^+)$ because of $\phi(1) = L^+$ and $P(1|1,(\phi,\Psi)(1)) = 1$. Now let $L^+>n_I$. Obviously $V_{R,j}(\sigma_R,\sigma_P)\leqq V_{R,j}(\phi,\Psi)$ $(j=5,6,7)$. Then (σ_R,σ_P) with $(\sigma_R,\sigma_P)(i) = (\phi,\Psi)(i)$ $(i=5,6,7)$ maximizes $V_{R,4}(\sigma_R,\sigma_P)$ if $(\sigma_R,\sigma_P)(4)=(\phi,\Psi)(4)$ under consideration of the compromise condition (C,R). Hence $V_{R,4}(\sigma_R,\sigma_P)\leqq V_{R,4}(\phi,\Psi)$. The final step of the backward iteration yields $V_R(\sigma_R,\sigma_P)$ $\leqq V_R(\phi,\Psi)$ for each pair $(\sigma_R,\sigma_P)\,\varepsilon\,\Sigma_R\times\Sigma_P$.

If Ψ is an optimal response to ϕ, i.e. $V_P(\phi,4) = \sup_{\Sigma_P} V_P(\phi,\sigma_P)$, it is not important to derive a hierarchical solution since the regulator can enforce his maximum payoff.

Definition: The payoff vector $(V_R(\sigma_R,\sigma_P),\ V_P(\sigma_R,\sigma_P))$ is Pareto-optimal if there is no other strategy pair $(\sigma_R',\sigma_P'\,\varepsilon\,\Sigma_R\times\Sigma_P)$ such that $V_j(\sigma_R,\sigma_P)\leqq V_j(\sigma_R',\sigma_P')$ $(j=R,P)$ and that at least one inequality is strict.

Theorem: Let $(\phi,\Psi)\,\varepsilon\,\Sigma_R\times\Sigma_P$ be defined as in the preceding theorem. Then Ψ is an optimal response to ϕ, i.e. $V_P(\phi,\Psi)\geqq V_P(\phi,\sigma_P)$ $(\sigma_P\varepsilon\Sigma_P)$, and $V_R(\phi,\Psi)\geqq$ $V_R(\sigma_R,\sigma_P)$ $(\sigma_R\varepsilon\Sigma_R,\ \sigma_P\varepsilon\Sigma_P)$ if one of the following conditions holds

(1) $L^+\leqq n_I$;

(2) $L^+>n_I$ and (C,R);

(3) $L^+>n_I$ and not only (C,R) but also (C,P) is violated.

In these cases $(V_R(\phi,\Psi),V_P(\phi,\Psi))$ is a Pareto-optimal payoff vector.

Sketched proof:

Case (1): Because of $\phi(1) = L^+\leqq n_I$ $P(1|1,(\phi,\sigma_P)(1)) = 1$. Hence $V_P(\phi,\sigma_P)=$ $\frac{1}{1-\rho}u_P(\sigma_P(1,L^+))$ where $\sigma_P(1,L^+)\leqq L^+$ is maximized by Ψ. In order to obtain a greater payoff $V_P(\sigma_R,\sigma_P)$ for one stage at least L^+ has to be replaced by $n>L^+$. But then the regulator's payoff is smaller because of $u_R(n)<$ $u_R(L^+)$.

Case (2): $P(5|4,(\phi,\sigma_P)(4)) = 1$ because of $\phi(4) = n_I$. By backward iteration evaluating $V_{P,i}(\phi,\sigma_P)$ (i=5,4,3,2,1) one immediately sees that Ψ maximizes $V_P(\phi,.)$. The proof of the Pareto-optimality relies on the fact that only strategies σ_R with $\sigma_R(i) = \phi(i)$ (i=1,..,5) give maximal payoff to the regulator. The verification of this fact requires a lengthy and uninstructive discussion which we therefore omit.

Case (3): Given ϕ the assessment $\Psi(i,1):=1$ (i=5,6,7) belongs to an optimal response for all values of the parameters. Because of $\phi(4) = L^+$ and $L^+>n_I$ a strategy σ_P maximizeing $V_{P,4}(\phi,.)$ takes either the value $\sigma_P(4,L^+) = n_I$ or the value L^+. Since (C,P) is violated the second assessment yields a larger utility. Hence Ψ maximizes $V_{P,4}(\phi,.)$. Then oviously Ψ maximizes $V_{P,i}(\phi,.)$ (i=3,2,1). The Pareto-optimality of $(V_R(\phi,\Psi), V_P(\phi,\Psi))$ can again be verified by changing some values of (ϕ,Ψ) (i) proving that they reduce the regulator's payoff.

If (C,P) holds and (C,R) is violated the strategy Ψ is generally not an optimal response of ϕ. The situation can arise where the regulator by reduction of his own payoff can force the maximizing producer to a no-compromise strategy. In order to keep the analytical part small we only treat a special case where this situation cannot arise.

Definiton: The vector $(\underline{n},n_I,n_R,\bar{n},\rho,p_6,p_7)$ satisfies the strict compromise condition (SC) if

$$u_P(\underline{n}) > \frac{1}{1-\rho(1-p_6-p_7)} \{(1-\rho)u_P(\bar{n}) + p_6\rho u_P(n_R) + p_7\rho[u_P(n_I) + c_P]\}$$

holds.

(SC) can be interpreted by the way that the utmost offer and threat of the regulator cannot match the value of a compromise for the producer.

Theorem: Let (SC) hold. A hierarchical solution (τ_R,τ_P) is given by $\tau_R=\phi$ and $\tau_P(\sigma_R) = \gamma\varepsilon\Sigma_P$ for each $\sigma_R\varepsilon\Sigma_R$ where

$$\gamma(i,1) := 1 \quad (i=1,2,3,5,6,7)$$

$$\gamma(4,1) := \min(1,n_I)$$

Sketched proof:
Because of (SC) the second component of $\sigma_P(4,\sigma_R(4))$ equals n_I for any optimal response σ_P of any $\sigma_R\varepsilon\Sigma_R$. By backward iteration one immediately sees that γ is an optimal response of each $\sigma_R\varepsilon\Sigma_R$, i.e. $V_P(\sigma_R,\sigma_P) \leq V_P(\sigma_R,\gamma)$. $V_{R,5}(.,\gamma)$ is maximized by ϕ and , more generally, $V_{R,i}(.,\gamma)$ (i=4,3,2,1) as one can see by backward iteration.

Remark:
In case of $L^+ > n_I$ and (SC) but violated (C,R) the regulator generally does not obtain the possible maximum payoff

$$V_R(\phi,\Psi) \underset{\neq}{>} V_R(\phi,\gamma) .$$

Part of the results can be given in a more illustrative way. In the case of $n_I < L \leq n_R$ let

$$c_j = -m_j(u_j(L^+) - u_j(n_I)) \quad (j=R,P)$$

m_j is assumed to be a constant positive factor. It specifies the weight of the severe consequences of a noise reducing sentence has to be regarded for all other noise-producing activities. A short calculation yields that (C,j) is equivalent to

$$m_j p_7 \rho > 1-\rho(1-p_6) \quad (j=R,P)$$

The second theorem implies that in the case of $n_I < L^+ \leqq n_R$ and $m_R p_7 \rho > 1-\rho(1-p_6)$ the regulator prefers the compromise: $\phi(4) = n_I$. In case of $n_I < L^+ \leqq n_R$ and $m_j p_7 \rho \leqq 1-\rho(1-p_6)$ (j=R,P), however, the lawsuit will result in a sentence: $(\phi,\Psi)(4) = (L^+,L^+)$.

An elementary calculation shows that the expected duration d of the lawsuit is $d = \frac{1}{p_6+p_7}$. Given d, condition (C,j) is equivalent to

$$\frac{1}{d} \geqq p_7 > \frac{1-\rho+\rho\frac{1}{d}}{\rho(m_j+1)} \qquad (j=R,P)$$

The following example illustratesthe relevance of the results. Let d=4 years, $\rho=0.9$, and $m_R=m_P=10$. Then (C,j) (j=R,P) is approximately given by $p_7 > 0.03$. Hence a lawsuit should only be executed if the probability for a sentence in favor of the residents in one year is not greater than three percent. If $p_7=0.03$ then the probability of such a sentence being pronounced at all is $dp_7=0.12$.

4. CONCLUSIONS

A main element of the model is the consideration of the impactees' reactions in standard setting. Under certain assumptions the model could identify the important areas in the decision process of the regulator and the producer. In particular the decision about offering and accepting or rejecting a compromise turned out to be a crucial importance. This decision could be determined as a function of the model parameters in which the subjective probabilities of the outcome of the court proceedings play a major role.

Model limitations include the "short-sightedness" of the impactees' response which only covers present standards an noise levels. Consequently the strategies of the regulator and the producer do not include commitments for later time periods, e.g. in the form of quality standards. The model results indicate, however, that such extensions are feasible, although at a substantially greater effort. For example, strategies could be in the form of long-term noise reduction plans instead of short-term standards, and impactees responses would take into account the nature of these plans.

REFERENCES

[1] von Winterfeldt, D., Standards Against Noise Pollution: The Case of Shinkansen Trains in Japan, Technical Report TR-4, Volkswagenwerk Foundation, International Institute for Applied Systems Analysis, Laxenburg, Austria,1978

[2] Owen, G., Game Theory, W.B. Sanders Co., Philadelphia, 1968.

[3] McKinsey, J.C.C., Introduction to the Theory of Games, Rand Corporation, McGraw-Hill, 1952.

[4] Hashimoto, M., Present Status of Noise Control in Japan, Inter Noise, 1975, 718-729.

[5] Yorino, T., Environmental Problems of the Shinkansen, in A. Straszak (ed.) Proceedings of the Shinkansen Conference. International Institute for Applied Systems Analysis, Laxenburg, Austria, 1978.

[6] Höpfinger, E. and R. Avenhaus, A Game Theoretic Framework for Dynamic Standard Setting Procedures, RM-78, International Institute for Applied Systems Analysis, Laxenburg, Austria, 1978.

[7] Hinderer, K., Foundations of Nonstationary Dynamic Programming with Discrete Time Parameter, in Lecture Notes in Operations Research and Mathematical Systems, Springer Verlag, Berlin/Heidelberg/New York, 1970.

GAME THEORY AND RELATED TOPICS
O. Moeschlin, D. Pallaschke (eds.)

ASYMPTOTIC VALUES OF MIXED GAMES*

Abraham Neyman
Department of Mathematics
University of California
Berkeley, California

1. INTRODUCTION

Since 1960 attention has focused more and more on games with large masses of players,[1/] i.e., where some of the participants are individually insignificant. Milnor and Shapley (1961), Shapiro and Shapley (1960), Shapley (1961) and Hart (1973), investigated value theories of "oceanic games," i.e., weighted majority games in which a sizable fraction of the total vote is controlled by a few large (major) players, and the rest is distributed among a large number of small (minor) voters. Shapiro and Shapley (1960), Milnor and Shapley (1961), and Shapley (1961), presented asymptotic results for the values of the major players, when the others become smaller and smaller. As for the minor ones, finding the limit of their values, turned out to be a much more difficult task; even in the case where there are no major players, this was an open problem for many years - only recently solved by the author (1979).

The main purpose of this paper is to settle the above question in general - i.e., for games with <u>both</u> major and minor voters. Intuitively, the result is that, for a coalition of small players, the (limit) value does not depend on its composition, but <u>only</u> on the total vote it has. More precisely, we consider two measures on the set of small players: the 'voting' and the 'value'. We prove that, as the largest minor vote tends to zero, the distance between the above two measures (defined as the bounded variation of the difference of their normalizations) also tends to zero.

The problem finds its natural and more general setting in the context of values of games with a continuum of players. The central interest is in those that are obtained as limits of values of finite approximants. The <u>asymptotic</u> value is the "strongest" possible such value in the sense, that if it exists for a

---

*This work was supported by National Science Foundation Grant SOC75-21820-A01 at the Institute for Mathematical Studies in the Social Sciences, Stanford University.

[1/]Kuhn and Tucker list fourteen outstanding research problems in |10|. The eleventh urged us "to establish significant asymptotic properties of n-person games, for large n" (|10|, p. xii).

particular game v, then any limiting value[2/] will exist for that game and equal the asymptotic one. It should be pointed out that the existence of the asymptotic value for v is essentially a strong statement on the limit of the values of games with finitely many players[3/] (which approximate v).

The asymptotic value of games with a continuum of players has been studied extensively, (|10|, |5|, |7|, |9| and |15|). The set of all games having an asymptotic value is denoted by ASYMP. It has long been known [Kannai (1966), Aumann-Shapley (1974)] that non-atomic games that are "sufficiently differentiable" (i.e., games in pNA) have asymptotic values (i.e., $pNA \subset ASYMP$). Recently [Neyman (1979)] established the existence of an asymptotic value for singular non-atomic games (i.e., proved that $bv'NA \subset ASYMP$). As for mixed games, i.e., games with finitely many large players it has been shown [Fogelman and Quinzii (1975)] that mixed games that are sufficiently differentiable are in ASYMP (i.e., that $pFL \subset ASYMP$). The main result of this paper is that singular mixed games are in ASYMP i.e. to prove that $bv'FL \subset ASYMP$.

2. STATEMENT OF THE MAIN RESULT

A game is a set-function $v\colon \mathcal{C} \to \mathbb{R}$, where $(I,\mathcal{C})$ is a (standard) measurable space (the player space) with $v(\emptyset) = 0$; it is called finite if $\mathcal{C}$ if finite. The Shapley value of a finite game v is the measure on $\mathcal{C}$ given by

$$\phi v(A) = E(v(P^{\mathcal{R}}_A \cup A) - v(P^{\mathcal{R}}_a)) \tag{2.1}$$

where $P^{\mathcal{R}}_A$ is the set of players (atoms of $\mathcal{C}$) preceding A (an atom of $\mathcal{C}$) in the order $\mathcal{R}$, and E is the expectation operation when each order has equal probability |18|. To define the asymptotic value for a game v that is not necessarily finite, one approximate it by finite games. Specifically, if Π is a finite subfield of $\mathcal{C}$, define a finite game v_Π on Π by $v_\Pi = v|\Pi$. Given an S in $\mathcal{C}$ (a "coalition"), an increasing sequence $\{\Pi_1,\Pi_2,\ldots\}$ of finite subfield of $\mathcal{C}$ is called S-admissible if $S \in \Pi_1$ and $\cup_i \Pi_i$ generates $\mathcal{C}$. An asymptotic value of v is a set function ψv on $\mathcal{C}$ such that for all coalitions S and all S-admissible sequence, we have

$$\lim \phi v_{\Pi_n}(S) = \psi v(S) \quad . \tag{2.2}$$

[2/] Like the μ-value |2|, |3| and the partition value |16|.

[3/] For example, in (1977) Aumann-Kurz used the μ-value as the underlying value concept, which restricted the conclusions of their model to democratic societies. Later, when their games were shown to have asymptotic value (cf. |15|), it led to more general results.

The set of all games having asymptotic values is denoted by ASYMP.

Let $(I,\mathcal{C})$ be a measurable space isomorphic to $([0,1],\mathcal{B})$ where $\mathcal{B}$ is the σ-field of Borel sets in $[0,1]$. We denote by FL the set of all measures on $(I,\mathcal{C})$ with finitely many atoms. The space of all real-valued functions f of bounded variation on $[0,1]$ that obey $f(0) = 0$ and are continuous at 0 and 1 is denoted bv'. The closed subspace of BV spanned by the set functions of the form $f \circ \mu$, where $\mu \in FL$ is a probability measure and $f \in bv'$ is denoted $bv'FL$.

Main Theorem: $bv'FL \subset ASYMP$.

3. VALUES FOR FINITE GAMES - PREPARATIONS FOR THE PROOF

We begin by recalling that a finite game in coalitional (or characteristic function) form (a finite game for short) is usually represented as a pair (N,v), where N is a finite set and v is a real-valued function on the family 2^N of all subsets of N, with $v(\emptyset) = 0$ (clearly, this is equivalent to our definitions in the previous section; N is the set of atoms of the finite field $\mathcal{C}$). We may consider ϕv as a measure on N.

The formula (2.1) for the Shapley value uses the finite probability space of the orders on N. It turns out that it is much more convenient and powerful to replace that discrete probability space by a continuous one.

This is done as follows: Let (Ω,Σ,P) be a probability space such that to every $i \in N$ corresponds a real-valued random variable X_i, defined on (Ω,Σ,P), having uniform distribution on $(0,1)$; furthermore, let the random variables X_i be mutually independent. This "continuous embedding" induces, for almost all $\omega \in \Omega$, an order $\mathcal{R}(\omega)$ on N by $i\mathcal{R}(\omega)j$ iff $X_i(\omega) < X_j(\omega)$, and for every order $\mathcal{R}$ on Ω Prob $(\mathcal{R}(\omega) = \mathcal{R}) = 1/|N|!$. In what follows we will use $\mathcal{P}_i^{\omega}$ or $\mathcal{P}_i$ instead of $\mathcal{P}_i^{\mathcal{R}(\omega)}$. Observe that the stochastic process $N^t\colon [0,1] \to 2^N$ defined by $N^t(\omega) = \{i\colon i \in N, X_i(\omega) \le t\}$, is nondecreasing, has stationary increments, which are sums (unions) of independent random variables, $N^0 = \emptyset$, and $N^1 = N$. In particular, if w is a measure on N, then $w(N^t)$ is a sum of <u>independent</u> (real-valued) random variables. For $\omega \in \Omega$, $i \in N$ and a game v on N we define

$$\Delta(i,\omega) = v(N^{X_i(\omega)}) - v(N^{X_i(\omega)} \setminus \{i\}) \quad .$$

Obviously $\mathcal{P}_i^\omega = N^{X_i(\omega)}\setminus\{i\}$ and therefore,

$$\phi v(i) = E(\Delta(i,\omega)) = \int_0^1 E(\Delta(i,\omega)|X_i = t)\cdot dt \quad . \tag{3.1}$$

A weighted majority game is one of the form $f_g \circ w$ where w is a non-negative measure on the players' set N (called: the voting measure), $0 < q < w(N)$, and $f_q(x) = 0$ or 1, according to $x < q$ or $x \geq q$. It is normalized if w is a probability measure, i.e., $w(N) = 1$. In the case of a finite (or countable) N, we shall sometimes use a more explicit symbol for the game, namely $[q; w_1, w_2, \ldots]$. Here $N = \{1,2,\ldots\}$ and w_i stands for $w(\{i\})$. We will also use M_i for $M\setminus\{i\}$; in particular $N_i^t = N^t\setminus\{i\}$.

Lemma 1: Let $v \equiv [q; w_1,\ldots,w_n]$ be a finite weighted majority game, with $0 < q < w(N)$. Then

$$\phi v(i) = \int_0^1 \text{Prob}\ (w(N_i^t) \in [q - w_i, q))\cdot dt$$

$$= E\left(\int_0^1 \chi_{[q-w_i,q)}(N_i^t)\cdot dt\right) \quad .$$

Proof:

$$\phi v(i) = E(\Delta(i,\omega)) = \text{Prob}\ (w(\mathcal{P}_i) \in [q - w_i, q))$$

$$= \int_0^1 \text{Prob}\ (w(\mathcal{P}_i) \in [q - w_i, q)|X_i = t)\cdot dt$$

$$= \int_0^1 \text{Prob}\ (w(N_i^t) \in [q - w_i, q))\cdot dt \quad ,$$

and apply Fubini's theorem.

Lemma 2: Let $v = [q; w_1,\ldots,w_n]$ be a weighted majority game, and let $T = \inf\{t: w(N^t) \geq q\}$. Then

$$\phi v(i) = \text{Prob}\ (T = X_i) \quad .$$

Proof: Follows easily from the finiteness of N.

We turn now to the "key" lemma, which is a reformulation, in terms of weighted majority games, of the main result of |14|.

Lemma 3: For every $\varepsilon > 0$ there exist $K > 0$ and $\delta > 0$ such that if $v = [q; w_1,\ldots,w_n]$ is a normalized weighted majority game with $w_i < \delta$ and $K\cdot w_i < q < 1 - K\cdot w_i$ for every $i \in N$, then

$$\sum_{i=1}^n |\phi v(\{i\}) - w_i| < \varepsilon \quad .$$

Consider a sequence of $(m + n_k)$-person normalized weighted majority games

$$v_k = [q; w_1^k, \ldots, w_m^k, w_{m+1}^k, \ldots, w_{m+n_k}^k]$$

such that

(3.2) $$\sum_{j=1}^{n_k} w_{m+j}^k \to \alpha > 0 \quad \text{as} \quad k \to \infty$$

(3.3) $$w_i^k \to w_i \qquad \text{as} \quad k \to \infty$$

and

(3.4) $$\max_j w_{m+j}^k = w_{max}^k \to 0 \quad \text{as} \quad k \to \infty \quad .$$

We shall use the following notations: M for the set $\{1,\ldots,m\}$ of "major" players; N_k for the set $\{m + 1,\ldots,m + n_k\}$ of "minor" players; w^k for the voting measure on $M \cup N_k$; and w for the 'limiting' voting measure on M. (Note that the players do not retain their identities from game to game in the sequence, nevertheless, in our continuous embedding, $X_1,\ldots,X_m$ will be independent of k.) We associate to that sequence of games the "ideal" stochastic process Z^t defined by $Z^t = w(M^t) + \alpha t$, and we define the <u>hitting time</u> of $[q,1]$ by $t_q = \inf\{t: Z^t \geq q\}$. The stochastic process Z^t and the hitting time t_q are "limits" of the "actual" stochastic processes $Z_k^t = w^k(M^t) + w^k(N_k^t)$, and hitting times $t_q^k = \inf\{t: Z_k^t \geq q\}$.

<u>Lemma 4</u>: For every $\varepsilon > 0$, $0 \leq t \leq 1$, and $0 < q < 1$

$$\text{Prob}\,(|Z_k^t - Z^t| > \varepsilon) \xrightarrow[k\to\infty]{} 0$$

and

$$\text{Prob}\,(|t_q - t_q^k| > \varepsilon) \xrightarrow[k\to\infty]{} 0 \quad .$$

<u>Proof</u>: The first part follows from (3.3) and Chebyshev's inequality (the weak law of large numbers) applied to $w^k(N^t)$ (which is a sum of independent random variables). The second part is then implied by the observation that $Z^{t_2} - Z^{t_1} \geq \alpha(t_2 - t_1)$ whenever $0 \leq t_1 < t_2 \leq 1$.

Let $A_i(\omega) = (w(M_i^{X_i}), w(M^{X_i})]$, and let $\phi_i = \text{Prob}\,(q \in A_i(\omega)) = \int_0^1 \text{Prob}\,(q \in A_i(\omega) | X_i = t)dt$.

<u>Lemma 5</u>: Let ϕ_i^k denote the value of the game v_k to the i-th major player. Then

$$\phi_i^k \to \phi_i \quad \text{as} \quad k \to \infty \quad .$$

Proof: This is Theorem 1 in Milnor-Shapley (1961) (alternatively, Theorem 1 in Shapiro-Shapley (1960)). For completeness we present here a short proof, based on our continuous embedding. Let $B_q = \{t\colon t\alpha + w(S) = q$ for some $S \subset M\}$. The set B_q is finite. For every $t \notin B_q$, $\min_S |w(S) + \alpha t - q| > 0$, and thus (by Chebyshev's inequality) we deduce that, for every $1 \leq i \leq m$

$$f_k(t) = \text{Prob}\ (w(M_i^t) + w(N_k^t) \in (q - w_i, q] \mid X_i = t)$$
$$\underset{k\to\infty}{\longrightarrow} \text{Prob}\ (q \in A_i(\omega) \mid X_i = t) \quad .$$

Using Lebesque dominated convergence theorem, we finally conclude that

$$\phi_i^k = \int_0^1 f_k(t) \cdot dt \underset{k\to\infty}{\longrightarrow} \int_0^1 \text{Prob}\ (q \in A_i(\omega) \mid X_i = t) \cdot dt = \phi_i \quad .$$

4. PROOF OF THE MAIN THEOREM

The notations are as in the previous sections.

Lemma 6: Let ϕ_i^k, for $1 \leq i \leq m + n_k$, denote the value of the game v_k to the i-th player, and let $\eta = 1 - \sum_{i=1}^m \phi_i$. Then,

$$(4.1) \qquad \sum_{i=m+1}^{m+n_k} |\alpha \cdot \phi_i^k - \eta \cdot w_i^k| \to 0 \quad \text{as} \quad k \to \infty \quad .$$

Proof: Let $t^k = \inf\{t\colon w^k(M^t \cup N_k^t) \geq q\}$; by Lemma 2, $\phi_i^k = \text{Prob}\ (X_i = t^k)$. In order to prove (4.1) it is enough to prove that for every $S_k \subset N_k$,

$$(4.2) \qquad \lim_{k\to\infty} \alpha(\phi v_k)(S_k) - \eta w^k(S_k) = 0 \quad .$$

By Lemma 5, and the efficiency of the Shapley value, $\phi v_k(N_k) \to \eta$, as $k \to \infty$, and therefore it will be enough to prove that for every $S_k \subset N_k$,

$$(4.3) \qquad \liminf_{k\to\infty} (\alpha\phi v_k(S_k) - \eta w^k(S_k)) \geq 0 \quad .$$

Let $S_k \subset N_k$ be given. Thus,

$$(4.4) \qquad \phi v_k(S_k) = \sum_{i\in S_k} \text{Prob}\ (t^k = X_i) = \text{Prob}\ \{t^k \in X(S_k)\}$$

where $X(S_k)$ is the random finite set $\{X_i\}_{i\in S_k}$. The idealization of t^k, namely t_q, is a function of $X_1,\ldots,X_m$ only. Let

$$\bar{t}^k = \inf \{t: w^k(N_k^t) + w^k(M^{t_q}) \geq q\} = \inf \{t: w^k(N_k^t) \geq q_k\} \quad ,$$

$$\text{where} \quad q_k = q - w^k(M^{t_q}) \quad .$$

Consider the weighted majority games (which depend on $X_1,\ldots,X_m$),

$$u_k = [q_k; w^k_{m+1},\ldots,w^k_{m+n_k}]$$

on the set of players N_k. Let $\bar{\phi}^k$ be its value, and let ϕ^k be the restriction to N_k of the value of v_k. We would like to approximate ϕ^k in terms of $\bar{\phi}^k$ (as a "sub average"), and then use Lemma 3.

Let H^ε denote the event that for all $1 \leq i \leq m$, $|t_q - X_i| > \varepsilon$; the event $q \notin A_i(\omega)$ for all $1 \leq i \leq m$ will be denoted by H. Obviously, $\eta = \text{Prob}\,(H)$. Now we claim that

$$(4.5) \qquad \text{Prob}\,(H^\varepsilon|H) \underset{\varepsilon\to 0}{\longrightarrow} 1 \quad .$$

Indeed, if $|t_q - X_i| < \varepsilon$ and $q \notin A_i(\omega)$ then $X_i \in B_q + (-\varepsilon,\varepsilon)$ where $B_q = \{t: q = t\alpha + w(S)$ for some $S \subset M\}$. Since B_q is finite, the Lebesque measure of $B_q + (-\varepsilon,\varepsilon)$ tends to zero as $\varepsilon \to 0$, which completes the proof of (4.5).

For every fixed $\varepsilon > 0$, if q^k satisfies

$$(4.6) \qquad \alpha(\tfrac{\varepsilon}{2}) < q_k < \alpha(1 - \tfrac{\varepsilon}{2}) \quad ,$$

then by Lemma 3,

$$(4.7) \qquad \lim_{k\to\infty} \alpha\phi u_k(S_k) - w^k(S_k) = 0 \quad .$$

Lemma 4 now implies $\text{Prob}\,(|t_q - t^k| > \varepsilon/2) \underset{k\to\infty}{\longrightarrow} 0$, hence,

$$(4.8) \qquad \text{Prob}\,(\bar{t}^k = t^k|H^\varepsilon) \underset{k\to\infty}{\longrightarrow} 1 \quad .$$

Now,

$$(4.9) \qquad \phi v_k(S_k) = \text{Prob}\,(t^k \in X(S_k)) \geq \text{Prob}\,(t^k = \bar{t}^k \wedge \bar{t}^k \in X(S_k) \wedge H) \quad .$$

Observe that for (sufficiently) small $\varepsilon > 0$, there is K (large enough), such that for every $k \geq K$, (4.6) is implied by H^ε. As $\phi u_k(S_k) = \text{Prob}\,(\bar{t}^k \in X(S_k))$, we can combine (4.5), (4.7), (4.8) and (4.9), using standard probability rules, to get (4.2) - which completes the proof of this lemma.

We proceed by proving that the "jump functions" are in ASYMP. We identify a finite subfield of C with the partition it induces. If v is any set function on (I,C), its dual v^* is defined by $v^*(S) = v(I) - v(I\setminus S)$. Observe that $v \in$ ASYMP iff $v^* \in$ ASYMP (easily shown by reversing order, see (|5|, p, 140)). If $0 < q < 1$ then $\bar{f}_q(x) = 0$ or 1 according to $x \leq q$ or $x > q$.

Lemma 7: Let λ be a probability measure in FL, and $0 < q < 1$. Then $f_q \circ \lambda \in$ ASYMP and $\bar{f}_q \circ \lambda \in$ ASYMP.

Proof: Let $(\Pi_k)_{k=1}^{\infty}$ be an increasing S-admissible sequence of finite subfield of C. Let $\{a_1,\ldots,a_m\}$ be the finite set of atoms of λ. As $(\Pi_k)_{k=1}^{\infty}$ is increasing and $\cup\Pi_k$ generates C, there exists K (large enough) such that for all $k \geq K$, each of the a_i's is in a different atom of Π_k. For such sequences $(\Pi_k)_{k=1}^{\infty}$, it is known that, as $k \to \infty$, we have $\lambda(A_k^i) \to \lambda(\{a_i\})$ whenever $a_i \in A_k^i \in \Pi_k$, and $\lambda(A_k) \to 0$ whenever $A_k \subset \Pi_k$ does not contain any of the atoms a_i $(1 \leq i \leq m)$. Therefore, the sequence of finite games $(f_q \circ \lambda)_{\Pi_k}$ are of the form $[q: w_1^k,\ldots,w_m^k,w_{m+1}^k,\ldots,w_{m+n_k}^k]$, and satisfy (3.2), (3.3), and (3.4) whenever λ contains a (non-trivial) non-atomic part. Thus by Lemma 6 we conclude that $f_q \circ \lambda \in$ ASYMP. If λ is purely atomic, then obviously $f_q \circ \lambda \in$ ASYMP.

Now observe that $\bar{f}_q \circ \lambda = (f_{1-q} \circ \lambda)^*$ and therefore $\bar{f}_q \circ \lambda \in$ ASYMP.

Lemma 8: Let $f \in bv'$ be right continuous, and let λ be a probability measure in FL. Then $f \circ \lambda \in$ ASYMP.

Proof: Let $S \in C$, and let $(\Pi_k)_{k=1}^{\infty}$ be an S-admissible sequence. By Lemma 3.4 of |15|,

$$\phi(f \circ \lambda)_{\Pi_k}(S) = \int_0^1 \phi(f_q \circ \lambda)_{\Pi_k}(S_k)df(q) \quad .$$

For every $0 < q < 1$, $f_q \in$ ASYMP (Lemma 7); thus as $k \to \infty$, $\phi(f_q \circ \lambda)_{\Pi_k}(S) \to \psi(f_q \circ \lambda)_{\Pi_k}(S)$ where ψ denotes the asymptotic value. Using Lebesque's dominated convergence theorem, we conclude that $\psi(f \circ \lambda)_{\Pi_k}(S) \to \int_0^1 \psi(f_q \circ \lambda)(S) \cdot df(g)$, hence $f \circ \lambda \in$ ASYMP and

$$\psi(f \circ \lambda)(S) = \int_0^1 \psi(f_q \circ \lambda)(S)df(q) \quad . \tag{4.10}$$

Theorem A: $bv'FL \subset$ ASYMP.

Proof: Theorem F of |5|, asserts that ASYMP is a closed symmetric linear subspace of BV-the space of all set functions of bounded variation. Thus, by Lemma 5 and Lemma 6, ASYMP contains every game of the form

(4.11) $$f = g + \sum_{i=1}^{\infty} \alpha_i \cdot \bar{f}_{q_i} \quad ,$$

with $g \in bv'$ right continuous, $q_i \in (0,1)$, and $\sum_{i=1}^{\infty} |\alpha_i| < \infty$. Since every $f \in bv'$ has such (i.e., (4.11)) a representation, we conclude, by recalling definition of $bv'FL$, and using again the closeness of ASYMP that $bv'FL \subset \text{ASYMP}$.

5. FURTHER RESULTS AND OPEN PROBLEMS

In this section we shall state few additional results, and present some open problems.

Denote by M the space of measures on the underlying mesurable space $(I,\mathcal{C})$, M_a will denote the subspace of M of all purely atomic measures, and $M_b = M \setminus M_a$, i.e., M_b is the set of all measures with a (non-trivial) non-atomic part; M^1 denotes the subset of M of all non-negative measures μ, with $\mu(I) = 1$, and FL^1, M_a^1, M_b^1 are similarly defined. The closed subspace of BV spanned by power of measures in M^1 is denoted by pM, and the one spanned by the set functions of the form $f \circ \mu$ where $f \in bv'$ and $\mu \in M^1$, is denoted by $bv'M$. In the same manner the spaces pFL, pM_a, pM_b, $bv'M_a$, $bv'M_b$ are defined. Let $\mathcal{A}$ denote the closed algebra generated by games of the form $f \circ \mu$, where $\mu \in FL^1$ and $f \in bv'$ is continuous. If Q_1 and Q_2 are two subspaces of BV we denote by $Q_1 * Q_2$ the minimal closed space which contain Q_1, Q_2 and $Q_1 \cdot Q_2$.

<u>Theorem B</u>: $\text{pFL} = \text{pM} \subsetneqq bv'FL \subsetneqq \text{pM} * bv'FL \subsetneqq \mathcal{A} * bv'FL \subsetneqq \text{ASYMP}$.

We would like to replace (in Theorem B) FL by M. This leads us to the following,

<u>Open Problem</u>: Is $bv'M \subset \text{ASYMP}$, or even is $bv'M_a \subset \text{ASYMP}$ or $bv'M_b \subset \text{ASYMP}$, or even is $f_q \circ \mu \in \text{ASYMP}$ for every $\mu \in M_b^1$, or for every $\mu \in M_a^1$.

This problem has proved to be very stubborn. The last part turns out to be equivalent to the open problem raised by Shapley (|20|) as to whether or not every weighted majority game with countable many players is regular. Observe, that our method implies, that if $\mu \in M^1$ and $f_q \circ \mu \in \text{ASYMP}$ for every $0 < q < 1$ than $f \circ \mu \in \text{ASYMP}$ for every $f \in bv'$, and thus these problems reduces to those of the jump functions. A positive solution to this question would imply in particular the existence of a partition value (in particular a value) on $bv'M$. Even this is unknown. Indeed we state

Open Problem: Does there exist a partition value on bv'M, or, even, does there exist a value on bv'M.

Along these lines, we can state the following:

Theorem C: (a) There exists a partition value on $bv'M_b$.

(b) If $f \in bv'$ and $\lambda \in M^1$, then for every subset B of the set of atoms of the measure λ, and every B-admissible sequence of partitions $(\Pi_k)_{k=1}^{\infty}$, the limit $\lim_{k\to\infty} \phi(f \circ \lambda)_{\Pi_k}(B)$ exists and is independent of the particular sequence $(\Pi_k)_{k=1}^{\infty}$.

We turn now to results concerning asymptotic pre-values; for each Borel measure λ on $[0,1]$ the λ-pre value ϕ_λ of the finite game (N,v) is the measure on N given by,

$$\phi_\lambda v(i) = \int_0^1 E(\Delta(i,\omega) | X_i = t) \cdot d\lambda(t) \quad .$$

The main conceptual interest is in those that are semi values, i.e., where λ is a probability measure (see |6|, |21|). However the pre values are mathematically convenient.

As with values, we may investigate the limiting λ-pre values, and we define (in the natural way) the space $ASYMP_\lambda$ to be the space of all games having an asymptotic λ-pre value. The proof of Lemma 5, reveals that the λ-pre values for the major player in the sequence of games v_k (we use notations from previous sections), converges to the limit, $\int f(t) \cdot d\lambda(t)$, whenever λ does not have an atom in the set B_q. (When λ has an atom in the set B_q the λ-pre value of the major player does not necessarily converge.) Thus if λ is non-atomic, the λ-pre values of the major players converge to a limit for every $0 < q < 1$. However, in order that the λ-pre values of the sequence of games v_k will converge (in the same manner that the values do) further assumptions (on λ) are needed.

Theorem D: (a) $bv'FL \subset ASYMP_\lambda$ iff λ is absolutely continuous with respect to Lebesque measure ℓ, and such that $d\lambda/d\ell$ is continuous.

(b) $pM \subset ASYMP_\lambda$ iff λ is absolutely continuous with respect to the Lebesque measure ℓ, and $d\lambda/d\ell \in L_\infty$.

References

|1| Z. Artstein: Values of games with denumerably many players, Int. J. Game Theory 1 (1971) 27-37.

|2| R. J. Aumann and M. Kurz: Power and taxes, Econometrica 45 (1977) 1137-1161.

|3| R. J. Aumann and M. Kurz: Power and taxes in a multi-commodity economy, J. Math. 27 (1977) 185-234.

|4| R. J. Aumann and M. Kurz: Power and taxes in a multi-commodity economy (updated), Journal of Public Economics 9 (1978) 139-161.

|5| R. J. Aumann and L. S. Shapley: Values of non-atomic games (Princeton University Press, Princeton, 1974).

|6| P. Dubey, A. Neyman and R. J. Weber: Value theory without efficiency, unpublished manuscript (1978).

|7| F. Fogelman and M. Quinzii: Asymptotic values of mixed games, unpublished (1975).

|8| S. Hart: Values of mixed games, Int. J. Game Theory 2 (1973) 69-85.

|9| S. Hart: Asymptotic values of games with a continuum of players, J. Math. Econ. 4 (1977) 57-80.

|10| Y. Kannai: Values of games with a continuum of players, Isr. J. Math. 4 (1966) 54-58.

|11| H. W. Kuhn and A. W. Tucker: Editors' preface to contributions to the theory of games (volume I), Annals of Mathematics Study 24 (Princeton University Press, Princeton, N.J., v-xiii, 1950)

|12| J. W. Milnor and L. S. Shapley: Values of large games II: Oceanic games, RM-2649, The Rand Corporation, Santa Monica, California. Basis of |13| (1961).

|13| J. W. Milnor and L. S. Shapley: Values of large games II: Oceanic games, Math. Oper. Res 3 (1978) 290-307. Based on |12| and |19|.

|14| A. Neyman: Renewal theory for sampling without replacement, TR-380, School of Oper. Res. and I.E., Cornell Univ., Ithaca, N.Y. (1978).

|15| A. Neyman: Singular games have asymptotic values, to appear in Math. Oper. Res. (1979).

|16| A. Neyman and Y. Tauman: The partition value, to appear in Math. Oper. Res. (1979).

|17| N. Z. Shapiro and L. S. Shapley: Values of large games, I: A limit theorem, RM-2648, The Rand Corporation, Santa Monica, California (1960), and Math. Oper. Res. 3 (1978) 1-9.

|18| L. S. Shapley: A value for n-person games, in Contribution to the Theory of Games, Vol. II, ed. by H. W. Kuhn and A. W. Tucker (Princeton University Press, Princeton, 1953) 307-317.

|19| L. S. Shapley: Values of large games, III: A Corporation with two large stockholders, RM-2650, The Rand Corporation, Santa Monica, California (1961). Included in |12|.

|20| L. S. Shapley: Values of games with infinitely many players, in Recent Advances in Game Theory, papers delivered at a meeting of the Princeton University Conference in October 1961, privately printed for members of the conference (1961) 113-118.

|21| R. J. Weber: Subjectivity in the valuation of games. To appear in this volume.

GAME THEORY AND RELATED TOPICS
O. Moeschlin, D. Pallaschke (eds.)

GAME THEORETIC ANALYSIS OF VOTING SCHEMES

Bezalel Peleg

Institute of Mathematics

The Hebrew University of Jerusalem

This paper consists of a survey of the results on consistent voting. First, in Section 2, we discuss the implications of the Gibbard-Satterthwaite theorem for the theory of committees. Then we proceed, in Sections 3-5, to describe the results on existence of strong representations for symmetric, weak and strong simple games.

1. INTRODUCTION

This paper is devoted to an exhaustive survey of known constructions of social choice functions (i.e., voting schemes), which are not distorted by manipulation of preferences by coalitions of voters. We start, in Section 2, with the formulation of the Gibbard-Satterthwaite theorem on the manipulability of voting schemes. Then we discuss its implications to the theory of committees (see Corollary 2.8). Motivated by the discussion in Section 2 we proceed in Section 3 to define strong representations of committees (see Definition 3.3). Then the known results on the existence of strong representations for symmetric and weak simple games (i.e., committees), are stated (see Theorems 3.7. and 3.9). In Section 4, using the result of Nakamura [1976], we provide an upper bound on the order of a strong representation of a simple game without veto players (see Theorem 4.4). The capacity (i.e., the maximum order of a strong representation), of strong simple games is determined by that bound (see Theorem 4.5). Finally, in Section 5 we report Polishchuk's results on monotonicity and uniqueness of strong representations of symmetric simple games.

From the point of view of presentation this paper is self contained. However, proofs of the results are not given or discussed, and the interested reader is referred to the original works.

2. The Gibbard-Satterthwaite Theorem and Its Implications

Let A be the set of alternatives. We assume that A contains at least two members. A linear order on A is a complete, reflexive, transitive and antisymmetric binary relation on A. We denote by L the set of all linear orders on A.

Let N be a finite set with n members. N is called a society, members of N are called voters, or players, and non-empty subsets of N are called coalitions. For a coalition S we denote by L^S the set of all functions from S to L.

Definition 2.1. A social choice function (SCF) is a function $F : L^N \to A$.

The strategic aspects of a voting situation are made precise by the following definition.

Definition 2.2. Let F be a SCF and let $R^N \in L^N$. The game associated with F and R^N is the n-person game in normal form $g(F,R^N)$ where

(2.1) L is the set of strategies for each player $i \in N$.

(2.2) F is the outcome function.

(2.3) R^i is the preference relation of player $i \in N$ on the outcome space A.

Definition 2.3. Let F be a SCF and let $R^N \in L^N$. $Q^N \in L^N$ is an equilibrium point (e.p.) of $g(F,R^N)$ if for each $i \in N$

$F(Q^N)R^iF(Q^{N-\{i\}},T^i)$ for all $T^i \in L$.

(Here $Q^{N-\{i\}}$ is the restriction of Q^N to $N - \{i\}$).

A SCF F is nonmanipulable if for each $R^N \in L^N$ R^N itself is an e.p. of $g(F,R^N)$. Let F be a SCF. The range of F is $\rho(F) = \{x \mid x = F(R^N)$ for some $R^N \in L^N\}$. A player $j \in N$ is a dictator of F if for all $R^N \in L^N$ and all $x \in \rho(F)$ $F(R^N)R^jx$. A fundamental result of Gibbard [1973] and Satterthwaite [1975] is the following.

Theorem 2.4. If a SCF F is nonmanipulable and $\rho(F)$ contains at least three alternatives, then there exists a dictator for F.

In order to understand the implications of Theorem 2.4 to the theory of committees we associate with each SCF a committee, i.e., a monotonic and proper simple game, in the following way. Let $F : L^N \to A$ be a SCF.

Definition 2.5. A coalition S is winning (with respect to F) if $[R^N \in L^N, x \in A$ and xR^iy for all $i \in S$ and $y \in A] \Rightarrow F(R^N) = x$.

Definition 2.6. The simple game associated with F is the game $G^*(F) = (N,W)$ where W is the set of winning coalitions with respect to F. Furthermore, F is called a representation of $G^*(F)$.

Remark 2.7. Since A contains at least two alternatives, $G^*(F)$ is proper (i.e., if $S \in W$ then $N - S \notin W$). Also, $G^*(F)$ is monotonic (i.e., if $S \in W$ and $T \supset S$ then $T \in W$).

A simple game $G = (N,W)$ is dictatorial if there exists $j \in N$ (a dictator) such that $[S \in W \Leftrightarrow j \in S]$. As a result of Theorem 2.4 we get:

Corollary 2.8. Let $F : L^N \to A$ be a SCF and let $G^*(F) = (N,W)$. If A contains at least three alternatives and if $N \in W$ and $G^*(F)$ is nondictatorial, then F is manipulable.

Thus, a non-trivial and nondictatorial committee which has to choose one alternative out of a set which contains at least three alternatives has no representative nonmanipulable voting procedure.

3. Strong Representations of Committees

In this section we proceed to investigate voting procedures for committees, i.e. SCF's, which, although they may be manipulable, are not distorted by manipulation. Where a SCF $F : L^N \to A$ "is not distorted by manipulation" if for each $R^N \in L^N$ $F(R^N)$ is the outcome of a strong e.p. of $g(F,R^N)$. The formal definitions are :

Definition 3.1. Let F be a SCF and let $R^N \in L^N$. $Q^N \in L^N$ is a strong e.p. (s.e.p.) of $g(F,R^N)$ if for every coalition S and for every $P^S \in L^S$, there exists a player $i \in S$ such that $F(Q^N)R^iF(Q^{N-S},P^S)$.(Here Q^{N-S} is the restriction of Q^N to N-S).

Definition 3.2. A SCF F is exactly and strongly consistent if there exists a function $E : L^N \to L^N$ such that for each $R^N \in L^N$ $E(R^N)$ is a s.e.p. of $g(F,R^N)$, and the composition $F \circ E = F$.

Thus, if F is an exactly and strongly consistent SCF and $R^N \in L^N$, then each coalition, and in particular each voter, can enforce $F(R^N)$ to be the outcome of $g(F,R^N)$. Note, however, that R^N may not be a s.e.p. of $g(F,R^N)$. In such a case $E(R^N) \neq R^N$, and therefore some of the players have to manipulate their votes in order to play a s.e.p. which yields $F(R^N)$ as the outcome of $g(F,R^N)$. However, such a manipulation is compatible with the rules of $g(F,R^N)$ (since each $R \in L$ is a legal strategy for each $i \in N$).

We proceed to the main definition.

Definition 3.3. Let $G = (N,W)$ be a (monotonic) simple game and let $A = \{x_1,\ldots,x_m\}$ be the set of alternatives. A SCF $F : L^N \to A$ is a strong representation (SR) of G of order m if $G^*(F) = G$ (see Definition 2.6), and F is exactly and strongly consistent. (Note that in Peleg [1978]$_b$ a SR is called a representation).

In order to state the known existence theorems for SR's we need to mention the following properties of SCF's.

Definition 3.4. A SCF F is monotonic if it satisfies: If $F(R^N) = x$, $R_1^N \in L^N$ and for all $a,b \in A - \{x\}$ and all $i \in N$, $aR^ib \Leftrightarrow aR_1^ib$ and $xR^ia \Rightarrow xR_1^ia$, then $F(R_1^N) = x$.

Let F be a SCF. A permutation π of N is a symmetry of F if for all $R^N = (R^1,\ldots,R^n)$ in L^N $F(R^N) = F(R^{\pi(1)},\ldots,R^{\pi(n)})$. The group of all symmetries of F is denoted by SYM(F).

Definition 3.5. A SCF F is faithful if SYM(F) = SYM(G*(F)), where SYM(G*(F)) is the symmetry group of the simple game G*(F).

Theorem 3.6. Every proper simple game has a monotonic and faithful SR of order 2. (See Theorem 4.3 in Peleg $[1978]_b$.)

A simple game is symmetric if it can be described completely by a pair (n,k of natural numbers, where n is the number of players and k is the size of a minimal winning coalition. For a game (n,k) we denote $b = n - k + 1$. Clearly, b is the size of a minimal blocking coalition in (n,k). (A coalition in a simple game is blocking if its complement is losing.)

Theorem 3.7. Let (n,k) be a proper symmetric simple game. Let further $t = [n/b] + 1$ if $n \equiv -1(b)$, and $t = [n/b]$ otherwise. Then (n,k) has a faithful SR of every order m such that $2 \le m \le t$. (See Theorem 5.1 in Peleg $[1978]_b$.)

Remark 3.8. Let (n,k) be a proper symmetric simple game and let $2 \le m \le t$. Let A be a set of m alternatives. By Theorem 3.7 there exists a SCF $F : L^N \to A$ which is a faithful SR of (n,k). Since F is faithful it is anonymous, i.e. $SYM(F) = S_n$, where S_n is the symmetric group of the n players. Furthermore, F has the following peculiar property. For each $R^N \in L^N$ one can choose a s.e.p. Q^N of $g(F,R^N)$ such that $F(Q^N) = F(R^N)$ (see Definition 3.2) and, in addition, for some $i \in N$ $Q^i = R^i$. (This last assertion can be verified by a careful examination of the proof of Lemma 4.1 in Peleg $[1978]_a$.) Hence, for each profile R^N of true preferences there exists a voter i who can enforce $F(R^N)$ to be the outcome of $g(F,R^N)$ by sticking to his true preference R^i.

Let $G = (N,W)$ be a simple game. G is weak if $V = \cap\{S \mid S \in W\} \neq \emptyset$. The members of V are called veto players.

Theorem 3.9. Let $G = (N,W)$ be a weak game. Then G has a monotonic and faithful SR of every order $m \ge 2$. (See Theorem 6.1 in Peleg $[1978]_b$.)

4. Capacities of Committees

Definition 4.1. Let G be a proper simple game. If G has a SR of maximum order m, then the capacity of G, $\mu(G)$, is defined as $\mu(G) = m$; otherwise (i.e., if G has no SR of maximum order), the capacity of G, $\mu(G)$, is defined as $\mu(G) = \infty$.

By Theorem 3.6 $\mu(G) \ge 2$. By Theorem 3.9 if G is weak then $\mu(G) = \infty$. Clearly, $\mu(G)$ is the maximum number of alternatives that the players of G can dea with simultaneously in a consistent way. A bound on the capacity of a proper simp le game without veto players can be determined in the following way.

Let $G = (N,W)$ be a simple game and let A be a set of alternatives. Let further $R^N \in L^N$ and $x,y \in A$ $x \neq y$. x dominates y with respect to R^N if $\{i \in N \mid xR^iy\}$ is winning. The core of A, with respect to G and R^N, is the set of undominated alternatives in A, and is denoted by $C(A,N,W,R^N) = C(R^N)$. The following simple result is true.

Theorem 4.2. Let $F : L^N \to A$ be an exactly and strongly consistent SCF and let $G = G^*(F) = (N,W)$ (see Definition 2.6). Then, for all $R^N \in L^N$ $F(R^N) \in C(A,N,W,R^N)$. (See Theorem 3.5 in Peleg $[1978]_b$.)

It is clear from Theorem 4.2 that the problem of existence of SR's for committees is closely connected to the problem of existence of a non-empty core for simple games. The latter problem has already been solved completely in Nakamura [1976]. We recapitulate here Nakamura's result: Let $G = (N,W)$ be a simple game without veto players. Nakamura's number of G, $\nu(G)$, is defined by

(4.1) $\nu(G) = \min \{|\sigma| \mid \sigma \subset W \text{ and } \cap\{S \mid S \in \sigma\} = \emptyset\}$,

where, here and in the sequel, if B is a finite set then $|B|$ denotes the number of members of B.

Theorem 4.3. (Nakamura [1976]). Let $G = (N,W)$ be a simple game without veto players and let A be a finite set of alternatives. Then $C(A,N,W,R^N) \neq \emptyset$ for all $R^N \in L^N$ iff $|A| < \nu(G)$.

Theorems 4.2. and 4.3 imply:

Theorem 4.4. If $G = (N,W)$ is a proper simple game without veto players then $\mu(G) < \nu(G)$ (see (4.1)). (See Theorem 4.5 in Peleg $[1978]_b$.)

A simple game $G = (N,W)$ is strong if

(4.2) $S \notin W \Rightarrow N - S \in W$.

Theorems 3.6 and 4.4 determine the capacity of strong games.

Theorem 4.5. If $G = (N,W)$ is an essential (i.e., non-dictatorial) proper and strong game then $\mu(G) = 2$. (See Theorem 4.6 in Peleg $[1978]_b$.)

5. Monotonicity and Uniqueness of SR's of Symmetric Games

The construction in Peleg $[1978]_a$ of SR's of symmetric games is essentially unique. Furthermore, it can be modified to yield monotonic SR's. These two results are due to Polishchuk [1978]. The details are as follows. Let $R \in L$ and $Y \subset A$. We denote by R/Y the restriction of R to A - Y. Let now $A = \{x_1,\dots,x_m\}$ be a finite set with m alternatives, and assume that $n \geq m - 1$. Let $\beta : A \to \{1,\dots,n\}$ satisfy $\sum_{i=1}^{m} \beta(x_i) = n + 1$.

Definition 5.1. Let $R^N \in L^N$. A feasible elimination procedure (f.e.p.) with respect to R^N is a sequence $(x_{i_1},C_1;\dots;x_{i_{m-1}},C_{m-1};x_{i_m})$, where $C_1,\dots,C_{m-1}$ are coalitions, which satisfies

(5.1) If $1 \leq s < t \leq m-1$ then $C_t \cap C_s = \emptyset$ and $i_s \neq i_t$;

(5.2) $|C_j| = \beta(x_{i_j})$, $j = 1,\dots,m-1$.

(5.3) x_{i_j} is the worst alternative with respect to

$$R^s / \{x_{i_1},\dots,x_{i_{j-1}}\} \text{ for all } s \in C_j,\ j = 1,\dots,m-1.$$

(5.4) $\{x_{i_m}\} = A - \{x_{i_1},\dots,x_{i_{m-1}}\}$.

A member $x \in A$ is called R^N-maximal if there exists a f.e.p. with respect to R^N $(x_{i_1},C_1;\ \dots;\ x_{i_{m-1}},C_{m-1};x_{i_m})$ such that $x = x_{i_m}$. We denote by $F^*(R^N)$ the set of all R^N-maximal members of A. Since $\sum_{i=1}^{n} \beta(x_i) = n + 1$, $F^*(R^N) \neq \emptyset$ for all $R^N \in L^N$.

The following definitions are needed for the description of the results.

Definition 5.2. Let $F : L^N \to A$ be a SCF. A coalition S is winning for $a \in A$ if:

$[R^N \in L^N$ and aR^ix for all $i \in S$ and all $x \in A] \Rightarrow F(R^N) = a$.

Definition 5.3. An anonymous SCF satisfies condition $W(h_1,\ldots,h_m)$ if for all $1 \le i \le m$ h_i is the size of a minimal winning coalition for x_i.

We denote by 2^A the set of all non-empty subsets of A.

Theorem 5.4. Let $\sigma : 2^A \to A$ satisfy $\sigma(B) \in B$ for all $B \in 2^A$. Then the composition $F = \sigma \circ F^*$ is an exactly and strongly consistent anonymous SCF which satisfies $W(h_1,\ldots,h_m)$, where $h_i = n + 1 - \beta(x_i)$, $i = 1,\ldots,m$. (See Lemma 4.1 in Peleg $[1978]_a$ and Theorem 4.6 in Polishchuk [1978].)

Theorem 5.4 implies Theorem 3.7. (See Peleg $[1978]_a$.)

The following interesting and important result implies that the construction of faithful SR's for (n,k) games is essentially unique.

Let h_i, $i = 1,\ldots,m$, satisfy

(5.5) $0 < h_i \le n$ and $\sum_{i=1}^{m} h_i = (m-1)(n+1)$.

Define $\beta(x_i) = n + 1 - h_i$, $i = 1,\ldots m$.

Theorem 5.5. Let F be an exactly and strongly consistent anonymous SCF which satisfies condition $W(h_1,\ldots,h_m)$ (where $h_1,\ldots,h_m$ satisfy (5.5)). Then for every $R^N \in L^N$ $F(R^N) \in F^*(R^N)$. (See Theorem 6.6 in Polishchuk [1978] where, actually, a somewhat stronger result is proved.)

A social decision function (SDF) is a function $H : L^N \to 2^A$.

Definition 5.6. A SDF H is monotonic if it satisfies : If $x \in H(R^N)$, $R_1^N \in L^N$ and for all $a,b \in A - \{x\}$ and all $i \in N$, $aR^ib \Leftrightarrow aR_1^ib$ and $xR^ia \Rightarrow xR_1^ia$, then $x \in H(R_1^N)$. (See condition (M) in Dutta and Pattanaik [1978].)

Clearly, F* as defined above is a monotonic SDF. However, unfortunately, a monotonic SDF may have no monotonic selection as the following example shows.

Example 5.7. Let $A = \{a,b,c\}$ and $N = \{1,2\}$. Denote $R_1 = (a,b,c)$, $R_2 = (a,c,b)$, $R_3 = (b,a,c)$, $R_4 = (b,c,a)$, $R_5 = (c,a,b)$ and $R_6 = (c,b,a)$. Define a SDF H by:

(5.6) $H(R_1,R_1) = H(R_1,R_2) = H(R_2,R_1) = H(R_2,R_2) = H(R_1,R_4) = H(R_4,R_1) = \{a\}$.

(5.7) $H(R_3,R_3) = H(R_3,R_4) = H(R_4,R_3) = H(R_4,R_4) = H(R_2,R_3) = H(R_3,R_2) = \{b\}$.

(5.8) $H(R_1,R_3) = H(R_3,R_1) = \{a,b\}$;

and $H(Q^1,Q^2) = \{c\}$ otherwise. Then H is monotonic, anonymous and Paretian (i.e., it satisfies conditions (M), (AN) and (WP) of Dutta and Pattanaik [1978]). However, as the reader can easily verify, H has no monotonic selection.

The following stronger version of monotonicity implies the existence of monotonic selections.

Definitions 5.8. A SDF H is strongly monotonic if it satisfies:

If $R^N, R_1^N \in L^N$, $x \in A$ and for all $a,b, \in A - \{x\}$ and all $i \in N$, $aR^ib \Leftrightarrow aR_1^ib$ and

$xR^i a \Rightarrow xR_1^i a$, then $H(R_1^N) \subset \{x\} \cup H(R^N)$.

Theorem 5.9. The SDF F^* is strongly monotonic. (See Theorem 5.7 in Polishchuk [1978] where, actually, a stronger result is proved.)

References

[1] Dutta, B. and P.K. Pattanaik [1978], "On Nicely Consistent Voting Systems," Econometrica, 46, 163-170.

[2] Gibbard, A. [1973], "Manipulation of Voting Schemes: A General Result," Econometrica, 41, 587-601.

[3] Nakamura, K. [1976], "The Vetoers in a Simple Game With Ordinal Preferences," to appear in International Journal of Game Theory.

[4] Peleg,B. [1978]$_a$, "Consistent Voting Systems," Econometrica, 46, 153-161.

[5] Peleg,B. [1978]$_b$, "Representation of Simple Games by Social Choice Functions," International Journal of Game Theory, 7, 81-94.

[6] Polishchuk, I. [1978], "Monotonicity and Uniqueness of Consistent Voting Systems," Center for Research in Mathematical Economics and Game Theory, The Hebrew University, Jerusalem.

[7] Satterthwaite, M. [1975], "Strategy-Proofness and Arrow's Conditions: Existence and Correspondence Theorems for Voting Procedures and Social Welfare Functions," Journal of Economic Theory, 10, 187-217.

GAME THEORY AND RELATED TOPICS
O. Moeschlin, D. Pallaschke (eds.)

EQUILIBRIUM PLANS FOR NON-ZERO-SUM MARKOV GAMES

U. Rieder

Institute of Mathematical Statistics
University of Karlsruhe

This paper considers non-cooperative countable-person Markov games with Borel state and action spaces and where the transition law is given by a bounded transition measure. We establish the existence and construction of Nash equilibrium plans for finite-stage games as well as for the limit of finite-stage Markov games.

1. Introduction and summary

In recent years several authors have investigated non-cooperative non-zero-sum Markov (stochastic) games. Mostly infinite-stage Markov games with a finite set of players and with a countable state space are studied (see the survey report of Parthasarathy and Stern (1977)). The present paper considers a non-cooperative non-zero-sum finite-stage Markov game with a countable set of players. The state and action spaces are Borel sets, the transition law is given by a bounded transition measure. Thus we may include models arising from semi-Markov processes and models transformed by so-called "bounding" functions. On the other hand, it should be noted that our model is in a certain sense equivalent to a Markov game with a stochastic transition law and with a state-and action-dependent discount factor.

We consider the important questions of the existence and of the construction of Nash equilibrium plans in finite-stage Markov games as well as in the limit of finite-stage games. The analysis is based on dynamic programming methods. A key tool is a measurable selection theorem of equilibrium functions.

In section 2 the non-cooperative dynamic game model to be considered is formulated. It is assumed that the sets of all admissible actions are compact and that the immediate return and the transition measure depend continuously on the actions. The existence of measurable equilibrium functions is proved in section 3. Section 4 contains the main results for finite-stage Markov games. They generalize results of van der Wal and Wessels (1977). The final section deals with infinite-stage Markov games. A convergence assumption is introduced which on the one hand seems to be rather weak, but on the other hand can be easily verified in practical situations. We establish the existence of ε-equilibrium plans. The only other results of this type seem to be in Himmelberg et. al. (1976) and in Whitt (1978). As noted by Federgruen (1978), the proof in Sobel (1973) is not valid. As an application of some characterizations we give a relatively elementary proof of the existence of a stationary equilibrium plan if the state space is countable. The existence result is due to Whitt (1978) who gave a different proof. Finally we remark that for two-person zero-sum Markov games there are stronger and more complete existence results (cf. Rieder (1978)).

2. The non-zero-sum Markov game model

In this paper we consider a non-cooperative non-zero-sum Markov game with a countable set of players specified by a tupel $(I,S,(A_i),(D_i),(q_i),(r_i),(V_i^0))$ of the following meaning:

(i) I is the countable set of players.

(ii) S is the *state space*. S is assumed to be a Borel space, i.e. S is a non-empty Borel subset of some Polish (complete, separable metric) space and is endowed with the σ-algebra of Borel subsets of S.

(iii) A_i is the *space of actions* for player $i \in I$. A_i is assumed to be a Borel space. Let $A = \times A_i$.

(iv) D_i is a measurable subset of $S \times A_i$, $i \in I$. It is assumed that D_i contains the graph of a measurable map from S into A_i. For any $s \in S$ the non-empty and measurable s-section $D_i(s)$ of D_i is called the *set of all admissible actions for player* $i \in I$, if the system is in state s. Define $D(s) = \times D_i(s)$, $s \in S$. We assume that

$$K = \{(s,a) \in S \times A:\ a \in D(s)\}$$

is a measurable subset of $S \times A$.

(v) q_i is a transition measure from K to S such that

$$0 < \rho = \sup_i \sup_{(s,a)} q_i(s,a,S) < \infty.$$

q_i specifies the *transition law* for player $i \in I$ during a single stage.

(vi) r_i is a measurable function on K, the so-called *reward function for player* $i \in I$. (r_i) is assumed to be uniformly bounded.

(vii) V_i^o is a measurable function on S, the so-called *terminal reward function for player* $i \in I$. (V_i^o) is assumed to be uniformly bounded.

Remark. The transition law q_i is allowed to depend on i. Thus our model includes non-cooperative discounted dynamic games where the discount factor depend on $i \in I$.

For $i \in I$ let $P(A_i)$ denote the set of all probability measures on A_i. Let $P(A_i)$ be topologized with the usual weak topology based on bounded continuous functions and let $PA := \times P(A_i)$ be endowed with the product topology. We recall that the σ-algebra of Borel subsets on $P(A_i)$ coincides with the smallest σ-algebra on $P(A_i)$ such that for each Borel subset B of A_i the map $p \to p(B)$ is measurable (cf. Rieder (1975)). Further, $P(A_i)$ and hence PA are Borel spaces (cf. Hinderer (1970)). For $s \in S$ let $P(D_i(s))$ be the set of all probability measures on A_i with support $D_i(s)$. Then it is not difficult to check that

$$C = \{(s,p) \in S \times PA : p \in \times P(D_i(s))\} \tag{2.1}$$

is a measurable subset of $S \times PA$. We observe that the s-section $C(s)$ of C can be identified as the set of all product probability measures on $D(s)$. Moreover, for any $s \in S$ and for any bounded continous function v on $D(s)$ it follows from Theorem 1.3.2 in Billingsley (1968), which is easily extended to a countable product of probability measures, that

$$p \to \int v(a)p(da) \quad \text{is continuous on } C(s). \tag{2.2}$$

We write F_i for the set of all decision rules for player $i \in I$, i.e. of all measurable maps $f_i : S \to P(A_i)$ such that $f_i(s) \in P(D_i(s))$ for all $s \in S$. Then F_i^N is the set of all (randomized Markovian) N-*stage policies* for player i. Finally, define $F = \times F_i$ and let $F^N = \times F_i^N$ represent the set of all N-*stage (randomized Markovian) plans*. We remark that F can be identified as the set of all measurable maps $f : S \to PA$ whose graphs belong to C. The elements of F will be called *decision plans*. We make use of the following abbreviations:

$$\overline{q_i}(s,p,\cdot) = \int q_i(s,a,\cdot)p(da) \qquad (s,p) \in C,\ i \in I$$
$$\overline{r_i}(s,p) = \int r_i(s,a)p(da) \qquad (s,p) \in C,\ i \in I.$$

It follows from Lemma 6.2 in Rieder (1975) that $\overline{q}_i$ is a transition measure from C to S and that $\overline{r}_i$ is a measurable function on C. If v is a bounded measurable function on S, then $Q_i(f)v$, for $f \in F$ and $i \in I$ is defined by

$$Q_i(f)v(s) = Q_{is}(f)v = \int \overline{q}_i(s, f(s), ds')v(s') \qquad s \in S$$

$$r_i(f)(s) = r(f)(i,s) = \overline{r}_i(s, f(s)) \qquad f \in F,\ i \in I,\ s \in S.$$

For $\pi = (f_o, \dots, f_{N-1}) \in F^N$, $i \in I$ and $s \in S$ let us write

$$Q^n_{is}(\pi) = Q_{is}(f_o)Q_i(f_1)\dots Q_i(f_{n-1}) \qquad 0 \leqq n \leqq N$$

where $Q^o_{is}(\pi)$ is the probability measure concentrated at the point s (independent of $i \in I, \pi \in F^N$).

Let $N \in \mathbb{N}$. Then for any plan $\pi \in F^N$, $i \in I$, $s \in S$ the functions

$$(2.3) \qquad V^N_\pi(i,s) = \sum_{n=o}^{N-1} Q^n_{is}(\pi) r_i(f_n) + Q^N_{is}(\pi)V^o_i \qquad (i,s) \in I \times S$$

are well-defined. $V^N_\pi(i,s)$ is the *"expected" total reward for player* $i \in I$ if the plan π is used and the system starts in $s \in S$.

For any $f \in F$ and $g_i \in F_i$ let $[f|g_i]$ denote the decision plan $f' \in F$ with $f'_j = f_j$ for $j \neq i$ and $f'_i = g_i$. Now we can define

$$(2.4) \qquad G^N_\pi(i,s) = \sup_{\sigma_i} V^N_{[\pi|\sigma_i]}(i,s) \qquad (i,s) \in I \times S$$

where $[\pi|\sigma_i]$ is defined as the N-stage plan $([f_o|g^i_o], \dots, [f_{N-1}|g^i_{N-1}])$ if $\pi = (f_o, \dots, f_{N-1}) \in F^N$ and $\sigma_i = (g^i_o, \dots, g^i_{N-1}) \in F^N_i$. $G^N_\pi(i,s)$ is the *maximal "expected" total reward for player* $i \in I$ if the system starts in s and the other players are using the policies $\pi_j \in F^N_j$, $j \neq i$. With these preparations we can now introduce the concept of optimality. A plan $\pi \in F^N$ is called an *equilibrium plan (in the sense of Nash)* if

$$(2.5) \qquad V^N_\pi(i,s) = G^N_\pi(i,s), \qquad (i,s) \in I \times S.$$

Hence, whenever the players choose an equilibrium plan, none of them can improve his own "expected" total reward by changing his policy. We will show that equilibrium plans exist if some continuity and compactness assumptions are fulfilled.

In this paper we do not consider non-Markovian policies. The justification for our restriction to the set F^N follows from the following theorem which can be proved by means of Theorem 8.1 in Rieder (1975).

THEOREM 2.1 If π is an equilibrium plan within F^N then π is an equilibrium plan within the broader set of not necessarily Markovian plans as well.

3. Existence of equilibrium functions

Formulations and proofs simplify considerably by the use of the following isotone operators. Denote by M the set of all bounded universally measurable functions v on I×S. The operators L, L_f and U_f are well-defined on M by

$$Lv(i,s,p) = \overline{r}_i(s,p) + \int \overline{q}_i(s,p,ds')v(i,s') \qquad i\in I,\ (s,p)\in C$$

$$L_f v(i,s) = Lv(i,s,f(s)) \qquad f\in F,\ (i,s)\in I\times S$$

$$U_f v(i,s) = \sup_{g_i\in F_i} Lv(i,s,[f|g_i](s)) \qquad f\in F,\ (i,s)\in I\times S$$

Let $v\in M$. The decision plan $f\in F$ is called an *equilibrium function of Lv*, if

$$(3.1) \qquad L_f v(i,s) = U_f v(i,s), \qquad (i,s)\in I\times S$$

i.e. if $L_f v=U_f v$.

In this paper we shall make use of the following condition

Condition (A)

(A1) $D_i(s)$ is a compact subset of A_i, $(i,s)\in I\times S$.

(A2) $q_i(s,a_n,\cdot)$ converges setwise to $q_i(s,a,\cdot)$ as $a_n\to a$, $a_n,a\in D(s)$, for all $(i,s)\in I\times S$.

(A3) $r_i(s,\cdot)$ is continuous on $D(s)$ for all $(i,s)\in I\times S$.

Remarks. (1) (A1) implies that the measurable set D_i contains the graph of a measurable map from S into A_i for all $i\in I$ (cf. the definition of the game).

(2) It is easily verified that (A2) is equivalent to condition

(A2)' For all $(i,s) \in I\times S$, $a_n, a \in D(s)$ $a_n \to a$ implies
$\int q_i(s,a_n,ds')v(i,s') \to \int q_i(s,a,ds')v(i,s')$
for all bounded measurable functions v on $I\times S$.

(3) (A2) and (A3) are both satisfied if I is finite and $D_i(s)$ is countable (and discrete) for all $i \in I$ and $s \in S$.

THEOREM 3.1 Assume (A). If v is a bounded measurable function on $I\times S$ then there exists an equilibrium function of Lv.

Proof. From (A1) it follows that $P(D_i(s))$ is a compact convex subset of $M(A_i)$, the set of all finite signed measures on A_i. Recall that $M(A_i)$ endowed with the weak topology, is a locally convex linear topological Hausdorff space. Define for $i \in I$, $s \in S$

$$\Phi_i(s) = \{p \in C(s): Lv(i,s,p)=L'v(i,s,p)\}$$
$$\Phi(s) = \times \Phi_i(s)$$

where $L'v(i,s,p)=\sup_{\mu_i} Lv(i,s,[p|\mu_i])$. By the existence theorem of equilibrium points for continuous games (see e.g. Glicksberg (1952)) $\Phi(s)$ is non-empty. The function Lv is measurable on $I\times C$ and by (2.2), (A2)' and (A3), $Lv(i,s,\cdot)$ is continuous on C(s) for all $(i,s) \in I\times S$. Hence L'v is measurable and $L'v(i,s,\cdot)$ is continuous for all (i,s). Then we conclude that Φ is a compact-valued measurable correspondence from S into PA. From the selection theorem of Kuratowski and Ryll-Nardzewski (1965) we get the existence of a measurable map $f:S\to PA$ such that $f(s) \in \Phi(s)$ for all $s \in S$, i.e. $f \in F$ and $L_f v = U_f v$. The proof is complete. □

4. Finite-stage Markov games

Throughout this section let $N \in \mathbb{N}$ be fixed. Let $V^o \in M$ be defined by $V^o(i,s)=V_i^o(i,s)$, $(i,s) \in I\times S$. By means of a dynamic program (e.g. a dynamic program with a state space which is the union of a fictitious absorbing state and the original state space) the relations (4.1) and (4.2) are easily derived from the results in Hinderer (1970) or Rieder (1975):

(4.1) $$V^N_{(f_o,\ldots,f_{N-1})} = L_{f_o}\cdots L_{f_{N-1}} V^o$$

(4.2) $$G^N_{(f_o,\ldots,f_{N-1})} = U_{f_o}\cdots U_{f_{N-1}} V^o$$

Concerning the construction of an equilibrium plan in finite-stage games we get the following theorem.

THEOREM 4.1 Let $v_o=V^o$. If f_n is an equilibrium function of Lv_{n-1} and $v_n:=U_{f_n}v_{n-1}$, $1\le n\le N$, then the plan $\pi=(f_N,\ldots,f_1)$ is an equilibrium plan and

$$v_N=V^N_\pi=G^N_\pi.$$

Proof. First we get $L_{f_1}V^o=U_{f_1}V^o=v_1$ and by induction

$$L_{f_n}\cdots L_{f_1}V^o=U_{f_n}\cdots U_{f_1}V^o=v_n.$$

Then the statements follow from (4.1) and (4.2). □

THEOREM 4.2 Assume (A). There exists an equilibrium plan.

Proof. From Theorem 3.1 we know that there exists an equilibrium function of Lv_{n-1}, since by induction v_{n-1} is a bounded measurable function on $I\times S$. In view of Theorem 4.1 the proof is complete. □

It is worth to remark that the existence of an equilibrium plan cannot be obtained directly via the well-known Nash theorem since the "expected" total rewards V^N_π are far from being linear in $\pi_j\in F^N_j$ for $j\in I$.

5. Infinite-stage Markov games

First we will introduce a rather weak convergence assumption which is closely related to the m-stage contraction assumption of Denardo (1967). Let the positive real numbers δ_n be defined by

(5.1) $$\delta_n = \sup_{i,s,\pi} Q^n_{is}(\pi)\mathbb{1}, \quad n\in\mathbb{N}+\{0\}$$

where $\mathbb{1}$ is the constant function on S equal to one. Then $\delta_o=1$ and $\delta_1=\rho$.

The proof of the following Lemma is not very difficult and is, therefore, not given here.

LEMMA 5.1 (a) $\lim_n \delta_n^{1/n}$ exists and is equal to $\inf_n \delta_n^{1/n}$.

(b) The following statements are equivalent:

(i) $\delta_m < 1$ for some $m \in \mathbb{N}$

(ii) $\inf_n \delta_n^{1/n} < 1$

(iii) $\sum_0^\infty \delta_n < \infty$

(iv) $\delta_n \to 0 \ (n \to \infty)$.

Throughout this section let the following convergence assumption (C) be satisfied.

(C) $$\delta_m < 1 \quad \text{for some } m \in \mathbb{N}.$$

The condition (C) means that a generalized discounted Markov game is considered. Let $\pi = (f_n) \in F^\infty$. Define

(5.2) $$V_\pi = \lim_n L_{f_1} \ldots L_{f_n} V^0$$

(5.3) $$G_\pi = \lim_n U_{f_1} \ldots U_{f_n} V^0$$

By Lemma 5.1 and (C), the limits in (5.2) and (5.3) exist and are independent of V^0. $V_\pi(i,s)$ may be regarded as the *"expected" discounted reward for player* $i \in I$ if the system starts in $s \in S$ and if the players use the infinite-stage plan $\pi \in F^\infty$. Analogously for G_π.

Let $\varepsilon > 0$. A plan $\pi \in F^\infty$ is called an *ε-equilibrium plan* if for all $(i,s) \in I \times S$ $V_\pi(i,s) \geqq G_\pi(i,s) - \varepsilon$. $\pi \in F^\infty$ is called an *equilibrium plan* if $V_\pi(i,s) = G_\pi(i,s)$ for all $(i,s) \in I \times S$. We will show that one can obtain an ε-equilibrium plan in the infinite-stage game by means of a sequence of equilibrium plans in the finite-stage games.

THEOREM 5.2 Assume (A). If condition (C) is satisfied then for any $\varepsilon > 0$, there exists an ε-equilibrium plan.

Proof. Let $\|\cdot\|$ denote the supremum norm. Given $\varepsilon > 0$, let n be such that $2b_n \leqq \varepsilon$, where

$$b_n = \delta_n (\sup_f \|r(f)\| \cdot \sum_0^\infty \delta_k + \|V^0\|).$$

By Theorem 4.2 there exists a plan $\pi_n=(f_o,\ldots,f_{n-1})\in F^n$ such that $V^n_{\pi_n}=G^n_{\pi_n}$. From (A1) we conclude that the measurable set C contains the graph of a measurable map g from S to PA. We will prove that $\pi=(f_o,\ldots,f_{n-1},g,g,\ldots)\in F^\infty$ is an ε-equilibrium plan. One verifies that

$$\|V_\pi - V^n_{\pi_n}\| \leqq b_n \quad \text{and} \quad \|G_\pi - G^n_{\pi_n}\| \leqq b_n .$$

Then we obtain

$$V_\pi \geqq V^n_{\pi_n} - \varepsilon/2=G^n_{\pi_n} - \varepsilon/2 \geqq G_\pi - \varepsilon . \quad \square$$

A *stationary plan* is a plan (f_n) where $f_n=f$ is independent of n. For such a policy we write f^∞. Whether there exists a stationary ε-equilibrium plan in the theorem above is a question which remains open. In the case of a two-person zero-sum Markov game there is a stationary ε-equilibrium plan, even a stationary equilibrium plan. Recently Whitt (1978) has proved the existence of a stationary ε-equilibrium plan under somewhat different assumptions. The following theorem contains characterizations of stationary equilibrium plans.

THEOREM 5.3 Assume (C) and let $f\in F$.
The following statements are equivalent:
(i) f^∞ is a stationary equilibrium plan.
(ii) f is an equilibrium function of LV_{f^∞}.
(iii) f is an equilibrium function of LG_{f^∞}.
If $\delta_m<1$, then (i)<=>(iv)<=>(v).
(iv) There exists $v\in M$ such that $v=L^m_f v$ and f is an equilibrium function of Lv.
(v) There exists $v\in M$ such that $v=U^m_f v$ and f is an equilibrium function of Lv.

The proof is straightforward by observing that V_{f^∞} and G_{f^∞} are the unique fixed points of L_f and U_f, respectively. If $\delta_m<1$, then $V_{f^\infty}(G_{f^\infty})$ is the unique fixed point of L^m_f (U^m_f). $\square$

If S is countable then it is well-known that there is a stationary equilibrium plan. Parthasarathy (1973) and Federgruen (1978) have used criteria (iii) and (ii) of Theorem 5.3, resp., whereas Whitt (1978) has taken a more direct approach. Here we will apply (iv).

The proof reduces to the proof of Takahashi (1964) when m=1 and I and S are finite.

THEOREM 5.4 Assume (A). If S is countable and condition (C) is satisfied then there exists a stationary equilibrium plan.

Proof. Let $\delta_m<1$. Then, for each $f\in F$, the operator L_f^m maps M_o into itself, where

$$M_o = \{v\in M: \|v\| \leqq (1-\delta_m)^{-1}\sup_f \|r(f)\|\cdot\sum_o^{m-1} \delta_k\}$$

Since S is countable and by (A) we may look upon $M_o\times F$ as a compact convex subset of some locally convex linear topological Hausdorff space (cf. proof of Theorem 3.1). Define for $i\in I$, $v\in M_o$ and $f\in F$

$$\Gamma_i(v,f)=\{g_i\in F_i: Lv(i,s,[f|g_i](s))=U_f v(i,s) \text{ for all } s\in S\}$$

$$\Gamma(v,f)=\bigtimes \Gamma_i(v,f).$$

Using Proposition 18 on p.232 in Royden (1968), we conclude that for each i and s, the functions

$$(v,f,g) \rightarrow Lv(i,s,[f|g_i](s))$$

and thus

$$(v,f) \rightarrow U_f v(i,s) \text{ and } (v,f) \rightarrow L_f^m v(i,s)$$

are continuous on $M_o\times F\times F$ and $M_o\times F$, respectively. Hence we obtain that

$$(v,f) \rightarrow (L_f^m v,\Gamma(v,f))$$

is a closed correspondence of $M_o\times F$ to $M_o\times F$ with compact convex values. Then by the Glicksberg-Fan fixed point theorem (cf. Glicksberg (1952), Fan (1952)), there exist $v\in M_o$ and $f\in F$ such that $v=L_f^m v$ and $f\in\Gamma(v,f)$, i.e. $L_f v=U_f v$. By Theorem 5.3(iv), f^∞ is a stationary equilibrium plan. □

References

Billingsley,P. (1968) Convergence of Probability Measures. Wiley, New York.

Denardo,E.V. (1967) Contraction mappings in the theory underlying dynaming programming. SIAM Review 9, 165-177.

Fan,K. (1952) Fixed-point and minimax theorems in locally convex topological linear spaces. Proc. Nat. Akad. Sci. 38, 121-126.

Federgruen,A. (1978) On N-person stochastic games with a denumerable state space. Adv. Appl. Prob. 10, 452-471.

Glicksberg,I. (1952) A further generalization of the Kakutani fixed point theorem with application to Nash equilibrium points. Amer. Math. Soc. 3, 170-174.

Himmelberg,C.J., Parthasarathy,T., Raghavan,T.E.S. and van Vleck,F. (1976) Existence of p-equilibrium and optimal stationary strategies in stochastic games. Proc. Amer. Math. Soc. 60, 245-251.

Hinderer,K. (1970) Foundations of Non-stationary Dynamic Programming with Discrete Time Parameter. Lecture Notes in Operations Research and Mathematical Systems 33, Berlin.

Kuratowski,K. and Ryll-Nardzewski,C. (1965) A general theorem on selectors. Bull. Acad. Polon. Sci. Ser. Sci. Math. Astronom. Phys. 13, 397-403.

Parthasarathy,T. and Stern,M. (1977) Markov games- a survey, 1-46 in: E. Roxin, P. Liu and R. Sternberg (eds.) Differential Games and Control Theory II, Marcel Dekker, New York-Basel.

Rieder,U. (1975) Bayesian dynamic programming. Adv. Appl. Prob. 7, 330-348.

Rieder,U. (1978) Semi-continuous dynamic games. Preprint Univ. Karlsruhe, to be published.

Royden,H.L. (1968) Real Analysis. Macmillan, New York.

Sobel,M.J. (1973) Continuous stochastic games. J. Appl. Prob. 10, 597-604.

Takahashi,M. (1964) Equilibrium points of stochastic non-cooperative n-person games. J. Sci. Hiroshima Univ. Ser.A 28, 95-99.

Wal,J. van der and Wessels,J. (1977) Successive approximation methods for Markov games, 39-55 in: H.C.Tijms and J.Wessels (eds.) Markov Decision Theory, Math. Centre Tracts 93, Amsterdam.

Whitt,W. (1978) Representation and approximation of noncooperative sequential games. Operations Research Center, Bell Laboratories, Holmdel, New Jersey.

GAME THEORY AND RELATED TOPICS
O. Moeschlin, D. Pallaschke (eds.)

SOLVING STOPPING STOCHASTIC GAMES BY MAXIMIZING A LINEAR FUNCTION SUBJECT TO QUADRATIC CONSTRAINTS

Uriel G. Rothblum*
School of Organization and Management
Yale University
New Haven, Connecticut

It is shown that the value and optimal strategies of a discounted stochastic game can be found by maximizing a linear function subject to quadratic constraints.

Consider a stochastic game as defined by Shapley (1953). Let $\{1,\dots,n\}$ is the set of positions and assume that there are p_k (resp., q_k) possible actions to player I (resp., player II) at state k. If state k is observed and the players take, respectively, their i-th and j-th action then the first player takes a^k_{ij} from the second player and the probability that state m will be observed at the next time epoch is p^{km}_{ij} where $\sum_{m=1}^n p^{km}_{ij} < 1$ for every state k and actions i and j. The game will stop with probability $1 - \sum_{m=1}^n p^{km}_{ij} > 0$.

For a given vector $\alpha \in R^n$, let $A^k(\alpha)$ be the $p_k \times q_k$ matrix whose ij-th element is $a^k_{ij} + \sum_{m=1}^n p^{km}_{ij}\alpha^m$. Let T be the operator on R^n defined by $T\alpha = \beta$, where β^k equals the value of the two persons zero sum game whose payoff matrix is $A^k(\alpha)$.

As observed by Shapley (1953), the unique solution to the fixed point equation $T\alpha = \alpha$ is a vector $v \in R^n$, where v^k is the value of the game with starting state k and the optimal stationary strategies for this game are the optimal strategies for the game whose payoff matrix is $A^k(v)$. Since T is monotone (i.e., $T\alpha \le T\beta$ whenever $\alpha \le \beta$) it follows (e.g., Denardo (1967)) that the unique fixed point of T is the unique solution to

*Research was supported by NSF Grant ENG-76-15599 and ONR contract N00014-77-C-0518. The author also wishes to acknowledge some useful discussions with Professor Matthew Sobel. The paper was revised while the author was visiting the Faculty of Industrial and Management Engineering, Technion, Haifa, Israel.

<u>Program 1</u>. $\max \sum_{k=1}^{n} w^k$

subject to: $Tw \geq w$.

The constraint $Tw \geq w$ is equivalent to the existence of probability vectors $x^k = (x_1^k,\ldots,x_{p_k}^k)$, for $k = 1,\ldots,n$, such that

$$\sum_{i=1}^{p_k} x_i^k(a_{ij}^k + \sum_{m=1}^{n} p_{ij}^{km}w^m) \geq w^k , \; k = 1,\ldots,n \text{ and}$$
$$j = 1,\ldots,q_k$$

Thus, Program 1 can be rewritten as:

<u>Program 2</u>. $\max \sum_{k=1}^{n} v^k$

subject to: $x_i^k \geq 0$, $k = 1,\ldots,n$ and $i = 1,\ldots,p_k$

$$\sum_{i=1}^{p_k} x_i^k = 1 , \qquad k = 1,\ldots,n$$

$$\sum_{i=1}^{p_k} x_i^k a_{ij}^k + \sum_{i=1}^{p_k} \sum_{m=1}^{n} x_i^k p_{ij}^{km} v^m \geq v^k ,$$

$$k = 1,\ldots,n \text{ and } j = 1,\ldots,q_k$$

It is easily seen that the optimal x_i^k's for the above program form an optimal stationary strategy.

It should be pointed out that every complementary problem as well as every integer program can be formulated as problems in which a linear objective is maximized subject to quadratic constraints. So, in order to develop an efficient computational procedure for solving program 2, it is expected that one would have to use the structure of the program. The study of this structure might relate to recent results of Bewley and Kohlberg (1976).

Developing algorithms for solving Program 2 in its general form seems to be a difficult problem. Therefore it is worthwhile to consider restricted cases in which the program simplifies. We next consider one such case. When the p_{ij}^{km}'s do not depend on i, i.e., when the transition probabilities depend only on the actions of the second player, Program 2 reduces to the following linear program:

Program 3. $\max \sum_{k=1}^{n} v^k$

subject to: $x_i^k \geq 0$, $k = 1,\ldots,n$ and $i = 1,\ldots,p_k$

$$\sum_{i=1}^{p_k} x_j^k = 1 , \qquad k = 1,\ldots,n$$

$$\sum_{i=1}^{p_k} a_{ij}^k x_i^k + \sum_{m=1}^{n} p_j^{km} v^m \geq v^k , \quad k = 1,\ldots,n \text{ and } j = 1,\ldots,q_k ,$$

where above $p_j^{km} \equiv p_{ij}^{km}$ for $i = 1,\ldots,p_k$. The fact that the value of a stochastic game in which one player controls the transition probabilities can be computed by a linear program was obtained by Parthasarathy and Raghavan (1977).

References

[1] T. Bewley and E. Kohlberg: The theory of stochastic games with zero stop probabilities, Technical Report No. 21, Harvard University, Cambridge, Massachusetts (1976) 94 pages.

[2] E.V. Denardo: Contraction mappings in the theory underlying dynamic programming, SIAM Review 9 (1967) 165-177.

[3] T. Parthasarathy and T.E.S. Raghavan: Finite algorithms for stochastic games, presented at the International Conference on Dynamic Programming, Vancouver, Canada (1977).

[4] L.S. Shapley: Stochastic games, Proceedings of the National Academy of Sciences 39 (1953) 1095-1100.

GAME THEORY AND RELATED TOPICS
O. Moeschlin, D. Pallaschke (eds.)

VALUE ON THE SPACE OF ALL SCALAR INTEGRABLE GAMES

Yair Tauman
C.O.R.E.
Université Catholique de Louvain
Belgium

We present the existence of a unique continuous value on the space spanned by all scalar non-atomic games $f \circ \mu$ where f is integrable on $[0, 1]$ and continuous at 0 and 1 and μ is non-atomic probability measure. The proof given here is also an alternative simple proof for the existence of a unique continuous value on bv'NA.

INTRODUCTION

Aumann and Shapley in their book "Values of Non Atomic Games" [1] proved the existence of a unique value on* pNA and extended this result to the space bv'NA. Recently, A. Neyman [3] has succeeded to prove that bv'NA contained in ASYMP. He also gave an example of a game of the form $f \circ \mu$ not in ASYMP where $\mu \in NA^1$ and f is a bounded function on $[0, 1]$, continuous at 0 and 1 which vanishes everywhere but for a countable number of points. This f is a bounded, measurable function but $f \circ \mu \notin BV$ (otherwise $f \circ \mu \in bv'NA$ and in particular $f \circ \mu \in ASYMP$). However, we will show here that one can prove the existence of the value on the larger space, the one spanned by all games of the form $f \circ \mu$ where f is integrable on $[0, 1]$ (with respect to Lebesgue measure) continuous at 0 and 1 and $\mu \in NA^1$.

Our proof of existence (which employes another method than the one used by Aumann and Shapley for bv'NA) gives an easier proof for the existence of a unique continuous value on bv'NA.

PRELIMINARIES

We use the definitions and notations of [1]. Let $(I, \mathcal{C})$ be a measurable space which is isomorphic to $([0, 1], \mathcal{B})$ where $\mathcal{B}$ is the σ-field of Borel sets on $[0, 1]$ (i.e., there is a one to one mapping from I onto $[0, 1]$ that is measurable in

---

* The notation is given in the following section.

both directions). A set function (or a game) is a real valued function v on C such that $v(\phi) = 0$. A game v is monotonic if for each $S, T \in C$, $S \subset T \Longrightarrow v(S) \leqslant v(T)$. If Q is a space of games, Q^+ denotes the subset of monotonic games in Q. A game v is of bounded variation if it is the difference between two monotonic games. The space of all games of bounded variation is called BV. The subspace of BV consisting of all bounded, finitely additive set functions is denoted by FA.

Let Q be a subspace of BV. A mapping of Q into BV is positive if it maps Q^+ into BV^+. Let G be the group of the automorphisms of (I, C) (i.e., one to one mappings of I onto itself that are measurable in both directions). Each $\Theta \in G$ induces a linear mapping Θ^* of BV onto itself, defined by $(\Theta^* v)(S) = v(\Theta S)$ for all $S \in G$. A subspace Q of BV is called symmetric if $\Theta^* Q = Q$ for all $\Theta \in G$.

Let Q be a symmetric subspace of games. A value on Q is a mapping φ from Q into FA that satisfies the following requirements :

(1) φ is linear

(2) φ is positive

(3) $\varphi(\Theta^* v) = \Theta^*(\varphi v)$ for all $v \in Q$ and $\Theta \in G$

(4) $(\varphi v)(I) = v(I)$ for all $v \in Q$.

For each game v in BV define $\|v\|_{BV} = \sup \|v\|_\Omega$ where the sup ranges over all chains of the form

$$\Omega : \phi = S_0 \subset S_1 \subset S_2 \subset \dots \subset S_n = I , \quad S_i \in C , \quad 0 \leqslant i \leqslant n$$

and $\|v\|_\Omega$ is defined by $\sum_{i=0}^{n-1} |v(S_{i+1}) - v(S_i)|$.

$\| \; \|_{BV}$ defines a norm on BV (Proposition 4.1 in [1]) which we call the variation norm. For $v \notin BV$ we write $\|v\|_{BV} = +\infty$.

Denote the set of all non-atomic measures on (I, C) by NA and by NA^1 the set of all probability measures in NA. pNA is the closed linear subspace of BV spanned by all powers of NA^1 measures. The space of all real valued functions of bounded variation on $[0,1]$ satisfying $f(0) = 0$ and being continuous at 0 and 1 is denoted by bv'. The closed symmetric subspace of BV spanned by the set functions of the form $f \circ \mu$ where $f \in bv'$ and $\mu \in NA^1$ is called bv'NA.

Let Q be a set of games (which might contain elements not belonging to BV) and let $\varphi : Q \to BV$ be an operator. We will write $\|\varphi\| \leqslant K$ $(K \in E^1)$ iff for each $v \in Q$ $\|\varphi v\|_{BV} \leqslant K \cdot \|v\|_{BV}$. We will say that φ is continuous if there exists $K \in E^1$ for which $\|\varphi\| \leqslant K$. By saying that Q is closed we mean that whenever $\|v_n - v\|_{BV} \xrightarrow[n\to\infty]{} 0$ and $v_n \in Q$ for all n, then, v is in Q.

Let λ be the Lebesgue measure on $[0,1]$ and let In' be the set of all integrable functions (with respect to λ) on $[0,1]$ which are continuous at 0 and 1 and satisfying $f(0) = 0$. In'NA is the closure of the smallest linear space of games containing all games of the form $f \circ \mu$ where f is in In' and $\mu \in NA^1$.

The purpose of this paper is to prove the following

THEOREM. There exists a unique continuous value φ on In'NA. The range of φ is NA and $\|\varphi\| \leqslant 1$. For every $f \circ \mu$ in In'NA

$$\varphi(f \circ \mu) = f(1) \cdot \mu.$$

Proof. For every $f \in$ In' and for every $0 < \delta < 1$ define the function f^δ on $[0,1]$ by

$$f^\delta(x) = \int_0^1 f((1-\delta)x + \delta\cdot y)\,d\lambda(y) - \int_0^1 f(\delta\cdot y)\,d\lambda(y).$$

LEMMA 1. For each $f \in$ In' and for each $0 < \delta < 1$

(1) f^δ is absolute continuous on $[0,1]$

(2) $f^\delta(0) = 0$ and $f^\delta(1) \xrightarrow[\delta\to 0]{} f(1)$

Proof. (1) f is integrable on $[0,1]$, therefore if $g(x) = \int_0^x f(t)\,dt$ then g is absolute continuous (a.c.) on $[0,1]$; and for every $a \neq 0$ and b, $g(ax+b)$ is a.c. on $\left[-\frac{b}{a}, \frac{1-b}{a}\right]$

$$f^\delta(x) = \int_0^1 f((1-\delta)x + \delta\cdot y)\,d\lambda(y) - \int_0^1 f(\delta\cdot y)\,d\lambda(y)$$

$$\frac{1}{\delta}\int_{(1-\delta)x}^{(1-\delta)x+\delta} f(t)\,d\lambda(t) - \int_0^1 f(\delta\cdot y)\,d\lambda(y).$$

For a fixed $0 < \delta < 1$ it is enough to prove that $\int_{(1-\delta)x}^{(1-\delta)x+\delta} f(t)\,d\lambda(t)$ is a.c. on $[0,1]$

$$\int_{(1-\delta)x}^{(1-\delta)x+\delta} f(t)\,d\lambda(t) = \int_0^{(1-\delta)x+\delta} f(t)\,d\lambda(t) - \int_0^{(1-\delta)x} f(t)\,d\lambda(t)$$

$$= g((1-\delta)x+\delta) - g((1-\delta)x)$$

$g((1-\delta)x+\delta)$ and $g((1-\delta)x)$ are a.c. on $\left[\frac{-\delta}{1-\delta}, 1\right]$ and $\left[0, \frac{1}{1-\delta}\right]$, respectively. Hence, $g((1-\delta)x+\delta) - g((1-\delta)x)$ is a.c. on $[0, 1]$.

(2) follows immediately from the definition of f^δ and from the facts that f is continuous at 0 and 1 and $f(0) = 0$.

Denote by A the set of all scalar games $f \circ \mu$ where $f \in In'$ and $\mu \in NA^1$, and let $L(A)$ be the linear space generated by A. The space In'NA is the closure of $L(A)$.

<u>LEMMA 2</u>. For every finite set of games in A $\{f_1 \circ \mu_1, \ldots, f_n \circ \mu_n\}$, for every $0 < \delta < 1$ and subchain Λ

$$\Lambda : S_0 \subset S_1 \subset \ldots \subset S_k$$

there exists a subchain $\Lambda^{y,\delta}$

$$\Lambda^y_\delta : S_0^{y,\delta} \subset S_1^{y,\delta} \subset \ldots \subset S_k^{y,\delta}$$

such that for every $1 \leqslant i \leqslant n$ and $0 \leqslant j \leqslant k-1$

$$(f_i^\delta \circ \mu_i)(S_{j+1}) - (f_i^\delta \circ \mu_i)(S_j) = \int_0^1 [(f_i \circ \mu_i)(S_{j+1}^{y,\delta}) - (f_i \circ \mu_i)(S_j^{y,\delta})]d\lambda(y$$

The proof of Lemma 2 makes use of the following result due to Dvoretzky - Wald - Wolfowitz [2].

<u>LEMMA 3</u>. Let η be a finite vector of NA measures. Let $g_1, g_2, \ldots, g_m$ be m measurable functions on I satisfying

(1) $0 \leqslant g_i(t) \leqslant 1$ for every $t \in I$ and $1 \leqslant i \leqslant m$

(2) $g_1 \leqslant g_2 \leqslant \cdots \leqslant g_m$.

Then there are m measurable sets

$$T_1 \subset T_2 \subset \ldots \subset T_m$$

such that

$$\eta(T_i) = \int_I g_i \, d\eta \text{ for every } 1 \leqslant i \leqslant m.$$

For the proof, see Lemma 22.1 of [1] (p.146).

Proof of Lemma 2. For every $1 \leqslant i \leqslant n$ and $0 \leqslant j \leqslant 1$ and for a given $0 < \delta < 1$

$$(f_i^\delta \circ \mu_i)(S_{j+1}) - (f_i^\delta \circ \mu_i)(S_j) = \int_0^1 [f_i((1-\delta)\mu_i(S_{j+1}) + \delta \cdot y) -$$

$$- f_i((1-\delta)\mu_i(S_j) + \delta \cdot y)]\, d\lambda(y).$$

For every $0 \leqslant y \leqslant 1$ and $0 \leqslant j \leqslant k$ define measurable functions $g_j^{y,\delta} : I \to [0,1]$ by

(1) $$g_j^{y,\delta}(t) = (1-\delta)\chi_{S_j}(t) + \delta \cdot y$$

where χ_{S_j} is the characteristic function of S_j. Obviously

(2) $$0 \leqslant g_j^{y,\delta}(t) \leqslant 1$$

(3) $$g_0^{y,\delta} \leqslant g_1^{y,\delta} \leqslant \dots \leqslant g_k^{y,\delta}.$$

Applying Lemma 3 for $\mu = (\mu_1, \dots, \mu_n)$ we deduce from (1), (2) and (3) the existence of a subchain $\Lambda^{y,\delta}$

$$\Lambda^{y,\delta} : S_0^{y,\delta} \subset S_1^{y,\delta} \subset \dots \subset S_k^{y,\delta}$$

for which

$$\mu(S_j^{y,\delta}) = (1-\delta)\mu(S_j) + \delta \cdot y \qquad 0 \leqslant j \leqslant k-1.$$

Thus, we get for every $1 \leqslant i \leqslant n$ and $0 \leqslant j \leqslant k-1$

$$(f_i^\delta \circ \mu_i)(S_{j+1}) - (f_i^\delta \circ \mu_i)(S_j) = \int_0^1 [(f_i \circ \mu_i)(S_{j+1}^{y,\delta}) - (f_i \circ \mu_i)(S_j^{y,\delta})]\, d\lambda(y)$$

and the proof of Lemma 2 is completed.

DEFINITION. For every $v \in L(A)$ if $v = \sum_{i=1}^{n} f_i \circ \mu_i$ where $f_i \circ \mu_i \in A$ $(1 \leqslant i \leqslant n)$ then the game v^δ is defined for every $0 < \delta < 1$ by $v^\delta = \sum_{i=1}^{n} f_i^\delta \circ \mu_i$.

LEMMA 4. For every $v \in L(A)$ and $0 < \delta < 1$, v^δ is well defined.

Proof. We have to prove that if $\sum_{i=1}^{n} f_i \circ \mu_i = \sum_{j=1}^{m} g_j \circ \nu_j$ ($f_i \circ \mu_i$ and $g_j \circ \nu_j$ in A) then $\sum_{i=1}^{n} f_i^\delta \circ \mu_i = \sum_{j=1}^{m} g_j^\delta \circ \nu_j$. It is enough to prove that if $\sum_{i=0}^{n} f_i \circ \mu_i = 0$ then $\sum_{i=1}^{n} f_i^\delta \circ \mu_i = 0$. Let $S \in \mathcal{C}$ and assume that $\sum_{i=1}^{n} f_i \circ \mu_i = 0$

$$\left(\sum_{i=1}^{n} f_i^\delta \circ \mu_i\right)(S) = \sum_{i=1}^{n} \int_0^1 f_i((1-\delta)\mu_i(S) + \delta \cdot y)\, d\lambda(y) - \sum_{i=1}^{n} \int_0^1 f_i(\delta \cdot y)\, d\lambda(y).$$

Applying Lemma 2 for the subchain $\Lambda : \phi \subset S$ and for $\mu = (\mu_1, \dots, \mu_n)$ we get for every $0 \leqslant y \leqslant 1$ and $0 < \delta < 1$ a set $S^{y,\delta} \in \mathcal{C}$ such that

$$\sum_{i=1}^{n} (f_i^\delta \circ \mu_i)(S) = \int_0^1 [\sum_{i=1}^{n} f_i(\mu_i(S^{y,\delta}))]\, d\lambda(y) - \int_0^1 [\sum_{i=1}^{n} f_i(\delta\cdot y)]\, d\lambda(y).$$

Hence

$$(1) \qquad \sum_{i=1}^{n} (f_i^\delta \circ \mu_i)(S) = -\int_0^1 [\sum_{i=1}^{n} f_i(\delta\cdot y)]\, d\lambda(y).$$

By Lyapunov theorem (which asserts that the range of any finite vector of NA measures is convex) we have for every $0 < \delta < 1$ and $0 \leq y \leq 1$ a set $S^{y,\delta} \in C$ such that

$$\mu_i(S^{y,\delta}) = \delta\cdot y \quad \text{for every } 1 \leq i \leq n.$$

Hence

$$0 = \sum_{i=1}^{n} (f_i \circ \mu_i)(S^{y,\delta}) = \sum_{i=1}^{n} f_i(\delta\cdot y)$$

and therefore by (1), $\sum_{i=1}^{n} (f_i^\delta \circ \mu_i)(S) = 0.$

<u>COROLLARY 5</u>. For every $0 < \delta < 1$, if $v \in L(A)$ then $v^\delta \in pNA$.

<u>Proof</u>. Theorem C in [1] (p.25) asserts that if f is absolute continuous on $[0, 1]$ and $f(0) = 0$ then $f \circ \mu \in pNA$ for every $\mu \in NA^1$. Applying Lemma 1 we get that for every $0 < \delta < 1$ and for every $f \circ \mu \in A$, $f^\delta \circ \mu \in pNA$. Since pNA is a linear space and each $v \in L(A)$ is a finite sum of games in A, v^δ is in pNA.

<u>COROLLARY 6</u>. Let v be in $L(A)$. If v is a monotonic game then for every $0 < \delta < 1$, v^δ is a monotonic game.

<u>Proof</u>. Let $v = \sum_{i=1}^{n} f_i \circ \mu_i$ and let S and T be any two coalitions in C with $S \subset T$. Let Λ be the subchain $\Lambda : S \subset T$. For every $0 < \delta < 1$ and for every $0 \leq y \leq 1$ Lemma 2 implies that there are coalitions $S^{y,\delta}$ and $T^{y,\delta}$ in C with $S^{y,\delta} \subset T^{y,\delta}$ such that for every $1 \leq i \leq n$

$$(f_i^\delta \circ \mu_i)(T) - (f_i^\delta \circ \mu_i)(S) = \int_0^1 [(f_i \circ \mu_i)(T^{y,\delta}) - (f_i \circ \mu_i)(S^{y,\delta})]\, d\lambda(y).$$

Hence

$$(*) \qquad v^\delta(T) - v^\delta(S) = \int_0^1 [v(T^{y,\delta}) - v(S^{y,\delta})]\, d\lambda(y).$$

Since v is a monotonic game the integrand in $(*)$ is non-negative for every $0 \leq y \leq 1$ and thus $v^\delta(T) - v^\delta(S) \geq 0$.

<u>COROLLARY 7</u>. For every $v \in L(A)$ and every $0 < \delta < 1$

$$\|v^\delta\|_{BV} \leq \|v\|_{BV}.$$

Proof. Let Ω be a chain

$$\Omega : \phi = S_0 \subset S_1 \subset \ldots \subset S_k = I$$

and let $0 < \delta < 1$. By Lemma 3 together with (*) for each $0 \leqslant y \leqslant 1$ there is a subchain $\Omega^{y,\delta}$

$$\Omega^{y,\delta} : S_0^{y,\delta} \subset S_1^{y,\delta} \subset \ldots \subset S_k^{y,\delta}$$

such that

$$\sum_{i=0}^{k-1} |v^{\delta}(S_{i+1}) - v^{\delta}(S_i)| = \sum_{i=0}^{k-1} |\int_0^1 [v(S_{i+1}^{y,\delta}) - v(S_i^{y,\delta})] d\lambda(y)|$$

Hence

$$\|v^{\delta}\|_{\Omega} \leqslant \int_0^1 [\sum_{i=0}^{k-1} |v(S_{i+1}^{y,\delta}) - v(S_i^{y,\delta})|] d\lambda(y).$$

The integrand is bounded by $\|v\|_{BV}$, therefore $\|v\|_{\Omega} \leqslant \|v\|_{BV}$. Since this inequality holds for every chain Ω, our proof is completed.

LEMMA 8. There exists a unique value φ on $L(A)$. The range of φ is NA and $\|\varphi\| \leqslant 1$.

Proof. Define the operator $\varphi : L(A) \rightarrow NA$ by

$$(1) \qquad \varphi(\sum_{i=1}^{n} f_i \circ \mu_i) = \sum_{i=1}^{n} f_i(1) \cdot \mu_i \qquad f_i \circ \mu_i \in A.$$

Assume for the time being that φ is well defined (i.e., that φv is independent of the special representation of v). From the definition of φ it is obvious that φ is linear, symmetric and efficient. It is left to be shown that φ is positive. Let $v = \sum_{i=1}^{n} f_i \circ \mu_i$ be a monotonic game in $L(A)$.

From Corollary 6 we know that v^{δ} is also monotonic. By Corollary 5 $v^{\delta} \in pNA$. Thus if φ_{pNA} is the value on pNA then $\varphi_{pNA}\, v^{\delta} \geqslant 0$. But

$$\varphi_{pNA}(f \circ \mu) = f(1) \cdot \mu \quad \text{for every } f \circ \mu \in pNA$$

(see Proposition 6.1 in [1], p.38), hence

$$\varphi_{pNA}\, v^{\delta} = \sum_{i=1}^{n} f_i^{\delta}(1) \cdot \mu_i \geqslant 0.$$

When $\delta \rightarrow 0$ we get, by Lemma 1, that

$$\sum_{i=1}^{n} f_i(1) \cdot \mu_i \geqslant 0.$$

Therefore $\varphi v \geqslant 0$, i.e., φ is positive.

Now we can easily prove that φ is well defined. For that purpose it is enough to show that whenever $v = 0$ then $\varphi v = 0$. But $v = 0$ implies that v and $-v$ are both monotonic games. Hence, by the positivity and linearity of φ, $\varphi v \geqslant 0$ and $-\varphi v \geqslant 0$ which implies that $\varphi v = 0$.

Thus we prove that φ is a value on $L(A)$. For the uniqueness of φ on $L(A)$ we need

LEMMA 8. Let Q be a symmetric space of games and let φ be a value on Q. If $\mu \in NA^1$ and $f \circ \mu$ is in Q then $\varphi(f \circ \mu) = f(1) \cdot \mu$.

The only difference between Lemma 8 here and Proposition 6.1 in [1] is that Proposition 6.1 requires Q to be contained in BV. But the proof does not make any use of that property.

It remains to prove that $\|\varphi\| \leqslant 1$. Let $v \in L(A)$. Applying Corollary 7 and the fact that $\|\varphi_{pNA}\| \leqslant 1$ (Theorem B in [1] we get for every $0 < \delta < 1$

$$\|\varphi_{pNA} v^{\delta}\|_{BV} \leqslant \|v^{\delta}\|_{BV} \leqslant \|v\|_{BV} .$$

But since $\|\varphi_{pNA} v^{\delta}\|_{BV} \xrightarrow[\delta\to 0]{} \|\varphi v\|_{BV}$ we get $\|\varphi v\|_{BV} \leqslant \|v\|_{BV}$.

Proof of the theorem. Let $v \in In'NA$. Then, there are games $(v_n)_{n=1}^{\infty}$ in $L(A)$ for which

$$\|v_n - v\|_{BV} \xrightarrow[n\to\infty]{} 0.$$

Since $\|\varphi v_n - \varphi v_m\| \leqslant \|v_n - v_m\|$, $(\varphi v_n)_{n=1}^{\infty}$ is a Cauchy sequence in NA. NA is a Banach space (Proposition 4.4 in [1], p.28) and so $(\varphi v_n)_{n=1}^{\infty}$ has a limit. (This limit is independent of the choice of $(v_n)_{n=1}^{\infty}$.) The definition $\varphi v = \lim_{n\to\infty} \varphi v_n$ extends φ from $L(A)$ to In'NA. It is easy to verify that the extended φ on In'NA is a linear, symmetric and efficient operator. We will show now that $\|\varphi\| \leqslant 1$. Let $v \in In'NA$. If $v_n \longrightarrow v$ $(v_n \in L(A))$ then

$$\|\varphi v_n\|_{BV} \leqslant \|v_n\|_{BV} \leqslant \|v_n - v\|_{BV} + \|v\|_{BV}$$

and letting $n \longrightarrow \infty$, we have $\|\varphi v\|_{BV} \leqslant \|v\|_{BV}$.

To complete the proof of the theorem we have to prove that φ is positive on In'NA. The proof is the same as given for Proposition 4.6 in [1] and it is as follows : Let v be a monotonic game in In'NA then $\|\varphi v\|_{BV} \leqslant \|v\|_{BV} = v(I)$.

Assume now that there is a coalition $S \in C$ with $(\varphi v)(S) < 0$, then

$$v(I) \geqslant \|\varphi v\|_{BV} \geqslant |(\varphi v)(S)| + |(\varphi v)(I) - (\varphi v)(S)| >$$

$$> |v(I) - (\varphi v)(S)| = v(I) - \varphi v(S) > v(I).$$

This contradiction establishes the theorem.

ACKNOWLEDGEMENT. I would like to thank Joseph Greenberg for a useful discussion and remarks.

REFERENCES

[1] Aumann R.J. and L.S. Shapley, Values of Non Atomic Games, Princeton University Press, 1974.

[2] Dvoretzky, A., A. Wald, and J. Wolfowitz, "Relations Among Certain Ranges of Vector Measures", Pac. J. Math., 1 (1951), pp.59-74.

[3] Neyman, A. "Singular Games Have Asymptotic Values", to appear in Math. Oper. Res..

GAME THEORY AND RELATED TOPICS
O. Moeschlin, D. Pallaschke (eds.)

POSITIVE MARKOV GAMES WITH STOPPING ACTIONS

J. van der Wal
Department of Mathematics
Eindhoven University of Technology
Eindhoven, the Netherlands

Abstract. In this paper we consider the two-person zero-sum Markov (stochastic) game with finite state and action spaces, where the immediate payoffs from player II to player I are strictly positive, but where player II has in each state an action which terminates the play immediately. This game is a special case of the positive Markov games considered by Kushner and Chamberlain. Making explicit use of the fact that player II can terminate the game immediately we derive with the method of successive approximations, an ε-band for the value of the game and stationary ε-optimal strategies for both players.

Introduction and notations. Consider a dynamic system with finite state space $S := \{1,2,\ldots,N\}$ to be observed at $t = 0,1,2,\ldots$. The behaviour of the system is influenced by two players, P_1 and P_2, having completely opposite aims. In each state $i \in S$ there exist two finite nonempty sets of actions, one for each player. The set for P_1 is denoted by K_i the one for P_2 by L_i. If at time t the system is observed in state i, then P_1 selects an action from K_i and P_2 on action from L_i. As a joint result of the state i and the two selected actions, k by P_1 and ℓ by P_2, P_1 will receive an amount $r(i,k,\ell)$ from P_2, the systems moves with probability $p(j|i,k,\ell)$ to state j and the play terminates with probability $1-\sum_j p(j|i,k,\ell)$.
We make the following two assumptions:

i) $r(i,k,\ell) \geq a > 0$ for all i, k and ℓ.

ii) in each state $i \in S$ there exists an action $\hat{\ell}_i \in L_i$ such that $\sum_{j\in S} p(j|i,k,\hat{\ell}_i) = 0$ for all $k \in K_i$.

So, as long as the play goes on P_2 looses at least an amount a in each step. But P_2 has in each state an action which terminates the play immediately, of course at some positive cost.
Both players are interested in maximizing their total expected reward.
A *policy* f for P_1 is any function which adds to each state $i \in S$ a probability distribution $f(i)$ on K_i. And $f(i,k)$ denotes the probability by which action k is taken in state i. $f(i)$ can be called a randomized action.
A (Markov) *strategy* π for P_1 is a sequence of policies $\pi = (f_0,f_1,\ldots)$. If at time n the system is observed in state i strategy π prescribes the randomized action $f_n(i)$. Note that we might also consider history dependent strategies but as one may show

(cf. [7]) it is sufficient to consider Markov strategies only.
A stationary strategy π is a strategy in which the policies f_n are identical. Nota-tion: $\pi = f^{(\infty)}$.
Similarly, we define policies g and strategies ρ for P_2.
For each pair of strategies π,ρ let $V(\pi,\rho)$ be the N-columnvector with i-th component equal to the total expected reward for P_1 (possibly ∞) if at time 0 the system is observed in state i and strategies π and ρ are used.
In [3] Maitra and Parthasarathy have considered positive Markov games, however th assume $V(\pi,\rho) < \infty$ for all π and ρ.
Kushner and Chamberlain [2] consider a more general positive stochastic game in their assumptions A_2 through A_4 than we do. They proved that this game has a valu and that the method of successive approximations converges.
In section 4 of this paper we show how the structure of the game (assumptions i and ii) can be exploited further to obtain, with the method of successive approximations, upper and lower bounds for the value of the game and nearly optimal stationary strategies. This we do using the fact that for sensible strategies for P_2 th probability that the play still goes on decreases exponentially.
Before we do this we give some preliminary results concerning the method of successive approximations for finite stage games in section 2. To make this paper self-contained we prove in section 3 that the game has a value, that stationary optimal strategies exist and that the method of successive approximations converges f any starting vector. Finally in section 5 we weaken assumption ii.
We conclude this section with some notations.
For a vector $v \in \mathbb{R}^N$ we define $\|v\|$, $\bar{v}$ and $\underline{v}$ by

$$\|v\| = \max_i |v(i)|, \quad \bar{v} = \max_i v(i), \quad \underline{v} = \min_i v(i) .$$

And $e \in \mathbb{R}^N$ is the N-columnvector with all components equal to one: $e = (1,1,\dots,1)$.

2. <u>Finite stage Markov games</u>. In section 3 we will show that - as in the discount Markov game (cf. Shapley [5]) - the value of the n-stage game may serve as an approximation for the ∞-stage game. Therefore we repeat in this section some results f finite stage Markov games.

We consider the n-stage Markov game with terminal payoff w. So we observe the system at $t = 0,1,\dots,n-1$, and if, as a result of the actions taken at time $n-1$, the system reaches state j at time n, then P_1 receives a terminal payoff $w(j)$ fro P_2. Let $W_n(\pi,\rho)$ be the total expected reward vector for P_1 in this game if strategies π and ρ are used.
Before we show that this game has a value and how to determine it we first give some more notations. Define the N-columnvector $r(f,g)$, the $N \times N$ matrix $P(f,g)$ an the operators $L(f,g)$ and U on $\mathbb{R}^N$ by

$$r(f,g)(i) := \sum_k \sum_\ell f(i,k)g(i,\ell)r(i,k,\ell), \quad i \in S$$

$$P(f,g)(i,j) := \sum_k \sum_\ell f(i,k)g(i,\ell)p(j|i,k,\ell), \quad i,j \in S$$

$$L(f,g)v := r(f,g) + P(f,g)v$$

$$Uv := \max_f \min_g L(f,g)v \ ,$$

where maxmin is taken componentwise.
So $(L(f,g)v)(i)$ is the expected payoff for P_1 if in the matrixgame with entries $r(i,k,\ell) + \sum_j p(j|i,k,\ell)v(j)$ the randomized actions $f(i)$ and $g(i)$ are taken. And $(Uv)(i)$ is the value of this matrixgame. $L(f,g)v$ is also the expected payoff for P_1 in the 1-stage game with terminal payoff v if policies f and g are used.
Clearly the operators $L(f,g)$ and U are monotone. I.e. if $v \geq w$ then $L(f,g)v \geq L(f,g)w$ and $Uv \geq Uw$. As we will treat the n-stage game by a dynamic programming approach it will be more convenient to renumber the observation points in reversed order: the initial time becomes $t = n$, the last actions are taken at $t = 1$ and the terminal payoff takes place at $t = 0$.
Define $w_0 := w$

$$(2.1) \qquad w_t = Uw_{t-1}, \quad t = 1,2,\ldots,n$$

and let $\pi_n = (f_n,\ldots,f_1)$ and $\rho_n = (g_n,\ldots,g_1)$ be n-stage strategies with f_t, g_t (to be applied at reverse time t) satisfying

$$(2.2) \qquad L(f,g_t)w_{t-1} \leq L(f_t,g_t)w_{t-1} = w_t \leq L(f_t,g)w_{t-1}$$

for all f and g. So $f_t(i)$ and $g_t(i)$ are optimal randomized actions in the matrix-games with entries $r(i,k,\ell) + \sum_j p(j|i,k,\ell)w_{t-1}(j)$. Then we have the following result.

Theorem 2.1. The n-stage stochastic game with terminal payoff w has value w_n $(= U^n w)$ and the strategies π_n and ρ_n are optimal in this game for P_1 and P_2 respectively:

$$W_n(\pi,\rho_n) \leq W_n(\pi_n,\rho_n) = w_n \leq W_n(\pi_n,\rho) \quad \text{for all } \pi \text{ and } \rho \ .$$

Proof. The proof is fairly straightforward using the monotonicity of the $L(f,g)$ and U operators, cf. [7]. □

An immediate consequence of this theorem is:

Corollary 2.2. Let $(f_n,\ldots,f_1)$ as defined by (2.2) be an optimal strategy for P_1 in the n-stage game with terminal payoff w, then $(f_m,\ldots,f_1)$ is optimal in the m-stage game with terminal payoff w $(m \leq n)$, and $(f_n,\ldots,f_{n-m+1})$ is optimal in the m-stage game with terminal payoff $U^{n-m}w$.

3. Value and stationary optimal strategies. In this section we show that the operator U has a unique fixed point v^*, that v^* is the value of the ∞-stage game and that the sequence $U^n w$ converges to v^* for all $w \in \mathbb{R}^N$. Further it will be shown that optimal policies for the 1-stage game with terminal payoff v^* give stationary optimal strategies for the ∞-stage game.

Let b be the N-columnvector with i-th component

$$b(i) := \max_{k \in K_i} r(i,k,\hat{\ell}_i) .$$

And let $p(\pi,\rho,n)$ be the N-columnvector with i-th component $p(\pi,\rho,n,i)$ equal to the probability that the play still continues at stage n if the game has started in state i and strategies π and ρ are used.

Further denote by $\rho_n(w)$ an optimal strategy for P_2 for the n-stage game with terminal payoff w, and by $\rho_n(\pi,w)$ a strategy which is optimal if it is already known that P_1 plays π, i.e.

$$W_n(\pi,\rho_n(\pi,w)) = \min_{\rho} W_n(\pi,\rho) .$$

Then we have the following basic lemma.

Lemma 3.1.

i) $U^n w \le b$ for all $w \in \mathbb{R}^N$ and $n \in \mathbb{N}$.

If $na + \underline{w} > 0$ then

ii) $p(\pi,\rho_n(w),n,i) \le b(i)/(na + \underline{w})$.

iii) $p(\pi,\rho_n(\pi,w),n,i) \le b(i)/(na + \underline{w})$.

Proof. i) Clearly P_2 can restrict his losses in any n-stage game to b by terminating the play immediately, so $U^n w \le b$.

ii) From i) and $\rho_n(w)$ being optimal we have

$$W_n(\pi,\rho_n(w)) \le w_n = U^n w \le b .$$

On the other hand the losses for P_2 are at least

$$p(\pi,\rho_n(w),n,i)(na + \underline{w})$$

as, if the play still continues at (reversed) time 0, P_2 will have lost at least at each stage and at time 0 he looses additionally a terminal amount of at least together at least $na + \underline{w}$.

So we have

$$p(\pi,\rho_n(w),n,i)(na + \underline{w}) \le W_n(\pi,\rho_n(w))(i) \le b(i)$$

from which ii) follows immediately.

iii) The proof of iii) is similar to the proof of ii) once we have observed that

$$W_n(\pi,\rho_n(\pi,w)) \le W_n(\pi,\rho_n(w)) \le w_n \ . \qquad \square$$

$p(\pi,\rho_n(w),n)$ behaves not only as $O(n^{-1})$ but even as $o(n^{-1})$. This can be easily obtained from $\sum_{m=0}^{n} p(\pi,\rho_n(w),m) = O(1)$. In section 4 we will see that $p(\pi,\rho_n(w),n)$ decreases even exponentially fast. Similarly $p(\pi,\rho_n(\pi,w),n)$.

Define $v_n := U^n 0$, $n = 0,1,\dots$, then we have:

<u>Theorem 3.2.</u> $\lim_{n\to\infty} v_n$ exists and is finite.

<u>Proof</u>. From $r(i,k,\ell) \ge a > 0$ we have $v_1 \ge v_0 = 0$, and from the monotonicity of U we get $v_{n+1} \ge v_n$, $n = 1,2,\dots$. So v_n is a nondecreasing sequence, which according to lemma 3.1(i) is bounded by b. So $\lim_{n\to\infty} v_n$ exists and does not exceed b. $\square$

Define $v^* := \lim_{n\to\infty} v_n$. Then as Uv is continuous in v $Uv^* = v^*$, so there exist policies f^* and g^* satisfying

$$L(f,g^*)v^* \le v^* \le L(f^*,g)v^* \quad \text{for all f and g} \ . \tag{3.1}$$

Now we get the following theorem.

<u>Theorem 3.3.</u>

i) v^* is the value of the ∞-stage game.

ii) The strategies $f^{*(\infty)}$ and $g^{*(\infty)}$, satisfying (3.1), are optimal for P_1 and P_2 respectively.

<u>Proof</u>. i) P_1 can get at least v_n by playing an optimal strategy for the n-stage game first and playing arbitrarily thereafter. So $\sup_\pi \inf_\rho V(\pi,\rho) \ge v_n$, and with $n \to \infty$ $\sup_\pi \inf_\rho V(\pi,\rho) \ge v^*$.

Further let $\pi_n = (f_n,\dots,f_1)$ (in reversed time notation) be an arbitrary n-stage strategy for P_1, and $V_n(\pi,\rho)$ denote the total expected reward for P_1 in the n-stage game with terminal payoff 0. Then we have by the monotonicity of the $L(f,g)$ operators and inequality (3.1)

$$V_n(\pi_n,g^{*(\infty)}) = L(f_n,g^*)\dots L(f_1,g^*)0 \le L(f_n,g^*)\dots L(f_1,g^*)v^* \le v^* \ .$$

So also for all π

$$V(\pi,g^{*(\infty)}) = \lim_{n\to\infty} V_n(\pi,g^{*(\infty)}) \le v^* \ . \tag{3.2}$$

Hence

$$\inf_\rho \sup_\pi V(\pi,\rho) \le \sup_\pi V(\pi,g^{*(\infty)}) \le v^* \ .$$

So v^* is the value of the ∞-stage game.

ii) From (3.2) we already have that $g^{*(\infty)}$ is optimal for P_2. Remains to show the optimality of $f^{*(\infty)}$. Consider again the n-stage game. Then we have for a strategy

$\rho_n(f^{*(\infty)},0) = (\tilde{g}_n,\ldots,\tilde{g}_1)$ (reversed time again) that

$$\begin{aligned} V_n(f^{*(\infty)},\rho_n(f^{*(\infty)},0)) &= L(f^*,\tilde{g}_n)\ldots L(f^*,\tilde{g}_1)0 \\ &\geq L(f^*,\tilde{g}_n)\ldots L(f^*,\tilde{g}_1)v^* - p(f^{*(\infty)},\rho_n(f^{*(\infty)},0),n)\|v^*\| \\ &\geq v^* - p(f^{*(\infty)},\rho_n(f^{*(\infty)},0),n)\|v^*\| \ . \end{aligned}$$

With lemma 3.1(iii) we get $p(f^{*(\infty)},\rho_n(f^{*(\infty)},0),n) \to 0$. Hence for all ρ

$$\lim_{n\to\infty} V_n(f^{*(\infty)},\rho) \geq v^*.$$

So $f^{*(\infty)}$ is an optimal strategy for P_1.

By definition we have $U^n 0 \to v^*$. But one might also want to start the successive approximation procedure with a scrapvector $w \neq 0$, for example $w = b$. That this also leads to a convergent algorithm is stated in the following theorem.

Theorem 3.4. $\lim_{n\to\infty} U^n w = v^*$ for all $w \in \mathbb{R}^N$.

Proof. The proof we give here will be similar to the proof in theorem 3.3 for the optimality of $f^{*(\infty)}$. Consider the n-stage game with terminal payoff w and let $W_n(\pi,\rho)$ denote again the total expected reward for P_1 in this game. Let P_1 play $f^{*(\infty)}$ and P_2 an optimal reply $\rho_n(f^{*(\infty)},w) = (\bar{g}_n,\ldots,\bar{g}_1)$. Then we have

$$\begin{aligned} U^n w \geq W_n(f^{*(\infty)},\rho_n(f^{*(\infty)},w)) &= L(f^*,\bar{g}_n)\ldots L(f^*,\bar{g}_1)w \\ &\geq L(f^*,\bar{g}_n)\ldots L(f^*,\bar{g}_1)v^* - p(f^{*(\infty)},\rho_n(f^{*(\infty)},w),n)\|w - v^*\| \\ &\geq v^* - p(f^{*(\infty)},\rho_n(f^{*(\infty)},w),n)\|w - v^*\| \ . \end{aligned}$$

From lemma 3.1(iii) we have $p(f^{*(\infty)},\rho_n(f^{*(\infty)},w),n) \to 0$, so $\liminf_{n\to\infty} U^n w \geq v^*$. To prove $\limsup_{n\to\infty} U^n w \leq v^*$ let P_2 play $g^{*(\infty)}$ and P_1 an optimal n-stage reply $\bar{\pi}_n = (\bar{f}_n,\ldots,\bar{f}_1)$ to $g^{*(\infty)}$, then we have

$$\begin{aligned} U^n w \leq W_n(\bar{\pi}_n,g^{*(\infty)}) &= L(\bar{f}_n,g^*)\ldots L(\bar{f}_1,g^*)w \\ &\leq L(\bar{f}_n,g^*)\ldots L(\bar{f}_1,g^*)v^* + p(\bar{\pi}_n,g^{*(\infty)},n)\|w - v^*\| \\ &\leq v^* + p(\bar{\pi}_n,g^{*(\infty)},n)\|w - v^*\| \ . \end{aligned}$$

Also with lemma 3.1(ii) $p(\bar{\pi}_n,g^{*(\infty)},n) \to 0$ as $g^{*(\infty)}$ is an optimal strategy in the stage game with terminal payoff v^* (cf. (3.1)). So also $\limsup_{n\to\infty} U^n w \leq v^*$. Together this gives us $U^n w \to v^*$ as $n \to \infty$.

4. Bounds on v^* and nearly optimal stationary strategies. In the previous section we have shown that the successive approximations $U^n w$ converge to v^* and that $p(\pi,\rho,n)$ tends to zero for all relevant strategies ρ for P_2 (cf. lemma 3.1(ii) and the proof of theorem 3.4). In order to derive bounds on v^* and ε-optimal strategies we need that $p(\pi,\rho,n) \to 0$ exponentially fast for all sensible strategies ρ.
Consider the approximations $U^n w$. And let $\rho_n(w) = (g_n,\dots,g_1)$ be an optimal strategy for P_2 in the n-stage game with terminal payoff w satisfying (2.2). Then by corollary 2.2 $\rho_n(w) = (g_n,\dots,g_{n-m+1})$ is optimal in the m-stage game with terminal payoff $U^{n-m}w = w_{n-m}$.
So with lemma 3.1(ii) assuming $ma + \underline{w}_{n-m} > 0$

$$p(\pi,\rho_n^m(w),m) \le b/(ma + \underline{w}_{n-m}) \ .$$

Define $\delta(m,w)$ by

$$\delta(m,w) := \min(1,\bar{b}/(ma + \underline{w})) \ .$$

Then we can write

$$\bar{p}(\pi,\rho_n^m(w),m) \le \delta(m,w_{n-m}) \ . \tag{4.1}$$

So if P_2 plays $\rho_n(w)$ then at reversed time $n-m$ the play will have terminated with a probability of at least $1-\delta(m,w_{n-m})$. And given that the (n-stage) play still goes on at stage m (reversed time $n-m$) and that P_2 continues to play $\rho_n(w)$ the play will terminate before "stage" $n+1$ (reversed time 0) with a probability of at least $1-\delta(n-m,w)$. From this reasoning we get

$$\bar{p}(\pi,\rho_n(w),n) \le \delta(m,w_{n-m})\delta(n-m,w) \ . \tag{4.2}$$

This leads us to

Lemma 4.1. Let $\rho_n(w)$ be optimal in the n-stage game with terminal payoff w, then if $Uw \ge w$

$$p(\pi,\rho_n(w),n) \le \delta(m,w)\delta(n-m,w) \qquad (m \le n) \ .$$

Proof. From $Uw \ge w$ we get with the monotonicity of U that $w_{n-m} \ge w$. So $\delta(m,w_{n-m}) \le \delta(m,w)$. Substituting this into (4.2) gives us the desired result. □

The lemma implies that $p(\pi,\rho_n(w),n)$ decreases exponentially fast

Corollary 4.2. If $Uw \ge w$ and $n = \alpha p + q$, $\alpha,p,q \in \mathbb{N}$ then

$$\bar{p}(\pi,\rho_n(w),n) \le \delta^{\alpha}(p,v)\delta(q,v) \ .$$

This enables us to obtain bounds on v^*.

Theorem 4.3. If $Uw \ge w$ and $p \in \mathbb{N}$ such that $\delta(p,w) < 1$ then

$$Uw \le v^* \le Uw + (1-\delta(p,w))^{-1} \sum_{q=1}^{p} \delta(q,w)\| Uw - w\| .e \ .$$

Proof. From $Uw \geq w$ we have $w_n := U^n w \geq Uw$, $n \geq 1$ so also $\lim_{n\to\infty} U^n w = v^* \geq Uw$. To derive the second inequality we observe that

$$v^* - Uw = \lim_{n\to\infty} (w_n - w_1) = \lim_{n\to\infty} [(w_n - w_{n-1}) + (w_{n-1} - w_{n-2}) + \ldots + (w_2 - w_1)] .$$

Let $\pi_{t+1}(w) = (f_{t+1},\ldots,f_1)$ and $\rho_{t+1}(w) = (g_{t+1},\ldots,g_1)$ be optimal strategies for the $(t+1)$-stage game with terminal payoff w, as defined by (2.2), then

$$\begin{aligned} w_{t+1} - w_t &= L(f_{t+1},g_{t+1})\ldots L(f_2,g_2)w_1 - L(f_t,g_t)\ldots L(f_1,g_1)w \\ &\leq L(f_{t+1},g_t)\ldots L(f_2,g_1)w_1 - L(f_{t+1},g_t)\ldots L(f_2,g_1)w \\ &\leq \delta(t,w)\| Uw - w\| e , \end{aligned}$$

where the first inequality follows from the optimality of π_{t+1} and ρ_{t+1} and the second inequality follows from lemma 4.1. Thus

$$v^* - Uw \leq \sum_{t=1}^{\infty} \delta(t,v)\| Uw - w\| e .$$

Using corollary 4.2 we get an upperbound for the sum at the right hand side:

$$\begin{aligned} \sum_{m=1}^{\infty} \delta(m,w) &= \sum_{\alpha=0}^{\infty} \sum_{q=1}^{p} \delta(\alpha p + q,w) \leq \sum_{\alpha=0}^{\infty} \sum_{q=1}^{p} \delta^{\alpha}(p,w)\delta(q,w) \\ &= (1-\delta(p,w))^{-1} \sum_{q=1}^{p} \delta(q,w) . \end{aligned}$$

So

$$v^* \leq Uw + (1-\delta(p,w))^{-1} \sum_{q=1}^{p} \delta(q,w)\| Uw - w\| . e .$$

[

Let $\hat{f}$ and $\hat{g}$ be optimal policies in the 1-stage game with terminal payoff w:

$$L(f,\hat{g})w \leq L(\hat{f},\hat{g})w = Uw \leq L(\hat{f},g)w \quad \text{for all f and g} .$$

Then we have the following result on the nearly optimality of $\hat{f}^{(\infty)}$ and $\hat{g}^{(\infty)}$ in the ∞-stage game.

Theorem 4.4.

i) If $Uw \geq w$ then $V(\hat{f}^{(\infty)},\rho) \geq Uw$ for all ρ.

If $w > 0$ and $Uw \leq w + \varepsilon e$ with $0 \leq \varepsilon < a$ then for some $\gamma < 1$

ii) $P(f,\hat{g})w \leq \gamma w$ for all f,

iii) $V(\pi,\hat{g}^{(\infty)}) \leq Uw + \hat{\varepsilon}\gamma(1-\gamma)^{-1}w$ for all π with $\hat{\varepsilon} = \varepsilon \max_{i\in S} \{1/w(i)\}$.

Proof. i) Let $\rho_n(\hat{f}^{(\infty)},0) = (g_n,\ldots,g_1)$ be an optimal reply to $\hat{f}^{(\infty)}$ in the n-stage game with terminal payoff 0, then we have for all ρ

$$\begin{aligned} V_n(\hat{f}^{(\infty)},\rho) &\geq V_n(\hat{f}^{(\infty)},\rho_n(\hat{f}^{(\infty)},0)) = L(\hat{f},g_n)\ldots L(\hat{f},g_1)0 \\ &\geq L(\hat{f},g_n)\ldots L(\hat{f},g_1)w - p(\hat{f}^{(\infty)},\rho_n(\hat{f}^{(\infty)},0),n)\| w\| . \end{aligned}$$

By lemma 3.1(iii) again $p(\hat{f}^{(\infty)},\rho_n(\hat{f}^{(\infty)},0),n) \to 0$. Further

$$L(\hat{f},g_n)\dots L(\hat{f},g_1)w \ge L(\hat{f},g_n)\dots L(\hat{f},g_2)Uw \ge L(\hat{f},g_n)\dots L(\hat{f},g_2)w \ge \dots \ge Uw ,$$

so for all ρ

$$V(\hat{f}^{(\infty)},\rho) = \lim_{n\to\infty} V_n(\hat{f}^{(\infty)},\rho) \ge Uw .$$

ii) From $Uw \le w + \varepsilon e$ we have for all f

$$P(f,\hat{g})w = L(f,\hat{g})w - r(f,\hat{g}) \le L(f,\hat{g})w - ae \le w + \varepsilon e - ae < w .$$

So for some $\gamma < 1$ we have $P(f,\hat{g})w \le \gamma w$ for all f.

iii) Let $\pi_n = (f_n,\dots,f_1)$ be an arbitrary n-stage strategy, then

$$\begin{aligned} L(f_n,\hat{g})\dots L(f_1,\hat{g})0 &\le L(f_n,\hat{g})\dots L(f_1,\hat{g})w \le L(f_n,\hat{g})\dots L(f_2,\hat{g})Uw \\ &\le L(f_n,\hat{g})\dots L(f_2,\hat{g})(w+\varepsilon e) = \\ &= L(f_n,\hat{g})\dots L(f_2,\hat{g})w + P(f_n,\hat{g})\dots P(f_2,\hat{g})\varepsilon e . \end{aligned}$$

With $\hat{\varepsilon} = \varepsilon \max_i \{1/w(i)\}$, we have $\varepsilon e \le \hat{\varepsilon}w$. So

$$P(f_n,\hat{g})\dots P(f_2,\hat{g})\varepsilon e \le \hat{\varepsilon}\gamma^{n-1}w .$$

Continuing in this way we get

$$L(f_n,\hat{g})\dots L(f_1,\hat{g})w \le Uw + \hat{\varepsilon}(\gamma+\gamma^2+\dots+\gamma^{n-1})w .$$

Letting $n \to \infty$ we get for all π

$$V(\pi,\hat{g}^{(\infty)}) \le \lim_{n\to\infty} [Uw + \hat{\varepsilon}(\gamma+\gamma^2+\dots+\gamma^{n-1})w] = Uw + \hat{\varepsilon}\gamma(1-\gamma)^{-1}w . \qquad \square$$

Clearly $Uw + \hat{\varepsilon}\gamma(1-\gamma)^{-1}w$ is also an upperbound on v^*. If we consider the successive approximations $U^n v$, then we know from theorem 3.4 that $U^{n+1}v - U^n v$ will tend to zero, so with $w = U^n v$ we have for n sufficiently large $Uw \le w + \varepsilon e$.
Theorem 4.4(ii) states that the function w is strongly excessive with respect to the set of transition matrices $P(f,\hat{g})$. So the resulting Markov decision process if $\hat{g}$ is fixed is contracting (cf. van Hee and Wessels [1] and van Nunen [4]). Once we have observed this we see that theorem 4.4(iii) is a standard result in contracting dynamic programming. From theorems 4.3 and 4.4 we see that the method of successive approximations yields stationary ε-optimal strategies and (arbitrary close) bounds on v^*. For the ε-optimal strategy for P_1 however we used the monotonicity ($Uw \ge w$).
Also if we do not have $Uw \ge w$ we can derive an interesting result on the ε-optimality of $\hat{f}^{(\infty)}$.

Theorem 4.5. If $Uw \geq w - \varepsilon e$ then for all ρ $V(\hat{f}^{(\infty)},\rho) \geq Uw - \varepsilon a^{-2}(\bar{b} - a)b$.

Proof. Let $\tilde{g}^{(\infty)}$ be an optimal reply to $\hat{f}^{(\infty)}$ in the ∞-stage game: $V(\hat{f}^{(\infty)},\tilde{g}^{(\infty)}) = \min_\rho V(\hat{f}^{(\infty)},\rho)$. That such a strategy exists is a result from negative dynamic programming (cf. Strauch [6], theorem 9.1). Define $\tilde{v} = V(\hat{f}^{(\infty)},\tilde{g}^{(\infty)})$, then $L(\hat{f},\tilde{g})\tilde{v} = \tilde{v}$. From this we get $P(\hat{f},\tilde{g})\tilde{v} \leq \tilde{v} - ae$, which with $ae \leq \tilde{v} \leq \bar{b}e$, hence $\tilde{v}/\bar{b} \leq e$, gives

$$P(\hat{f},\tilde{g})\tilde{v} \leq \tilde{v} - a\tilde{v}/\bar{b} = (1 - a/\bar{b})\tilde{v} .$$

One may argue again that $p(\hat{f}^{(\infty)},\tilde{g}^{(\infty)},n) \to 0$ $(n \to \infty)$, hence

$$V(\hat{f},\tilde{g}^{(\infty)}) = \lim_{n\to\infty} L^n(\hat{f},\tilde{g})w .$$

Further we have

$$\begin{aligned} L^n(\hat{f},\tilde{g})w &\geq L^{n-1}(\hat{f},\tilde{g})Uw \geq L^{n-1}(\hat{f},\tilde{g})(w - \varepsilon e) \\ &\geq L^{n-1}(\hat{f},\tilde{g})w - P^{n-1}(\hat{f},\tilde{g})\varepsilon e \geq \ldots \geq Uw - \varepsilon(P(\hat{f},\tilde{g}) + \ldots + P^{n-1}(\hat{f},\tilde{g}))e . \end{aligned}$$

And with $e \leq a^{-1}\tilde{v}$ (from $\tilde{v} \geq ae$) we get

$$P^k(\hat{f},\tilde{g})e \leq a^{-1}P^k(\hat{f},\tilde{g})\tilde{v} \leq a^{-1}(1 - a/\bar{b})^k\tilde{v} \leq a^{-1}(1 - a/\bar{b})^k b .$$

Combining this yields

$$V(\hat{f}^{(\infty)},\tilde{g}^{(\infty)}) \geq Uw - \varepsilon a^{-1} \sum_{k=1}^{\infty} (1 - a/\bar{b})^k b = Uw - \varepsilon a^{-2}(\bar{b} - a)b .$$ □

5. Some final remarks. In this paper we made the rather restrictive assumption that P_2 can terminate the play immediately. This assumption can be weakened to:
ii') there exists a strategy $\hat{\rho}$ for P_2 such that

$$c = \max_\pi V(\pi,\hat{\rho}) < \infty .$$

Lemma 3.1 with b replaced by c then clearly holds for $w \leq c$ as $U^n w \leq U^n c \leq c$. So for $w \leq c$ we obtain exactly the same results as here with b replaced by c. And if $w \geq c$ lemma 3.1 holds with b replaced by $c + \|w - c\|e$. Again all results carry over with in theorem 4.5 b replaced by c.

Note that if we start the successive approximation procedure with b, then clearly $Ub \leq b$, hence $U^{n+1}b \leq U^n b$ for all n. So if we take $w = U^n b$ in theorem 4.4(iii) then we get $V(\pi,\hat{g}^{(\infty)}) \leq U^{n+1}b$. In this case also the result in theorem 4.5 can be sharpened as we also have $\tilde{v} \leq U^{n+1}b$. Thus we can replace b by $U^{n+1}b$.

References

[1] Hee, K.M. van and Wessels, J., Markov decision processes and strongly excessive functions, to appear in Stoch. Proc. Appl.

[2] Kushner, H.J. and Chamberlain, S.G., Finite state stochastic games: Existence theorems and computational procedures, IEEE Trans. AC. 14 (1969), 248-255.

[3] Maitra, A. and Parthasarathy, T., On stochastic games, II, J. Opt. Theory Appl. 8 (1971), 154-160.

[4] Nunen, J.A.E.E., Contracting Markov decision processes, Mathematica Centre Tract 71, Mathematical Centre, Amsterdam 1976.

[5] Shapley, L.S., Stochastic games. Proc. Nat. Acad. Sci. USA 39 (1953), 1095-1100.

[6] Strauch, R.E., Negative dynamic programming, Ann. Math. Statist. 37 (1966), 871-890.

[7] Wal, J. van der, and Wessels, J., On Markov games, Statistica Neerlandica 30 (1976), 51-71.

GAME THEORY AND RELATED TOPICS
O. Moeschlin, D. Pallaschke (eds.)

SUBJECTIVITY IN THE VALUATION OF GAMES

Robert James Weber
Yale University
New Haven, Connecticut U.S.A.

The recent axiomatic study of probabilistic values of games has clarified the relationship between various valuation methods and the players' subjective perceptions of the coalition-formation process. This has important bearing upon the increasingly-common use of the Banzhaf value in measuring the apportionment of power among the players in voting games. The incompatibility of the players' hypothesized subjective beliefs (under the Banzhaf valuation scheme) leads to the strange phenomenon of "pitfall" points (points of value discontinuity) in weighted majority games with several major players and an ocean of minor players. Such results argue against the use of the Banzhaf value (or indeed, of any value other than the Shapley-Shubik index) in the measurement of power in weighted voting systems.

Introduction. Let N be a finite set of players. A simple game v is a (0-1)-function on the subsets (coalitions) of N, which satisfies $v(\emptyset) = 0$ and $v(S) \geq v(T)$ for all $S \supset T$. The collection of all simple games on N is denoted $SG(N)$. We often think of a simple game as representing the decision rule of a political body: coalitions S for which $v(S) = 1$ are said to be winning, and those for which $v(S) = 0$ are said to be losing.

The weighted voting game $[q: w_1,\ldots,w_n]$ is a simple game defined on the player set $N = \{1,\ldots,n\}$, in which a coalition S is winning if and only if $w(S) = \Sigma_{i \in S} w_i \geq q$. The quantity q is the quota of the game, and $w_1,\ldots,w_n$ are the players' voting weights.

Many organizations are formally administered as weighted voting games. (The stockholders of a corporation, for example, are traditionally accorded voting weights equal to the number of shares they own.) Other institutions use decision rules which, while not explicitly formulated as weighted voting games, are nevertheless equivalent to such games. (The United Nations Security Council passes measures with the approval of at least nine of its fifteen members; each of the five permanent members has the right to veto any measure. This situation is fully represented by a weighted voting game in which the quota is 39, the permanent members have voting weights of 7 each, and the remaining ten members have weights of 1 each.)

Legislative bodies in which the legislators represent districts of unequal populations are frequently organized as weighted voting games. How should the voting weights of the various legislators be determined? An obvious approach is to make the weights proportional to the district populations. However, this can lead to highly unsatisfactory results. For example, assume that four districts respectively contain 30, 30, 30, and 10 percent of the total population. An assignment of weights yielding a game such as [51: 30,30,30,10] will leave the residents of the fourth district without effective representation; their legislator will never be an essential member of a winning coalition. Again, assume that three districts respectively contain 45, 45, and 10 percent of the population. Although the third district is much smaller than the other two, the proportionally-weighted voting game [51: 45,45,10] is actually symmetric; all coalitions of two or more legislators are winning.

Situations analogous to those just given (although perhaps a bit more subtle) have arisen in a number of municipal legislative bodies in the United States (for example, see [5]). The courts have generally ruled that such situations violate the "equal representation" principles of the U.S. Constitution, and have required that voting weights be reallocated in order to provide a more equitable distribution of influence.

These same courts have shown a willingness to accept the idea of a measure of "power" of the players in a simple game. If the relative power of the legislators in a weighted-voting legislative body is roughly proportional to the population of the legislators' districts, then the situation is deemed satisfactory.

In order to facilitate comparisons among various measures of power, we shall present several properties which might be desired of such measures. Two particular measures, the Shapley-Shubik and Banzhaf power indices, have received much attention. We will find that both of these reside within a common axiomatic framework which provides a natural interpretation of them in terms of the players' subjective perceptions of the process of coalition formation.

Probabilistic values. As a fixed player $i \in N$ varies his attention over the games in $SG(N)$, he will perceive himself as having greater influence in some games than in others. A value for i on $SG(N)$ is a real-valued function $\mu_i : SG(N) \to R$ which indicates the subjective assessment, by player i, of his power in the various games.

For any $i \in N$, consider the simple game on N in which the winning coalitions are precisely those containing i. Player i is a dictator in this game; it is difficult to imagine him in a more powerful position. On the other hand, since all games in $SG(N)$ are monotonic (if $S \supset T$ and T is winning, then S is also winning), player i's membership can never hurt a coalition. Therefore, his weakest position arises when he cannot contribute anything to any coalition. In this case, when $v(S\cup i) = v(S)$ for all $S\subset N|i$, we say that i is a zero-dummy of v. (For notational convenience, we will often omit the braces when indicating one-element sets.) Generally, if $v(S\cup i) = v(S) + v(i)$ for all $S\subset N|i$, then i is a strategic dummy in v; both dictators and zero-dummies are strategic dummies.

Combining these observations, we impose the following normalization requirement upon a value μ_i.

(P1) For any $v\in SG(N)$, $0\leq\mu_i(v)\leq 1$. If i is a strategic dummy in v, then $\mu_i(v) = v(i)$.

A simple game on N is completely characterized by its collections of winning and losing coalitions. Hence, a particularly elementary measure of player i's power in a game would arise from simply tallying the winning and losing coalitions to which he belongs. This idea is embodied in the next requirement.

(P2) There are constants $\{a_T:T\subset N\}$, $\{b_T:T\subset N\}$ such that for every $v\in SG(N)$, $\mu_i(v) = \sum_{\substack{T \text{ wins} \\ \text{in } v}} a_T + \sum_{\substack{T \text{ loses} \\ \text{in } v}} b_T$.

There are a number of equivalent formulations of this requirement, some less transparent than others. For example, for any games u and w in $SG(N)$ define the games $u\vee w$ and $u\wedge w$ by $(u\vee w)(S) = \max(u(S),w(S))$ and $(u\wedge w)(S) = \min(u(S),w(S))$. These two new games are also in $SG(N)$. Recently, attention has been given to the requirement that a value satisfy $\mu_i(u) + \mu_i(w) = \mu_i(u\vee w) + \mu_i(u\wedge w)$ for all games u and w under consideration; this is a lattice-theoretic analogue of linearity. In essence, this says that the transfer of winning coalitions from one game to another does not affect the total value of the two games. (P2) is a direct consequence of this.

A probabilistic value is a value satisfying (P1) and (P2). These two properties together imply that probabilistic values have a very special form.

Theorem. A value μ_i for i on $SG(N)$ is a probabilistic value if and only if there is a collection $\{P_T:T\subset N|i\}$ of nonnegative constants satisfying $\Sigma P_T = 1$, such that for every $v\in SG(N)$,

$$\mu_i(v) = \sum_{T\subset N|i} P_T[v(T\cup i)-v(T)] .$$

Proof. It follows from (P2) that for any v,

$$\mu_i(v) = \sum_{v(T)=1} a_T + \sum_{v(T)=0} b_T = \sum_{T\subset N} (a_T - b_T)v(T) + \sum_{T\subset N} b_T .$$

For any $T\subset N$, let $\hat{v}_T\in SG(N)$ be the game in which the winning coalitions are precisely those S for which $S\supsetneq T$. If T is nonempty, let $v_T\in SG(N)$ have as winning coalitions those S for which $S\supset T$. The game $\hat{v}_N$ is identically zero. Player i is a zero-dummy in this game; it follows from (P1) that $\Sigma b_T=0$. For every $T\subset N$, define $c_T = a_T - b_T$. Then $\mu_i(v) = \Sigma c_T v(T)$.

Let T be any nonempty coalition in $N|i$. Player i is a zero-dummy in v_T. Therefore, $\mu_i(v_T)=0=\Sigma_{T\subset S\subset N|i}(c_{S\cup i}+c_S)$. By induction, beginning with $T=N|i$ and proceeding to successively smaller coalitions, it follows that $c_{T\cup i}+c_T=0$. Define $P_T = c_{T\cup i} = -c_T$, and also define $P_\emptyset = c_i$. Then $\mu_i(v) = \Sigma_{T\subset N|i} P_T[v(T\cup i)-v(T)]$.

From (P1), it follows that $\Sigma P_T = \mu_i(v_i) = 1$. Furthermore, for any $T\subset N|i$, $P_T = \mu_i(\hat{v}_T) \geq 0$. This establishes the formula in the theorem. It is easily verified that any value defined in this manner indeed satisfies both (P1) and (P2). □

Our principle concern is with an interpretation suggested by the representation in the theorem. Player i can view the coefficients P_T as subjective probabilities. Coalitions form through a process of accretion. At some point, a coalition in $N|i$ will approach i and invite him to join; the probability that he is approached by T is P_T. The quantity $\mu_i(v)$ is then the probability that i is pivotal, converting the coalition he joins from losing to winning. Hence, the normalization and simplicity assumptions lead directly to a subjective model of coalition formation, in which a player's sole concern is with his marginal contribution to the coalition he joins. We shall hold this model in mind throughout the remainder of this paper.

Semivalues. When developing a measure to compare the influence of the various players in a game, it seems reasonable to adopt a symmetric point of view. In addition, it is desirable that the measurement method be applicable to all finite-player games (rather than merely to games on a fixed player set).

Let U be an infinite set, the universe of players. A simple game v on U is a monotonic (0,1)-function on the subsets of U which satisfies $v(\emptyset) = 0$. Any $N\subset U$ such that $v(S) = v(S\cap N)$ for all $S\subset U$ is a carrier of v. A finite simple game is a simple game with a finite carrier; the set of all such games is $SG(U)$. A value Ψ_i for $i\in U$ is a real-valued function on $SG(U)$. For any finite $N\subset U$, $SG(N)$ can be embedded in $SG(U)$ by treating the players in $U|N$ as zero-dummies. A value Ψ_i is said to satisfy (P1) and (P2) if its restriction to each SG(N) satisfies these conditions. A <u>permutation</u> $\pi : U \to U$ is a one-to-one onto mapping. For any permutation π and game $v \in SG(U)$, define $\pi v \in SG(U)$ by $(\pi v)(S) = v(\pi S)$ for all $S \subset U$. A <u>semivalue</u> $\Psi = (\Psi_i)_{i \in U}$ is a collection of values satisfying (P1), (P2), and the following symmetry condition:

(P3) $\Psi_i(\pi v) = \Psi_{\pi i}(v)$ for all $i \in U$, all permutations π of U, and all games $v \in SG(U)$.

Let ξ be a probability distribution on [0,1]. The value Ψ^{ξ} on SG(U) is defined for all $v \in SG(N)$ and $i \in N \subset U$ by

$$\Psi_i^{\xi}(v) = \sum_{S \subset N|i} P_s^n \, [v(S \cup i) - v(S)] \, ,$$

where $P_s^n = \int_0^1 t^s (1-t)^{n-s-1} d\xi(t)$. (Here, n and s generically denote the cardinalities of N and S.) Note that the definition of $\Psi_i^{\xi}(v)$ is independent of the selection of a carrier N of v. It is not difficult to verify that Ψ^{ξ} is a semivalue. The following result is derived in [3].

<u>Theorem</u>. Let Ψ be a semivalue on SG(U). Then there is a probability distribution ξ on [0,1] such that $\Psi = \Psi^{\xi}$.

Adopting the interpretation of probabilistic values given in the previous section, we can view a semivalue in the following manner. Given $v \in SG(N)$, the players in $N|i$ are assigned random positions on [0,1] which are chosen independently and uniformly. The position of i is chosen according to the distribution ξ. Then $\Psi_i^{\xi}(v)$ is the expected marginal contribution of i (or equivalently, the probability that i is pivotal) when he joins the coalition of players whose positions precede his.

A semivalue Ψ^{ξ} gives each player in turn a distinguished treatment when his value is computed. A fully-symmetric treatment arises when ξ is the uniform distribution on [0,1]. This yields the <u>Shapley value</u> [8,9], with $P_s^n = s!(n-s-1)!/n!$. On the other hand, if ξ is the probability distribution concentrated at the point ½ then the i-subjective viewpoint associated with Ψ_i^{ξ} is highly idiosyncratic; i considers himself likely to hold a central position among the players. This yields the <u>Banzhaf value</u> [1], with $P_s^n = 1/2^{n-1}$.

<u>Consistency</u>. The use of a semivalue Ψ to compare the relative influence of the players of a game is not affected by rescaling. Hence, one may work instead with the normalized value $\bar{\Psi}$, defined for all $v \in SG(U)$ by $\bar{\Psi}(v) = \Psi(v)/\Sigma\Psi_i(v)$. Historically, most applications of the Banzhaf value have employed this normalization.

A conceptual difficulty with this approach is that the sum in the normalizing factor combines the subjective probabilities of different individuals, a probabilistic analogue of adding apples and oranges.

Theorem. The Shapley value is the only semivalue Ψ for which $\Sigma\Psi_i(v) = 1$ for every nonzero $v \in SG(U)$.

This characterization of the Shapley value in the context of simple games first appeared in [2]. For an alternative derivation, let λ denote the Lebesque measure on [0,1] and let $\xi \neq \lambda$ be any other probability measure. Select $x \in [0,1]$ such that $(D\xi)(x) > 1$, or such that x is in the support of the component of ξ singular with respect to λ . Let v_k be a k-player game in which all coalitions of more than xk players win. Then for $x \neq 1$, sufficiently large values of k can be found so that $\Sigma\Psi_i^\xi(v_k) > 1$. (If $x=1$, the same result holds when v_k is a k-player unanimity game.)

The theorem provides a particular distinction to the Shapley value: it is the unique semivalue arising from consistent expectations. The effect of inconsistency on normalized values is discussed in the next section.

Games with many minor players. We consider weighted voting games consisting of a set $M = \{1,2,\ldots,m\}$ of "major" players, and a large number k of "minor" players of total voting weight $\alpha > 0$. Let $v_k = [q:w_1,\ldots,w_m,\frac{\alpha}{k},\ldots,\frac{\alpha}{k}]$, and let $v_k' = (q:w_1,\ldots,w_m,\frac{\alpha}{k},\ldots,\frac{\alpha}{k})$; a coalition wins in the latter game if it has total weight strictly greater than q . The following results concern the semivalues (unnormalized and normalized) of the major players when k is large.

Let Ψ^t denote the semivalue associated with the probability distribution concentrated at t . Define $z_t = [q-t\alpha:w_1,\ldots,w_m]$ and $z_t' = (q-t\alpha:w_1,\ldots,w_m)$; if $q-t\alpha \leq 0$ then $z_t=0$, and if $q-t\alpha<0$ then $z_t'=0$.

Theorem. For all $0<t<1$ and $i \in M$,

$$\lim \Psi_i^t(v_k) = \lim \Psi_i^t(v_k') = \tfrac{1}{2}\,\Psi_i^t(z_t) + \tfrac{1}{2}\,\Psi_i^t(z_t') .$$

Theorem. For all $0<t<1$ and $i \in M$,

$$\lim \bar{\Psi}_i^t(v_k) = \lim \bar{\Psi}_i^t(v_k') = \begin{cases} \bar{\Psi}_i^t(z_t) = \bar{\Psi}_i^t(z_t') & \text{if } t \notin P \\ 0 & \text{if } t \in P , \end{cases}$$

where $P = \{t : \text{for some } S \subset M ,\ w(S)+t\alpha=q\}$.

Both of these results follow in a straightforward manner from the central limit theorem, or from a few judicious applications of Stirling's formula. As k becomes large, each major player considers the total weight of the minor players in the coalition he will eventually join to be equally likely to be slightly more or slightly less than $t\alpha$; the first theorem is an immediate consequence of this. Fix a minor player j , and let L be the (binomially-distributed) number of other minor players in the coalition which j eventually joins. If $w(S)+t\alpha=q$, then $\psi_j^t(v_k) \geq t^s(1-t)^{m-s}\Pr(L<tk\leq L+1)$; this probability is asymptotically proportional to $k^{-\frac{1}{2}}$. On the other hand, if $\epsilon = \min\{|t-t'|:t'\epsilon P\}>0$, then $\psi_j^t(v_k) \leq \Pr(L<(t-\epsilon)k \text{ or } (t+\epsilon)k\leq L+1)$; this probability is asymptotically proportional to $k^{-\frac{1}{2}}r^k$, where $r = \exp(-\epsilon^2/[2t(1-t)])<1$. Consequently, in one case the sum of the values of the k minor players increases without bound, and in the other case the sum approaches zero. Hence, the normalized semivalues behave as indicated in the second theorem. (The cases $t=0$ and $t=1$ require separate treatment, because the distribution of L is degenerate. However, this degeneracy makes direct computation of the values ψ^0 and ψ^1 trivial.)

For any probability distribution ξ on [0,1], and any $v\epsilon SG(U)$ and $i\epsilon U$, $\psi_i^\xi(v) = \int_0^1 \psi_i^t(v)d\xi(t)$. Since the integrand is nonnegative and bounded, the first limit theorem carries over to general distributions in an obvious manner.

The possible behavior of normalized values, as indicated in the second theorem, is best illustrated by an example. Let β denote the normalized Banzhaf value, and consider the games $v_k = [55: 40+\epsilon, 30, 20, \frac{10-\epsilon}{k}, \ldots, \frac{10-\epsilon}{k}]$ for various values of ϵ . If $\epsilon=0$, $(\beta_1,\beta_2,\beta_3)(v_k)\to(0,0,0)$; that is, the minor players share essentially all the influence in the game. For any small $\epsilon>0$, $(\beta_1,\beta_2,\beta_3)(v_k)\to(\frac{3}{5},\frac{1}{5},\frac{1}{5})$. This discontinuity may seem unsurprising, since the voting weight of a major player has been increased at the expense of the minor players. But for any small $\epsilon<0$, we have $(\beta_1,\beta_2,\beta_3)(v_k)\to(\frac{1}{3},\frac{1}{3},\frac{1}{3})$; according to the normalized Banzhaf value, a sacrifice in voting weight can benefit the major players! (It can be shown that for a general distribution ξ and a game $v\epsilon SG(U)$, the occurrence of this type of "pitfall discontinuity" is related to the existence of a $t\epsilon P\cap(0,1)$ such that $d\xi/d\lambda$ either does not exist or is unbounded in every neighborhood of t .)

Of course, this strange behavior results from the inconsistency of the players' subjective expectations. If $\epsilon=0$, each minor player considers himself relatively more likely to be pivotal than any of his fellow minor players; for $\epsilon\neq 0$, the opposite situation holds. And the normalized values of the major players depend critically upon this collective optimism or pessimism of the minor players.

References. Probabilistic values were first characterized in [11], and semivalues in [3]. Limiting results concerning weighted voting games are given for the Shapley value in [7], and for the Banzhaf value in [4]. These two papers allow for the unequal distribution of voting weights among the minor players in a game; the techniques used in these papers can be adapted to the problems treated here. Asymptotic properties of the relative values of the minor players are investigated in the paper by Neyman appearing elsewhere in this volume; see also [6]. A comparison of the Shapley and Banzhaf values, based on a probabilistic model different from that presented here, can be found in [10].

[1] J.F. Banzhaf, III. One man, 3.312 votes: A mathematical analysis of the Electoral College, Villanova Law Review 13 (1968), 304-332.

[2] P. Dubey. On the uniqueness of the Shapley value, Int. J. Game Theory 4 (1975), 131-139.

[3] P. Dubey, A. Neyman, and R.J. Weber. Value theory without efficiency, Cowles Foundation Discussion Paper No. 513, Yale University.

[4] P. Dubey and L.S. Shapley. Mathematical properties of the Banzhaf index, Math. of Oper. Res. (to appear).

[5] W.F. Lucas. Measuring power in weighted voting systems, Case Studies in Applied Mathematics, Mathematical Association of America, 1976, 42-106.

[6] A. Neyman. Singular games have asymptotic values, Math. of Oper. Res. (to appear).

[7] N.Z. Shapiro and L.S. Shapley. Values of large games, I: A limit theorem, Math. of Oper. Res. 3 (1978), 1-9.

[8] L.S. Shapley. A value for n-person games, Ann. Math. Study 28 (1953), 307-317.

[9] L.S. Shapley and M. Shubik. A method for evaluating the distribution of power in a committee system, Amer. Pol. Sci. Rev. 48 (1954), 787-792.

[10] P.D. Straffin. Homogeneity, independence and power indices, Public Choice 30 (1977), 107-118.

[11] R.J. Weber. Probabilistic values for games, Cowles Foundation Discussion Paper No. 471R, Yale University.

Acknowledgement. The preparation of this paper was supported in part by grants from the U.S. Office of Naval Research and the National Science Foundation.

GAME THEORY AND RELATED TOPICS
O. Moeschlin, D. Pallaschke (eds.)

AN ALGORITHM TO DETERMINE ALL EQUILIBRIUM POINTS OF A BIMATRIX GAME

H.-M. Winkels
Ruhr-Universität Bochum
Institut für Unternehmungsführung und
Unternehmensforschung

It is a classical problem of game theory to construct the set $\mathbb{E}$ of all equilibrium points for a bimatrix game (S, T, A, B). The purpose of this paper is to show that $\mathbb{E}$ can be characterized as the union of its NASH components and to give a general method for constructing $\mathbb{E}$ without the assumption of non-degeneracy made in the known complementarity approaches. The algorithm has two essential steps: the determination of all extreme equilibrium points and the construction of all NASH components. The method is demonstrated by an example of a rather degenerated game.

1. INTRODUCTION

We consider a bimatrix game (S, T, A, B) where

$S = \{s_1,\ldots,s_m\}$ is the set of pure strategies of player p_1,

$T = \{t_1,\ldots,t_n\}$ is the set of pure strategies of player p_2 and

$A = (a_{ij})$ and $B = (b_{ij})$ are (m×n)-matrices formed by the (cardinal) utilies a_{ij} and b_{ij} of p_1 and p_2, respectively, when p_1 uses strategy s_i and p_2 strategy t_j. Let

$M(S) = \{x\in\mathbb{R}^m : x\cdot 1_m = 1;\ x \geqq 0\}$ and

$M(T) = \{y\in\mathbb{R}^n : 1_n\cdot y = 1;\ y \geqq 0\}$ denote the set of mixed strategies for p_1 and p_2 where 1_q means $(1,\ldots,1)\in\mathbb{R}^q$. In a natural way we consider s_i and t_j as elements of M(S) and M(T), respectively.

A pair (x^*,y^*) of mixed strategies is called an equilibrium point if and only if

$$x^*Ay^* \geqq xAy^* \qquad \forall x\in M(S)$$
$$x^*By^* \geqq x^*By \qquad \forall y\in M(T).$$

Then x^* is an equilibrium strategy of p_1 and y^* of p_2. By $\mathbb{E}$, $\mathbb{E}_1$ and $\mathbb{E}_2$ we denote the set of all equilibrium points and the sets of all equilibrium strategies for p_1 and p_2, respectively. The Theorem of NASH [10] guarantees that these sets are not empty.

In the case $b_{ij} = -[\alpha\cdot a_{ij} + k]\ \forall i,j : \alpha > 0;\ k\in\mathbb{R}$ the theory of matrix games shows $\mathbb{E}_1 = \{x\in M(S) : xA \geqq val(A)\cdot 1_n\}$,

$\mathbb{E}_2 = \{y\in M(T) : Ay \leqq val(A)\cdot 1_m\}$

and $\mathbb{E} = \mathbb{E}_1 \times \mathbb{E}_2$.

Here val(A) = minimax A = maximin A is the value of A.
We want to give an analogous characterization of $\mathbb{E}$ for the general case and a method for its construction without assuming any structural properties of the game. Recently it has been tried to find $\mathbb{E}$ by means of complementary algorithms. But it is not as yet certain whether these methods yield all elements of $\mathbb{E}$, especially in degenerate cases, see [1], [3], [4], [7], [11] and [12]. Our procedure combines the approaches of VOROB'EV [13], KUHN [6], and MANGASARIAN [8]. The essential steps will be the determination of all extreme equilibrium points and a combinatorical search of all NASH components.

2. PROPERTIES OF EQUILIBRIUM POINTS

2.1 THE PREVENTIVE VALUES.
For $x^*\in M(S)$ we denote by
$\sigma(x^*) := \max\{x^*By : y\in M(T)\} = \max\{x^*Bt_j : j = 1,\dots,n\}$
the preventive value for p_2 to x^*, i.e. the value of the best response to x^* measured in utility units of p_2. The same holds for
$\tau(y^*) := \max\{xAy^* : x\in M(S)\} = \max\{s_iAy^* : i = 1,\dots,m\}$
with respect to p_1 and $y^*\in M(T)$.

The next result is well-known:
2.2 THEOREM. If $x^*\in M(S)$ and $y^*\in M(T)$ then the following conditions are equivalent:
(1) (x^*, y^*) is an equilibrium point
(2) $\tau(y^*) = x^*Ay^*$; $\sigma(x^*) = x^*By^*$
(3) $(x^*Ay^*)\cdot 1_m \geqq Ay^*$; $(x^*By^*)\cdot 1_n \geqq x^*B$.

2.3 THE ASSOCIATED POLYHEDRA. We call the solution sets
$P(B) := \{(x, \xi)\in\mathbb{R}^{m+1} : x\cdot 1_m = 1;\ x \geqq 0;\ xB \leqq \xi\cdot 1_n\}$ and
$P(A) := \{(y, \eta)\in\mathbb{R}^{n+1} : 1_n\cdot y = 1;\ y \geqq 0;\ Ay \leqq \eta\cdot 1_m\}$
the associated polyhedra of the bimatrix game.
For $x\in M(S)$ and $y\in M(T)$ we have
$(y, \tau(y))\in P(A)$ and $(x, \sigma(x))\in P(B)$.

The following Theorem is due to MANGASARIAN [8] and shows the connections between bimatrix games and quadratic programming. We use a slight modification which can easily be verified:

2.4 THEOREM. For $x^*\in M(S)$ and $y^*\in M(T)$ the following statements are equivalent:
(1) (x^*, y^*) is an equilibrium point
(2) There exist $\xi^*, \eta^*\in\mathbb{R}$ such that $(x^*, \xi^*; y^*; \eta^*)$ is an optimal solution of the quadratic optimization problem:

$$g(x, \xi, y, \eta) := x(A+B)y - \xi - \eta \to \max!$$
$$(x, \xi)\in P(B);\ (y, \eta)\in P(A).$$

(3) There exist ξ^*, $\eta^* \in \mathbb{R}$ such that

$$x^*(A+B)y^* - \xi^* - \eta^* = 0;$$

$$(x^*, \xi^*) \in P(B);\ (y^*, \eta^*) \in P(A).$$

(4) There exist ξ^*, $\eta^* \in \mathbb{R}$ such that

$$(x^*B - \xi^* \cdot 1_n)y^* = 0;\ (x^*, \xi^*) \in P(B)$$

$$x^*(Ay^* - \eta^* \cdot 1_m) = 0;\ (y^*, \eta^*) \in P(A).$$

If one of these conditions holds then we necessarily have $\xi^* = \sigma(x^*)$ and $\eta^* = \tau(y^*)$ in (2), (3) and (4).

Proof: To show (3) $\Leftrightarrow$ (4) use the property:

$$x^*(A+B)y^* - \xi^* - \eta^* = x^*(Ay^* - \eta^* \cdot 1_m) + (x^*B - \xi^* \cdot 1_n)y^*$$

For the main steps of the proof see [9]. □

3. THE STRUCTURE OF THE EQUILIBRIUM SET $\mathbb{E}$

3.1 CORRESPONDENCE BETWEEN EQUILIBRIUM STRATEGIES. For non-empty subsets X of $\mathbb{E}_1$ and Y of $\mathbb{E}_2$ we define

$$K(X) := \{y \in \mathbb{E}_2 : (x, y) \in \mathbb{E}\ \ \forall x \in X\}$$

and $\quad L(Y) := \{x \in \mathbb{E}_1 : (x, y) \in \mathbb{E}\ \ \forall y \in Y\}$.

Instead of $K(\{x\})$ or $L(\{y\})$ we simply write $K(x)$ and $L(y)$.

We now formulate only properties of the "K-operator". The formulations for "L" are analogous.

(1) $K(X)$ is empty or convex $\qquad \forall X \subseteq \mathbb{E}_1$

(2) $K(X_1 \cup X_2) = K(X_1) \cap K(X_2) \qquad \forall X_1, X_2 \subseteq \mathbb{E}_1$

(3) $X_1 \subseteq X_2 \Rightarrow K(X_1) \supseteq K(X_2) \qquad \forall X_1, X_2 \subseteq \mathbb{E}_1$

(4) $K(X_1) = K(X_3) \Rightarrow K(X_1) = K(X_2) = K(X_3) \qquad \forall X_1 \subseteq X_2 \subseteq X_3 \subseteq \mathbb{E}_1$

(5) $K(X) \neq \emptyset \Rightarrow X \subseteq L(K(X)) \qquad \forall X \subseteq \mathbb{E}_1$

(6) $K(X) \neq \emptyset \Rightarrow K[L(K(X))] = K(X) \qquad \forall X \subseteq \mathbb{E}_1$

(7) $K(\langle X \rangle) = K(X) \qquad \forall X \subseteq \mathbb{E}_1$

where $\langle \ldots \rangle$ denotes the convex closure.

Proof: (1) - (5) are immediate consequences of the definitions. (6): With (3) and (5) we have $K(X) \supseteq K[L(K(X))]$. Consider $y^* \in K[L(K(X))]$. We have to show: $(x, y^*) \in \mathbb{E}\ \forall x \in L(K(X))$. Take $x \in L(K(X))$. Then $(x, y) \in \mathbb{E}\ \forall y \in K(X)$. Now remember that $y^* \in K(X)$ by the first inclusion. (7): Because of $X \subseteq \langle X \rangle$ we have $K(X) \supseteq K(\langle X \rangle)$. Now let $y \in K(X)$ and $x = \Sigma_{i=1}^{r} \lambda_i x^i$ be a convex combination of elements in X. $L(y)$ is convex and therefore $x \in \langle x^1, \ldots, x^r \rangle \subseteq L(y)$. Thus $(x, y) \in \mathbb{E}$ and finally $y \in K(\langle X \rangle)$. □

The following Lemma plays the essential role for the combination of the VOROB'EV/KUHN and the MANGASARIAN approaches.

3.2 LEMMA. Let (x^*, y^*) be an equilibrium point and $(x^i, \xi_i) \in P(B)$, $i = 1,\dots,r$, such that $(x^*, \sigma(x^*)) = \Sigma_{i=1}^r \lambda_i (x^i, \xi_i) : \lambda_i > 0\ \forall i;\ \Sigma_{i=1}^r \lambda_i = 1$.
Then (x^i, y^*) is an equilibrium point and $\xi_i = \sigma(x^i)$ for $i = 1,\dots,r$.
An analogous result holds when y^* is a convex combination of elements in $P(A)$.

Proof: We use Theorem 2.4(4). With $\xi^* := \sigma(x^*)$ the first equation of 2.4(4) gives: $0 = (x^*B - \xi^* \cdot 1_n) \cdot y^* = [(\Sigma_{i=1}^r \lambda_i x^i)B - (\Sigma_{i=1}^r \xi_i) \cdot 1_n] \cdot y^*$
$= \Sigma_{i=1}^r \lambda_i [x^i B - \xi_i \cdot 1_n] \cdot y^*$.
Because of $\lambda_i [x^i B - \xi_i \cdot 1_n] \cdot y \leqq 0$ and $\lambda_i > 0$ for $i = 1,\dots,r$ this implies
$(x^i B - \xi_i \cdot 1_n) \cdot y^* = 0$, $i = 1,\dots,r$. (a)
In an analogous way we get from the second equation of 2.4(4) with $\eta^* = \tau(y^*)$:
$x^i \cdot (Ay^* - \eta^* \cdot 1_m) = 0$, $i = 1,\dots,r$. (b)
(a), (b) and Theorem 2.4 imply that (x^i, y^*) is an equilibrium point and $\xi_i = \sigma(x^i)$ for $i = 1,\dots,r$. □

3.3 EXTREME EQUILIBRIUM POINTS. (x^*, y^*) is called extreme equilibrium point if (x^*, y^*) is an equilibrium point such that $(x^*, \sigma(x^*))$ is an extreme point (vertex) of $P(B)$ and $(y^*, \tau(y^*))$ is an extreme point of $P(A)$.
In this case x^* and y^* are called extreme equilibrium strategies of p_1 and p_2, respectively, and the finite sets of all such strategies are denoted by $\mathbb{E}_1^e$ for p_1 and by $\mathbb{E}_2^e$ for p_2.

3.4 DETERMINATION OF EXTREME EQUILIBRIUM POINTS. Let $x \in M(S)$ and $y \in M(T)$. By Theorem 2.4 (x, y) is an extreme equilibrium point if and only if there exist $\xi, \eta \in \mathbb{R}$ such that (x, ξ) is an extreme point of $P(B)$, (y, η) is an extreme point of $P(A)$, $x \cdot (Ay - \eta \cdot 1_m) = 0$, and $(xB - \xi \cdot 1_n) \cdot y = 0$.
If one (and therefore both) of these conditions holds we have $\xi = \sigma(x)$ and $\eta = \tau(y)$. This equivalence can be used to find $\mathbb{E}_1^e$ and $\mathbb{E}_2^e$.

The next theorem is due to Mangasarian [8] and establishes the central role of $\mathbb{E}_1^e$ and $\mathbb{E}_2^e$.

3.5 THEOREM. For each $x^* \in \mathbb{E}_1$ there exist $x^1,\dots,x^r$ in $\mathbb{E}_1^e$ such that $(x^*, \sigma(x^*)) = \Sigma_{i=1}^r \lambda_i (x^i, \sigma(x^i)) \quad : \quad \lambda_i > 0\ \ \forall i;\ \Sigma_{i=1}^r \lambda_i = 1$.
Therefore $\mathbb{E}_1 \subseteqq \langle \mathbb{E}_1^e \rangle$.
The analogous result holds for $y^* \in \mathbb{E}_2$.
We give a proof for this theorem because Mangasarian's argumentation has a small gap.

Proof: There exists $y^* \in \mathbb{E}_2$ such that (x^*, y^*) is an equilibrium point. Theorem 2.4 implies that $(x^*, \sigma(x^*))$ is an optimal solution to the following linear optimization problem:

$$f(x, \xi) = x(A + B)y^* - \xi - \tau(y^*) \to \max!$$
$$(x, \xi) \in P(B).$$

If we denote by $P^e(B)$ the set of all extreme points of $P(B)$ a theorem of Goldman [5] shows that
$P(B) = \langle P^e(B)\rangle + \{\mu\cdot(0, 1): \mu \geqq 0\}$.
Therefore (here Mangasarian's proof is incomplete) each optimal solution is a strict (i.e. $\lambda_i > 0$ $\forall i$) convex combination of optimal solutions (x^i, ξ_i), $i = 1,\ldots,r$, in $P^e(B)$. By Theorem 2.4 we have $(x^i, y^*)\in\mathbb{E}$ for $i = 1,\ldots,r$. We now repeat the same argumentation where we change the roles of "x*" and "y*" and get $(y^1, \eta_1),\ldots,(y^s, \eta_s)$ in $P^e(A)$ such that
$(y^*, \tau(y^*))$ is a strict convex combination of all of them. By Lemma 3.2 we have $(x^i, y^j)\in\mathbb{E}$ for $i = 1,\ldots,r$ and $j = 1,\ldots,s$. Hence $x^i\in\mathbb{E}_1^e$ for $i = 1,\ldots,r$. □

3.6 CORRESPONDENCE WITH EXTREME EQUILIBRIUM STRATEGIES. Let X be a non-empty subset of $\mathbb{E}_1$ and Y be a non-empty subset of $\mathbb{E}_2$. Using the notations

$k(X) := \{y\in\mathbb{E}_2^e : (x, y)\in\mathbb{E}\ \ \forall x\in X\}$ and

$l(Y) := \{x\in\mathbb{E}_1^e : (x, y)\in\mathbb{E}\ \ \forall y\in Y\}$ we have

(1) $\langle k(X)\rangle = K(X)$

(2) $k(X)$ is the set of all extreme points of $K(X)$.

Proof: (1): Because of $k(X) \subseteqq K(X)$ and the convexity of $K(X)$ we obtain $\langle k(X)\rangle \subseteqq K(X)$. Consider $y\in K(X)$. Theorem 3.5 shows that there exist $y^1,\ldots,y^r$ in $\mathbb{E}_2^e$ such that
$(y, \tau(y)) = \Sigma_{j=1}^s\mu_j(y^j, \tau(y^j))$: $\mu_j > 0$ $\forall j$; $\Sigma_{j=1}^s\mu_j = 1$.
By Lemma 3.2 we get $y^j\in k(X)$ for $j = 1,\ldots,s$. Hence $y\in\langle k(X)\rangle$.
(2): Consider $y^*\in k(X)$ and suppose that y^* is a strict convex combination of other elements in $k(X)$: $y^* = \Sigma_{j=1}^s\mu_j y^j$. Because there exists an element $x^*\in X$, such that $(x^*, y^*)\in\mathbb{E}$ and $(x^*, y^j)\in\mathbb{E}$ $\forall j = 1,\ldots,r$ we may conclude
$\tau(y^*) = x^*Ay^* = \Sigma\mu_j(x^*Ay^j) = \Sigma\mu_j\tau(y^j)$.
But $(y^*, \tau(y^*))$ and $(y^j, \tau(y^j))$, $j = 1,\ldots,r$ are now different extreme points of $P(A)$. Contradiction! □

The next theorem gives the same representation of $\mathbb{E}$ as the Theorem of VOROB'EV [9], [13] with the difference that it is more constructive.

3.7 THEOREM. $\mathbb{E} = \cup\{\langle X\rangle \times \langle k(X)\rangle : \emptyset \neq X \subseteqq \mathbb{E}_1^e\}$

$\mathbb{E} = \cup\{\langle l(Y)\rangle \times \langle Y\rangle : \emptyset \neq Y \subseteqq \mathbb{E}_2^e\}$.

Proof: We only prove the first equality.
1. Let $\emptyset \neq X \subseteqq \mathbb{E}_1^e$ and $K(X) \neq \emptyset$. By 3.1(5) we have $X \subseteqq L(K(X))$ and by the convexity $\langle X\rangle \subseteqq L(K(X))$. Hence $\langle X\rangle \times K(X) \subseteqq \mathbb{E}$ and with $K(X) = \langle k(X)\rangle$ we obtain $\langle X\rangle \times \langle k(X)\rangle \subseteqq \mathbb{E}$.
2. Let $(x, y)\in\mathbb{E}$. Theorem 3.5 and Lemma 3.2 guarantee the existence of $x^1,\ldots,x^r$ in $\mathbb{E}_1^e$ such that $x\in\langle x^1,\ldots,x^r\rangle$ and $y\in K(x^1,\ldots,x^r) = \langle k(x^1,\ldots,x^r)\rangle$. □

4. NASH COMPONENTS

It is now our aim to reduce the number of subsets for the characterization of $\mathbb{E}$.

4.1 NASH PAIRS AND NASH COMPONENTS. Let $X_0 \subseteq \mathbb{E}_1^e$ and $Y_0 \subseteq \mathbb{E}_2^e$, both non-empty. Then we call (X_0, Y_0) a NASH pair and $\langle X_0 \rangle \times \langle Y_0 \rangle$ a NASH component if $k(X_0) = Y_0$ and $l(Y_0) = X_0$.

4.2 MAXIMALITY OF NASH COMPONENTS. Let $X \times Y \subseteq \mathbb{E}$ be non-empty. Then the following conditions are equivalent:

(1) $X \times Y$ is a NASH component.

(2) $K(X) = Y$ and $L(Y) = X$.

(3) There do not exist X' and Y' such that

$$X \times Y \subsetneqq X' \times Y' \subseteq \mathbb{E}.$$

Proof: (1) $\Rightarrow$ (2): Obvious. (2) $\Rightarrow$ (1): The pair $(l(Y), k(X))$ has the property $\langle l(Y) \rangle \times \langle k(X) \rangle = X \times Y$. We show that it is a NASH pair: $k(l(Y)) = k(L(Y)) = k(X)$ and $l(k(X)) = l(K(X)) = l(Y)$. (2) $\Rightarrow$ (3): Suppose there exist X' and Y' such that $X \times Y \subseteq X' \times Y' \subseteq \mathbb{E}$. Then $Y' \subseteq K(X) = Y$ and $X' \subseteq L(Y) = X$. (3) $\Rightarrow$ (2): We have $X \times Y \subseteq L(K(X)) \times K(X) \subseteq \mathbb{E}$ by 3.1(5). Hence by assumption: $Y = K(X)$ and $K = L(K(X)) = L(Y)$. □

4.3 OVERLAPPING PROPERTY OF NASH COMPONENTS. Two distinct NASH components have at most boundary points in common.

Proof: Let (X_0, Y_0) and (X', Y') be NASH pairs of distinct NASH components, $x^1,\ldots,x^r$ be the different elements of X_0, and $y^1,\ldots,y^s$ the different elements of Y_0. 4.2 implies $Y' \setminus Y_0 \neq \emptyset$ or $X' \setminus X_0 \neq \emptyset$ because (X_0, Y_0) is a NASH pair. Suppose $y' \in Y' \setminus Y_0$ and $(x^*, y^*) \in \langle X_0 \rangle \times \langle Y_0 \rangle \cap \langle X' \rangle \times \langle Y' \rangle$ with $x^* = \Sigma_{i=1}^r \lambda_i x^i : \lambda_i > 0 \;\; \forall i;\; \Sigma_{i=1}^r \lambda_i = 1$.
Since (X', Y') is a NASH pair, too, we have $(x^*, y') \in \mathbb{E}$, but Lemma 3.2 implies $y' \in k(X_0) = Y_0$. Contradiction! □

4.4 GENERATION OF NASH PAIRS. Let X be a non-empty subset of $\mathbb{E}_1^e$ with $K(X) \neq \emptyset$. Then (X_0, Y_0) defined by $Y_0 = k(X)$ and $X_0 = l(Y_0)$ is a NASH pair and $\langle X \rangle \times \langle k(X) \rangle \subseteq \langle X_0 \rangle \times \langle Y_0 \rangle$.
We call (X_0, Y_0) the NASH pair generated by X.
In the same way a non-empty subset Y of $\mathbb{E}_2^e$ with $L(Y) \neq \emptyset$ generates the NASH pair $(l(Y), k[l(Y)])$.

Proof: We only show the first part. 3.1(6) gives $K[L(K(X))] = K(X)$ and 3.6(2) implies $k[L(K(X))] = k(X) =: Y_0$. Hence, by 3.1(7) and 3.6(1): $Y_0 = k[L(K(X))] = k[l(k(X))] = k[l(Y_0)] = k(X_0)$. Thus $\langle Y_0 \rangle = \langle k(X) \rangle$ and $X \subseteq L(K(X)) = L(k(X)) = L(Y_0) = \langle l(Y_0) \rangle = \langle X_0 \rangle$. □

4.5 COROLLARY. $\mathbb{E}$ is the union of the NASH components:

$$\mathbb{E} = \cup\{\langle X_o\rangle \times \langle Y_o\rangle : (X_o, Y_o) \text{ is a NASH pair}\}.$$

We now indicate three properties of NASH pairs which are very useful if one wants to find all such pairs.

4.6 COLLECTING PROPERTY OF NASH PAIRS. Let (X_o, Y_o) be the NASH pair generated by $X \subseteqq \mathbb{E}_1^e$ and $X \subseteqq X' \subseteqq X_o$. Then X' generates (X_o, Y_o), too, and in the case of $X' \neq X_o$ there exists no NASH pair of the form (X', Y') with $Y' \subseteqq \mathbb{E}_2^e$.
An analogous result can be formulated with respect to $Y \subseteqq \mathbb{E}_2^e$.

Proof: $X \subseteqq X' \subseteqq X_o$ implies $k(X) \supseteqq k(X') \supseteqq k(X_o) = Y_o = k(X)$. Therefore (X_o, Y_o) is generated by X', too. In the case of $X' \neq X_o$ we suppose that there is a NASH pair (X', Y') with $Y' \subseteqq \mathbb{E}_2^e$. Then we have $Y' = k(X') = Y_o$ and therefore $X' = l(Y') = l(Y_o) = X_o$. Contradiction! □

4.7 ELIMINATION PROPERTY OF NASH PAIRS. Let (X_o, Y_o) be a NASH pair generated by $X \subseteqq \mathbb{E}_1^e$ and $x \in \mathbb{E}_1^e$. If $k(x) \cap Y_o = \emptyset$ then there exists no NASH pair (X', Y') with $\{x\} \cup X \subseteqq X'$.
An analogous result holds with respect to $Y \subseteqq \mathbb{E}_2^e$ and $y \in \mathbb{E}_2^e$.

Proof: Trivial! □

Our last property shows how a condensation into equivalence classes might be advantageous.

4.8 EQUIVALENCE CLASSES ON $\mathbb{E}_1^e$ AND $\mathbb{E}_2^e$. Two elements x and x' in $\mathbb{E}_1^e$ are called interchangeable, and we write $x \sim_1 x'$, if $k(x) = k(x')$. With respect to $\mathbb{E}_2^e$ we have $y \sim_2 y'$ if $l(y) = l(y')$. $\sim_1$ and $\sim_2$ are equivalence relations on $\mathbb{E}_1^e$ and $\mathbb{E}_2^e$, respectively.

Let $\mathbb{E}_1^\sim = \{x^1,\ldots,x^\alpha\}$ and $\mathbb{E}_2^\sim = \{y^1,\ldots,y^\beta\}$ be sets of representatives for $\sim_1$ and $\sim_2$, respectively. We denote by $X^i := \{x \in \mathbb{E}_1^e : x \sim_1 x^i\}$, $i = 1,\ldots, \alpha$, and $Y^j := \{y \in \mathbb{E}_2^e : y \sim_2 y^j\}$, $j = 1,\ldots,\beta$, the corresponding equivalence classes. For $X \subseteqq \mathbb{E}_1^\sim$ and $Y \subseteqq \mathbb{E}_2^\sim$ we define

$$\bar{k}(X) := \{y \in \mathbb{E}_2^\sim : (x, y) \in \mathbb{E} \quad \forall x \in X\} \text{ and}$$
$$\bar{l}(Y) := \{x \in \mathbb{E}_1^\sim : (x, y) \in \mathbb{E} \quad \forall y \in Y\}.$$

4.9 THE EQUIVALENCE CLASS PROPERTY. Each NASH pair (X_o, Y_o) is generated by a subset $X \subseteqq \mathbb{E}_1^\sim$. In this case

$$Y_o = \cup\{Y^j : y^j \in \bar{k}(X)\} \text{ and}$$
$$X_o = \cup\{X^i : x^i \in \bar{l}(\bar{k}(X))\}.$$

An analogous result holds with respect to $\mathbb{E}_2^\sim$.

Proof: $X_o \cap \mathbb{E}_1^\sim$ generates (X_o, Y_o). The rest can be proven in the usual way. □

5. THE ALGORITHM

STEP 1: THE DETERMINATION OF $\mathbb{E}_1^e$ AND $\mathbb{E}_2^e$.

This step can be done by first calculating the extreme points of P(A) and P(B) and then applying 3.4. The extreme points may be found, for example, by the algorithm of BALINSKI [2], by a modified version of BURGER's algorithm [15], or by geometric considerations in the case of only two pure strategies for one player. For the calculation the following Tableau contains all necessary information :

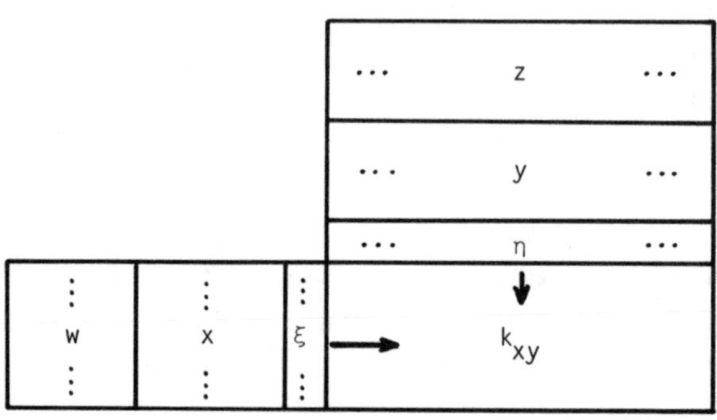

Tab. 5.1

The $(x|\xi)$-block contains (as rows) all extreme points of P(B), the w-block the corresponding values of $xB - \xi\cdot 1_n$, the $(y|\eta)$-block (as columns) all extreme points of P(A), and the z-block the corresponding values of $Ay - \eta\cdot 1_m$.
Then $k_{xy} = 1$ if and only if $w\cdot y = 0$ and $x\cdot z = 0$, and $k_{xy} = 0$ otherwise. Therefore, $k_{xy} = 1$ and $(x, y)\in\mathbb{E}$ are equivalent. Hence the non-zero k-rows yield all elements of $\mathbb{E}_1^e$ and the non-zero k-columns all elements of $\mathbb{E}_2^e$. x and x' are interchangeable if their k-rows are identical.

It is also possible to calculate $\mathbb{E}_1^e$ and $\mathbb{E}_2^e$ directly with another modification of BURGER's algorithm which yields all complementary extreme points of $P(A) \times P(B)$ in the sense of 2.4(4), see [15].
For non-degenerated problems (in the terminology of complementarity algorithms) the method of [7] will give the desired result.

STEP 2: THE SEARCH FOR ALL NASH COMPONENTS.

For the determination of all NASH components all necessary informations are contained in Tab. 5.2.

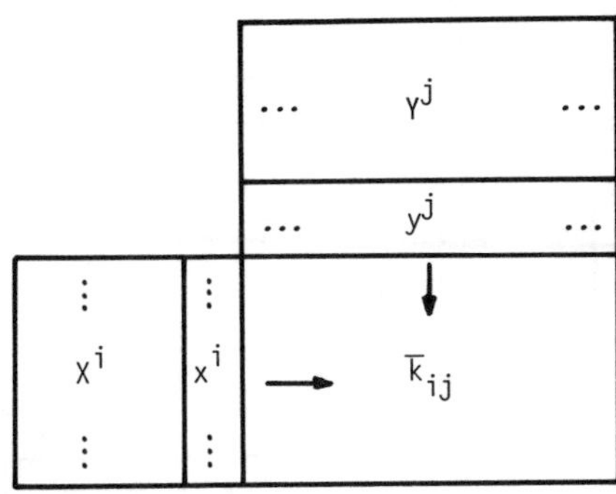

Tab. 5.2

Here $\mathbb{E}_1^{\sim} = \{x^1,\ldots,x^{\alpha}\}$ and $\mathbb{E}_2^{\sim} = \{y^1,\ldots,y^{\beta}\}$ are sets of representatives and $\overline{k}_{ij} = 1 \Leftrightarrow (x^i, y^j) \in \mathbb{E}$; $\overline{k}_{ij} = 0$ otherwise.
Note: $\overline{k}(X) = \{y^j : \overline{k}_{ij} = 1 \quad \forall x^i \in X\} \quad \forall X \subseteqq \mathbb{E}_1^{\sim}$
and $\quad \overline{T}(Y) = \{x^i : \overline{k}_{ij} = 1 \quad \forall y^j \in Y\} \quad \forall Y \subseteqq \mathbb{E}_2^{\sim}$.
Tab. 5.2 can be directly obtained from Tab. 5.1.

Now it is appropriate to choose the smaller one of $\mathbb{E}_1^{\sim}$ and $\mathbb{E}_2^{\sim}$. Suppose this is $\mathbb{E}_1^{\sim}$. Then all NASH pairs can be found in the following way.

(α) Start with $\mathbb{M}$ as the set of all non-empty subsets of $\mathbb{E}_1^{\sim}$.
(β) Choose X in $\mathbb{M}$ minimal with respect to "$\subseteqq$".
(γ) Construct and store the NASH pair generated by X.
(δ) Eliminate in $\mathbb{M}$ all subsets X' with $X' \subseteqq X_0$ or $\{x\} \cup X \subseteqq X'$ for $x \in \mathbb{E}_1^{\sim} \setminus X$ with $\overline{k}(X) \cap \overline{k}(x) = \emptyset$.
(ε) If $\mathbb{M} \neq \emptyset$ go to (β), otherwise all NASH pairs are found.

6. EXAMPLE

Consider the bimatrix game with the matrices A and B, given at the upper left of Tab. 6.1 which corresponds to Tab. 5.1 for this example. The extreme points of P(B) and P(A) were found by the modified BURGER algorithm and by geometric considerations, respectively. We get $\mathbb{E}_1^e = \{x^1,\ldots,x^8\}$ and $\mathbb{E}_2^e = \{y^1,\ldots,y^4\}$ and see that only x^7 and x^8 are interchangeable.

A:

$$\begin{vmatrix} 1 & 3 \\ 1 & 3 \\ 3 & 1 \\ 3 & 1 \\ \frac{5}{2} & \frac{5}{2} \\ \frac{5}{2} & \frac{5}{2} \end{vmatrix}$$

B:

$$\begin{vmatrix} 1 & 2 \\ 0 & -1 \\ -2 & 2 \\ 4 & -1 \\ -1 & 6 \\ 6 & -1 \end{vmatrix}$$

	y^1	y^2	y^3	y^4
z_1	0	0	-1	-2
z_2	0	0	-1	-2
z_3	-2	-1	0	0
z_4	-2	-1	0	0
z_5	$-\frac{1}{2}$	0	0	$-\frac{1}{2}$
z_6	$-\frac{1}{2}$	0	0	$-\frac{1}{2}$
y_1	0	$\frac{1}{4}$	$\frac{3}{4}$	1
y_2	1	$\frac{3}{4}$	$\frac{1}{4}$	0
η	3	$\frac{5}{2}$	$\frac{5}{2}$	3

	w_1	w_2	x_1	x_2	x_3	x_4	x_5	x_6	ξ \ η	y^1	y^2	y^3	y^4
	0	-1	0	1	0	0	0	0	0	0	0	0	0
x^1	0	-5	0	0	0	1	0	0	4	0	0	0	1
	0	-7	0	0	0	0	0	1	6	0	0	0	0
x^2	0	0	$\frac{1}{2}$	$\frac{1}{2}$	0	0	0	0	$\frac{1}{2}$	1	1	0	0
	0	0	$\frac{5}{6}$	0	0	$\frac{1}{6}$	0	0	$\frac{3}{2}$	0	0	0	0
	0	0	$\frac{7}{8}$	0	0	0	0	$\frac{1}{8}$	$\frac{13}{8}$	0	0	0	0
x^3	-1	0	1	0	0	0	0	0	2	1	0	0	0
	0	0	0	$\frac{4}{5}$	$\frac{1}{5}$	0	0	0	$-\frac{2}{5}$	0	0	0	0
x^4	0	0	0	0	$\frac{5}{9}$	$\frac{4}{9}$	0	0	$\frac{2}{3}$	0	0	1	1
x^7	0	0	0	0	$\frac{7}{11}$	0	0	$\frac{4}{11}$	$\frac{10}{11}$	0	0	1	0
	-4	0	0	0	1	0	0	0	2	0	0	0	0
x^5	0	0	0	$\frac{7}{8}$	0	0	$\frac{1}{8}$	0	$-\frac{1}{8}$	0	1	0	0
x^8	0	0	0	0	0	$\frac{7}{12}$	$\frac{5}{12}$	0	$\frac{23}{12}$	0	0	1	0
x^6	0	0	0	0	0	0	$\frac{1}{2}$	$\frac{1}{2}$	$\frac{5}{2}$	0	1	1	0
	-7	0	0	0	0	0	1	0	6	0	0	0	0

Tab. 6.1

Tab. 6.2 corresponds to Tab. 5.2

	y^1	y^2	y^3	y^4
x^1	0	0	0	1
x^2	1	1	0	0
x^3	1	0	0	0
x^4	0	0	1	1
x^5	0	1	0	0
x^6	0	1	1	0
x^7, x^8	0	0	1	0

Tab. 6.2

Using this information and the procedure described above with respect to $\mathbb{E}_2^{\sim} = \mathbb{E}_2^e$ we obtain the following NASH components: $\langle x^2, x^3\rangle \times \{y^1\}$; $\langle x^2, x^5, x^6\rangle \times \{y^2\}$, $\langle x^4, x^6, x^7, x^8\rangle \times \{y^3\}$; $\langle x^1, x^4\rangle \times \{y^4\}$; $\{x^2\} \times \langle y^1, y^2\rangle$, $\{x^6\} \times \langle y^2, y^3\rangle$; $\{x^4\} \times \langle y^3, y^4\rangle$.

7. FINAL REMARKS

Further details and numerical examples can be found in [14]. A general method for determining all solutions of a linear complementarity problem is presented in [15]. In particular, this has the important consequence that all equilibrium points of a polymatrix game can also be constructed.

References

[1] Aggarwal, V. (1973) On the Generation of All Equilibrium Points for Bimatrix Games Through the Lemke-Howson-Algorithm. Mathematical Programming 4, 233 - 234.

[2] Balinski, M. L. (1961) An Algorithm for Finding all Vertices of Convex Polyhedral Sets. SIAM Journal on Applied Mathematics 9, 72 - 88.

[3] Bastian, M. (1976) Another Note on Bimatrix Games. Mathematical Programming 11 (3), 299 - 300.

[4] Bastian, M. (1976) Lineare Komplementärprobleme im Operations Research und in der Wirtschaftstheorie. Verlag Anton Hain, Meisenheim.

[5] Goldman, A. K. (1956) Resolution And Separation Theorems For Polyhedral Convex Sets, 41 - 71. In: Kuhn, H. W.; and A. W. Tucker; (Eds.) (1956) Linear Inequalities And Related Systems, Princeton University Press, New Jersey.

[6] Kuhn, H. W. (1961) An Algorithm for Equilibrium Points in Bimatrix Games. Proceedings of the National Academy of Sciences, Vol. 47, 1656 - 1662.

[7] Lemke, C. E.; and S. J. Grotzinger (1976) On Generalizing Shapley's Index Theory to Labelled Pseudomanifolds. Mathematical Programming 10 (2), 245 - 262.

[8] Mangasarian, O. L. (1964) Equilibrium Points of Bimatrix Games. SIAM Journal of Applied Mathematics 12 (4), 778 - 780.

[9] Parthasarathy, T.; and T. E. S. Raghavan (1971) Some Topics in Two-Person Games. Elsevier, New York.

[10] Nash, J. (1951) Non-Cooperative Games. Annals of Mathematics, Vol. 54 (2), 286 - 295.

[11] Todd, M. J. (1976) Comments On A Note By Aggarwal. Mathematical Programming 10, 130 - 133.

[12] Todd, M. J. (1978) Bimatrix-Games - an Addendum. Mathematical Programming 14, 112 - 115.

[13] Vorob'ev, N. N. (1958) Equilibrium Points in Bimatrix Games. (Transl. by R. DeMarr) Theory of Probability And Its Applications, Vol. III (3), 297 - 309.

[14] Winkels, H.-M. (1978) Die Menge aller Gleichgewichtspunkte eines Bimatrixspieles - Ihre Struktur und ihre Berechnung. Working Paper No. 14, Institut für Unternehmungsführung und Unternehmensforschung; Ruhr-Universität Bochum, D-4630 Bochum (July).

[15] Winkels, H.-M. (1979) Die Bestimmung aller Lösungen eines linearen Restriktionssystems und Anwendungen. Working Paper No. 15; Institut für Unternehmungsführung und Unternehmensforschung; Ruhr-Universität Bochum, D-4630 Bochum (January).

PART II
FIXED-POINT AND OPTIMIZATION THEORY

GAME THEORY AND RELATED TOPICS
O. Moeschlin, D. Pallaschke (eds.)

FIXED-POINT AND RELATED THEOREMS FOR NON-COMPACT CONVEX SETS*

Ky Fan

Department of Mathematics
University of California, Santa Barbara

1. Introduction.

The purpose of this talk is to discuss generalizations of some fixed point and related theorems in our earlier work [6], [7], [11] by relaxing the compactness condition. Due to the limitation of space, we shall not go into the details of the proofs, but shall confine ourselves to explain the mutual relations between the results.

2. Generalizations of Knaster-Kuratowski-Mazurkiewicz's theorem.

Among the various applications of the celebrated combinatorial lemma of Sperner [17], an important one is Knaster-Kuratowski-Mazurkiewicz's theorem [13] which was used in their simple proof of Brouwer's fixed point theorem. In [6], Knaster-Kuratowski-Mazurkiewicz's theorem (which applies only to the finite dimensional case) was extended to the following form:

LEMMA 1. *Let* X *be an arbitrary set in a Hausdorff topological vector space* E. *To each* $x \in X$, *let a closed set* $F(x)$ *in* E *be given such that the following two conditions are satisfied:*

(a) *The convex hull of every finite subset* $\{x_1, x_2, \dots, x_n\}$ *of* X *is contained in the corresponding union* $\bigcup_{i=1}^{n} F(x_i)$.

(b) $F(x)$ *is compact for at least one* $x \in X$.

Then $\bigcap_{x \in X} F(x) \neq \emptyset$.

Lemma 1 was used in [6], [7], [11] to prove fixed point and minimax theorems in topological vector spaces. Generalizations of Lemma 1 have been given by Brézis-Nirenberg-Stampacchia [2] and Dugundji-Granas [4]. Here, by relaxing the compactness condition, we extend Lemma 1 to the following result.

THEOREM 1. *In a Hausdorff topological vector space, let* Y *be a convex set and* $\emptyset \neq X \subset Y$. *For each* $x \in X$, *let* $F(x)$ *be a relatively closed subset of* Y *such that the convex hull of every finite subset* $\{x_1, x_2, \dots, x_n\}$ *of* X *is contained in the corresponding union* $\bigcup_{i=1}^{n} F(x_i)$. *If there is a nonempty subset* X_o *of* X *such that the intersection* $\bigcap_{x \in X_o} F(x)$ *is compact and* X_o *is contained in a compact convex subset of* Y, *then* $\bigcap_{x \in X} F(x) \neq \emptyset$.

* Work supported in part by the National Science Foundation.

For certain applications, it is convenient to reformulate Theorem 1 in the following form.

THEOREM 2. *Let X be a nonempty convex set in a Hausdorff topological vector space. Let B be a subset of $X \times X$ having the following properties:*

(a) *For each $x \in X$, the section $\{y \in X: (x,y) \in B\}$ is relatively closed in X.*

(b) *For every finite subset $\{x_1, x_2, \ldots, x_n\}$ of X and for any n positive numbers α_i with $\sum_{i=1}^{n} \alpha_i = 1$, we have $\left(x_k, \sum_{i=1}^{n} \alpha_i x_i\right) \in B$ for at least one index k.*

Then for every nonempty compact convex subset K of X, either there exists $y_1 \in X$ such that $(x, y_1) \in B$ for all $x \in X$; or there exists $y_2 \in X \setminus K$ such that $(x, y_2) \in B$ for all $x \in K$.

3. *Fixed point theorems.*

As a first direct consequence of Theorem 2, we have the following result generalizing [10, Theorem 2].

THEOREM 3. *Let X be a convex set in a normed vector space E, and let $f: X \to E$ be a continuous map. Let K be a nonempty compact convex subset of X. If for every $y \in X \setminus K$ there is a point $x \in K$ such that*

$$\|x - f(y)\| < \|y - f(y)\|,$$

(Observe that this condition is fulfilled in case X is compact and $K = X$.) *then there exists a point $\hat{y} \in X$ satisfying*

$$\|\hat{y} - f(\hat{y})\| = \min_{x \in X} \|x - f(\hat{y})\|$$

(In particular, if $f(\hat{y}) \in X$, then $\hat{y}$ is a fixed point of f.)

The next result is also proved by using Theorem 2.

THEOREM 4. *Let X be a nonempty paracompact convex set in a Hausdorff topological vector space. Let Φ be a nonempty convex set of lower semi-continuous convex functions on X. Let S be a subset of $X \times \Phi$ having the following properties:*

(a) *For each $\varphi \in \Phi$, the section $S(\varphi) = \{x \in X: (x, \varphi) \in S\}$ is open in X.*

(b) *For each $x \in X$, the section $S(x) = \{\varphi \in \Phi: (x, \varphi) \in S\}$ is nonempty and convex.*

Then for every nonempty compact convex subset K of X, either there exists $(y_1, \varphi_1) \in S$ such that $y_1 \in K$ and $\varphi_1(y_1) = \min_{x \in X} \varphi_1(x)$; or there exists $(y_2, \varphi_2) \in S$ such that $y_2 \in X \setminus K$ and $\varphi_2(y_2) \leq \varphi_2(x)$ for all $x \in K$.

We shall see that Theorem 4 includes a fixed point theorem as a special case. Let X be a set in a real Hausdorff topological vector space E. Let f be a set-valued mapping defined on X such that for each $x \in X$, $f(x)$ is a nonempty subset of E. As in [10], [11], we shall say that f is *upper demi-continuous* on X if for every $x_o \in X$ and any open half-space H in E containing $f(x_o)$, there is a neighborhood N of x_o in X such that $f(x) \subset H$ for all $x \in N$. We shall denote by E' the dual space of E (i.e., the vector space of all continuous linear functionals on E), and recall that an open half-space H in E is a set of the form $H = \{x \in E: \varphi(x) < t\}$, where $0 \neq \varphi \in E'$ and t is a real number. The boundary of X will be denoted by $\mathrm{Bd}\, X$. The following fixed point theorem is a special case of Theorem 4.

THEOREM 5. *Let X be a paracompact convex set in a real, locally convex, Hausdorff topological vector space E, and let K be a nonempty compact convex subset of X. Let f, g be upper demi-continuous set-valued mappings defined on X such that the following conditions are satisfied:*

(a) *For each $x \in X$, $f(x)$ and $g(x)$ are nonempty closed convex subsets of E, at least one of which is compact.*

(b) *For any $x \in K \cap \mathrm{Bd}\, X$ and $\varphi \in E'$ such that $\varphi(x) \leq \varphi(y)$ for all $y \in X$, there exist $u \in f(x)$ and $v \in g(x)$ with $\varphi(u) \geq \varphi(v)$.*

(c) *For any $x \in X \setminus K$ and $\varphi \in E'$ such that $\varphi(x) \leq \varphi(y)$ for all $y \in K$, there exist $u \in f(x)$ and $v \in g(x)$ with $\varphi(u) \geq \varphi(v)$.*

Then there exists a point $\hat{x} \in X$ such that $f(\hat{x}) \cap g(\hat{x}) \neq \emptyset$.

To see that Theorem 5 is a special case of Theorem 4, we define $S \subset X \times E'$ as follows: $(x, \varphi) \in S$ iff there is a real number t such that $\varphi(u) < t$ for all $u \in f(x)$, and $\varphi(v) > t$ for all $v \in g(x)$. Conditions (b), (c) in Theorem 5 imply that the conclusion of Theorem 4 is false. Then by Theorem 4, there is a point $\hat{x} \in X$ such that the section $S(\hat{x})$ is empty. Because E is locally convex, condition (a) of Theorem 5 and $S(\hat{x}) = \emptyset$ imply that $f(\hat{x}) \cap g(\hat{x}) \neq \emptyset$.

In case X is compact and $K = X$, Theorem 5 reduces to an earlier result [11, Theorem 5]. As another special case of Theorem 5, we mention the following fixed point theorem.

THEOREM 6. *Let X, K be convex sets in a locally convex Hausdorff topological vector space E such that $X \supset K \neq \emptyset$, X is paracompact and K is compact. Let f be an upper demi-continuous set-valued mapping defined on X such that for each $x \in X$, $f(x)$ is a nonempty closed convex subset of E. If $f(x) \cap X \neq \emptyset$ for every $x \in K \cap \mathrm{Bd}\, X$, and $f(x) \cap K \neq \emptyset$ for every $x \in X \setminus K$, then there exists a point $\hat{x} \in K$ such that $\hat{x} \in f(\hat{x})$.*

4. Minimax inequality.

We recall that a real-valued function f defined on a convex set X is said to be *quasi-concave* if for every real number t the set $\{x \in X: f(x) > t\}$ is convex. The following result was given in [11] together with several applications.

THEOREM 7. *Let X be a compact convex set in a Hausdorff topological vector space. Let f be a real-valued function defined on $X \times X$ such that:*

(a) *for each fixed $x \in X$, $f(x,y)$ is a lower semi-continuous function of y on X.*

(b) *For each fixed $y \in X$, $f(x,y)$ is a quasi-concave function of x on X.*

Then the minimax inequality $\min_{y \in X} \sup_{x \in X} f(x,y) \leq \sup_{x \in X} f(x,x)$ *holds.*

Generalizations of Theorem 7 and their connections with variational inequalities have been given by Brézis-Nirenberg-Stampacchia [2], Mosco [16], Allen [1], Tarafdar-Thompson [19], Dugundji-Granas [4], Lassonde [14]. We state here the following generalization due to Allen [1].

THEOREM 8 (Allen). *Let* X *be a nonempty convex set in a Hausdorff topological vector space. Let* f *be a real-valued function defined on* $X \times X$ *such that:*

(a) *For each fixed* $x \in X$, $f(x,y)$ *is a lower semi-continuous function of* y *on* X.

(b) *For each fixed* $y \in X$, $f(x,y)$ *is a quasi-concave function of* x *on*

(c) $f(x,x) \leq 0$ *for all* $x \in X$.

(d) *There is a nonempty compact convex subset* K *of* X *such that for each* $y \in X \setminus K$ *there is an* $x \in K$ *with* $f(x,y) > 0$.

Then there exists $\hat{y} \in X$ *such that* $f(x, \hat{y}) \leq 0$ *for all* $x \in X$.

Observe that in case of a compact X, condition (d) in Theorem 8 is satisfied by $K = X$. Theorem 8 reduces to Theorem 7 in case of a compact X. Theorem 8 can be easily derived from Theorem 2.

5. *A matching theorem for open coverings of convex sets.*

The following result is obtained by simply restating Theorem 1 in terms of the complements $A(x)$ of $F(x)$ in Y.

THEOREM 9 (Matching Theorem). *In a Hausdorff topological vector space, let* Y *be a convex set and let* $\emptyset \neq X \subset Y$. *For each* $x \in X$ *let* $A(x)$ *be a relatively open subset of* Y *such that* $\bigcup_{x \in X} A(x) = Y$. *If there is a nonempty subset* X_o *of* X *such that the complement of* $\bigcup_{x \in X_o} A(x)$ *in* Y *is compact or empty, and* X_o *is contained in a compact convex subset of* Y, *then there exists a finite subset* $\{x_1, x_2, \ldots, x_n\}$ *of* X *such that the convex hull of* $\{x_1, x_2, \ldots, x_n\}$ *contains a point of the corresponding intersection* $\bigcap_{i=1}^{n} A(x_i)$.

An immediate consequence of Theorem 9 is the following useful result.

THEOREM 10. *Let* X *be a nonempty convex set in a Hausdorff topological vector space. Let* $A \subset X \times X$ *be such that:*

(a) *For each* $x \in X$, $\{y \in X : (x, y) \in A\}$ *is open in* X.

(b) *For each* $y \in X$, $\{x \in X : (x, y) \in A\}$ *is nonempty and convex.*

(c) *There is a nonempty compact convex set* $K \subset X$ *such that*

$$K \cap \{x \in X : (x, y) \in A\} \neq \emptyset$$

for every $y \in X \setminus K$.

Then there exists a point $\hat{x} \in X$ *such that* $(\hat{x}, \hat{x}) \in A$.

Note that in the case of a compact X, condition (c) in Theorem 10 is satisfied by $K = X$. Thus for a compact X, Theorem 10 becomes a known result which has several applications [3], [6], [10], [11], [12], [20]. Using Theorem 10 one can easily prove

THEOREM 11. *Let* $X_1, X_2, \ldots, X_n$ *be* n (≥ 2) *nonempty convex sets each in a Hausdorff topological vector space, and let* $X = \prod_{i=1}^{n} X_i$. *Let* $A_1, A_2, \ldots, A_n$ *be* n *subsets of* X *such that:*

(a) *For each* i *and for every point* $x_i \in X_i$, *the section* $A_i(x_i)$ *formed by all points* $(x_1, \ldots, x_{i-1}, x_{i+1}, \ldots, x_n)$ *of* $\prod_{j \neq i} X_j$ *such that* $(x_1, \ldots, x_{i-1}, x_i, x_{i+1}, \ldots, x_n) \in A_i$ *is open in* $\prod_{j \neq i} X_j$.

(b) *For each* i *and for every point* $(x_1, \ldots, x_{i-1}, x_{i+1}, \ldots, x_n)$ *of* $\prod_{j \neq i} X_j$, *the section* $A_i(x_1, \ldots, x_{i-1}, x_{i+1}, \ldots, x_n)$ *formed by all* $x_i \in X_i$ *such that* $(x_1, \ldots, x_{i-1}, x_i, x_{i+1}, \ldots, x_n) \in A_i$ *is nonempty and convex.*

(c) *There is a compact convex set* $K \subset X$ *such that*

$$K \cap \prod_{i=1}^{n} A_i(x_1, \ldots, x_{i-1}, x_{i+1}, \ldots, x_n) \neq \emptyset$$

for every $(x_1, x_2, \ldots, x_n)$ *in* $X \setminus K$.

Then the intersection $\bigcap_{i=1}^{n} A_i$ *is nonempty.*

Theorem 11 generalizes an earlier result [7], which is the case of compact $X_1, X_2, \ldots, X_n$. As we have seen in [7], [8], this special case of Theorem 11 leads directly to Nash's equilibrium point theorem, minimax theorem and other applications.

6. The 1978 model of Sperner's lemma.

Theorem 9 can be used to prove the following combinatorial result which may be called "the 1978 model of Sperner's lemma".

LEMMA 2. *Let* T *be a triangulation of an* $(n-1)$*-simplex* $a_1 a_2 \cdots a_n$. *Let* ψ *be a function which assigns to each vertex* v *of* T *an integer* $\psi(v) \in \{1, 2, \ldots, n\}$. *Then there exists a* $(k-1)$*-simplex* $v_1 v_2 \cdots v_k$ *in the triangulation* T *(for some* k *between* 1 *and* n*) and* k *indices* $i_1 < i_2 < \cdots < i_k$ *(between* 1 *and* n*) such that* $v_1 v_2 \cdots v_k \subset a_{i_1} a_{i_2} \cdots a_{i_k}$ *and* $\psi(v_j) = i_j$ *for* $j = 1, 2, \ldots, k$.

Theorem 9 will cease to be valid if the word "open" in its statement is replaced by "closed". However one can use Lemma 2 to prove the following result for a finite covering by relatively closed subsets.

THEOREM 12. *Let* Y *be a convex set in a Hausdorff topological vector space, and let* $A_1, A_2, \ldots, A_n$ *be a finite number* n *of relatively closed subsets of* Y *such that* $\bigcup_{i=1}^{n} A_i = Y$. *Then for any* n *points* $x_1, x_2, \ldots, x_n$ (not necessarily distinct) <u>of</u> Y, <u>there exist</u> k <u>indices</u> (for some k) $i_1 < i_2 < \cdots < i_k$ (between 1 and n) <u>such that the convex hull of</u> $\{x_{i_1}, x_{i_2}, \ldots, x_{i_k}\}$ <u>contains a point of</u> $\bigcap_{j=1}^{k} A_{i_j}$.

For other combinatorial results related to Sperner's lemma, the reader is referred to [9], [15], [18].

References

[1] G. Allen: Variational inequalities, complementarity problems, and duality theorems, J. Math. Anal. Appl. 58 (1977) 1-10.

[2] H. Brézis, L. Nirenberg and G. Stampacchia: A remark on Ky Fan's minimax principle, Boll. Unione Mat. Ital. (4) 6 (1972) 293-300.

[3] F. E. Browder: The fixed point theory of multi-valued mappings in topological vector spaces, Math. Ann. 177 (1968) 283-301.

[4] J. Dugundji and A. Granas: KKM maps and variational inequalities, to appear in Ann. Scu. Norm. di Pisa.

[5] J. Dugundji and A. Granas: Fixed point theory (book in press).

[6] K. Fan: A generalization of Tychonoff's fixed point theorem, Math. Ann. 142 (1961) 305-310.

[7] K. Fan: Sur un Théorème minimax, C. R. Acad. Sci. Paris, Groupe 1, 259 (1964) 3925-3928.

[8] K. Fan: Applications of a theorem concerning sets with convex sections, Math. Ann. 163 (1966) 189-203.

[9] K. Fan: Simplicial maps from an orientable n-pseudomanifold into S^m with the octahedral triangulation, J. Combinatorial Theory 2 (1967) 588-602.

[10] K. Fan: Extensions of two fixed point theorems of F. E. Browder, Math. Z. 112 (1969), 234-240.

[11] K. Fan: A minimax inequality and applications, Inequalities III (Edited by O. Shisha, Academic Press, New York and London, 1972), 103-113.

[12] I. S. Iohvidov: On a lemma of Ky Fan generalizing the fixed-point principle of A. N. Tihonov (Russian), Dokl. Akad. Nauk SSSR 159 (1964) 501-504. English transl.: Soviet Math. Dokl. 5 (1964) 1523-1526.

[13] B. Knaster, C. Kuratowski and S. Mazurkiewicz: Ein Beweis des Fixpunktsatzes für n-dimensionale Simplexe, Fund. Math. 14 (1929) 132-137.

[14] M. Lassonde: Multiapplications KKM en analyse non linéaire, Thèse Ph. D. Math. Univ. de Montréal, 1978.

[15] P. Mani: Zwei kombinatorisch-geometrische Sätze vom Typus Sperner-Tucker-Ky Fan, Monatsh. Math. 71 (1967) 427-435.

[16] U. Mosco: Implicit variational problems and quasi variational inequalities, Lecture Notes in Mathematics, 543: Nonlinear operators and the calculus of variations, Bruxelles 1975 (Edited by J. P. Gossez, E. J. Lami Dozo, J. Mawhin, L. Waelbroeck, Springer-Verlag, Berlin-Heidelberg-New York, 1976, 83-156.

[17] E. Sperner: Neuer Beweis für die Invarianz der Dimensionszahl und des Gebietes, Abhandl. Math. Sem. Univ. Hamburg 6 (1928) 265-272.

[18] E. Sperner: Kombinatorik bewerteter Komplexe, Abhandl. Math. Sem. Univ. Hamburg 39 (1973) 21-43.

[19] E. Tarafdar and H. B. Thompson: On Ky Fan's minimax principle, J. Austral. Math. Soc. (Series A) 26 (1978) 220-226.

[20] G. Wittstock: Über invariante Teilräume zu positiven Transformationen in Räumen mit indefiniter Metrik, Math. Ann. 172 (1967) 167-175.

GAME THEORY AND RELATED TOPICS
O. Moeschlin, D. Pallaschke (eds.)

METHODS OF ASYMPTOTIC FIXED-POINT THEORY APPLIED TO SEMI-DYNAMICAL SYSTEMS

Christian C. Fenske*

Institute of Mathematics

University of Gliessen

Our investigations are motivated by effects observed in the study of functional differential equations. Suppose, one is given an autonomous functional differential equation which admits the trivial solution and this solution is unstable. Then methods of asymptotic fixed point theory can often be applied to conclude the existence of a nontrivial periodic solution (e.g., [4; p.251]). In the simplest case, one might think of an autonomous retarded differential equation. Such an equation generates a semi-flow which gets locally compact when the parameter increases past the time lag. So we consider the following situation: (X,d) is a metric space and $\Phi = (\phi_t)_{t\in\mathbb{R}_+}$ is a semiflow on X such that ϕ_T is locally compact for some $T \geq O$. A *semiflow* is a continuous mapping $\Phi: X \times [O,\infty) \to X$, $x,t \mapsto \phi_t x$

such that $\phi_O = id$ and $\phi_t\phi_s = \phi_{t+s}$ for $t,s \geq O$, and by ϕ_T being *locally compact* we mean that each $x \in X$ possesses a neighbourhood that is mapped into a compact set. We need a global compactness assumption and require Φ to be of compact attraction:

D e f i n i t i o n : Let X be a topological space, Φ a semiflow on X. A set $A \subset X$ is called an *attractor* if for each $x \in X$ and each neighbourhood U of A there is $t \geq O$ with $\phi_t x \in A$. A set $A \subset X$ is *invariant* if $\phi_t(A) \subset A$ for each $t \geq O$, and an attractor $A \subset X$ is *uniform* if it has arbitrarily small invariant neighbourhoods.

The following proposition is well-known if $T = O$, i.e., X is locally compact ([1; p.90]); it is much more difficult if one does not assume local compactness of X but only local compactness of ϕ_t for all $t > O$. In fact, one can do even better by assuming local compactness only for t bigger than some $T \geq O$:

---

*The author gratefully acknowledges support by the Deutsche Forschungsgemeinschaft - SFB 72 an der Universität Bonn.

P r o p o s i t i o n 1 : *Let (X,d) be a metric space, Φ a semiflow on X. Assume that ϕ_T is locally compact for some $T \geq 0$ and that Φ has a compact attractor, B. Then Φ has a uniform compact attractor.*

Before we give the proof, we need a simple lemma:

L e m m a 1 : *Under the hypotheses of Proposition 1, let U be a neighbourhood of B such that $\overline{\phi_T(U)}$ is compact. For $x \in \partial U$ let $\tau_x := \inf\{t > 0 \mid \phi_t x \in U\}$. Then $\tau_U := \sup\{\tau_x \mid x \in \partial U\}$ is finite.*

Proof: The proof is almost as in [2; p.64]. Suppose, there are $x_n \in \partial U$, $t_n \to \infty$ such that $\phi_t x_n \notin U$ for $t \in [0,t_n]$. We may assume $\phi_T x_n \to x'$. B is an attractor, so there are $t \geq 0$ with $\phi_t x' \in U$, a neighbourhood V of x' with $\phi_t(V) \subset U$, and $n_0 \in \mathbb{N}$ such that $\phi_T x_n \in V$ if $n \geq n_0$. Hence $t_n \leq t+T$ for $n \geq n_0$ - a contradiction. *q.e.d.*

Proof of Proposition 1: 1) Let $B' := \bigcup_{n=0}^{\infty} \phi_{nT}(B)$, $B_1 := \overline{B'}$. We claim that B_1 is compact; so $A_1 := \overline{\Phi(B_1 \times \mathbb{R}_+)}$ will be a compact invariant attractor. Choose $\rho > 0$ so small that if $V := \{x \mid d(x,B) < \rho\}$ we have that $\overline{\phi_T(V)}$ is compact. Put then $V_m := \{x \mid d(x,B) < \frac{\rho}{m}\}$, $\tau_m := \tau_{V_m}$, where τ_{V_m} is obtained from Lemma 1. Now let (x_n) be a sequence in B', we want to show that (x_n) has a convergent subsequence (in X). For $m,n \in \mathbb{N}$ let $t_n^{(m)} := \inf\{t > 0 \mid x_n \in \phi_t(V_m)\}$. It is easy to see that for each m the set $\{t_n^{(m)} \mid n \in \mathbb{N}\}$ is bounded. There are then two cases:
Case 1: There is m such that $\{n \mid t_n^{(m)} \geq T\}$ is infinite. Then we may assume $t_n^{(m)} \xrightarrow[n\to\infty]{} t \geq T$. Choose $y_n \in \partial V_m$ and $\sigma_n \in [t_n^{(m)}, t_n^{(m)} + \frac{1}{n}]$ with $x_n = \phi_{\sigma_n} y_n$. We may assume $\phi_T y_n \to y$, hence $x_n = \phi_{\sigma_n - T}\phi_T y_n \to \phi_{t-T} y$.
Case 2: For all m the set $\{n \mid t_n^{(m)} \geq T\}$ is finite. Then for $m \in \mathbb{N}$ choose $n(m) \leq n(m+1)$ with $t_{n(m)}^{(m)} \leq T$, $y_m \in \partial V_m$ and $\sigma_m \in [t_{n(m)}^{(m)}, t_{n(m)}^{(m)} + \frac{1}{m}]$ with $x_{n(m)} = \phi_{\sigma_m} y_m$, where we may assume $y_m \to y \in B$ and $\sigma_m \to \sigma \leq T$, hence $x_{n(m)} \to \phi_\sigma y$.

2) For $M \subset X$ one defines $D(M)$, the *positive prolongation* of M, to be the set of all $y \in X$ such that there exist sequences (x_n) in X and (t_n) in $\mathbb{R}_+$ with $d(x_n,M) \to 0$ and $\phi_{t_n} x_n \to y$. We now claim that $A := D(A_1)$ is a compact invariant attractor. It is certainly an attractor since $A_1 \subset A$, and A is invariant, since A is invariant. So let (y_n) be a sequence in A; we have to exhibit a convergent subsequence. The case $\inf\{d(y_n,A_1) \mid n \in \mathbb{N}\} = 0$ being trivial, we assume that $\inf\{d(y_n,A_1) \mid n \in \mathbb{N}\} =: \varepsilon > 0$. Let $U_\varepsilon := \{x \mid d(x,A_1) < \varepsilon\}$, choose a neighbourhood V of A such that $\Phi(\overline{V} \times [0,T]) \subset U_{\varepsilon/2}$ and $\delta > 0$ such that $U_\delta := \{x \mid d(x,A_1) < \delta\} \subset V$ and such that $\overline{\phi_T(U_\delta)}$ is compact. According to Lemma 1, $\tau := \tau_{U_\delta} < \infty$. We consider $y \in D(A_1) \setminus U_\varepsilon$. There are sequences $x_n \to x \in A_1$ and (t_n) in $\mathbb{R}_+$ such that $\phi_{t_n} x_n \to y$. We may assume that $x_n \in U_\delta$ and $\phi_{t_n} x_n \notin U_{\varepsilon/2}$ for all n. There is $\tau_n \in (0,t_n)$ such

that $\phi_{\tau_n} x_n \in \partial U_\delta$ and $\phi_t x_n \notin U_\delta$ as long as $t \in [\tau_n, t_n]$. But this implies $0 \leq T < t_n - \tau_n < \tau$, hence $\phi_{t_n} x_n = \phi_{t_n - \tau_n - T} \phi_T \phi_{\tau_n} x_n \subset \Phi(\overline{\phi_T(U_\delta)} \times [0, \tau - T])$. So there is a compact set containing all y_n, hence there is a convergent subsequence.
3) An easy argument like the one in [2; p.60, Lemma 1.10] now shows that $D(A) = D(D(A_1)) = D(A_1) = A$. We now claim that A is *stable*, which means that each neighbourhood U of A contains a neighbourhood V of A such that $\phi_t(V) \subset U$ for all $t \geq 0$: Choose $\varepsilon > 0$ such that the closure of $U_1 := \{x \mid d(x,A) < \varepsilon\}$ is contained in U and $\overline{\phi_T(U_1)}$ is compact. Call $U_n := \{x \mid d(x,A) < \frac{\varepsilon}{n}\}$. If our claim were false there would be sequences $t_n > 0$ and $x_n \in U_n$ with $\phi_{t_n} x_n \notin U$. Since $d(x_n, A) \to 0$ we may assume $x_n \to x \in A$. Since (t_n) cannot contain a convergent subsequence, we may assume $t_n \to \infty$. But then we will show that $(\phi_{t_n} x_n)$ has a convergent subsequence the limit of which has to be an element of $D(A) = A$ - a contradiction. In fact, Lemma 1 shows that $\tau := \tau_{U_1} < \infty$. Call τ_n the supremum of all $t \geq 0$ such that $\phi_t x_n \in \overline{U}_1$ and $\phi_t x_n \notin U_1$ for all $\tau \in [t, t_n]$. Observe that $\tau_n < t_n < \tau_n + \tau$, so there is n_0 such that $\tau_n > T$ whenever $n \geq n_0$. But for $n \geq n_0$ it is easily verified that $\phi_{t_n} x_n = \phi_{t_n - \tau_n} \phi_T \phi_{\tau_n - T} x_n \subset \Phi(\overline{\phi_T(U_1)} \times [0, 2\tau])$ which is a compact set.
4) We complete the proof by showing that A is a uniform attractor. Let W be a neighbourhood of A. Choose neighbourhoods $V \subset U \subset W$ of A such that $\overline{\phi_T(U)}$ is compact while $\phi_t(V) \subset U$ for all $t \geq 0$. Let then Ω denote the set of all $x \in U$ such that $\phi_t x \in U$ for all $t \geq 0$. Then Ω is, of course, invariant and contained in W. Moreover, it is easily seen that there is $T_0 > 0$ such that for each $x \in U$ there is $t \leq T_0$ with $\phi_t x \in V$. But this implies that Ω equals the set of those $x \in U$ such that $\phi_t x \in U$ for all $t \in [0, T_0]$, and now it is routine to show that $X \setminus \Omega$ is actually closed, so Ω is an invariant neighbourhood contained in W. *q.e.d.*

Devotees of filters or nets may wish to remove the metrizability assumption from Proposition 1, but we need it for other reasons, namely in addition to the hypotheses in Proposition 1 we now assume that X is an *ANR*, i.e., X is homeomorphic to a retract of an open set in a normed linear space. By H_* we denote singular homology with rational coefficients. Then it is shown in [3] that for each $t \geq T$ the generalized Lefschetz number of ϕ_t in the sense of Leray, $\Lambda(\phi_t)$, is defined. Now Φ itself provides a homotopy between ϕ_t and the identity map ϕ_0. But this implies that $\Lambda(\phi_0) = \Lambda(id)$ is defined, hence $\chi(X)$, the Euler number of X, must be finite. Moreover, Proposition 1 implies that for $t \geq T$ each ϕ_t is a map of compact attraction as defined in [3], and according to [3] there exists a fixed point index for such maps. Returning now to the situation discussed in the beginning of our paper we obtain the following theorem:

T h e o r e m 1 : *Let X be an ANR and let $\Phi = (\phi_t)_{t \in \mathbb{R}_+}$ be a semiflow on X. Assume that Φ has a compact attractor and that ϕ_T is locally compact for some $T \geq 0$. Then $\chi(X) < \infty$. If, in addition, there exists a closed invariant set $E \subset X$ and a neighbourhood U of E such that for each $x \in \overline{U} \setminus E$ there is $t > 0$ with $\phi_t x \in X \setminus \overline{U}$, then*

i) $\phi_t(X \setminus E) \subset X \setminus E$ for each $t \geq 0$; i.e., Φ restricts to a semiflow on $X \setminus E$.

ii) $\Phi | (X \setminus E) \times \mathbb{R}_+$ has a compact attractor, hence

iii) $\chi(X \setminus E) < \infty$.

iv) $\mathrm{ind}(X, \phi_t, U) = \chi(X) - \chi(X \setminus E)$ for all $t \geq T$.

If, in addition, $\chi(X \setminus E) \neq 0$ then there exists a stationary point in $X \setminus E$ or there is a periodic orbit of period t for each $t \geq T$.

Proof: The initial statement was proved before the theorem. Assertion i) is trivial, and to prove ii) let A denote a uniform compact attractor. Then $(X \setminus U) \cap A$ is a compact attractor for $\Phi | (X \setminus E) \times \mathbb{R}_+$. The discussion preceding the theorem then proves iii), and iv) is proven as in [3]. The final statement follows from the Lefschetz fixed point theorem: The assumption implies that $\Lambda(\phi_t | X \setminus E) \neq 0$, so for $t \geq T$ each ϕ_t possesses a fixed point. If this is true for all $t \geq 0$ there must be a stationary point in $X \setminus E$ (cf. [3]). If for some $t \in (0,T)$ there are no fixed points of ϕ_t the result means that for each $t \geq T$ there will be a periodic orbit of (not necessarily minimal) period t (this was first observed by Peitgen in [5]). *q.e.d.*

Thinking of semiflows generated by autonomous functional differential equations it is usually easy to see whether there are staionary points. So one might be interested in the case when there are no stationary points in $X \setminus E$. So we let A denote a uniform compact attractor for Φ on $X \setminus E$, and we denote by $\mathbb{P}$ the set of periodic points. It is obvious that $\mathbb{P} \subset A$, so $\overline{\mathbb{P}}$ is compact. Assuming that there are no stationary points in $X \setminus E$ amounts to assuming that there is a positive lower bound for the periods of Φ in $X \setminus E$. Then, on $\overline{\mathbb{P}}$, there are no start points of Φ; in fact Φ has *global negative existence* on $\overline{\mathbb{P}}$, i.e., for $x \in \overline{\mathbb{P}}$ and $t > 0$ the set $\phi_{-t}(x) := \phi_t^{-1}(\{x\}) \cap \overline{\mathbb{P}}$ is non-empty. So we may define a *generalized flow* $\Phi: \overline{\mathbb{P}} \times \mathbb{R} \to 2^{\overline{\mathbb{P}}}$ by $\Phi(x,t) := \phi_t x$. We then have $\phi_t \phi_s = \phi_{t+s}$ if $t \cdot s \geq 0$. In order to investigate the structure of $\mathbb{P}$ we shall construct a substitute for the Poincaré-map. Since the result may be of independent interest we describe it in a somewhat more general situation. If x is a periodic point we denote the minimal period by $p(x)$. The following proposition is easy and well-known [6; theorems 5.1,5.2,5.3]:

Proposition 2: *Let (Y,d) be a compact metric space, denote the set of nonempty compact subsets of Y by $K(Y)$, let Φ be a semiflow on Y, and assume that Φ has global negative existence. Then the generalized flow $\Phi: Y \times \mathbb{R} \to K(Y)$ generated by Φ is upper semicontinuous, and $t \mapsto \phi_t x$ is continuous for $x \in X$ fixed (with respect to the Hausdorff metric).*
Hence for $x,y \in Y$ the mapping $t \mapsto d(x,\phi_t y)$ is continuous and for $x,y \in Y$, $J \subset \mathbb{R}$ compact and $\varepsilon > 0$ there is $\delta > 0$ such that $d(x,\phi_t y) - d(x,\phi_t y') < \varepsilon$ whenever $t \in J$ and $d(y,y') < \delta$.

We retain the assumptions of Proposition 2 and proceed as in [2; pp. 50-51]. Let x be a periodic point and define a mapping $\psi_x: Y \times \mathbb{R} \to \mathbb{R}$

$$y,t \mapsto \int_t^{t+p(x)/2} d(x,\phi_s y)\,ds.$$

Then we obtain:

1) For $y \in Y$, $t \in \mathbb{R}$, and $\varepsilon > 0$ there is $\delta > 0$ such that $d(y,y') < \delta$ implies $\psi_x(y,t) - \psi_x(y',t) < \varepsilon$.

2) Let $\varepsilon > 0$. There is $\delta > 0$ such that $\Phi(\{x\} \times [-\delta,\delta]) \subset B(x;\varepsilon) := \{y \mid d(x,y) < \varepsilon\}$.

3) Choose ε_0 such that $D_2\psi_x(y,0) > 0$ for $y \in \overline{B}(x;\varepsilon_0) := \{y \mid d(x,y) \leq \varepsilon_0\}$, where D_2 denotes the partial derivative with respect to the second variable. This is possible, since $D_2\psi_x(y,0) = d(x,\phi_{p(x)/2}y) - d(x,y)$. So it is sufficient to choose $\varepsilon_0 < \frac{1}{2}d(x,\phi_{p(x)/2}x)$ such that $y \in \overline{B}(x;\varepsilon_0)$ implies $d(x,\phi_{p(x)/2)}x) - d(x,\phi_{p(x)/2}y) < \frac{1}{2}d(x,\phi_{p(x)/2}x)$.

4) Choose $\tau_0 > 0$ such that $\Phi(\{x\} \times [-\tau_0,\tau_0]) \subset B(x;\varepsilon_0)$ and $D_2\psi_x(y,t) > 0$ for $y \in \overline{B}(x;\varepsilon_0)$ and $t \in [-\tau_0,\tau_0]$. Then $\sup\{\psi_x(y,0) \mid y \in \phi_{-t}x\} < \psi_x(x,0) < \psi_x(x,t)$ for $t \in [0,\tau_0]$. This is possible, since it is easily seen that for $y \in \overline{B}(x;\varepsilon_0)$ there are $\delta(y) > 0$ and $t(y) > 0$ such that $D_2\psi_x(y',t') > 0$ whenever $y' \in B(y;\delta(y))$ and $|t'| < t(y)$. Hence the second condition can be fulfilled since $\overline{B}(x;\varepsilon_0)$ is compact. Moreover, for $t \in (0,\tau_0]$ and $y \in \phi_{-t}x$ we have

$$\psi_x(y,0) < \psi_x(y,t) = \int_t^{t+p(x)/2} d(x,\phi_s y)\,ds = \int_0^{p(x)/2} d(x,\phi_s x)\,ds = \psi_x(x,0)$$

while the second inequality is obvious.

5) Choose $\zeta > 0$ such that
a) $\overline{B}(\phi_{\tau_0}x;\zeta) \cup \overline{B}(\phi_{-\tau_0};\zeta) \subset B(x;\varepsilon_0)$.
b) $\psi_x(y,0) > \psi_x(x,0)$ whenever $y \in \overline{B}(\phi_{\tau_0}x;\zeta)$.
c) $\psi_x(y,0) < \psi_x(x,0)$ whenever $y \in \overline{B}(\phi_{-\tau_0}x;\zeta)$.
It is obvious that a) and b) can be satisfied. So suppose, there is a sequence (y_n) with $d(y_n,\phi_{-\tau_0}x) \to 0$ and $\psi_x(y_n,0) \geq \psi_x(x,0)$. Now 4) implies that $\varepsilon := \frac{1}{2}[\psi_x(x,0) - \sup\{\psi_x(y,0) \mid y \in \phi_{-\tau_0}x\}]$ is positive. So for each $z \in \phi_{-\tau_0}x$ we

may choose $\eta_z > 0$ such that $|\psi_x(y,0) - \psi_x(z,0)| < \varepsilon$ whenever $y \in B(z;\eta_z)$ (note that $\psi_x(\cdot,0)$ is continuous). Now there is n_0 such that $y_n \in \bigcup_{z \in \phi_{-\tau_0}x} B(z;\eta_z)$ for $n \geq n_0$. So fix $n \geq n_0$ and choose $z \in \phi_{-\tau_0}x$ with $y_n \in B(z;\eta_z)$. But then $\psi_x(y_n,0) - \psi_x(z,0) \geq \psi_x(x,0) - \sup\{\psi_x(y,0) | y \in \phi_{-\tau_0}x\} = 2\varepsilon$.

6) Choose $\delta > 0$ such that

a) $\phi_{\tau_0}(\overline{B}(x;\delta)) \subset B(\phi_{\tau_0}x;\zeta)$.

b) $\phi_{-\tau_0}(\overline{B}(x;\delta)) \subset B(\phi_{-\tau_0}x;\zeta)$.

c) $\Phi(\overline{B}(x;\delta) \times [-\tau_0,\tau_0]) \subset B(x;\varepsilon_0)$.

7) If $y \in B(x;\delta)$ and $\psi_x(y,0) < \psi_x(x,0)$ then there is a unique $\tau(y) \in (0,\tau_0)$ such that $\psi_x(\phi_{\tau(y)}y,0) = \psi_x(x,0)$. If $\psi_x(y,0) > \psi_x(x,0)$ then there exist $\tau(y) \in (0,\tau_0)$ and $y_1 \in \phi_{-\tau_0}y$ such that $\psi_x(y_1,0) = \psi_x(x,0)$.

8) Choose ρ_1 such that $\phi_{p(x)}(B(x;\rho_1)) \subset B(x;\delta)$ and define $S_x := \{y \in B(x;\delta) | \psi_x(y,0) = \psi_x(x,0)\}$. Then $y \in S_x \cap B(x;\rho_1)$ implies the existence of $\tau \in (p(x) - \tau_0, p(x) + \tau_0)$ such that $\phi_\tau y \in S_x$.

9) Choose $\rho \in (0,\rho_1]$ such that for $y \in B(x;\rho)$ the following implication holds: if $\phi_\tau y \in S_x$ for some $\tau \in (0,p(x)+\tau_0)$ then $\phi_\tau y \in B(x;\frac{\delta}{2})$.

We summarize:

P r o p o s i t i o n 3 : *Let Y and Φ be as in Proposition 2, let x be a periodic point of Φ, choose ρ as in 9), define S_x as in 8), and let $\Sigma_x := S_x \cap B(x;\rho)$. Define $\tau_x\colon \Sigma_x \to \mathbb{R}$ by $\tau_x(y) := \inf\{\tau > 0 | \phi_\tau y \in S_x\}$. Then*

i) $y \in \Sigma_x$ implies $\phi_{\tau_x} y \in S_x$

ii) $\tau_x(y) \geq \tau_0$ if $y \in \Sigma_x$, where τ_0 is chosen as in 4).

iii) $\tau_x\colon \Sigma_x \to S_x$ is continuous.

Proof: We need only prove iii). Let $y_n \to y$ in Σ_x. We may assume $\tau_x(y_n) \to t$, hence $\phi_t y \in S_x$. Now $t < \tau_x(y)$ would imply $t = 0$, so we have to exclude $t > \tau_x(y)$. If $\mu := t - \tau_x(y)$ were positive, we might choose τ' such that $0 < \tau_x(y) - \tau' < \frac{\mu}{4}$ and $\psi_x(\phi_{\tau'}y,0) < \psi_x(x,0)$. Then choose $\tau'' \in (0,\frac{\mu}{4})$ such that $\psi_x(\phi_{\tau'+\tau''}y,0) > \psi_x(x,0)$ and finally $\sigma > 0$ such that for $z \in B(\phi_{\tau'}y;\sigma)$ there is $\tau \in [0,\tau'']$ with $\phi_\tau z \in S_x$. Then there is n_0 such that $n \geq n_0$ implies $\phi_{\tau'}y_n \in B(\phi_{\tau'}y;\sigma)$ and $|\tau_x(y_n) - t| < \frac{\mu}{4}$. But this implies $\tau_x(y_n) \leq \tau' + \tau'' \leq \tau_x(y) + \frac{\mu}{4}$ contradicting $\tau_x(y_n) > t - \frac{\mu}{4} = \tau_x(y) + \frac{3\mu}{4}$. *q.e.d.*

Now we are in a position to describe the structure of $\mathbb{P}$ in the situation discussed after Theorem 1, if we make the additional assumption that there is exactly one periodic orbit of period t for each $t \geq T$. Since we are only interested in $X \setminus E$ we forget about E and call the ANR $X \setminus E$ just X now.

T h e o r e m 2 : *Let (X,d) be a metric space and let Φ be a semiflow on X. Assume that Φ has a compact attractor and that ϕ_T is locally compact for some $T \geq 0$. Assume further that there is a positive lower bound t_0 for the periods of Φ and that there is exactly one orbit of period t for each $t \geq t_0$. Then the space $\mathbb{P}$ of periodic points is homeomorphic to the simplicial complex obtained as follows: Take a pentagon as drawn beneath, identify both top segments so that they form the boundary of a Moebius strip which is then to be identified with the bottom segment. Then identify the vertical segments as indicated. (The flow is parallel to the horizontal segments.)*

(Note: When we say that each period $t \geq t_0$ occurs just once, we do not mean that it appears as a minimal period.)

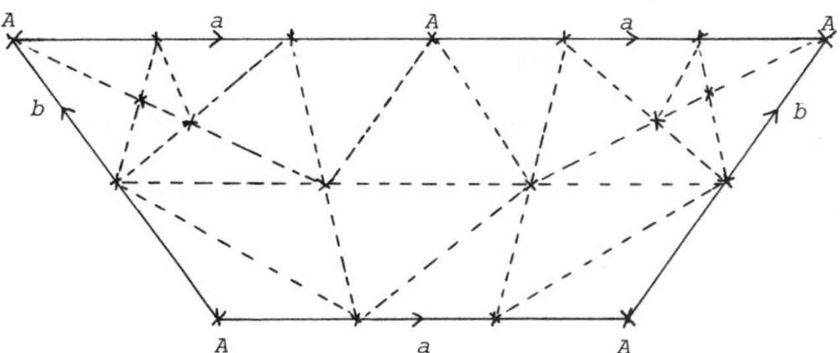

Proof: For $x \in \mathbb{P}$ let $p(x)$ denote again the minimal period. Since each $t \geq t_0$ occurs exactly once as a period, the minimal periods fill the interval $[t_0, 2t_0)$. Then it is easy to see that $\mathbb{P}$ is compact and connected. Denote by P_0 the orbit with period t_0. Then $p|\mathbb{P} \setminus P_0$ is continuous. Consider now a point $x \notin P_0$, choose $\tau_x\colon \Sigma_x \to S_x$ as above, where we may choose S_x and Σ_x so small that $S_x \cap P_0 = \emptyset$ and that τ_x is just $p|\Sigma_x$: it is obvious that $\tau_x(y) \leq p(y)$ for $y \in \Sigma_x$. Suppose that there is a sequence $x_n \to x$ such that $\tau_x(x_n) < p(x_n)$. Then $\tau_x(x_n) \to \tau_x(x)$ and $p(x_n) \to p(x)$. There is n_0 such that $\phi_{\tau_x(x_n)} x_n \in \Sigma_x$ whenever $n \geq n_0$, hence $p(x_n) \geq \tau_x(\tau_x(x_n)) + \tau_x(x_n)$. But this would imply $p(x) \geq \tau_x(\tau_x(x)) + \tau_x(x) = 2p(x)$ which is absurd. Now it is easy to see that for each $x \notin P_0$ we may choose an open interval $J_x \subset (t_0, 2t_0)$ such that

$$p^{-1}(J_x) \to J_x \times S^1, \quad y \mapsto (p(y), \exp(2\pi i t/p(y)))$$

is a homeomorphism where $t \in [0, p(y)]$ is such that $\phi_t y \in S_x$. These homeomorphisms are easily glued together and we see that $\mathbb{P} \setminus P_0$ is homeomorphic to $(t_0, 2t_0) \times S^1$. Moreover, for $x \in P_0$ we have a similar homeomorphism $p^{-1}([t_0, t_0+\varepsilon)) \to [t_0, t_0+\varepsilon) \times S^1$ for some $\varepsilon > 0$, and now we see that $\mathbb{P}$ is in fact the complex described above. Note that indeed $\chi(\mathbb{P}) = 0$ as required by Theorem 1, which is seen by triangulating $\mathbb{P}$ (for example, there is a triangulation with 10 vertices, 33 edges, and 23 faces). *q.e.d.*

R e f e r e n c e s

[1] Bhatia, N.P., O. Hajek: Local semi-dynamical systems. Lecture Notes in Mathematics 90, Berlin, Heidelberg, New York: Springer 1969

[2] Bhatia, N.P., G.P. Szegö: Stability theory of dynamical systems. Berlin, Heidelberg, New York: Springer 1970

[3] Fenske, C.C., H.-O. Peitgen: On fixed points of zero index in asymptotic fixed point theory. Pacific J. Math. 60 (1976) 391- 410

[4] Hale, J.: Theory of functional differential equations. New York, Heidelberg, Berlin: Springer 1977

[5] Peitgen, H.-O.: Methoden der topologischen Fixpunkttheorie in der nichtlinearen Funktionalanalysis. Habilitationsschrift Bonn (1976) (unpublished)

[6] Roxin, E.: Stability in general control systems. J. Diff. Eq. 1 (1976) 115-150

GAME THEORY AND RELATED TOPICS
O. Moeschlin, D. Pallaschke (eds.)

COMBINATORIAL FIXED-POINT ALGORITHMS

Lidia Filus

Department of Mathematics, Institute of Econometrics
Central School of Planning and Statistics
Warsaw, Poland

Abstract

In this paper we treat the fixed point algorithms of H. Scarf, H.W. Kuhn, "the homotopy method" and "the sandwich method" as the procedures to obtain sequences of almost complete sets which end up with a complete set.
We give necessary and sufficient conditions for the existence of an odd number of the sequences mentioned above. We also give necessary and sufficient conditions for the existence of an odd number of complete sets in some family of sets.

1. Introduction

In purpose to compute the economic equilibria many methods have been proposed for computing approximately fixed points of the continuous mapping.
The best known of them are algorithms of H. Scarf [10], H.W. Kuhn [8], B.C. Eaves [4] and "the sandwich method" [9].

The background of these three algorithms in [8], [4], [9] is Sperner's Lemma [12] on existence, in arbitrary simplicial subdivision of the n-simplex S of simplices complete with respect to some label function defined on the set of vertices of this subdivision with values in $\{0,1,\dots,n\}$.

It has been shown by B. Knaster, C. Kuratowski and S. Mazurkiewicz in [7] that the computation of approximately fixed points of the continuous mapping $f:S\to S$ can be reduced to the search of complete (with respect to some labelling function depending on f) simplices in an arbitrary fine simplicial subdivision of the simplex S.

The three algorithms mentioned above make easier the search of complete simplices in the simplicial subdivision of the n-simplex S.
For example in H.W. Kuhn's algorithm the objective is reached by:

a) pointing the starting simplex Z_o (which belongs to some extension of the simplicial subdivision of S), which is almost complete, which means that the set of labels of the vertices of Z_o is an n-element subset of $\{0,1,\ldots,n\}$; say $\{0,1,\ldots,n-1\}$,

b) an induction procedure to obtain the next simplex Z_1 and to obtain the unique simplex Z_j, when we know the last two simplices from the sequence $(Z_o, Z_1, \ldots, Z_{j-1})$ for $j \geq 2$.

The procedure is defined so that the obtained sequence of almost complete simplices ends up with a complete simplex.

In this procedure we pass from Z_{j-1} $(j=1,2,\ldots)$ to the neighbouring simplex Z_j passing through their common (n-1)-dimensional face, whose vertices have all labels from the set $\{0,1,\ldots,n-1\}$.

A similar procedure is used in "the sandwich method" and in "the homotopy method", the difference is that in "the sandwich method" we embed the n-dimensional simplex S with its regular simplicial subdivision into an (n+1)-dimensional simplex with its regular subdivision; the labelling function is extended in such a way that the search of a complete simplex in the simplicial subdivision of n-simplex becomes equivalent to the search of a complete simplex in the simplicial subdivision of the (n+1)-simplex.

In "the homotopy method" we have a sequence of simplicial subdivisions of an n-simplex S, these subdivisions are embedded into the simplicial subdivision of the cylinder $S \times [a,b]$, where a,b are reals, $a<b$.

In H. Scarf's algorithm, instead of the simplicial subdivision, the family of (n+1)-element primitive subsets of a set Y composed of all (n-1)-dimensional faces of the n-simplex S and a finite set of points in the interior of the simplex S are considered. H.E. Scarf defines the replacement operation for primitive sets and gives an algorithm to find a complete primitive set with respect to the labelling function $l: Y \rightarrow \{0,1,\ldots,n\}$.

This algorithm starts with some primitive set Z_o, which is either complete or almost complete and then an induction procedure is defined, by means of the replacement operation, to obtain a sequence of primitive sets $(Z_o, Z_1, \ldots, Z_k)$ such that $Z_o,\ldots,Z_{k-1}$ are almost complete and Z_k is a complete set.

In this paper we give a uniform description of these algorithms which covers the algorithms of H.E. Scarf, H.W. Kuhn, B.C. Eaves and "the sandwich method".

For this purpose we leave out the various geometrical constructions used in these algorithms. Instead of the family of primitive subsets in H.E. Scarf's algorithm or the family of n-simplices in the simplicial subdivision in algorithms of H.W. Kuhn, B.C. Eaves and "the sandwich method", we consider an arbitrary family P of (n+1)-element sets.

The labelling function is an arbitrary function

$$l:\bigcup P \to \{0,1,\dots,n\}.$$

In this way we can treat the various procedures used in those algorithms as a property of the binary relation R (the replacement relation) defined on the family P.

In this way we obtain a generalization of known theorems on sequences of simplices or primitive sets appointed by the algorithms mentioned above, looking at them as at procedures to obtain sequences $(Z_o,Z_1,\dots,Z_k)$ such that $(Z_j,Z_{j+1})\in R$, Z_j is an almost complete set for $j=0,1,\dots,k-1$, Z_k is a complete set.
We give necessary and sufficient conditions for the existence of an odd number of sequences $(Z_o,Z_1,\dots,Z_k)$ in P; we also give necessary and sufficient conditions for the existence of an odd number of complete sets in P.

Similar combinatorial results are presented in [5] and [6] (theorem 2.6), but they only give sufficient condition. In [6] it has been shown how the above mentioned algorithms follow from the combinatorial results.

A uniform combinatorial structure for the algorithms was also obtained by H. Tuy in [13]. In contrary to the results of H. Tuy, the replacement relation defined in [6] and in this paper is symmetric and irreflexive, which makes it possible to treat the members of P as vertices of some non-oriented graph and the replacement relation R as the set of arcs of the graph.

In this way we are able to obtain the results which are more general and include more possible cases of algorithms than the combinatorial results described above. In the next section we give only those graph results which can be used as lemmas in proving the combinatorial theorems.

2. Graph lemmas

We consider the *non-oriented graph* $\Gamma=(W,Q)$; it means $Q\subset W^2$ is a symmetric and irreflexive relation. The elements of W are called *vertices* and elements of Q are called *arcs* of the graph Γ.

For an arbitrary set $K\subset W$

$$K(x) = \{y\in K \mid xQy\}.$$

For any set K we denote by $\# K$ the cardinality of K.

We assume that W is finite.
An arbitrary set $V \subset W$ is fixed throughout the section.

From the symmetry of the relation R we obtain the following

Lemma 2.1

The number $\sum_{x \in V} \# V(x)$ is even.

It is easy to see that the lemma and the equality

$$\sum_{x \in W} \# W(x) = \sum_{x \in W \setminus V} \# W(x) + \sum_{x \in V} \# W(x)$$

imply:

Lemma 2.2

If the number $\# W(x)$ is odd for every $x \in W \setminus V$, then

[$\# (W \setminus V)$ is odd if and only if $\# \{x \in V \mid \# W(x) \text{ is odd}\}$ is an odd number].

By the path in Γ we mean an arbitrary sequence $(x^{o}, \dots, x^{k})$ of vertices such that $(x^{i}, x^{i+1}) \in Q$ for $i=0,1,\dots,k-1$.

The following theorem holds true:

Theorem 2.3

If $\# W(x) \leq 2$ for every $x \in V$, then the number of all paths $(x^{o}, \dots, x^{k})$ in Γ such that

1) $x^{j} \in V$ for $j=0,1,\dots,k-1$

2) $\# W(x^{o}) = 1$

3) $x^{k} \in W \setminus V$

4) $x^{j-1} \neq x^{j+1}$ for $j=1,\dots,k-1$

is odd if and only if $\# \{x \in V \mid \# W(x)=1\}$ is odd.

Similarly we can obtain:

Corollary 2.4

If $\# W(x) \leq 2$ for every $x \in V$ and if $\# V(x) \leq 1$ for every $x \in W \setminus V$ then every two paths which satisfy the conditions 1)-4) of theorem 2.3 are either disjoint or identical.

3. Combinatorial fixed point algorithms

The construction of fixed point algorithms in [4], [8], [9], [10] can be reduced to the following situation.
We are given a finite family P of (n+1)-element sets and a labelling function

$$1:\bigcup P \rightarrow \{0,1,\dots,n\} = N.$$

Our objective is to deduce, from some information about almost complete sets in P (i.e. sets $Z\in P$ for which $\# 1(Z)=n$) conclusions about the existence of a complete set in P (Z is complete if $1(Z)=N$). Moreover we want to find a finite sequence $(Z_o,Z_1,\dots,Z_k)$ of sets from P such that

i) Z_j is almost complete for $j=0,\dots,k-1$

ii) Z_k is complete, $k=1,2,\dots$

iii) for every $j=0,1,\dots,k-1$ the set $(Z_j\setminus Z_{j+1})\cup(Z_{j+1}\setminus Z_j)$ has exactly two elements.

In this section we prove a theorem on the existence of this sequence by means of the graph results of the previous section.

Assume that $Z,Z'\in P$. We say that Z,Z' are in the replacement relation if there exist elements z,z' such that $z\in Z$, $z'\in Z'$, $z\neq z'$ and $Z'=(Z\setminus\{z\})\cup\{z'\}$.
If this happens and $Z'=(Z\setminus\{z\})\cup\{z'\}$, $z\in Z$, $z'\in Z'$ and $z\neq z'$ then we say that the element z is _replaceable_ in Z by z'; we write in this case

$$Z' = Z(z/z').$$

If for $z\in Z$, there exists at most one $z'\neq z$ such that $Z(z/z')\in P$ then we say that z is _uniquely replaceable_ in Z.

The replacement of $z\in\bigcup P$ is _unique in_ some subfamily $P_o\subset P$ if z is replaceable uniquely in every set $Z\in P_o$ which contains z.

Every element $z\in Z$ such that $\#\{z'\in Z \mid 1(z')=1(z)\}>1$ is called _multiple_ in Z

For $m=0,1,\dots,n$ we define

$$V_m = \{Z\in P \mid 1(Z) = N\setminus\{m\}\}$$

and

$$W_m = \{Z\in V_m \mid \text{exactly one multiple element is replaceable in } Z\}.$$

Theorem 3.1

If for some m=0,1,...,n the replacement of multiple elements in V_m is unique and if every element z from the complete set Z such that l(z)=m is replaceable in Z and this replacement is unique, then the number of complete sets in P is odd if and only if $\# W_m$ is an odd number.

Proof: We define a relation R in P in the following way:

$$ZRZ' \quad \text{if} \quad Z' = Z(z/z') \quad \text{for some} \quad z \in Z\,,\ z' \notin Z.$$

Since $R \subset P^2$ is symmetric and irreflexive, thus (P,R) is a non-oriented graph. We choose an m which satisfies the assumptions of theorem 3.1 and define a subgraph $\Gamma=(W,Q)$ of the graph (P,R) in the following way:

$$W = \{Z \in P \mid l(Z) \supset N \setminus \{m\}\}$$

and

$$Q = \{(Z,Z') \in R \mid l(Z \cap Z') = N \setminus \{m\}\}.$$

Let $\quad V = V_m.$

Then

$$\{Z \in V \mid \# W(Z)=1\} = W_m$$

and

W\V is the subfamily of P composed of complete sets and containing no other sets.

From the assumptions of theorem 3.1 it follows that $\# W(Z)=1$ for every $Z \in W \setminus V$; hence V and W satisfy the assumptions of lemma 2.2.

Therefore

$\#(W \setminus V)$ is odd if and only if $\#\{Z \in V \mid \# W(Z)=1\}$ is odd, which means that the number of complete sets in P is odd if and only if $\# W_m$ is odd, which completes the proof of theorem 3.1.

Theorem 3.2

If for some m=0,1,...,n the replacement of multiple elements in V_m is unique, then the number of finite sequences $(Z_o,Z_1,\dots,Z_k)$ of sets from P such that

1) $Z_j \in V_m$ for $j=0,\dots,k-1$

2) $Z_o \in W_m$

3) Z_k is a complete set

4) $Z_{j-1} \neq Z_{j+1}$ for $j=1,\dots,k-1$

5) $Z_j = Z_{j-1}(z/z')$ for $j=1,\dots k$, z is multiple in Z_{j-1},

is odd if and only if $\#\mathcal{W}_m$ is odd.

Moreover every two sequences which satisfy 1)-5) are either disjoint or identical.

Proof: Let $(\mathcal{P},R)$ be the graph defined in the same way as in the proof of theorem 3.1.
Choose an m which satisfies the assumption of theorem 3.2 and define a subgraph $\Gamma=(W,Q)$ of the graph $(\mathcal{P},R)$ in the following way:

$$W = \{Z\in\mathcal{P}\,|\,l(Z)\supset N\setminus\{m\}\},$$

$$Q = \{(Z,Z')\in R\,|\,l(Z\cap Z') = N\setminus\{m\}\}.$$

We will show that $\Gamma=(W,Q)$ satisfies the assumptions of theorem 2.3 for $V=\mathcal{V}_m$. Then we show that for every $Z\in V$ the set $W(Z) = \{Z'\in W|ZQZ'\}$ has at most two elements. Suppose that $\#W(Z)\geq 3$ for some $Z\in V$. Let Z_1,Z_2,Z_3 be distinct sets from $W(Z)$. Then

$$Z_1 = Z(z'/z_1) \quad , \quad z' \text{ is multiple in } Z ,$$

$$Z_2 = Z(z''/z_2) \quad , \quad z'' \text{ is multiple in } Z ,$$

$$Z_3 = Z(z'''/z_3) \quad , \quad z''' \text{ is multiple in } Z.$$

Since every set $Z\in\mathcal{V}_m$ has exactly two multiple elements, therefore at least two of elements z',z'',z''' are equal. Assume that $z' = z''$.
Then we have $Z_1 = Z(z/z_1)$, $Z_2 = Z(z/z_2)$ and z is multiple in Z, $z_1 \neq z_2$.
It follows that z is replaceable in $Z\in\mathcal{V}_m$ by z_1 and by z_2 and $z_1\neq z_2$, which contradicts the assumption about the uniqueness of replacements in $\mathcal{V}_m$.
Therefore for every $Z\in V$ the set $W(Z) = \{Z'\in W|ZQZ'\}$ has at most two elements.
From theorem 2.3 we obtain the first part of theorem 3.2.
Let us note that to prove theorem 3.2 we also have to show that the assumptions of corollary 2.4 are satisfied, which means that $V(Z)\leq 1$ for every $Z\in W\setminus V$.
Suppose in contrary that $\#V(Z)>1$ for some $Z\in W\setminus V$ and let $V(Z)\supset\{Z',Z''\}$.
Since $Z'\in V$, $Z''\in V$ and $Z'QZ$, $Z''QZ$ we have $Z' = Z(z/z')$ and $Z'' = Z(z/z'')$ and z' is multiple in Z', z'' is a multiple element in Z'', $z'\neq z''$. It follows that $z'\in Z'$ is replaceable by $z''\in Z''$ and by $z\in Z$, which contradicts the assumption, that the replacement of multiple elements in $\mathcal{V}_m$ is unique. Therefore $\#V(Z)\leq 1$ for every $Z\in W\setminus V$ which completes the proof of theorem 3.2.

Theorem 3.2 is related to the fixed point algorithms mentioned in the first part of the paper; it can be shown in a similar way as it has been done in [6].

Theorem 3.1 is related to the Sperner's Lemma.

References

[1] P.S. Alexandroff, Combinatorial Topology, Vol.1, Graybock Press, Rochester, New York, 1956

[2] L.E.Y. Brouwer, Über Abbildung von Mannigfaltigkeiten, Mathematische Annalen, 71(1912), 97-115

[3] D.I.A. Cohen, On Sperner Lemma, Journal of Combinatorial Theory 2 (1967), 585-587

[4] B.C.Eaves, Homotopies for Computation of Fixed Points, Mathematical Programming 3 (1972), 1-22

[5] L. Filus, A Combinatorial Lemma Related to the Search of Fixed Points, Bull. Acad. Polon. Sci. Ser. Sci. Math. Astron. Phys., 7, 1977, 615-616

[6] L. Filus, A Combinatorial Lemma for Fixed Points Algorithms, Nonlinear Programming 3, Academic Press, 1978, 407-427

[7] B. Knaster, C. Kuratowski, S. Mazurkiewicz, Ein Beweis des Fixpunktsatzes für n-dimensionale Simplexe, Fundamenta Mathematicae 14(1929), 132-137

[8] H.W. Kuhn, Simplicial Approximation of Fixed Points, Proc. Nat. Acad. Sci. 61 (1968), 1238-1242

[9] H.W. Kuhn, J.G. Mac Kinnon, Sandwich Method for Finding Fixed Points, Journal of Optimization Theory and Applications, Vol.17 (1975), 189-204

[10] H.E. Scarf, The Approximation of Fixed Points of a Continuous Mapping, SIAM J. Appl. Math. 15 (1967), 1328-1343

[11] H.E. Scarf with the collaboration of T. Hansen, The Computation of Economic Equilibria, Yale University Press, New Haven, Connecticut 1973

[12] E. Sperner, Neuer Beweis für die Invarianz der Dimensionszahl und des Gebietes, Abh. Math. Sem. Univ. Hamburg 6 (1928)

[13] H. Tuy, Pivotal Methods for Computing Fixed Points, A Unified Approach, Proceedings of the IX International Symposium on Mathematical Programming, Budapest, 1976

GAME THEORY AND RELATED TOPICS
O. Moeschlin, D. Pallaschke (eds.)

DUALITY IN MONOIDAL CATEGORIES AND APPLICATIONS TO FIXED-POINT THEORY

Dieter Puppe *

Institute of Mathematics
University of Heidelberg

Exploiting the notion of duality in monoidal categories a new proof of the classical Lefschetz Hopf fixed point theorem is given. The same method proves an analogous result for continuous families of maps. In this case one obtains new invariants which sometimes ensure the existence of fixed points when the classical fixed point index fails. The notion of transfer is defined which, among other things, allows to investigate how the fixed points in a family depend on the parameter.

INTRODUCTION

We consider a notion of strong duality in monoidal categories C which specializes to S-duality in the sense of Spanier and J.H.C.Whitehead [Switzer 1975, 14.20] if C = Stab is stable homotopy theory and to fibre-wise S-duality in the sense of Becker and Gottlieb [1976] if $C = \mathrm{Stab}_B$ is stable homotopy theory over some fixed parameter space B. For any endomorphism of a strongly dualizable object we define a trace which coincides with the usual trace if $C = \mathrm{Mod}_R$ is the category of modules over some commutative ring R and with the Lefschetz number if C is the category of chain complexes $\partial\text{-}\mathrm{Mod}_R$ or the category of graded modules $\mathrm{Gr}\text{-}\mathrm{Mod}_R$.

If K is a compact neighborhood retract in $\mathbb{R}^n$ then K is strongly dualizable in Stab and the dual is essentially the complement $\mathbb{R}^n\text{-}K$ (Th. 2.1). The classical Lefschetz-Hopf fixed point theorem follows easily by using the general notion of trace (§ 3). All this has a straightforward generalization to the parametrized case (§ 4) containing part of the results of Dold [1974a, 1974b, 1976] and Becker-Gottlieb [1976]. In particular one obtains a fixed point index in $\mathrm{Stab}(B,S^0)$ which sometimes ensures the existence of fixed points in a continuous family of maps although there are no essential fixed points in each single map of the family (4.2).

---

* Report on joint work with A.Dold.

There is also a general notion of <u>transfer</u> in monoidal categories which gives a powerful tool in $Stab_B$ [Dold 1976, Becker-Gottlieb 1976]. In particular it allows to investigate how the fixed points in a family of maps depend on the parameter (§ 5).

Some aspects of what we indicate here are treated more extensively in [Dold-Puppe].

1. DUALITY IN MONOIDAL CATEGORIES

Let C be a (symmetric) monoidal category with multiplication $\otimes$ and neutral object I. That means that $\otimes$ is a bifunctor $(A,B) \longmapsto A \otimes B$ of C into itself and we have given coherent natural equivalences

$$A \otimes (B \otimes C) \cong (A \otimes B) \otimes C$$

$$I \otimes A \cong A \cong A \otimes I$$

$$\gamma = \gamma_{AB} : A \otimes B \xrightarrow{\cong} B \otimes A .$$

One knows by experience that it will do no harm to replace the three equivalences in the first two lines by equalities. In the third line one has to be a little more careful because $\gamma_{AA} : A \otimes A \longrightarrow A \otimes A$ is not the identity in general.

In order to understand this concept and the following abstract notions the reader is advised to keep always in mind the example that C is the category Mod_R of modules over a commutative ring R, $\otimes$ is the usual tensor product and $I = R$.

<u>1.1 Definition</u>. Let A,B be objects of C. B is called a <u>strong dual</u> of A if there are morphisms

$$\eta : I \longrightarrow A \otimes B \quad \text{and} \quad \varepsilon : B \otimes A \longrightarrow I$$

in C such that the following compositions are the identity morphisms of A and B resp.

$$(1.2) \quad \begin{aligned} id_A &: A = I \otimes A \xrightarrow{\eta \otimes id_A} A \otimes B \otimes A \xrightarrow{id_A \otimes \varepsilon} A \otimes I = A \\ id_B &: B = B \otimes I \xrightarrow{id_B \otimes \eta} B \otimes A \otimes B \xrightarrow{\varepsilon \otimes id_B} I \otimes B = B . \end{aligned}$$

We have taken this formulation from [Lindner 1978] except that he calls A an adjoint of B. In [Lindner 1978], [Pareigis 1976], [Ligon 1978] and [Dold-Puppe] one can find other characterizations and some general properties of strong duality. Here we just mention that strong duality between A and B is obviously symmetric in A,B and that it is equivalent to the condition that the map

$$C(X, Y\otimes B) \longrightarrow C(X\otimes A, Y)$$

$$\left(X \xrightarrow{f} Y\otimes B \right) \longmapsto \left\{ X\otimes A \xrightarrow{f\otimes id_A} Y\otimes B\otimes A \xrightarrow{id_Y\otimes\varepsilon} Y\otimes I = Y \right.$$

is a bijection for all X,Y [Dold-Puppe Th. 1.3]. Letting $Y = I$ one obtains in particular

$$C(X,B) \cong C(X\otimes A, I)$$

which means that B is a dual of A in the ordinary sense.

In Mod_R an object A is strongly dualizable if and only if it is a finitely generated projective module [Dold-Puppe 1.4].

2. STABLE HOMOTOPY, S-DUALITY

We define the stable homotopy category as follows (cf. [Dold-Puppe § 3] for more details): Objects are the pairs (X,n) where X is a well-pointed compactly generated space and $n \in \mathbf{Z}$. The set of morphisms from (X,n) to (Y,m) is

$$Stab((X,n),(Y,m)) = \operatorname*{colim}_{k\to\infty} [S^{n+k} \wedge X, X^{m+k} \wedge Y] ,$$

where $\wedge$ is the smash product in the category of (pointed) compactly generated spaces, the brackets $[\ ,\ ]$ denote the set of homotopy classes and the bonding maps are given by smashing with S^1. Composition of morphisms is the obvious one.

A monoidal structure on $Stab$ is given by

$$(X,n)\otimes(Y,m) = (X\wedge Y\ ,\ n+m) .$$

A neutral object is $I = (S^o,0)$. Forming the $\otimes$-product with (S^o,p) as a left factor defines a functor Σ^p of Stab into itself which sends (X,n) into $(S^o \wedge X, p+n) = (X, p+n)$. These functors Σ^p are defined for all $p \in \mathbb{Z}$ and satisfy

$$\Sigma^p \circ \Sigma^q = \Sigma^{p+q}$$

$$\Sigma^o = \text{identify functor} .$$

We sometimes abbreviate $(X,0)$ by X . Then we may write $\Sigma^n X$ instead of (X,n) .

If $p \geq 0$ then there is an obvious isomorphism

$$(S^o,p) \cong (S^p,0)$$

representend by the identity of S^p . This shows that Σ^p is equivalent to smashing with S^p , i.e. to ordinary p-fold suspension.

Sometimes an object of Stab will be represented by a pair of spaces. We make the convention that if $i : X' \longrightarrow X$ is any map between (unpointed) compactly generated spaces then (X,X') denotes the object $C_i = (C_i,0)$ of Stab where C_i is the mapping cone of i with the vertex of the cone as base point. Which map i we mean will always be clear from the context. Usually i will be an inclusion $X' \subset X$. Note that $(X,\emptyset) = X^+$, i.e. X with an additional isolated point as base point.

2.1 Theorem. *Let K be a compact subset of $\mathbb{R}^n$ and a neighborhood retract. Then $(K,\emptyset)$ and $\Sigma^{-n}(\mathbb{R}^n,\mathbb{R}^n - K)$ are strongly dual in Stab.*

If K is a subcomplex of a simplicial decomposition of $\mathbb{R}^n$ then the above theorem reduces to Lemma 5.1 of [Spanier 1959]. Apart from proving a somewhat more general theorem our point is that our proof is independent and quite different from former proofs. It consists of defining explicitely the morphisms ε ("evaluation") and η ("coevaluation") of Definition 1.1 and verifying directly the conditions (1.2). We need two lemmas.

2.2 Lemma. *Let X be metrizable and let X' and U be subspaces of X whose interiors cover X . Let $U' = U \cap X'$. Then the inclusion $(U,U') \subset (X,X')$ is an isomorphism in Stab .*

Now let $X' \subset X$ and $Y' \subset Y$. As usual we define

$$(X,X') \times (Y,Y') = (X \times Y,\ X' \times Y \ \cup\ X \times Y')$$

2.3 Lemma. *If X and Y are metric, then there is a canonical morphism*

$$(X,X') \otimes (Y,Y') \longrightarrow (X,X') \times (Y,Y')$$

in Stab . *It is an isomorphism if* X', Y' *are open in* X, Y *resp.*

The proofs are given in [Dold-Puppe] where the lemmas have the numbers 3.4 and 3.5 resp.

Now we define the evaluation

$$\varepsilon : \Sigma^{-n}(\mathbb{R}^n,\mathbb{R}^n-K) \otimes (K,\emptyset) \longrightarrow S^0$$

for an arbitrary subset K of $\mathbb{R}^n$ by the commutativity of

$$\begin{array}{ccc}
(\mathbb{R}^n,\mathbb{R}^n-K) \otimes (K,\emptyset) & \overset{\Sigma^n\varepsilon}{\dashrightarrow} & \Sigma^n S^0 \\
\downarrow & & \Vert\wr \\
(\mathbb{R}^n,\mathbb{R}^n-K) \times (K,\emptyset) & & S^n \\
\Vert & & \Vert\wr \\
(\mathbb{R}^n\times K,(\mathbb{R}^n-K) \times K) & \xrightarrow{\text{diff}} & (\mathbb{R}^n,\mathbb{R}^n-0) \\
(x,k) & \longmapsto & x - k
\end{array} \tag{2.4}$$

In order to define the coevaluation

$$\eta : S^0 \longrightarrow (K,\emptyset) \otimes \Sigma^{-n}(\mathbb{R}^n,\mathbb{R}^n-K)$$

we do need the hypotheses that K is compact and that there is a neighborhood V of K in $\mathbb{R}^n$ and a retraction $r : V \longrightarrow K$. We also choose a closed ball D in $\mathbb{R}^n$ (with center 0) such that $K \subset D$. Now we define $\Sigma^n\eta$ by

$$\begin{array}{ccc}
\Sigma^n S^0 & \overset{\Sigma^n\eta}{\dashrightarrow} & (K,\emptyset) \otimes (\mathbb{R}^n,\mathbb{R}^n-K) \\
\Vert\wr & & \downarrow p \\
(\mathbb{R}^n,\mathbb{R}^n-0) & & (K,\emptyset) \times (\mathbb{R}^n,\mathbb{R}^n-K) \\
i \Uparrow\wr & & \Vert \\
(\mathbb{R}^n,\mathbb{R}^n-D) \subset (\mathbb{R}^n,\mathbb{R}^n-K) & & (K \times \mathbb{R}^n, K \times (\mathbb{R}^n-K)) \\
j \Uparrow\wr & & \uparrow r \times \mathrm{id} \\
(V,V-K) & \xrightarrow{\text{diag}} & (V \times \mathbb{R}^n, V \times (\mathbb{R}^n-K)) \\
v & \longmapsto & (v,v)
\end{array} \tag{2.5}$$

i is an isomorphism in Stab because it is represented by a pair of homotopy equivalences. j and p are isomorphisms in Stab by Lemma 2.2 and 2.3 resp.

For the proof of the two identities (1.2) the reader is referred to [Dold-Puppe].

3. THE HOPF-LEFSCHETZ FIXED POINT THEOREM

Let again C be a monoidal category and A an object of C which has a strong dual B.

3.1 Definition. The trace σf of an endomorphism $f : A \longrightarrow A$ is the composition

$$(3.2) \qquad \sigma f : I \xrightarrow{\eta} A \otimes B \xrightarrow[\cong]{\gamma} B \otimes A \xrightarrow{\mathrm{id} \times f} B \otimes A \xrightarrow{\varepsilon} I .$$

If $C = \mathrm{Mod}_R$ then $\sigma f : R \longrightarrow R$ may be identified with $(\sigma f)(1)$, which is the usual trace of f.

For another example let $C = \text{Gr-Mod}_R$ be the category of graded modules over R. Objects are sequences of modules $(A_n \mid n \in \mathbb{Z})$ and morphisms are sequences of linear maps $(f_n : A_n \longrightarrow B_n \mid n \in \mathbb{Z})$. The monoidal structure is given by

$$(A \otimes B)_n = \bigoplus_{p+q=n} A_p \otimes B_q$$

and the equivalence $\gamma : A \otimes B \longrightarrow B \otimes A$ by

$$A_p \otimes B_q \longrightarrow B_q \otimes A_p$$

$$a \otimes b \longmapsto (-1)^{pq}\, b \otimes a$$

(Without the sign one would also get a monoidal category but it would not serve our purpose.) The neutral object I of Gr-Mod_R may be identified with R $(I_o = R,\ I_n = 0$ for $n \neq 0)$. Then $\sigma f = (\sigma f)(1)$ is the Lefschetz number

$$\sum_{n \in \mathbb{Z}} (-1)^n \sigma f_n .$$

The sign comes from the sign in the definition of γ.

Now we come to the Lefschetz-Hopf theorem. As in Theorem 2.1 let K be a compact subset of $\mathbb{R}^n$, V a neighborhood of K in $\mathbb{R}^n$ and $r : V \longrightarrow K$ a retraction. Let $f : K \longrightarrow K$ be a continuous map with fixed point set F. Let f^+ be the corresponding map of $K^+ = (K, \emptyset)$. Inserting the definitions (2.4) of ε and (2.5) of η into the definition (3.2) of σf one obtains the commutative diagram in Stab

$$
(3.3)\qquad
\begin{array}{ccccc}
\Sigma^n S^o & & \xrightarrow{\quad \Sigma^n \sigma f^+ \quad} & & \Sigma^n S^o \\
\Vert\wr & & & & \Vert\wr \\
(\mathbb{R}^n, \mathbb{R}^n - 0) & & & & (\mathbb{R}^n, \mathbb{R}^n - 0) \\
\cong\downarrow U & & & & \uparrow v - frv \\
(\mathbb{R}^n, \mathbb{R}^n - D) & \subset & (\mathbb{R}^n, \mathbb{R}^n - K) & & \uparrow \;\; \updownarrow v \\
 & & \cong\downarrow U & & \\
 & & (v, V-K) & \subset & (V, V-F)
\end{array}
$$

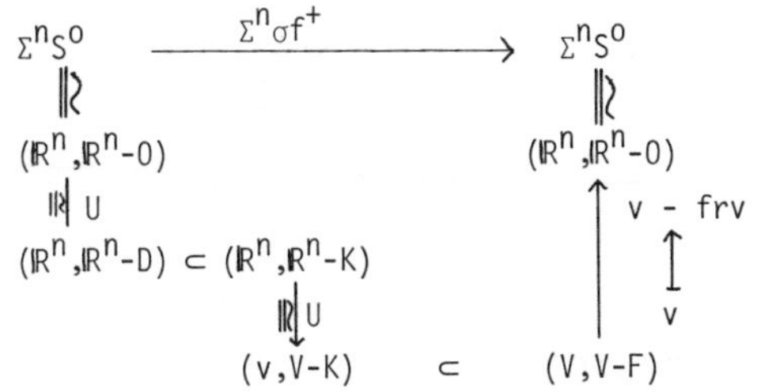

Going around the lower part of this diagram and applying homology $H_n(\ ;\mathbb{Z})$ one gets a homomorphism which is multiplication by the fixed point index of f. This is the definition of the index in [Dold 1972, p.202]. Hence

$$\text{index } f = H_n(\Sigma^n \sigma f^+;\mathbb{Z}) = \tilde{H}_o(\sigma f^+;\mathbb{Z}) = \tilde{H}_o(\sigma f^+;\mathbb{Q}) \ .$$

Reduced homology with rational coefficients is a functor from Stab to $\text{Gr-Mod}_{\mathbb{Q}}$ which takes $\wedge$ into $\otimes$ and S^o into $\mathbb{Q}$. Our definitions 1.1 and 3.1 imply immediately that such a functor is compatible with forming strong duals and trace. Hence

$$\tilde{H}_o(\sigma f^+;\mathbb{Q}) = \sigma\tilde{H}(f^+;\mathbb{Q}) = \sigma H(f;\mathbb{Q}) \ .$$

Thus we have shown that the fixed point index of f equals the Lefschetz number of $H(f;\mathbb{Q})$.

4. PARAMETRIZED FIXED POINT THEORY

One advantage of § 3 over other proofs of the Lefschetz-Hopf fixed point theorem is that it has straightforward generalizations to spaces with additional structure.

For a fixed "base space" B let us consider (compactly generated) "spaces over B" , i.e. continuous maps $p : E \longrightarrow B$. One should think of them mainly as families $(E_b \mid b \in B)$ of the "fibres" $E_b = p^{-1}b$, parametrized by B . The topology of E has just the purpose to make it meaningful to say that something happening in the family (E_b) depends continuously on b . Guided by this principle one can translate the whole content of § 2 and § 3 to the "parametrized" case.

In particular we construct a monoidal category Stab_B of stable homotopy over B whose objects are (up to formal suspension and desuspension Σ^n, $n \in \mathbb{Z}$) "well-sectioned" spaces over B , i.e. commutative diagrams

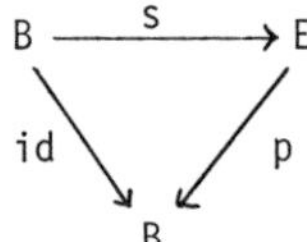

such that s is a cofibration in the category of spaces over B . If $E' \longrightarrow E$ is a fibre-wise mapping of spaces over B one can form the fibre-wise mapping cone and consider it as an object of Stab_B . We denote it by $(E,E')_B$ and abbreviate $(E,\emptyset)_B$ by $\overline{E}$. $\overline{E}$ is just the topological sum of E and B , and the inclusion $B \subset \overline{E}$ is the section of base points. By an almost literal translation of § 2 we get

4.1 Theorem. *Let B be metric, $K \subset B \times \mathbb{R}^n$ and $p : B \times \mathbb{R}^n \longrightarrow B$ the projection onto the first factor. Let $p|K : K \longrightarrow B$ be proper and K an ENR_B , i.e. there is a neighborhood V of K in $B \times \mathbb{R}^n$ and a fibre-wise retraction $r : V \longrightarrow K$ $\big((p|K) \circ r = p|V\big)$. Then $\overline{K} = (K,\emptyset)_B$ and $(B \times \mathbb{R}^n, (B \times \mathbb{R}^n)-K)_B$ are strongly dual in Stab_B .*

Now consider a fibre-wise map $f : K \longrightarrow K$ (where K is as in the above Theorem). Then a diagram analogous to (3.3) shows that the trace $\sigma\overline{f}$ of $\overline{f} : \overline{K} \longrightarrow \overline{K}$ equals the (parametrized) fixed point index of f as an element of $\mathrm{Stab}_B(B \times S^0, B \times S^0)$. The elements of this group are represented by maps g such that

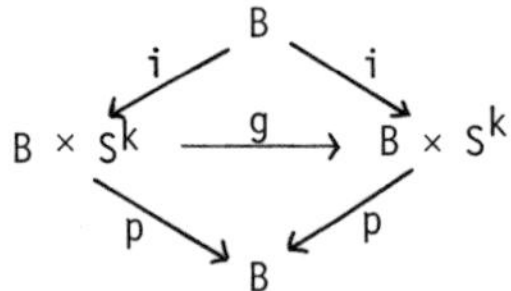

commutes, which in turn are in 1-1-correspondence to pointed maps

$$B^+ \wedge S^k = (B \times S^k)/(B \times \{x_0\}) \longrightarrow S^k .$$

In this way one shows that

$$\begin{aligned} \mathrm{Stab}_B(B \times S^0, B \times S^0) &\cong \mathrm{Stab}(B^+,S^0) \\ &\cong \mathrm{Stab}(S^0,S^0) \oplus \mathrm{Stab}(B,S^0) \end{aligned}$$

where, for the last group, one chooses some (non-degenerate) base point b in B. By these isomorphisms $\sigma\bar{f}$ corresponds to $(\sigma\bar{f}_b, \tilde{\sigma}f)$ where $\sigma\bar{f}_b \in \mathrm{Stab}(S^0,S^0) \cong \mathbb{Z}$ is the trace of the restriction of $\bar{f}$ to the fibre $E_b = p^{-1}b$ and $\tilde{\sigma}f$ is some element of $\mathrm{Stab}(B,S^0)$, which is an additional fixed point invariant. It cannot be seen by looking at the fibres separately, it reflects a relation between the fibres which comes from the continuity of the family. If $\tilde{\sigma}f \neq 0$ then f_x has a fixed point for at least one $x \in B$ even if the ordinary fixed point index of f_x vanishes for each $x \in B$.

4.2 Example. Let S^1 be the unit circle in $\mathbb{C} = \mathbb{R}^2$ and consider the fibre-wise map

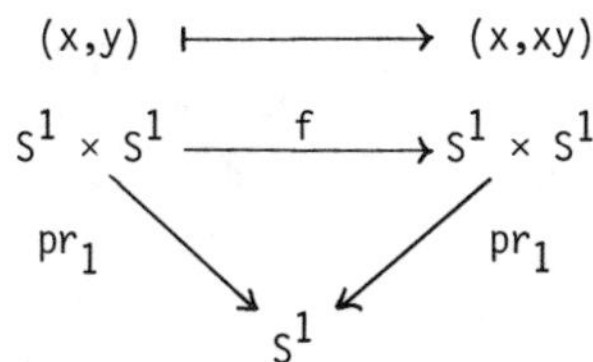

Thus in the notation above we take $B = S^1$, $K = S^1 \times S^1 \subset S^1 \times \mathbb{R}^2$. For each $x \in B$, the restriction f_x of f to $\{x\} \times S^1$ is a rotation of S^1. The Lefschetz number of f_x which equals the trace $\sigma\bar{f}_x$ vanishes, corresponding to the fact that f_x has no essential fixed points. If $x \neq 1$ then f_x has no fixed point at all. f_1 is the identity of S^1 which is homotopic to a map without fixed points by an arbitrarily small deformation.

But the whole family $f = (f_x \mid x \in S^1)$ does have essential fixed points. One can show (Dold [1974a] 5.3, [1976] 7.6, [1978] 5.14) that $\tilde{\sigma}f$ is the non-zero element of $\mathrm{Stab}(S^1,S^0) \cong \mathbb{Z}/2\mathbb{Z}$, which implies that every map in the fibre-wise homotopy class of f has fixed points.

5. TRANSFER

Let again C be a monoidal category and A an object of C which has a strong dual B. In addition let $d : A \longrightarrow A \otimes A$ and $c : A \longrightarrow I$ be given such that

(5.1)

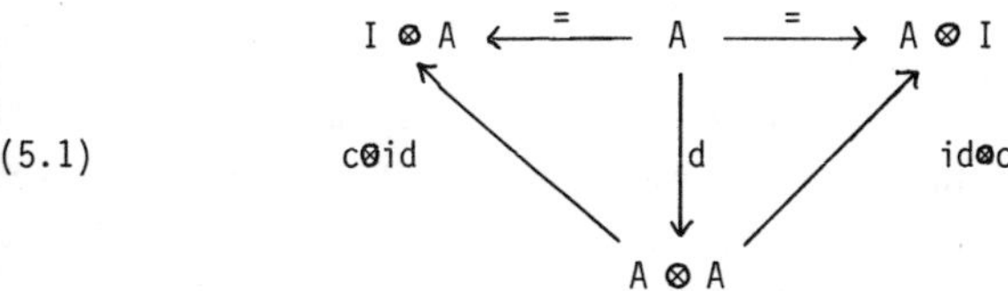

commutes. Then (A,d) or A is called a <u>coalgebra</u> with counit c.

<u>5.2 Definition</u>. The <u>transfer</u> τf of an endomorphism $f:A\to A$ is the composition

$$\begin{aligned}\tau f : I \xrightarrow{\eta} A \otimes B \xrightarrow[\cong]{\gamma} B \otimes A \xrightarrow{id_B \otimes d} B \otimes A \otimes A \\ \xrightarrow{id_B \otimes f \otimes id_A} B \otimes A \otimes A \xrightarrow{\varepsilon \otimes id_A} I \otimes A = A .\end{aligned} \tag{5.3}$$

<u>5.4 Proposition</u>. <u>Under the above hypotheses the composition</u> $c \circ (\tau f)$ <u>equals the trace</u> σf.

Using the right triangle of (5.1) this is an immediate consequence of the definitions of τf and σf (3.1).

In the case $C = \text{Stab}$ (cf. § 2) every object of the form $(X,\emptyset)$ has a canonical structure of a coalgebra with counit: The morphism d is the usual diagonal

$$(X,\emptyset) \longrightarrow (X \times X, \emptyset) = (X,\emptyset) \otimes (X,\emptyset)$$

$$x \longmapsto (x,x)$$

and c is the unique map $(X,\emptyset) \longrightarrow (P,\emptyset) = S^0$, where P is a one point space.

In the parametrized case $C = \text{Stab}_B$ (§ 4) the same construction can be applied fibre-wise. For each space $p : E \longrightarrow B$ over B this gives to the object $\overline{E} = (E,\emptyset)_B$ of Stab_B the structure of a coalgebra with counit

$$c = \overline{p} : \overline{E} \longrightarrow \overline{B} = B \times S^0 .$$

If $\overline{E}$ is strongly dualisable in Stab_B then for every map $f : E \longrightarrow E$ over B we have a transfer morphism

$$\tau\overline{f} : B \times S^0 = \overline{B} \longrightarrow \overline{E}$$

in Stab_B such that $\overline{p} \circ (\tau\overline{f}) = \sigma\overline{f}$ (by 5.4). This has important consequences for the properties of p which have been used e.g. by Becker-Gottlieb [1975, Th. 7.1] for a proof of the famous Adams conjecture [Adams 1963, Conjecture (1.2)]. We do not follow this line here because it would lead away from fixed point theory. For fixed point theory it is interesting to note that the transfer can be refined in such a way that its target is not E but - roughly speaking - the fixed point set F of f in E. More precisely:

5.5 Theorem. *Let B be metric and $K \subset B \times \mathbb{R}^n$. Let the projection $K \longrightarrow B$ be proper, let $f : K \longrightarrow K$ be a map over B with fixed point set F and let K be an ENR_B. Then for every neighborhood U of F in K there is a $\tau_U \in Stab_B(\overline{B},\overline{U})$ such that $\tau\overline{f}$ is τ_U followed by the inclusion $\overline{U} \subset \overline{K}$.*

This can be proved by inserting the definitions (2.4) of ε and (2.5) of η into the definition (5.3) of $\tau\overline{f}$. One obtains a diagram similar to (3.3) from which one may read off the result. In a slightly different formulation the result is contained in [Dold 1976, § 3].

As an example of an application we prove

5.6 Corollary. *Let K be a compact subset of $\mathbb{R}^n$ and a neighborhood retract. Let $f_t : K \longrightarrow K$ be a continuous family of maps and let F be the set of points $(t,x) \in I \times K$ such that x is a fixed point of f_t. Then the fixed point index λ of f_t equals the Lefschetz number of $H(f_t ; \mathbb{Q})$ and is independent of t. If $\lambda \neq 0$ then F has a connected component which meets $0 \times K$ and $1 \times K$.*

Proof. The first assertion is well known and was reproved in § 3. The family $(f_t \mid t \in I)$ gives rise to a map f over I defined by

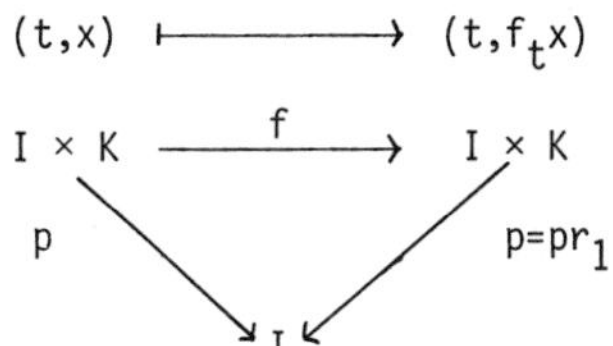

Let U be any neighborhood of F in $I \times K$. Then by Th. 5.5 there is a morphism τ_U such that the following diagram in $Stab_I$ commutes

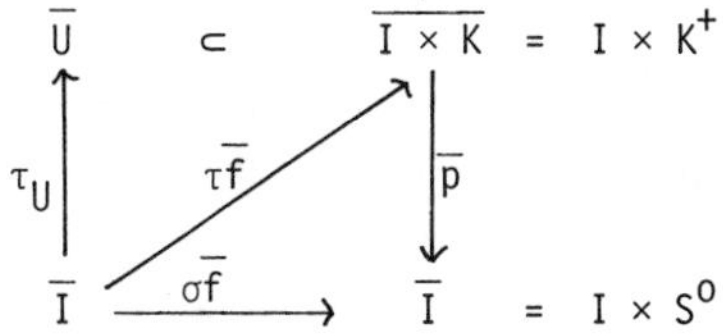

Since I is contractible, restriction to one of the fibres gives

$$Stab_I(I \times S^0, I \times S^0) \cong Stab(S^0,S^0) \cong \mathbb{Z} ,$$

and $\sigma\overline{f}$ is just multiplication by λ. In $Stab$ the above diagram induces

another commutative diagram

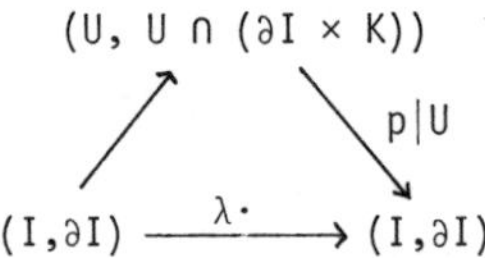

where $\partial I = \{0, 1\}$ is the boundary of I . Since $\lambda \neq 0$ it follows that in integral singular homology the map

$$(p|U)_* : H_1(U, U \cap (\partial I \times K)) \longrightarrow H_1(I,\partial I)$$

is non zero. Hence there exists a path in U which connects $0 \times K$ to $1 \times K$. Considering a descending sequence of closed neighborhoods of F in $I \times K$ with intersection F it is not hard to prove that F has a connected subset which meets $0 \times K$ and $1 \times K$.

REFERENCES

Adams, J.F. [1963]: On the groups J(X)-I. Topology 2, 181-195

Becker, J.C., Gottlieb, D.H. [1975]: The transfer map and fibre bundles. Topology 14, 1-12

Becker, J.C., Gottlieb, D.H. [1976]: Transfer maps for fibrations and duality. Compositio math. 33, 107-133

Dold, A. [1974a]: The fixed point index of fibre-preserving maps. Inventiones math. 25, 281-297

Dold, A. [1974b]: Transfert des points fixes d'une famille continue d' applications. C.r.Acad.Sci.Paris, Sér. A 278, 1291-1293

Dold, A. [1976]: The fixed point transfer of fibre-preserving maps. Math. Zeitschr. 148, 215-244

Dold, A. [1978]: Geometric cobordism and the fixed point transfer. In: Algebraic Topology Proceedings, Vancouver 1977, Editors: P.Hoffmann, R.Piccinini and D.Sjerve. Lecture Notes in Mathematics Vol. 673, 32-87, Springer-Verlag

Dold, A., Puppe D.: Duality, trace and transfer. To appear in the Proceedings of the Conference on Geometric Topology, Warsaw 24.8.-2.9.1978

Ligon, T.S. [1978]: Galois-Theorie in monoidalen Kategorien. Algebra-Berichte Nr. 35, Uni-Druck München

Lindner, H. [1978]: Adjunctions in monoidal categories. Manuscripta math. 26, 123-139

Pareigis, B. [1976]: Non-additive ring and module theory IV. The Brauer group of a symmetric monoidal category. In: Brauer groups, Proceedings of the Conference held at Evanston, October 11-15, 1975. Editor: D.Zelinsky. Lecture Notes in Mathematics Vol. 249, 112-133, Springer-Verlag

Spanier, E.H. [1959]: Function spaces and duality. Ann. of Math. II. Ser. 70, 338-378

Switzer, R.M. [1975]: Algebraic Topology - Homotopy and Homology. Die Grundlehren der mathematischen Wissenschaften in Einzeldarstellungen Band 212, Springer-Verlag

GAME THEORY AND RELATED TOPICS
O. Moeschlin, D. Pallaschke (eds.)

ON OPTIMAL STOCHASTIC CONTROL PROBLEMS WITH CONSTRAINTS

Maurice ROBIN
IRIA
B.P. 105
78150 LE CHESNAY
FRANCE

In many applications, one needs to consider stochastic control problems with more than one payoff.

The literature is almost exclusively devoted to the one-criterium problem. We consider here control problems for markov processes where we have to minimize some global cost $J(v)$, expectation of a functional of the process depending on the control v, under the constraint $I(v) \leq$ constant, where I is a functional similar to J. This is of course related to the bi-criteria problem with I and J. The stopping-time problem and some special cases of impulsive problems are studied.

1. INTRODUCTION

An extensive literature deals with optimal stochastic control problem where the objective is to minimize the mathematical expectation of a functional depending on the state of a Markov process. Under this category of problems, special cases are optimal stopping problem ([5] for example) and impulsive control problems ([1]. [4]). A typical optimal stopping example is to minimize

$$J_x(\tau) = E_x \int_0^\tau e^{-\alpha s} f(x_s)ds \quad , \qquad \alpha \in \mathbb{R}_+$$

on stopping times τ.

In this paper we are interested in control problems where, in addition to the cost $J_x(\tau)$, there is a constraint on admissible control, this constraint taking the form of another functional to be kept less than some constant. For example

$$I_x(\tau) = E_x \int_0^\tau e^{-\alpha s} g(x_s)ds \leq K$$

Such a situation is frequent in applications and is related to the fact that there are, in general, several criteria to be simultaneously minimized. Here we will re trict ourselves to one constraint. Very few publications have delt with this kinc of problems for stochastic control. Frid [3] gave some results for the control of a Markov chain and showed that there exists an ε-optimal element in the class of randomized strategies.

We will, see here, for optimal stopping and impulsive control problems, that one can prove the existence of an optimal control (which will be randomized in some cases) and we will obtain an algorithm, inspired by the Frid's method to compute ε-optimal control. The existence of an optimal control for the impulsive control of a one-dimensional diffusion process has been proved by Safonov [6] based on ex plicit computations.

2. OPTIMAL STOPPING PROBLEMS

2.1. Classical optimal stopping problem and results

Let $X = (\Omega, \mathcal{F}_t, x_t, P_x)$ be a homogeneous Markov process with values in a space E Let us assume that E is a compact metric space, endowed with its Borel σ-field. X is a right continous Feller process ; that is, if $\Phi(t)$ is the semi-group of X and C the Banach space of continous functions on E,

$$(2.1)\qquad \Phi(t)f \in C \;,\quad \forall f \in C$$

$$\lim_{t\downarrow 0} \Phi(t)\, f(x) = f(x) \quad \forall f \in C, \quad \forall X \in E$$

Now assume $f \in C$ is given, we define

$$(2.2)\qquad J_x(\tau) = E_x \int_0^\tau e^{-\alpha s} f(x_s)\, ds$$

for any stopping time τ w.r.t. $\{\mathcal{F}_t\}$.

Defining

$$(2.3)\qquad u(x) = \operatorname*{Inf}_{\tau} J_x(\tau),$$ we can state the following result:

Theorem 2.1. Under assumption (2.1.), u is the maximum element of the set of fun tions w such that

$$\begin{cases} w \le e^{-\alpha t}\Phi(t)w + \int_0^t e^{-\alpha s}\Phi(s)\, f\, ds \\ w \le 0 \\ w \in C \end{cases}$$

Moreover the stopping time

$$\tau = \mathrm{Inf}(s \geq 0,\ u(x_s) = 0) \text{ is optimal.}$$

We refer to [4] for proof of such type of results. Similar results for continuous processes appear in [1] and, excepted for $u \in G$ in [5] but with very different methods.

Extension to randomized stopping time

Let us now consider the previous problem with randomized stopping times. These stopping times are obtained in the following way.

Let $(\Omega', \mathfrak{F}', P')$ a probability space in which we assume that variables $\{\xi_p\}_{p\in[0,1]}$ are defined where ξ_p takes value 1 with probability p and 2 with probability 1-p.

Now we consider

$$\Omega_1 = \Omega \times \Omega',\ \mathfrak{F}_t^1 = \mathfrak{F}_t \otimes \mathfrak{F}'\ ,\ P_x^1 = P_x \otimes P'$$

Then $X^1 = (\Omega_1, \mathfrak{F}_t^1, x_t, P_x^1)$ is still a Markov process with semi-group $\Phi(t)$.

Given τ^1, τ^2, $\mathfrak{F}_t$-stopping times, one can define

$$\tau = \tau^1\ \chi_{\xi_p=1} + \tau^2\ \chi_{\xi_p=2} \quad \text{for any p.} \quad (\chi_A \text{ indicator function of the set } A)$$

Then τ is an $\mathfrak{F}_t^1$-stopping time. It is clear that this construction means that τ^1 is choosen "with probability p" and τ^2 with probability (1-p).

Now if

$$J_x(\tau) = E_x^1 \int_0^\tau e^{-\alpha s} f(x_s) ds \quad \text{for any } \mathfrak{F}_t^1\text{-stopping time,}$$

It is easy to see, that, due to independance of ξ_p from the process x_t, the previous theorem is still valid and that, because the class of randomized stopping time contains the $\mathfrak{F}_t$-stopping times, that $\tau = \mathrm{Inf}(s \geq 0\ u(x_s) = 0)$ is an optimal solution for J_x.

2.2. Stopping problem with constraint

For reasons which will be made clear later on, we will take the following particular problem : given $x \in E$,

$$P_0 \ :\ \text{minimize}\ \ J_x(\tau)$$

on the class V of randomized stopping times such that

$$E_x^1 \, e^{-\alpha\tau} \leq \theta \qquad \theta \in [0,1[$$

To this "conditional problem" is associated an "inconditional" one, namely

Π_0 : minimize $J_x(\tau)$ over the whole set of $\mathcal{F}_t^1$-stopping times (including the randomized ones).

We know that there exists a non-randomized optimal solution, that is

$$\tau^0 = \text{Inf } (s \geq 0 \quad u^0(x_s) = 0)$$

where $u^0(x_s) = \underset{\tau}{\text{Inf}} \; J_x(\tau)$

In order to get a non trivial problem P_0, we will assume that

$$E_x^1 \; e^{-\alpha\tau^0} > \theta$$

[Otherwise, of course, τ^0 is a solution of P_0].
To study the problem P_0, we introduce for $\lambda \in R^+$, the problem

$$\Pi_\lambda \; : \text{minimize } J_x^\lambda(\tau) = E_x^1\{ \int_0^\tau e^{-\alpha s} \, f(x_s)ds + \lambda e^{-\alpha\tau}\}$$

(over the class of randomized stopping times).

Let $\quad u^\lambda(x) = \underset{\tau}{\text{Inf}} \; J_x^\lambda(\tau)$

$$\tau^\lambda = \text{Inf } (s \geq 0, \; u^\lambda(x_s) = \lambda)$$

We know from general theory that the non randomized stopping time τ^λ is optimal for u^λ. (see [4]).

<u>Lemma 2.1.</u> If $E_x^1 \; e^{-\alpha\tau^\lambda} = \theta$ then τ^λ is optimal for P_0.

<u>Proof</u> :
By definition of τ^λ, we have

$$E_x^1[\int_0^{\tau^\lambda} e^{-\alpha s} \, f(x_s)ds + \lambda \, e^{-\alpha\tau^\lambda}] \leq E_x^1[\int_0^\tau e^{-\alpha s} \, f(x_s)ds + \lambda e^{-\alpha\tau}]$$

for any τ (eventually randomized).

From $E_x^1 \; e^{-\alpha\tau^\lambda} = \theta$, we get

$$E_x^1 \int_0^{\tau^\lambda} e^{-\alpha s} f(x_s)ds \leq E_x^1[\int_0^{\tau} e^{-\alpha s} f(x_s)ds + \lambda(e^{-\alpha\tau} - \theta)]$$

and for any τ such that $E_x^1 e^{-\alpha\tau} \leq \theta$, this implies

$$E_x^1 \int_0^{\tau^\lambda} e^{-\alpha s} f(x_s)ds \leq E_x^1 \int_0^{\tau} e^{-\alpha s} f(x_s)ds ,$$

therefore, τ^λ is optimal for P_0.

In the following, we look for some λ such that $E_x^1 e^{-\alpha\tau^\lambda} = \theta$ in order to get a solution of P_0.

Lemma 2.2. $||u^\lambda - u^{\lambda'}|| \leq |\lambda - \lambda'|$ where $||.||$ denotes the sup-norm on G

Proof :

$$|J_x^\lambda(\tau) - J_x^{\lambda'}(\tau)| \leq |\lambda - \lambda'| \qquad \forall x,\tau .$$

Lemma 2.3. $\lambda \to \tau^\lambda$ is increasing.

Proof :

Let $\lambda' = \lambda + \mu \quad \mu \geq 0$

We have

$$\tau^{\lambda+\mu} = \text{Inf}(s \geq 0, u^{\lambda+\mu}(x_s) = \lambda+\mu),$$

$$\tau^\lambda = \text{Inf}\ (s \geq 0, u^\lambda(x_s) = \lambda)$$

It is clear that $\lambda \to u^\lambda$ is increasing in the sense that $\lambda' \geq \lambda \to u^{\lambda'}(x) \geq u^\lambda(x)\ \forall x$.

Now, by the optimality of τ^λ

$$u^{\lambda+\mu}(x) \leq E_x^1\{ \int_0^{\tau^\lambda} e^{-\alpha s} f(x_s)ds + (\lambda+\mu)\ e^{-\alpha\tau^\lambda}\}$$

therefore

$$u^{\lambda+\mu}(x) \leq u^\lambda(x) + \mu E_x^1\ e^{-\alpha\tau^\lambda} .$$

Then if $x \in \{u^\lambda < \lambda\}$,

we get

$$u^{\lambda+\mu}(x) < \lambda + \mu ,$$

which implies

$$\{u^{\lambda} < \lambda\} \subset \{u^{\lambda+\mu} < \lambda+\mu\}$$

Therefore, because τ^{λ} (respectively $\tau^{\lambda+\mu}$) is the exit time from the open set $\{u^{\lambda} < \lambda\}$ (resp $\{u^{\lambda+\mu} < \lambda+\mu\}$) we have

$$\tau^{\lambda} \leq \tau^{\lambda+\mu} \quad .$$

Lemma 2.4. When $\lambda' \nearrow \lambda$, $\tau^{\lambda'} \nearrow \tau^{\lambda}$ (i.e., $\lambda \to \tau^{\lambda}$ is continuous from the left).

Proof :

For $\delta > 0$, let

$$\tau_{\delta}^{\lambda} = \operatorname{Inf}(s \geq 0,\ u^{\lambda}(x_s) \geq \lambda - \delta)$$

We have $\tau_{\delta}^{\lambda} \leq \tau^{\lambda}$. Moreover, for ε small enough $\tau_{\delta}^{\lambda} \leq \tau^{\lambda-\varepsilon}$.

In fact, on $[0,\tau_{\delta}^{\lambda}[$

$$u^{\lambda}(x_s) < \lambda - \delta$$

but we have $||u^{\lambda} - u^{\lambda-\varepsilon}|| \leq \varepsilon$,

therefore, if for example $\varepsilon \leq \frac{\delta}{2}$, we will have

$$u^{\lambda-\varepsilon}(x_s) \leq u^{\lambda}(x_s) + \varepsilon < \lambda - \frac{\delta}{2} < \lambda - \varepsilon$$

on $[0,\tau_{\delta}^{\lambda}[$, implying $\tau^{\lambda-\varepsilon} \geq \tau_{\delta}^{\lambda}$.

Let us prove now that $\tau_{\delta}^{\lambda} \nearrow \tau^{\lambda}$ for $\delta \to 0$.

τ_{δ}^{λ} is increasing, $\tau_{\delta}^{\lambda} \nearrow \hat{\tau}$ which is an $\mathfrak{F}_t^1$ stopping time and $\hat{\tau} \leq \tau^{\lambda}$

But,

$$u^{\lambda}(x_{\tau_{\delta}^{\lambda}}) \geq \lambda - \delta ,$$

u^{λ} is continuous, and from (2.1) x_t is quasi left continuous, that is ([1])

$$x_{\tau_{\delta}^{\lambda}} \to x_{\hat{\tau}} \quad P_x^1 \text{ a.s on } \{\hat{\tau} < +\infty\} .$$

Hence

$$u^{\lambda}(x_{\hat{\tau}}) \geq \lambda \ P_x^1 \text{ a.s on } \{\hat{\tau} < +\infty\}$$

implying $\hat{\tau} = \tau^{\lambda}$ at least when $\hat{\tau} < \infty$.

([1]) See Dynkin [2]. If E is not compact additional assymptions on Φ are needed, but the results can be extended.

But, if $\hat{\tau} = +\infty$, then $\tau^\lambda = +\infty$. Therefore, we have $\hat{\tau} = \tau^\lambda$ a.s., and the proof is completed.

It is clear now, that if we could prove that $\lambda \to \tau^\lambda$ is continuous or at least, that

$\lambda \to E_x^1 \, e^{-\alpha\tau^\lambda}$ is continuous, we could find λ such that $E_x^1 \, e^{-\alpha\tau^\lambda} = \theta$.

The difficulty lies in the fact that it is not true in general that $\tau^{\lambda'} \searrow \tau^\lambda$ when $\lambda' \searrow \lambda$, even for a very regular process (a diffusion for example), because this involves the smoothness of the boundary of $\Gamma_\lambda = \{u^\lambda < \lambda\}$. The problem is illustrated by the following picture.

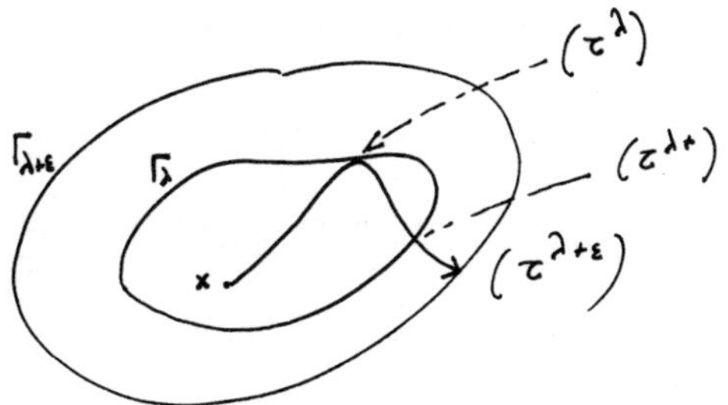

For a diffusion process and a smooth boundary of Γ_λ, the process exits from $\bar{\Gamma}_\lambda$ at τ^λ with probability one. But, in general, Γ_λ is not smooth enough, and $\{\tau^{\lambda+} > \tau^\lambda\}$ has a strictly positive probability.

Nevertheless, we can obtain the following result :

<u>Theorem 2.2.</u> Under the assumptions of § 2.1, there exists an optimal randomized stopping time for the problem P_0.

<u>Proof</u> :

Let us define

$$\lambda_0 = \sup [\lambda \geq 0 \mid E_x^1 \, e^{-\alpha\tau^\lambda} > \theta]$$

As a consequence of lemmas 2.3 and 2.4, $\lambda \to E_x^1 \, e^{-\alpha\tau^\lambda}$ is decreasing and continuous from the left.

Let

$$\theta_0 = E_x^1 \, e^{-\alpha\tau^{\lambda_0}} ,$$

we have : $\theta_0 \geq \theta$ and the two possible situations are illustrated by the figures 1 and 2 below.

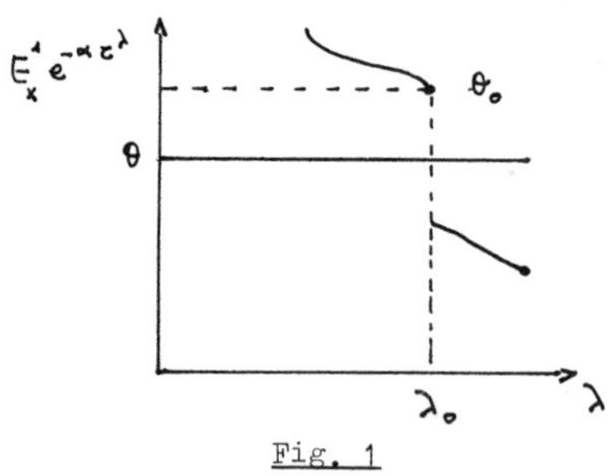

Fig. 1

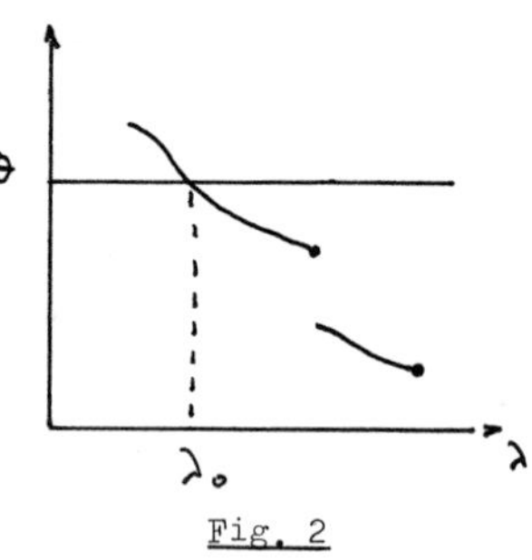

Fig. 2

It is clear that if λ_0 is a point of continuity of $E_x^1\, e^{-\alpha\tau^\lambda}$, then τ^{λ_0} is a non randomized stopping time such that

$E_x^1\, e^{-\alpha\tau^{\lambda_0}} = \theta$ and by lemma 2.1, is an optimal solution for P_0.

This is true also when λ_0 is the right end of a continuity interval of $E_x^1\, e^{-\alpha\tau^\lambda}$.

If $\tau^{\lambda_0+\varepsilon} \searrow \hat{\tau}$ and $P_x(\hat{\tau} > \tau^{\lambda_0}) > 0$ then either $\theta_0 > \theta$ or $E_x^1\, e^{-\alpha\hat{\tau}} < \theta$. In both cases

$$E_x^1[\, e^{-\alpha\tau^{\lambda_0}} - e^{-\alpha\hat{\tau}}] > 0$$

Now, the important fact is that $\hat{\tau}$ is optimal for Π_{λ_0} as well as τ^{λ_0} : actually, we have, for $\varepsilon > 0$

$$u^{\lambda_0+\varepsilon}(x) = E_x^1 \left[\int_0^{\tau^{\lambda_0+\varepsilon}} e^{-\alpha s} f(x_s)ds + (\lambda_0+\varepsilon)\, e^{-\alpha\tau^{\lambda_0+\varepsilon}}\right]$$

We know that $u^{\lambda_0+\varepsilon} \searrow u^{\lambda_0}$ uniformly and $\tau^{\lambda_0+\varepsilon} \searrow \hat{\tau}$ (by definition of $\hat{\tau}$). Moreover, the continuity from the right of $\mathfrak{F}_t^1$ implies that $\hat{\tau}$ is an $\mathfrak{F}_t^1$-stopping time. Therefore

$$u^{\lambda_0}(x) = E_x^1\left[\int_0^{\hat{\tau}} -e^{\alpha s} f(x_s)ds + \lambda_0\, e^{-\alpha\hat{\tau}}\right] = J_x^{\lambda_0}(\hat{\tau})$$

and that means that $\hat{\tau}$ is also an optimal stopping time for Π_{λ_0} (non randomized).

Then, from $E_x^1\, (e^{-\alpha\tau^{\lambda_0}} - e^{-\alpha\hat{\tau}}) > 0$, one can choose $p \in [0,1]$ such that

satisfies $\tau^* = \tau^{\lambda_0}\chi_{\xi_p = 1} + \hat{\tau}\chi_{\xi_p = 2}$

$\tau^{\lambda_0}\quad \chi_{\xi_p = 1} + \chi_{\xi_p = 2}$

$$E_x^1 \, e^{-\alpha\tau^*} = \theta$$

(take $p = \dfrac{\theta - E_x^1 \, e^{-\alpha\hat{\tau}}}{\theta_0 - E_x^1 \, e^{-\alpha\hat{\tau}}}$). This completes the proof of the theorem, since one can now use lemma 2.1.

Remark 2.1. There is no difficulty in generalizing the previous result to

$$J_x(\tau) = E_x \int_0^{\tau} e^{-\alpha s} f(x_s)\,ds + E\, e^{-\alpha\tau} \Psi(x_\tau)$$

where Ψ is a continuous function (If Ψ is in the domain of the infinitisimal generator of Φ, i.e. very smooth, then a translation reduces the problem to theorem 2.2).

Remark 2.2. The same type of results can also be obtained for the more general constraint

$$E_x^1 \, e^{-\alpha\tau} \Phi(x_\tau) \leq \text{constant}$$

where Φ is a positive function.

Remark 2.3. Theorem 2.2 given an existence result but the proof does not give an algorithm in order to compute a solution. The next section will give some idea on that point.

3. ε-OPTIMAL SOLUTIONS

Using a method similar to the one in [3], we give here a proof of existence of ε optimal solution.

As in §. 2 we will assume , u being the optimal cost defined by (2.3).

(3.1.) $\quad u(x) < u^0(x)$

which means that the constant is non trivial.
As usual, we will say that τ is (λ,ε)-optimal if

(3.2) $\quad J_x^\lambda(\tau) \leq u^\lambda(x) + \varepsilon$

Now let τ^0 be $(0,\delta)$-optimal for a small $\delta > 0$ such that

$$0 < \delta < u(x) - u^0(x)$$

(We know from §. 2.1 that such τ^0 exists).
By definition of τ^0

$$J_x(\tau^0) \leq u^0(x) + u(x) - u^0(x) \leq u^0(x)$$

Therefore, necessarily

$E_x^1 \, e^{-\alpha\tau^0} > \theta$ and this is also true for any τ which is $(0,\delta)$-optimal.

Let Λ be large enough such that

$$(3.2) \qquad E_x^1 \, e^{-\alpha\tau} \leq \theta$$

as soon as τ is (Λ,δ) optimal.
This is possible because we have

$$J_x(\tau) + \lambda \, E_x^1 \, e^{-\alpha\tau} \leq J_x(\tau^\lambda) + \lambda \, E_x^1 \, e^{-\alpha\tau^\lambda} + \delta$$

for any τ (λ,δ)-optimal.
Therefore

$$E_x^1 \, e^{-\alpha\tau} \leq \frac{J_x(\tau^\lambda) - J_x(\tau)}{\lambda} + E_x^1 \, e^{-\alpha\tau^\lambda} + \frac{\delta}{\lambda} .$$

Moreover

$$E_x^1 \, e^{-\alpha\tau^\lambda} \leq \frac{u^\lambda(x) - J_x(\tau^\lambda)}{\lambda} .$$

Then

$E_x^1 \, e^{-\alpha\tau^\lambda} \leq \frac{K}{\lambda}$ for a suitable constant K.

Hence

$E_x^1 \, e^{-\alpha\tau} \leq \frac{K' + \delta}{\lambda}$ for a suitable constant K', and for any τ (λ,δ)-optim

Now, from the continuity of u^λ w.r.t. λ, we can divide the interval $[0,\Lambda]$:

$$\lambda_0 = 0, \; \lambda_1 < \lambda_2 < \ldots < \lambda_n = \Lambda$$

in order to get

$$(3.3) \qquad |u^{\lambda_i}(x) - u^{\lambda_{i+1}}(x)| \leq \frac{\delta}{3} ,$$

and

$$(3.4) \qquad |\lambda_i - \lambda_{i+1}| \leq \frac{\delta}{3} .$$

Then, the main remark is that : if τ^i is $(\lambda_i, \frac{\delta}{3})$-optimal, then τ^i is also (λ_{i+1},δ)-optimal.

Indeed :

$$\begin{aligned} |u^{\lambda_{i+1}}(x) - J_x^{\lambda_{i+1}}(\tau^i)| \leq & |J_x^{\lambda_i}(\tau^i) - u^{\lambda_i}(x)| \\ & + |\lambda_i - \lambda_{i+1}| \, E_x^1 \, e^{-\alpha\tau^i} \\ & + |u^{\lambda_{i+1}}(x) - u^{\lambda_i}(x)| \end{aligned}$$

and the right hand side is less than δ by (3.3), (3.4) and the assumption that τ^i is $(\lambda_i, \frac{\delta}{3})$-optimal.

Now for each λ_i, one can associate a $(\lambda_i, \frac{\delta}{3})$-optimal $-$ τ^i . [Notice that τ^i is also (λ_i,δ)-optimal].

For this sequence, we have

$$E_x^1 \, e^{-\alpha\tau^0} > \theta \quad \text{and} \quad E_x^1 \, e^{-\alpha\tau^n} \leq \theta .$$

Therefore, there exists an index j such that

$$E_x^1 \, e^{-\alpha\tau^j} > \theta \quad \text{and} \quad E_x^1 \, e^{-\alpha\tau^{j+1}} \leq \theta$$

with τ^j, τ^{j+1} respectively $(\lambda_j, \frac{\delta}{3})$-optimal and $(\lambda_{j+1}, \frac{\delta}{3})$-optimal.

Consequently, τ^j and τ^{j+1} are both (λ_{j+1}, δ)-optimal from what we have just proved. Notice also that all the stopping times τ^j can be choosen as non-randomized.

One can then choose $p \in [0,1]$ such that the randomized stopping time τ_p associated to (p,τ^j,τ^{j+1}) satisfies

$$E_x^1 \, e^{-\alpha\tau_p} = \theta$$

and therefore τ_p is δ-optimal for the problem P_0 (see the proof of lemma 2.1 which is valid for δ-optimality as well).

4. AN IMPULSIVE CONTROL PROBLEM

Let us consider a replacement problem as follows :

x_t is a Markov process representing the degradation of a machine ; at any time, one can replace the machine by a new one, for which the state is $\bar{x}$, by paying $k > 0$. Moreover, the operation of the machine costs $f(x)$ per unit of time, when the state is x ($f > 0$, $t \in \mathbb{R}_+$).

General results about such problems are contained in [1], [4]. The problem here is simplified by the fact that we replace by a machine in a fixed state $\bar{x}$.

Let us give a precise statement of the problem (we use the notations of §. 2). In the following θ_t will be the translation operator on Ω .

We define the class V_1 of sequences v of $\mathcal{F}_t$-stopping times such that

$$(4.1) \qquad v = (\tau^n) \quad n \geq 1$$

$$\tau^n = \tau^{n-1} + \sigma^n \circ \theta_{\tau_{n-1}} \,,, \quad \tau^1 = \sigma^1 \,, \; \tau^n \nearrow \infty$$

where σ^n is any $\mathcal{F}_t$-stopping time. (This means that τ^n is independent of the "past before τ^{n-1}" see Dynkin [2]).

For any $v \in V$, we define the payoff

$$(4.2) \qquad J^k_{\bar{x}}(v) = E_{\bar{x}} \left[\int_0^{\sigma^1} e^{-\alpha s} f(x_s)ds + k\, e^{-\alpha\sigma^1} \right]$$

$$+ E_{\bar{x}}\, e^{-\alpha\sigma^1}\, E_{\bar{x}} \left[\int^{\sigma^2} e^{-\alpha s} f(x_s)ds + k\, e^{-\alpha\sigma^2} \right]$$

$$+ \left(\prod_{j=1}^{n-1} E_{\bar{x}}\, e^{-\alpha\sigma^j} \right) E_{\bar{x}} \left[\int_0^{\sigma^n} e^{-\alpha s} f(x_s)ds + k\, e^{-\alpha\sigma^n} \right]$$

in other words :

$$(4.3) \qquad J^k_{\bar{x}}(v) = \sum_{n \geq 1} \left(\prod_{j=1}^{n-1} E_{\bar{x}}\, e^{-\alpha\sigma^j} \right) E_{\bar{x}} \left(\int_0^{\sigma^n} e^{-\alpha s} f(x_s)ds + k\, e^{-\alpha\sigma^n} \right)$$

For any x, one defines $J_x(v)$ by

$$J^k_x(v) = \sum_{n \geq 2} \left(E_x\, e^{-\alpha\sigma^1} \right) \left[\prod_{j=2}^{n-1} E_{\bar{x}}\, e^{-\alpha\sigma^j} \right] E_{\bar{x}} \left(\int_0^{\sigma^n} e^{-\alpha\sigma} f(x_s)ds + k\, e^{-\alpha\sigma^n} \right)$$

$$+ E_x \left(\int_0^{\sigma^1} e^{-\alpha s} f(x_s)ds + k\, e^{-\alpha\sigma^1} \right)$$

[of course for $n = 2$, $\prod_{j=2}^{n-1}$ is taken equal to 1]

It can be shown (see [4]) that this can be written as

$$E_x^v \left(\int_0^\infty \bar{e}^{\alpha s} f(y_s) ds + k \sum_{j \geq 1} \bar{e}^{\alpha \tau^j} \right)$$

for a suitable process y_s and measure P_x^v .
Using general results from [4] one can state the following.

Theorem 4.1. Under the assumptions (2.1), and for $f \in C$,

(4.4) $u^k(x) = \operatorname{Inf}_{v \in V} J_x^k(v)$ is the maximum element of the set of functions w

satisfying

$$w(x) \leq k + w(\bar{x}) ,$$

$$(4.5) \quad w(x) \leq \bar{e}^{\alpha t} \Phi(t)\, w(x) + \int_0^t \bar{e}^{\alpha s} \Phi(s)\, f(x) ds ,$$

$$w \in C .$$

Moreover, an optimal control is given by

$$(4.6) \quad \tau^1 = \sigma^1 = \operatorname{Inf}\,(s \geq 0,\ u^k(x_s) = k + u^k(\bar{x}))$$

$$\tau^n = \tau^{n-1} + \sigma^1 \circ \theta_{\tau^{n-1}}$$

and u^k is the unique solution in C, of the equation

$$(4.7) \quad u^k(x) = \operatorname{Inf}_{\tau} E_x\left[\int_0^\tau \bar{e}^{\alpha s} f(x_s) ds + \bar{e}^{\alpha\tau}[k + u^k(\bar{x})]\right] .$$

The problem we want to solve now is

(4.8) P_0 : minimize $J_{\bar{x}}^0(v)$ i.e. with $k = 0$

for $v \in V$ such that

$$\sum_{n \geq 1} \left[\prod_{j=1}^{n} E_{\bar{x}}\, \bar{e}^{\alpha\sigma^j} \right] \leq C_0 \quad \text{(a given constant } > 0)$$

As for the optimal stopping problem we will introduce the problem

(4.9) Π_λ : minimize $J_x^\lambda(v)$

on impulsive controls v consisting of sequences of stopping times satisfying (4.1), eventually randomized.

We know that, for the problem Π_λ, an optimal (non-randomized) strategy is given by

$$\tau^\lambda = \text{Inf } (s \geq 0,\ u^\lambda(x_s) = \lambda + u^\lambda(\bar{x})).$$

We begin with

Lemma 4.1.

(i) $\lambda \to u^\lambda$ is increasing

(4.10) $||u^\lambda - u^{\lambda'}|| \leq |\lambda - \lambda'| \,.\, K(\lambda,\lambda')$

where

$$K(\lambda,\lambda') = \frac{||f||}{\alpha . \min(\lambda,\lambda')}$$

(ii) $\lambda \to \tau^\lambda$ is increasing.

Proof (i) $\lambda \to u^\lambda$ increasing is obvious.

(4.10) comes from the fact that for $\lambda > 0$, one can always restrict admissible controls to those such that

$$\lambda E_x \sum_{i \geq 1} e^{-\alpha\tau^i} \leq \frac{||f||}{\alpha}$$

because the right hand side is surely greater than $||u^\lambda||$.

(ii) The proof is similar to the one in lemma 2.3. : we know that $\tau^{\lambda+\mu}(\mu>0)$ is optimal for the stopping problem

$$\text{Inf}_\tau\ E_x^1 \left[\int_0^\tau e^{-\alpha s} f(x_s)ds + [\lambda+\mu + u^{\lambda+\mu}(\bar{x})]\, e^{-\alpha\tau} \right].$$

Therefore

$$u^{\lambda+\mu}(x) \leq E_x^1 \left[\int_0^{\tau^\lambda} e^{-\alpha s} f(x_s)ds + [(\lambda+\mu) + u^{\lambda+\mu}(\bar{x})]\, e^{-\alpha\tau^\lambda} \right].$$

Hence

$$u^{\lambda+\mu}(x) \leq u^\lambda(\bar{x}) + \mu E_x^1\, e^{-\alpha\tau^\lambda} + [u^{\lambda+\mu}(\bar{x}) - u^\lambda(\bar{x})]\, E_x^1\, e^{-\alpha\tau^\lambda}$$

$$u^{\lambda+\mu}(x) - u^{\lambda+\mu}(\bar{x})\, E_x^1\, e^{-\alpha\tau^\lambda} \leq u^\lambda(\bar{x}) - u^\lambda(\bar{x}) E_x^1\, e^{-\alpha\tau^\lambda} + \mu E_x^1\, e^{-\alpha\tau^\lambda}.$$

Then

$$u^{\lambda+\mu}(x) - u^{\lambda+\mu}(\bar{x}) \leq u^\lambda(\bar{x}) - u^\lambda(\bar{x})\, E_x^1\, e^{-\alpha\tau^\lambda} + \mu E_x^1\, e^{-\alpha\tau^\lambda}.$$

Therefore, if

$$u^{\lambda}(x) < \lambda + u^{\lambda}(\bar{x}) \text{ , we get}$$

$$u^{\lambda+\mu}(x) - u^{\lambda+\mu}(\bar{x}) < \lambda + \mu .$$

This implies

$$\{x \mid u^{\lambda}(x) < \lambda + u^{\lambda}(\bar{x})\} \subset \{x \mid u^{\lambda+\mu}(x) < \lambda + \mu + u^{\lambda+\mu}(\bar{x})\}$$

and $\tau^{\lambda+\mu} \geq \tau^{\lambda}$, which was to prove.

Now the rest of the results follows by reducing the problem to stopping problem. Indeed, knowing that for Π_{λ}, an optimal control is given by

$$\tau_n^{\lambda} = \tau_{n-1}^{\lambda} + \tau^{\lambda} o \theta_{\tau_{n-1}^{\lambda}} .$$

If we set

$$\beta_{\lambda} = E_{\bar{x}} \, e^{-\alpha\tau^{\lambda}} ,$$

the constraint we want to be satisfied will be expressed by

$$\frac{\beta_{\lambda}}{1 - \beta_{\lambda}} \leq c_0$$

(4.11) or

$$\beta_{\lambda} \leq \frac{c_0}{1 + c_0}$$

Furthermore, (4.7) gives, using Dynkin's formula

$$(4.12) \quad u^{\lambda}(x) = u^{\lambda}(\bar{x}) + \operatorname{Inf}_{\tau} E_x^1 \left[\int_0^{\tau} e^{-\alpha s} \, [f(x_s) - \alpha u^{\lambda}(\bar{x})] ds + e^{-\alpha\tau} \right] .$$

Therefore the results of lemma 2.4 and Theorem 2.2 are valid for

$$v^{\lambda}(x) = u^{\lambda}(x) - u^{\lambda}(\bar{x}) \text{ with the constraint (4.11)}$$

We obtain the following

<u>Theorem 4.2.</u> Under the assumptions of theorem 4.1. the problem P_0 (see(4.8)) has an optimal randomized solution.

Remark : We studied above a very special case of impulsive control. There is not yet any result on more general impulsive control problem as those described in [1] or [4].

REFERENCES

[1] A. BENSOUSSAN, J.L. LIONS, Temps d'arrêt optimal et inéquations variationnelles, Dunod - Paris - 1978.

[2] E. DYNKIN, Markov Processes. Springer Verlag, 1965 (Vol. 1).

[3] E. FRID, On optimal strategies in control problems with constraints. Theor. of Probability and Applications. Vol. , n° 1, 1972.

[4] M. ROBIN, Contrôle impulsionnel des processus de Markov. Doctoral Thesis. Paris, 1978.

[5] N.SHYRIAIEV,Sequential statistical analysis. A.M.S. Publication. Providence, 1973.

[6] M. SAFONOV, Control of a Wiener process when the number of switches is bounded Theor. Prob. and Applic. Vol. 21, n° 3 - 1976.

This list of references cannot pretend to be an exhaustive one. Extensive bibliographies can be found in [1], [4] (and also [5] for optimal stopping problems).

GAME THEORY AND RELATED TOPICS,
O. Moeschlin, D. Pallaschke (eds.)

MULTIFUNCTIONS AND OPTIMIZATION

S.Rolewicz
Institute of Mathematics
Polish Academy of Sciences

Let X,Y be two metric spaces. Let Γ be a multifunction $\Gamma : Y \longrightarrow 2^X$ and let f be a real valued function $f : X \longrightarrow R$. By the general optimization problem we mean the following problem.

$$(1) \quad f(x) \longrightarrow \inf, \quad x \in \Gamma y_o$$

The present above denotation is now in a common use in engineering optimization. In fact it contains several problems, for example

a) to find $\inf\limits_{x \in \Gamma y_o} f(x)$

b) to find x_o realizing this infimum

c) to find effective algorithms for calculations and so on.

In the present talk the problems a) and b) will be considered only.

Dolecki and Kurcyusz [1] have introduced for problem (1) a notion of generalized Lagrangean

$$(2) \quad L(x,\varphi,y_o) = f(x) - \sup_{y \in \Gamma^{-1}x} \varphi(y) + \varphi(y_o),$$

where φ belongs to a fixed class Φ of real valued functions defined on Y. About Φ we shall make only one assumption, that it is invariant under addition of a constant, $\Phi + c = \Phi$.

We shall recall that $\Gamma^{-1}x = \{y : x \in \Gamma y\}$ Dolecki and Kurcyusz have proved

THEOREM 1. [1] Let

$$(3) \quad \overline{f\Gamma}(y) \overset{df}{=} \inf_{x \in \Gamma y} f(x)$$

be so called primal functional. Then

(4) $\inf_{x \in \Gamma y_o} f(x) = \sup_{\varphi \in \Phi} \inf_{x \in X} L(x,\varphi,y_o)$

if and only if

(5) $\overline{f\Gamma}(y) = \text{sub}\{\varphi \in \Phi : \varphi \leq \overline{f\Gamma}\}$

THEOREM 2 [1]. There is $\varphi_o \in \Phi$ such that

(6) $\inf_{x \in \Gamma y_o} f(x) = \inf L(x,\varphi_o,y_o)$

if and only if $\overline{f\Gamma}(y)$ satisfies (5) and moreover there is φ_o such that

(7) $\overline{f\Gamma}(y) - \overline{f\Gamma}(y_o) \geq \varphi_o(y) - \varphi_o(y_o)$.

Till this moment I was not speaking about topological properties of f and Γ. All topological difficulties were shifted to investigations of properties of primal functional $\overline{f\Gamma}(y)$.

It would be interested to find properties warrantying (5) and (7) for different classes of functions Φ. The most important are

a) class of affine functionals giving classical duality theory for convex problems

b) classes $\Phi_i = \{\varphi : \varphi(y) = c - r\,\rho^i(y,\bar{y})$ c-real, r-real positive, $\bar{y}$ fixed element of $Y\}$ $1 \leq i < +\infty$

Essential role play classes Φ_2 and Φ_1. The class Φ_2 is very convenient for numerical consideration in Hilbert spaces. The class is not so convenient from nümerical point of view, but a lot of theorems were proved for it.

In the present talk we shall concentrate ourselves on the class Φ_1.

Observe that if f is Lipschitzian and Γ satisfies Lipschitz condition i.e. there is $K > 0$ such that

(8) $\Gamma(B(y_o,r)) \leq B(\Gamma y_o,Kr)$

where $B(A,\varepsilon) = \{y : \rho(A,y) \leq \varepsilon\}$ then $\overline{f\Gamma}(y)$ satisfies the Lipschit condition, too. Thus there is $\varphi_o \in \Phi_1$ satisfying (7).

In many problem we are interested in local minima. Let $x_o \in \Gamma y_o$. If there is a neighbourhood Q of x_o such that f is Lipshitzian on Q and $\Gamma y \cap Q$ is Lipschitzian, then using preceding consideration we trivially obtain the existence of φ_o satisfying (7).

Unfortunately the fact that Γ is Lipschitzian does not imply that $\Gamma y \cap Q$ is Lipschitzian, too. (see [2]). The main problem is that the intersection of two Lipschitzian multifunction is need not be Lipschitzian.

We say that a set A has $CP(c,x_o)$ property if for all $x \in A$ and all α, $0 < \alpha < 1$,

$$(9) \quad B(\alpha x_o + (1-\alpha)x,\ c\|x - x_o\|\alpha) \subset A$$

We say that a set A is $SC(d,x_o)$ if for all $x \in A$ and α, $0 \le \alpha \le 1$, there is x_α such that

$$(10) \quad \|(\alpha x_o + (1-\alpha)x) - x_\alpha\| < d\|x - x_o\| \cdot \alpha.$$

THEOREM 3. ([4]). Let Γ_i $i=1,2$ be Lipschitzian at y_o. Let $x_o \in \Gamma_1 y_o \cap \Gamma_2 y_o$. Let $\Gamma_1 y_o$ have $CP(c,x_o)$ property and $\Gamma_2 y_o$ be $SC(d,x_o)$. If $d < c$, then $\Gamma_1 y \cap \Gamma_2 y$ is Lipschitzian at y_o.

Theorem 3 requests of certain type regulatity conditions. Without this assumption we are able only to prove.

THEOREM 4. ([3]). Let Γy be Lipschitzian at y_o with constant K. Let $x_o \in \Gamma y_o$. Then for every neighbourhood Q of x_o and $c > 1$ there is a neighbourhood Q_o of x_o such that the multifunction $Q_o \cap \Gamma y$ is Lipschitzian at y_o with a constant cK.

Basing on Theorem 4 we proved

THEOREM 5. [3]. Let X,Y be as before. Suppose that Y has the following property

(*) for all $0 < \alpha < 1$, there is $k(\alpha)$, $0 < k(\alpha) < 1$ such that for all $y_1, y_2 \in Y$, there is $\bar{y}$ such that

(a) $\rho(y_1,\bar{y}) \le \alpha\rho(y_1,y_2)$

$$\rho(y_2,\bar{y}) \leq (1-k(\alpha)\alpha)\rho(y_1,y_2)$$

Moreover, let $f(x)$ be a locally Lipschitzian at x_o and a multifunction $\Gamma : Y \longrightarrow 2^X$ satisfies the following conditions.

(i) Γy is closed for all $y \in Y$

(ii) there is a neighbourhood Q of x_o and a constant K and a neighbourhood W of y_o, and $r_o > 0$ such that

$$Q \cap \Gamma(B(y,r)) \subset B(\Gamma y, Kr)$$

for $r < r_o$ and $y \in W$.

(iii) for all $r > 0$ there is $s > 0$ such that

$$\Gamma(B(y_o,r)) \supset B(x_o,s)$$

Then there is a $\varphi_o \in \Phi_1$ and a neighbourhood Q_o of x_o such that all local minima in Q_o of functions $f(x)$ and $L(x,\varphi_o,y_o)$ are those same.

It is interesting to known which class of metric spaces satisfies (*). Is this class topologically the same as the class of convex spaces ?

References

[1] S.Dolecki - S.Kurcyusz, On Φ-convexity in extremal problems, SIAM Jour. of Control, 16(1978), pp. 277-300.

[2] S.Dolecki - S.Rolewicz, Metric characterization of semicontinuity - preserving multifunctions, Jour. of Math. Anal and Appl. (to appear).

[3] S.Dolecki - S.Rolewicz, Exact penalty for local minima (to appear).

[4] S.Rolewicz, On intersections of multifunctions, Mathematische Operationsforschung und Statistik - Series Optimization (to appear).

GAME THEORY AND RELATED TOPICS
O. Moeschlin, D. Pallaschke (eds.)

EIN KOMBINATORISCHER UMSCHLIESSUNGSSATZ NEBST ANWENDUNGEN

E. Sperner [1)]

University of Hamburg

Die Kombinatorik ist nicht arm an Aussagen, die, obgleich selbst rein kombinatorischen Charakters, Anwendung in anderen Gebieten der Mathematik gefunden haben. Viele kombinatorischen Sätze sind überhaupt erst als Hilfsmittel für andere Fragen entstanden. Bei manchen von ihnen hat sich später der Umkreis der Anwendungen noch erheblich erweitert.

Ein Beispiel dafür ist jenes Lemma, das der Verfasser 1928 für den Beweis tiefliegender und damals keineswegs leicht zugänglicher topologischer Fragen aufgestellt und benutzt hat [2)](vgl.[1]!). Es hat sich teils bald danach(vgl. [2], [3] !) teils viele Jahre später für weitere Anwendungen als nützlich erwiesen (vgl. z.B. [7] bis [25]!).

Außerdem traten später in einer Parallelentwicklung (A.W. TUCKER 1945, KY FAN 1952, vgl. [4] bis [6]!) verwandte Resultate auf, die mit ähnlichen Beweismethoden gewonnen werden konnten und ihrerseits neuen topologischen Anwendungen dienten.

Die damit erreichte Vielfalt von Aussagen, welche gleichartiger Beweistechnik zugänglich waren, weckte das Bedürfnis nach einer einheitlichen Grundlegung. Der Versuch dazu wurde von mehreren Autoren unternommen (KY FAN [26], P. MANI [27], E. SPERNER [28]). Dabei blieb aber eine Weiterentwicklung noch unberücksichtigt, welche die erwähnten Resultate, die zunächst für "Komplexe o h n e Orientierung" (siehe unten § 8) aufgestellt wurden, auf solche "m i t Orientierung" ausdehnte (A.B. BROWN, S.S. CAIRNS [29], KY FAN [30], [31]).

Es ist das Ziel dieser Mitteilung, die allgemeine Grundlage, wie sie z.B. in [28] erreicht wurde, so zu erweitern, daß auch die Resultate, die sich auf orientierte Komplexe beziehen, mit erfaßt werden. Das Ergebnis ist ein kombinatorischer "Umschließungs"-Satz (§ 6) von großer Allgemeinheit und mit weitverzweigter Anwendbarkeit.

Die nachfolgenden Ausführungen haben zwar Beziehungen zur kombinatorischen (algebraischen) Topologie, die Darstellung ist aber völlig unabhängig davon und setzt keinerlei Kenntnis davon voraus.

§ 1. Orientierung auf Mengen

Wir beschäftigen uns mit Mengen, denen in bestimmter Weise eine Orientierung aufgeprägt wird. Das hierbei benutzte Verfahren soll am Beispiel einer (endlichen oder unendlichen) Menge

(1) $$A = \{a_i\}_{i \in J}, \qquad J = \text{Indexmenge}, \ |J| \le \infty$$

beschrieben werden. Wir betrachten die geordneten endlichen Mengen [3)]

1) Prof. Dr. E. Sperner, Rosenweg 1, D-7811 Sulzburg - Laufen, Fed.-Rep. of Germany.
2) Die Nummern in eckigen Klammern beziehen sich auf das Literaturverzeichnis am Ende der Abhandlung.
3) Mit $\mathbb{N}$ bezeichnen wir die Menge der ganzen Zahlen ≥ 0.

(2) $S^m = (a_1, a_2,\dots,a_m)$, $m \in \mathbb{N}$, $a_i \in A$, $a_i \neq a_k$ für $i \neq k$

Ein solches S^m nennen wir ein m-Eck auf A. (Man beachte, daß wir die m Elemente $a_1, a_2, \dots a_m$ in S^m stets alle als verschieden voraussetzen!) Sodann denken wir uns für jedes feste $m \in \mathbb{N}$ eine Indexmenge $\mathbb{P}_m$ gegeben, deren Mächtigkeit mit der der (ungeordneten) Teilmengen von A, welche m verschiedene Elemente aus A enthalte übereinstimmt, also

$$|\mathbb{P}_m| = |\{X \subset A : |X| = m\}|.$$

Schließlich greifen wir für festes m aus der G e s a m t h e i t $\{S^m\}$ der Mengen (2) eine Untermenge (*"Repräsentantenmenge"*)

(3) $\mathfrak{A}_m = \{S^m_\rho\}_{\rho \in \mathbb{P}_m}$

v o n d e r A r t heraus, daß sie der folgenden Forderung Genüge leistet:

F: *Zu jedem* S^m *der Art (2) gibt es* g e n a u e i n $\rho \in \mathbb{P}_m$, $\rho = \rho(S^m)$, *für welches* S^m_ρ *eine Permutation* $\pi(S^m)$ *von* S^m *darstellt.*

Die damit n a c h der Wahl von $\mathfrak{A}_m$ eindeutig erklärte Abbildung

(4) $S^m \to \rho = \rho(S^m) \in \mathbb{P}_m$

ist für das folgende grundlegend; dazu gehört weiter die Festsetzung:

Wenn für ein S^m gilt: $S^m_\rho = \pi\,(S^m)$ mit $\rho = \rho(S^m)$ gemäß (4), so setzen wir

(5) $S^m = \eta S^m_\rho$ mit $\begin{cases} \eta = +1, \text{ wenn } \pi \text{ eine gerade Permutation ist,} \\ \eta = -1, \text{ wenn } \pi \text{ eine ungerade Permutation ist.} \end{cases}$

Mit den Abbildungen (4), die für jedes $m \in \mathbb{N}$ zu definieren sind, und der jeweils zugehörigen Festsetzung (5) werden wir wiederholt zu operieren haben. Man beachte daß die Abbildungen (4) wohldefiniert sind, sobald die Repräsentantenmengen $\mathfrak{A}_m$ für alle $m \in \mathbb{N}$ gewählt sind!

Die zugrunde gelegte Menge A behalten wir bei und nennen sie im folgenden "*Grundmenge*". Wir benötigen aber noch eine zweite (beliebig wählbare, aber dann festzulegende) Menge

(1') $B = \{b_j\}_{j \in J'}$, J' = Indexmenge, $|J'| \leq \infty$,

welche "*Bewertungsmenge*" heißen soll.

Der Menge B prägen wir mit genau demselben Verfahren wie bei A eine Orientierung auf, nur verwenden wir zur besseren Unterscheidung andere Buchstaben. Entsprechend (2) betrachten wir die m-Tupel:

(2') $T^m = (b_1, b_2, \dots, b_m)$, $m \in \mathbb{N}, b_i \in B, b_i \neq b_k$ für $i \neq k$.

Die den $\mathbb{P}_m$ entsprechenden Indexmengen sollen jetzt mit Γ_m $(m \in \mathbb{N})$ und die Repräsentantenmengen (3) mit

(3') $\mathfrak{B}_m = \{T^m_\sigma\}_{\sigma \in \Gamma_m}$

bezeichnet werden. Auch die $\mathfrak{B}_m$ sollen der Forderung F (mit entsprechend geänderten Bezeichnungen) genügen, so daß dadurch wieder die (4) analogen Abbildungen

(4') $T^m \to \sigma = \sigma(T^m) \quad \in \Gamma_m$

und ebenso π aus $T^m_\sigma = \pi(T^m_\sigma)$ eindeutig definiert sind und wir entsprechend (5) setzen können:

(5') $$T^m = \eta T^m_\sigma \quad \text{mit } \eta \begin{cases} = +1, \text{ wenn } \pi \text{ eine gerade Permutation ist} \\ = -1, \text{ wenn } \pi \text{ eine ungerade Permutation ist.} \end{cases}$$

§ 2. Zwei Vektorräume.

Die in § 1 für die zwei Mengen A, B gebildeten Repräsentantenmengen $\mathfrak{A}_m$, $\mathfrak{B}_m$ benutzen wir nun für jedes $m \in \mathbb{N}$ als Basissysteme zweier Vektorräume. Die "Komponenten" der zu bildenden "Vektoren" beziehen wir aus einem beliebig wählbaren (nicht notwendig kommutativen) Ring[4] G mit Einselement, den wir von jetzt an fest gegeben denken.

Die Elemente K^m des ersten Vektorraums $\mathcal{K}_m$ schreiben wir in der Form

(6) $$K^m = \sum_{\rho \in \mathbb{P}_m} \alpha_\rho S^m_\rho ,$$

wobei

(6a) $\alpha_\rho \in G, \quad S^m_\rho \in \mathfrak{A}_m, \quad |\{\rho \in \mathbb{P}_m : \alpha_\rho \neq 0\}| < \infty$

sein soll und die $\mathbb{P}_m$ die in § 1 in (3) eingeführten Indexmengen sind. Die letzte Bedingung in (6a) verlangt, daß jeder "Vektor" K^m nur höchstens endlich viele "Komponenten" α_ρ besitzt, die $\neq 0$ sind.

Gerade das letzte setzen wir n i c h t voraus für die Elemente C^m des zweiten Vektorraumes $\mathcal{L}_m$, die wir mit Hilfe der Basis $\mathfrak{B}_m$ folgendermaßen schreiben:

(7) $$C^m = \sum_{\sigma \in \Gamma_m} \beta_\sigma T^m_\sigma , \qquad \beta_\sigma \in G,\ T^m_\sigma \in \mathfrak{B}_m.$$

Hierin sind die Γ_m wieder die schon in (3') benutzten Indexmengen.

Für $\mathcal{K}_m$ und $\mathcal{L}_m$ führen wir die üblichen Rechenoperationen ein, nämlich zunächst eine *Addition*: für ein Element der Form (6) und ein zweites

$$'K^m = \sum_{\rho \in \mathbb{P}_m} \alpha'_\rho S^m_\rho$$

durch die Festsetzung

(8) $$K^m + {'K^m} = \sum_{\rho \in \mathbb{P}_m} (\alpha_\rho + \alpha'_\rho)\, S^m_\rho .$$

Genauso wird natürlich in $\mathcal{L}_m$ ein Element (7) und $'C^m = \sum_{\sigma \in \Gamma_m} \beta'_\sigma T^m_\sigma$ addiert gemäß

---

4) Von G setzen wir also voraus, daß G eine nicht leere Menge ist mit
1. einer kommutativen und assoziativen Addition
2. einer assoziativen, beidseitig distributiven Multiplikation mit neutralem Element 1.

(9) $$C^m + {}'C^m = \sum_{\sigma\in\Gamma_m} (\beta_\sigma + \beta'_\sigma)\, T^m_\sigma .$$

Sodann wird die *skalare Links-Multiplikation* von (6) bzw. (7) mit einem Element $\lambda \in G$ definiert durch

(10) $$\lambda\, K^m = \sum_{\rho\in \mathbb{P}_m} (\lambda\alpha_\rho)\, S^m_\rho , \qquad \lambda\, C^m = \sum_{\sigma\in \Gamma_m} (\lambda\beta_\sigma)\, T^m_\sigma .$$

Wir nennen [5] ein

Element der Art (6) einem *m-Mengenkomplex,*

Element der Art (7) einem *m-Wertekomplex.*

Schließlich treffen wir noch folgende

<u>Konvention:</u> *Es sei* $\rho_0 \in \mathbb{P}_m$ *fest gewählt. Dann wird für dasjenige* K^m*, für welches in (6)*

$$\alpha_\rho = 0 \text{ für alle } \rho \neq \rho_0, \text{ aber } \alpha_{\rho_0} = 1$$

gilt, gesetzt:

(11) $$K^m = \sum_{\rho\in \mathbb{P}_m} \alpha_\rho S^m_\rho \equiv S^m_{\rho_0} .$$

D.h. In diesem Falle wird K^m mit $S^m_{\rho_0}$ identifiziert.

§ 3. Randoperator.

Den Begriff des *Randoperators* erklären wir (wie in der kombinatorischen Topologie) als Abbildung $\partial : \mathcal{K}_m \to \mathcal{K}_{m-1}$ in mehreren Schritten. Zunächst wird für ein einzelnes[6] $S^m_\rho = (a_1, a_2, \ldots, a_m)$ definiert:

(12) $$\partial S^m_\rho = \sum_{k=1}^{m} (-1)^{k+1} (a_1, \ldots, a_{k-1}, a_{k+1}, \ldots, a_m).$$

Um hierin die Glieder der rechten Seite als Linearkombination der Basis darzustellen, bestimmen wir für jedes

(13) $$S^{m-1}_k = (a_1, \ldots, a_{k-1}, a_{k+1}, \ldots, a_m) , \quad k=1,2,\ldots,m,$$

das durch (4) eindeutig erklärte $\rho_k \in \mathbb{P}_{m-1}$, für welches $\rho_k = \rho(S^{m-1}_k)$ gilt. Dazu gehört dann gemäß der Vorschrift (5) ein wieder eindeutig bestimmtes Element $\eta_k = \pm\, 1 \in G$, für welches

$$S^{m-1}_k = \eta_k\, S^{m-1}_{\rho_k}$$

wird. Trägt man dies für jedes Glied (13) in (12) ein, so erhält man mit der Abkürzung $\varepsilon_k = (-1)^{k+1}\eta_k$ weiter

5) Obwohl bekanntlich gleichartige Bildungen der algebraischen Topologie "Ketten" heißen, ziehen wir hier die obigen Namen vor, da sie unserem Gebrauch dieser Gebilde besser entsprechen.

6) Vgl. (11)!

(14) $\partial S^m_\rho = \sum_{k=1}^{m} \varepsilon_k S^{m-1}_{\rho_k}$.

Gemäß (11) ist damit ∂S^m_ρ als ein Element $K^{m-1} \in \mathcal{K}_{m-1}$ dargestellt.
Schließlich definieren wir ∂ für jedes Element der Form (6) durch

(15) $\partial K^m = \sum_{\rho \in \mathbb{P}_m} \alpha_\rho \partial S^m_\rho$.

Da hierin (wie in (6)) nur für höchstens endlich viele Glieder $\alpha_\rho \neq 0$ gilt, ist (15) nach (8) wohl definiert. Damit ist ∂ als Abbildung von $\mathcal{K}_m$ in $\mathcal{K}_{m-1}$ vollständig erklärt.

§ 4. Bewertungen.

Unter einer "*Bewertung*" verstehen wir eine Abbildung

(16) $\varphi : A \to B$.

Bei der durch

(17) $\varphi(S^m) = (\varphi(a_1), \varphi(a_2), \ldots, \varphi(a_m))$

beschriebenen Wirkung eines solchen φ auf ein m-Eck der Art (2) ist zu beachten, daß die rechte Seite von (17) in unserer Sprechweise nur dann wieder ein m-Eck auf B darstellt, wenn stets $\varphi(a_i) \neq \varphi(a_k)$ ist für $i \neq k$ (vgl. (2')!); in diesem Fall wollen wir $\varphi(S^m)$ "*regulär*" nennen, sonst "*irregulär*".

Jede Bewertung φ induziert gewisse Folgeabbildungen, nämlich eine Abbildung $\varphi'_m : \mathbb{P}_m \to \Gamma_m$ und eine weitere $\bar{\varphi}_m : \mathbb{P}_m \to G$ (für jedes $m \in \mathbb{N}$).

Die Abbildung φ'_m wird folgendermaßen erklärt. Man wähle ein festes $\sigma_m \in \Gamma_m$ und setze

(18a) $\varphi'_m(\rho) = \sigma_m$, wenn $\varphi(S^m_\rho)$ irregulär ist.

Wenn aber $\varphi(S^m_\rho)$ regulär, also ein m-Eck T^m auf B ist, so gehört dazu nach (4') ein $\sigma_\rho = \sigma(\varphi(S^m_\rho))$ eindeutig, mit dem wir dann setzen

(18b) $\varphi'_m(\rho) = \sigma_\rho$, wenn $\varphi(S^m_\rho)$ regulär ist.

Für d i e s e s σ_ρ gilt übrigens nach (5'):

(19) $\varphi(S^m_\rho) = \eta^{(m)}_\rho T^m_{\sigma_\rho}$, $\eta^{(m)}_\rho = \pm 1 \in G$.

Damit definieren wir weiter $\bar{\varphi}_m : \mathbb{P} \to G$, indem wir setzen:

(20) $$\bar{\varphi}_m(\rho) = \begin{cases} \eta^{(m)}_\rho, & \text{wenn } \varphi(S^m_\rho) \text{ regulär ist,} \\ 0, & \text{wenn } \varphi(S^m_\rho) \text{ irregulär ist.} \end{cases}$$

Die Abbildungen $\varphi'_m, \bar{\varphi}_m$ benötigen wir, um einen weiteren, für unsere Zwecke wichtigen Begriff zu erklären. Wir denken uns ein Komplex-Paar(K^m, C^m) in der Form (6) und (7) gegeben, sowie eine Bewertung φ: $A \to B$. Dann bilden wir mit den α_ρ aus (6), den β_σ aus (7) und den eben erklärten φ'_m, $\bar{\varphi}_m$ den Ausdruck

$$(21)\qquad \Phi\begin{matrix}C^m\\K^m\end{matrix} := \sum_{\rho\in\mathbb{P}_m} \alpha_\rho \beta_{\varphi'_m(\rho)} \bar{\varphi}_m(\rho).$$

Da nur endlich viele α_ρ von Null verschieden sind, stellt (21) ein wohldefiniertes Ringelement dar, welches wir den "*Effekt*" der Bewertung φ bezüglich (K^m, C^m) nennen wollen.

In die Definition (21) gehen zwar die Repräsentantenmengen $\mathfrak{A}_m$, $\mathfrak{B}_m$ ein, der Wert von (21) ist aber von der Wahl der $\mathfrak{A}_m, \mathfrak{B}_m$ unabhängig und allein durch φ, K^m, C^m bestimmt. Denn ändert man bei festgehaltenen φ, K^m, C^m die $\mathfrak{A}_m$, $\mathfrak{B}_m$ (im Rahmen der in § 1 dafür aufgestellten Bedingungen), so ändern sich zwar möglicherweise die α_ρ, β_σ, $\bar{\varphi}_m(\rho)$, aber jedes Produkt $\alpha_\rho \beta_{\varphi'_m(\rho)} \bar{\varphi}_m(\rho)$ bleibt dabei invariant und damit auch die Summe in (21). (vgl. [33]!).

Übrigens verhält sich (21) additiv sowohl in K^m *als auch in* C^m, *d.h. für beliebige* K^m, K_1^m, K_2^m, C^m, C_1^m, C_2^m *gelten die Gleichungen*

$$(22)\qquad \Phi\begin{matrix}C^m\\K_1^m\end{matrix} + \Phi\begin{matrix}C^m\\K_2^m\end{matrix} = \Phi\begin{matrix}C^m\\K_1^m + K_2^m\end{matrix}, \qquad \Phi\begin{matrix}C_1^m\\K^m\end{matrix} + \Phi\begin{matrix}C_2^m\\K^m\end{matrix} = \Phi\begin{matrix}C_1^m + C_2^m\\K^m\end{matrix}.$$

Der einfache Beweis fußt auf den Distributivgesetzen von G.

§ 5. Δ - Operator

Wir benötigen noch eine Abbildung, welche in der Bewertungsmenge B operiert und jeden Vektorraum $\mathcal{L}_m$ (dessen Elemente durch (7) dargestellt werden) in den Vektorraum $\mathcal{L}_{m+1}$ abbildet. Bedeutung und Zweck dieser Abbildung, welche wir mit Δ bezeichnen und Δ-*Operator* nennen wollen, wird im Anschluß an unseren Hauptsatz in den §§ 6 - 8 erkennbar werden.

Wir gehen aus von einem m-Wertekomplex

$$(23)\qquad C^m = \sum_{\sigma\in\Gamma_m} \beta_\sigma^{(m)} T_\sigma^m, \qquad \beta_\sigma^{(m)} \in G,\ T_\sigma^m \in \mathfrak{B}_m,$$

dem, wie gesagt, durch den Δ-Operator eindeutig ein (m+1)-Wertekomplex

$$(24)\qquad \Delta\, C^m = \sum_{\sigma\in\Gamma_{m+1}} \beta_\sigma^{(m+1)} T_\sigma^{m+1}, \qquad \beta_\sigma^{(m+1)} \in G,\ T_\sigma^{m+1} \in \mathfrak{B}_{m+1},$$

zugeordnet werden soll. Die Koeffizienten $\beta_\sigma^{(m)}$ in (23) betrachten wir als g e g e b e n, die $\beta_\sigma^{(m+1)}$ in (24) als g e s u c h t. Dementsprechend definieren wir Δ durch Angabe eines Verfahrens zur Berechnung der Koeffizienten $\beta_\sigma^{(m+1)}$ aus den $\beta_\sigma^{(m)}$. Zu dem Zweck greifen wir aus der Summe in (24) ein einzelnes Glied heraus und schreiben dafür

$$(25)\qquad T_\sigma^{m+1} = (b_0, b_1, b_2, \ldots, b_m), \qquad \sigma \in \Gamma_{m+1}.$$

Daraus bilden wir weiter die folgenden m-Wertekomplexe

(26) $T_k^m = (b_o, \ldots, b_{k-1}, b_{k+1}, \ldots b_m), \qquad k=o,1,\ldots,m,$

welche aus (25) durch S t r e i c h e n einer Komponente b_k, $0 \le k \le m$, hervorgehen. Zu jedem solchen T_k^m gehört nach (4') ein eindeutig bestimmtes $\sigma_k \in \Gamma_m$, zu welchem nach (5') ein (für jedes k wieder eindeutig bestimmtes) $\eta_k(\sigma) \equiv \eta_k = \pm 1 \in G$ gehört, so daß eine Gleichung

(27) $T_k^m = \eta_k T_{\sigma_k}^m, \qquad k = o,1,2,\ldots,m,$

für jedes k besteht.

Auf diese Weise haben wir mittels der Festsetzungen (25) bis (27) zu jedem $\sigma \in \Gamma_{m+1}$ genau (m+1) wohlbestimmte Elementepaare (σ_k, η_k), $k = o,1,2,\ldots,m$, $\sigma_k \in \Gamma_m$, $\eta_k = \pm 1$, erhalten, mit denen wir folgendermaßen weiter operieren. Wir greifen aus (23) zu jedem σ_k den Koeffizienten $\beta_{\sigma_k}^{(m)}$ heraus und bilden mit diesen $\beta_{\sigma_\kappa}^{(m)}$ den Ausdruck

(28) $\beta_\sigma^{(m+1)} = \sum_{k=o}^{m} (-1)^k \eta_k \beta_{\sigma_k}^{(m)}.$

Durch (28) sind die Koeffizienten in (24) in eindeutiger Weise definiert und damit ebenso der Operator Δ.

Es ist noch anzumerken, daß zur Definition und Berechnung von Δ zwar feste Repräsentantenmengen $\mathscr{L}_m$ benutzt werden müssen, der Wert von Δ aber von der Wahl der $\mathscr{L}_m$ nicht abhängt. Denn ändert man bei festhaltenem C^m die Wahl der Mengen $\mathscr{L}_m$, (im Rahmen der in § 1 festgelegten Bedingungen), so ändern sich zwar i.A. auch die Elemente η_k, $\beta_{\sigma_k}^{(m)}$, $\beta_\sigma^{(m+1)}$, aber ΔC^m bleibt dabei ungeändert. (Für den Beweis dieser Invarianzaussage vgl. (33)!)

Auch Δ *verhält sich wieder additiv in* C^m, *d.h. es gilt für beliebige Wertekomplexe* C_1^m, C_2^m

(29) $\Delta(C_1^m + C_2^m) = \Delta C_1^m + \Delta C_2^m.$

§ 6. Ein Umschließungssatz

Nunmehr haben wir alle Begriffe und Bezeichnungen bereit, um unser Hauptresultat kurz formulieren zu können. Es lautet:

Satz 1. *Für beliebige Wahl von* K^{m+1}, C^m, φ *gilt:*

(30) $\Phi_{\partial K^{m+1}}^{C^m} = \Phi_{K^{m+1}}^{\Delta C^m}.$

Bei der Gleichung (30) handelt es sich um eine reine Identität, deren Beweis man durch Rückgang auf die Definitionen der in (30) auftretenden Begriffe gewinnt. Dabei bedient man sich mit Vorteil des Additionstheorems (22). Denn dadurch läßt sich (30) zurückführen auf den einfachen Sonderfall, daß von den Koeffizienten α_ρ des Komplexes K^{m+1} nur ein einziger $\neq 0$ ist, daß also K^{m+1} ein (m+1)-Eck ist (im

Sinne von (11)). Bezüglich der Einzelheiten des Beweises verweisen wir auf [33].

Wir nennen Satz 1 deshalb einen "Umschießungssatz", weil damit (von links nach rechts gelesen) von einer Aussage über das Verhalten des Randes, d.i. der "Umschließung" des Komplexes K^{m+1} auf ein entsprechendes Verhalten des Komplexes K^{m+1} selbst geschlossen werden kann. Hierbei ist dann ein wesentlicher Schritt, daß zu gegebenem C^m das ihm zugeordnete ΔC^m bestimmt wird. Beispiele hierfür sind in den nächsten Paragraphen enthalten.

§ 7. Spezialisierung der Bewertungen.

Die große Allgemeinheit von Satz 1 bedingt die Weite seines Anwendungsbereiches. Bei den verschiedenen Anwendungen hat man jeweils die zugrunde liegenden Mengen und Begriffe in geeigneter Weise zu spezialisieren.
Als Beispiel behandeln wir eine Spezialisierung der Bewertungsmenge B aus (1') und des Bewertungskomplexes C^m, welche an das Lemma von 1928 anknüpft (vgl.[1]). Wir setzen nämlich:

(31) $$B = \{0,1,2,\ldots,m\}, \qquad C^m = (1,2,\ldots,m) \equiv T_0^m.$$

Die Bewertungsmenge B besteht also jetzt nur aus den m+1 Ziffern 0,1,...,m, während C^m nur aus einem einzigen m-Tupel besteht, d.h. daß in (7) alle Koeffizienten verschwinden bis auf den von T_0^m welcher = 1 ist. Alle übrigen Mengen und Begriffe behalten vorläufig ihre bisherige Allgemeinheit. Es folgt:

(31a) $$\Delta C^m = (0,1,2,\ldots,m).$$

Um den Inhalt von Satz 1 bei dieser Spezialisierung deutlicher hervortreten zu lassen, setzen wir

$$K^{m+1} = \sum_{\rho\in\mathbb{P}_{m+1}} \alpha_\rho S_\rho^{m+1}, \qquad \partial K^{m+1} = \sum_{\rho\in\mathbb{P}_m} \alpha'_\rho S_\rho^m,$$

und bedienen uns der folgenden Indexmengen[7)]

(32) $$I = \{\rho\in \mathbb{P}_{m+1} : \alpha_\rho \neq 0,\ \bar\varphi_{m+1}(\rho) \neq 0\}, \qquad I' = \{\rho\in\mathbb{P}_m : \alpha'_\rho \beta_{\varphi'_m(\rho)} \bar\varphi_m(\rho) \neq 0\}$$

um die beiden Seiten von (30) in der folgenden Form zu schreiben

(33) $$\Phi^{C^m}_{\partial K^{m+1}} = \sum_{\rho\in I'} \alpha'_\rho \eta_\rho^{(m)}, \qquad \Phi^{\Delta C^m}_{K^{m+1}} = \sum_{\rho\in I} \alpha_\rho \eta_\rho^{(m+1)}.$$

Die Schreibweise (33) für die in Frage stehenden Effekte ergibt sich aus der Definition (21), wenn man nur die Summen-Glieder aufschreibt , welche ≠ 0 sind und insbesondere zu regulären $\varphi(S_\rho^m)$, $\varphi(S_\rho^{m+1})$ gehören. Eben deswegen können auch nach (20) für die Werte $\bar\varphi_m(\rho)$, $\bar\varphi_{m+1}(\rho)$ die $\eta_\rho^{(m)}$, $\eta_\rho^{(m+1)}$ eingesetzt werden.

7) Die in (32) vorkommenden $\varphi'_m(\rho)$, $\bar\varphi_m(\rho)$, $\bar\varphi_{m+1}(\rho)$ sind die in § 4, Formel (18) und (20), erklärten, aus φ abgeleiteten Abbildungen!

Man mache sich nun die Bedeutung der Summen in (33), deren Gleichheit ja gerade durch den Satz 1 ausgedrückt wird, klar! Bei der ersten wird genau über diejenigen $\rho \in \mathbb{P}_m$ summiert, zu denen "Rand"-m-Ecke S^m_ρ mit $\varphi(S^m_\rho) = \pm (1,2,\dots,m)$ gehören, in der zweiten entsprechend über alle jene $\rho \in \mathbb{P}_{m+1}$, zu denen (m+1)-Ecke S^{m+1}_ρ mit $\varphi(S^{m+1}_\rho) = \pm (0,1,\dots,m)$ gehören.

Sätze über Abzählungen solcher Art sind in der Literatur schon wiederholt aufgetreten (vgl. [29], S. 133!), allerdings nur für Punktmengen des R^n. Um diesen Sätzen näherzukommen, hat man daher weiter zu spezialisieren:

A= Punktmenge des R^n,

$G = Z \equiv$ Ring der ganzen Zahlen,

K^m = Pseudomannigfaltigkeit[8].

Bei geeigneter (kohärenter!) Orientierung von K^m und des Randes ∂K^m nehmen dann die Koeffizienten α_ρ in (33) alle den Wert 1 an, und die Effekte gehen in *die algebraischen Summen* der Vorzeichen η^m_ρ, bzw. η^{m+1}_ρ, genau *der oben beschriebenen m-Ecke, bzw.* (m+1)-*Ecke* über, welche dann allerdings (m-1)-dimensionale, bzw. m-dimensionale Simplexe darstellen.

Auch Sätze der Art, wie sie von Ky Fan in [30] und [31] aufgestellt und bewiesen wurden, lassen sich als Sonderfälle unseres Satzes 1 gewinnen, wobei a n d e r e Spezialisierungen von B und C^m gebraucht werden. Da hier der Raum für die Darstellung dieser Verhältnisse nicht ausreicht, müssen wir auf [33] verweisen.

§ 8. Spezialisierung modulo 2.

Eine andere Spezialisierung, nämlich eine des Ringes G, hat andere Konsequenzen. Um diese aufzuzeigen, setzen wir jetzt voraus, daß

(34) $\quad G$ = *dem Primkörper der Charakteristik 2*

ist, während die Grundmenge A und die Bewertungsmenge B ihre frühere a l l g e m e i n e Bedeutung behalten sollen.

Zunächst stellt man fest, daß für die m-Ecke S^m, T^m jetzt jegliche Orientierung entfällt, weil in G ja (-1) = (+1) ist. Insbesondere gilt für jede Permutation $(\nu_1, \nu_2, \dots, \nu_m)$ von $(1,2,\dots,m)$ stets

(35) $\quad (a_1, a_2,\dots,a_m) = (a_{\nu_1}, a_{\nu_2}, \dots, a_{\nu_m}).$

Eine Gleichung $S^m_1 = S^m_2$ bedeutet jetzt immer, daß S^m_1 und S^m_2 als Mengen übereinstimmen ohne Rücksicht auf die Anordnung ihrer Elemente.

Auf Grund der Voraussetzung (34) sind unsere Grundbegriffe und ihre gegenseitigen Beziehungen einer veränderten Deutung fähig, die zuerst an den Vielecken und Komplexen

8) Zum Begriff "Pseudomannigfaltigkeit" vergleiche z.B. [26], [30], [31]!

$$(36)\quad \begin{cases} S^m = (a_1, a_2,\dots,a_m), & S^{m-1} = (a'_1, a'_2,\dots,a'_{m-1}) \\ K^m = \sum\limits_{\rho\in\mathbb{P}_m} \alpha_\rho S^m_\rho , & C^m = \sum\limits_{\sigma\in\Gamma_m} \beta_\sigma T^m_\sigma \end{cases}$$

erklärt werden soll. Hierfür definieren wir nämlich [9])

$$(37)\quad \begin{cases} S^m_\rho \in K^m :\Leftrightarrow \alpha_\rho = 1 , \\ T^m_\sigma \in C^m :\Leftrightarrow \beta_\sigma = 1, \\ S^{m-1} \subset S^m :\Leftrightarrow \exists\ k,\ 1 \le k \le m : S^{m-1} = (a_1,\dots,a_{k-1}, a_{k-1},\dots,a_m). \end{cases}$$

Durch die Definitionen (37) erhalten auch die Operationen ∂, Δ neue Bedeutungen, die sich folgendermaßen beschreiben lassen

$$(38)\qquad \partial K^m = \{S^{m-1} : |\{\rho \in \mathbb{P}_m : S^m_\rho \supset S^{m-1}\}| \equiv 1 \pmod 2\},$$

$$(39)\qquad \Delta C^m = \{T^{m+1} = (b_0, b_1,\dots,b_m) : \nu(T^{m+1}) \equiv 1 \pmod 2\}$$

mit $\qquad \nu(T^{m+1}) = |\{ k\in\mathbb{N} : 0 \le k\le m, (b_0,\dots,b_{k-1}, b_{k-1},\dots b_m) \in C^m\}|.$

Schließlich erhält auch der Effekt einer Bewertung $\varphi : A \Rightarrow B$ jetzt eine neue Bedeutung, nämlich

$$(40)\qquad \Phi^{C^m}_{K^m} = |\{S^m_\rho \in K^m : \varphi(S^m_\rho) \in C^m\}|.$$

Unser obiger Satz 1 kann jetzt natürlich auch in der folgenden Form geschrieben werden

Satz 2. *Wenn für den Ring G aus § 2 der Primkörper der Charakteristik 2 gewählt wird und für die Begriffe* K^m, C^m, ∂, Δ, Φ *die neuen Bedeutungen (36) bis (40) genommen werden, so gilt wieder für beliebige Wahl von* K^m, C^m, φ:

$$(41)\qquad \Phi^{C^m}_{\partial K^{m+1}} \equiv \Phi^{\Delta C^m}_{K^{m+1}} \pmod 2 .$$

Dieser Satz 2 ist eine leichte Verallgemeinerung von Satz 1 aus (28). Die Verallgemeinerung liegt darin, daß für die Bewertungsmenge B jetzt irgend eine Menge genommen werden kann. Dieser Umstand könnte neue Anwendungen ermöglichen. Insbesondere scheint der Fall, daß für B eine geeignete Menge von Vektoren eingesetzt wird, auch für numerische Anwendungen von Interesse zu sein.

Schlußbemerkung

Die Ähnlichkeit der Formeln (30) und (41) mit dem GAUSS'schen oder STOKES'schen Integralsatz ist nicht nur äußerlich sondern auch inhaltlich bedingt. In der Tat läßt sich durch geeignete (starke) Spezialisierung von A,B,G,K^{m+1}, C^m z.B. erreichen, daß die beiden Seiten von (30) Näherungssummen für die beiden Integrale des GAUSS'schen Integralsatzes darstellen; ein Grenzübergang führt dann zu dem Satz selbst. Dieser Zusammenhang wird an anderer Stelle erörtert werden (vgl. [33])

9) Man beachte, daß die Koeffizienten α_ρ,β_σ in (36) wegen (34) jetzt nur die Werte 0 oder 1 annehmen können!

Literaturhinweise

[1] Sperner E. (1928) *Neuer Beweis für die Invarianz der Dimensionszahl und des Gebietes,* "Abh. a. d. Math. Sem. d. Univ. Hamburg" 6, 265-272

[2] Menger K. (1928)-*Dimensionstheorie* (Leibzig - Berlin), p. 245 ff.

[3] Knaster B., Kuratowski C. und Mazurkiewicz S. (1929) - *Ein Beweis des Fixpunktsatzes für n-dimensionale Simplexe,* "Fund. Math." 14, 132-137.

[4] Tucker A.W. (1945) - *Some Topological Properties of Disk and Sphere,* "Proc. First Can. Math. Congress", Montreal, 285-309.

[5] Lefschetz S. (1949) - *Introduction to Topology* (Princeton), pp. 117, 138,139.

[6] Fan Ky (1952) - *A Generalization of Tucker's Combinatorial Lemma with Topological Applications,* "Annals of Mathematics", 56, 431-437.

[7] Bagemihl F. (1953) - *An Extension of Sperner's Lemma, with Applications to Closed-set Coverings and Fixed Points,* "Fund. Math.", 40, 3-12.

[8] Alexandrov P.S. (1956) - *Combinatorial Topology* (Rochester, N.Y.) Vol I, pp. 120 ff., 155 ff.

[9] Fan Ky (1958) - *Topological Proofs for Certain Theorems on Matrices with Non- Negative Elements,* "Monatshefte f. Mathematik", 62, 219-237.

[10] Burger E. (1959) -*Einführung in die Theorie der Spiele* (Berlin).

[11] Kuhn H.W. (1960) - *Some Combinatorial Lemmas in Topology,* "IBM Journ. Res. Develop.", 4, 518-524.

[12] Tompkins CH.B. (1964) - *Sperner's Lemma and Some Extensions,* from Beckenbach E.F., *Applied Combinatorial Mathematics* "University of California Engineering and Physical Sciences Extension Series" New York.

[13] Krasnosel'skii M.A. (1964) - *Topological Methods in the Theory of Nonlinear Integral Equations* , <Pergamon Press>, S. 81ff.

[14] Cohen D.I.A. (1967) - *On the Sperner Lemma,* <J.Comb. Theory>, 2,585-587.

[15] Fan Ky (1968) - *A Covering Property of Simplexes,* <Math. Scand.> 22, 17-20.

[16] Scarf H. (1967) - *The Approximation of Fixed Points of a Continuous Mapping,* <Siam J. Appl. Math.>, 15, 1328-1343.

[17] Kuhn H.W. (1968) - *Simplicial Approximation of Fixed Points,* <Proc. Nat. Acad. Sc.>, 61, 1238-1242.

[18] Kuhn H.W. (1969) - *Approximate Search For Fixed Points,* <Computing Methods in Optimization Problems> (New York) Vol. 2, 199-211.

[19] Allgower E.L. and Keller C.L. (1971) - *A Search Routine of a Sperner Simplex,* <Computing> 8, 157-165.

[20] Allgower E.L., Keller C.L. and Reeves T.E. (1971) - *A Program for the Numerical Approximation of a Fixed Point of an Arbitrary Continous Mapping of the n-Cube or n-Simplex into itself,* <Aerospace Research Laboratories>, 71-0257

[21] Eaves B.C. (1972) - *Homotopies for Computation of Fixed Points.* <Mathematical Programming> 3, 1-22.

[22] Yoseloff M. (1974) - *Topologic Proofs of some Combinatorial Theorems,* <J. Comb. Theory (A)>, 17, 95-111.

[23] Karamardian S. (1977) - *Fixed Points, Algorithms and Applications,* (Academic Press, Inc., New York).

[24] Prüfer M. (1978) - *Sperner Simplices and the Topological Fixed Point Index,* <Manuskript>.

[25] Prüfer M. (1978) - *Der topologische Fixpunktindex und die simpliziale Behandlung von nichtlinearen Eigenwertproblemen.* <Dissertation> , Bonn.

---

[26] Fan Ky (1960) - *Combinatorial Properties of Certain Simplicial and Cubical Vertex Maps,* <Archiv d. Math.> , 11. 368-377.

[27] Mani P. (1967) - *Zwei kombinatorisch-geometrische Sätze vom Typus Sperner-Tucker-Ky Fan,* <Monatshefte f. Mathematik>, 71, 427-435.

[28] Sperner E. (1973) - *Kombinatorik bewerteter Komplexe,* <Abh. a.d. Math. Sem. d. Univ. Hamburg>, 39, 21-43.

---

[29] Cairns S.S. (1961) - *Introductory Topology,* (Ronald Press, New York), 133 f

[30] Fan Ky, (1967) - *Simplicial Maps from an Orientable n-Pseudonanifold into S with the Octahedral Triangulation,* < J. Comb. Theory >, 2, 588-602.

[31] Fan Ky, (1970) - *A Combinatorial Property of Pseudomanifolds and Covering Properties of Simplexes.* <J. of Math. Analysis and Applications>, 31, 68-80.

[32] Kuhn H.W., (1974) - *A New Proof of the Fundamental Theorem of Algebra,* <Math. Programming Study>, 1, 148-158.

[33] Sperner E. (1979) - *Eine kombinatorische Identität,* <Resultate d. Math.> (erscheint demnächst).

PART III
MEASURE THEORETIC CONCEPTS AND OTHER TOOLS

GAME THEORY AND RELATED TOPICS
O. Moeschlin, D. Pallaschke (eds.)

EXTENDING PREMEASURES WITH COMPACT CONVEX VALUES IN A FRÉCHET SPACE

Ernst-Erich Doberkat

Department of Mathematics
University of Hagen

The aim of this paper is to establish an extension theorem of Hopf-Kolmogoroff's type for nondegenerate set valued premeasures. It is shown that it suffices to prove this for monotone premeasures, and this in turn is done by means of an extension procedure for modular functionals defined on a lattice and taking their values in a uniform semigroup.

§ 1

Given a real locally convex vector space E, and a measurable space $(X, \mathcal{A})$, $M: \mathcal{A} \longrightarrow C(E) := \{C;\ \emptyset \neq C \subset E \text{ compact and convex}\}$ is said to be a <u>set valued measure</u> iff given a sequence (A_n) of mutually disjoint sets in $\mathcal{A}$, <u>the following conditions</u> hold:

(i) every sequence (x_n) such that $\forall n \in \mathbb{N}: x_n \in M(A_n)$ is summable,

(ii) $M(\bigcup_{n\in\mathbb{N}} A_n) = \{\sum_{n\in\mathbb{N}} x_n;\ \forall n \in \mathbb{N}: x_n \in M(A_n)\}^-$

(the bar denotes closure).

These conditions imply that for any $x' \in E'$, the topological dual of E, $A \longmapsto S(x';M(A))$ is a finite signed measure, where S denotes the support function. This fact is used in |9| for the definition of a set valued measure, and is generalized in |5|: if F,G are real vector spaces in duality, F is endowed with the weak, G with the Mackey topology, a mapping M from $\mathcal{A}$ into the nonvoid closed, convex subsets of F then is said to be a weak set valued measure iff $A \longmapsto S(y;M(A))$ is a signed measure for any $y \in G$. In |16| as well as in |1| E is assumed to be finite dimensional; a countably additive correspondence, or a set valued measure, respectively, is a mapping M from $\mathcal{A}$ into the nonvoid subsets of E such that $M(\bigcup_{n\in\mathbb{N}} A_n)$ equals the set of all $\sum_{n\in\mathbb{N}} x_n$ such that (x_n) is summable, and $\forall n \in \mathbb{N}: x_n \in M(A_n)$. These measures are of interest e.g. in Mathematical Economics, see |11|, II. 4.2.

In this paper, E is assumed to be a Fréchet space. If $\mathcal{A}_o$ is an algebra over X, $M_o: \mathcal{A}_o \longrightarrow C(E)$ is said to be a <u>premeasure</u> iff the conditions (i), and (ii) hold for every disjoint sequence in $\mathcal{A}_o$ that has its union in $\mathcal{A}_o$. M_o is said to be <u>monotone</u> iff $A \subset B$ implies $M_o(A) \subset M_o(B)$, or, equivalently, iff $0 \in M_o(A) \forall A \in \mathcal{A}_o$. It is easily established (see 3.2) that a monotone premeasure is completely characterized by additivity, monotony, and continuity under monotone convergence, where C(E) is endowed with the Hausdorff metric.

The aim of this paper is to prove the following

1.1 Extension Theorem: If $\mathcal{A}_o$ is an algebra over X, and $M_o: \mathcal{A}_o \longrightarrow C(E)$ is a nondegenerate premeasure, there exists a uniquely determined measure M defined on the σ-algebra $s(\mathcal{A}_o)$ generated by $\mathcal{A}_o$ with values in $C(E)$ that extends M_o.

Here M_o is said to be nondegenerate iff there exists $A \in \mathcal{A}_o$ such that $\mathrm{card}(M_o(A)) > 1$; a monotone premeasure is either identical $\{0\}$, or nondegenerate. If M_o is monotone, 1.1 is referred to as the Monotone Extension Theorem, and the problem of proving 1.1 can be reduced to prove this theorem in its monotone version. Thus, let $M_o: \mathcal{A}_o \longrightarrow C(E)$ be a nondegenerate premeasure. The idea in reducing is to find a selective vector valued premeasure $m_o | \mathcal{A}_o$ for M_o, extending $M_o - m_o$, and m_o separately, and to construct an extension for M_o from those extensions. m_o is constructed in analogy to a proof of the Krein-Milman Theorem (|7|, V. 8.2). It is noted that a mapping N from $\mathcal{A}_o$ to $C(E)$ such that $\forall A \in \mathcal{A}_o : N(A) \subset M_o(A)$ is a premeasure iff N is additive, and if $A_n \downarrow \emptyset$ implies $N(A_n) \longrightarrow \{0\}$ with respect to the Hausdorff metric.

1.2 Lemma: There exists a vector valued premeasure $m_o: \mathcal{A}_o \longrightarrow E$ such that $m_o(A)$ is an extremal point of $M_o(A)$ for every $A \in \mathcal{A}_o$.

Proof:

1. Let $\mathcal{M}$ be the set of all premeasures P on $\mathcal{A}_o$ such that P(A) is an extremal subset of $M_o(A)$ for every $A \in \mathcal{A}_o$. Then evidently $M_o \in \mathcal{M}$. Given $P_i \in \mathcal{M}$, define $P_1 \le P_2$ iff $\forall A \in \mathcal{A}_o$: $P_2(A) \subset P_1(A)$. If $\mathcal{Q} \subset \mathcal{M}$ is a chain, $P := \bigcap \mathcal{Q}$ is additive: let A_1, A_2 be disjoint, then $P(A_1) + P(A_2) \subset P(A_1 \cup A_2)$ is easily established; if $x \in P(A_1 \cup A_2)$, there exists, given $p \in \mathcal{Q}$ $x_{p,i} \in p(A_i)$ such that $x = x_{p,1} + x_{p,2}$. Since $M_o(A_i)$ is compact, the ultrafilter generated by $\{\{x_{p,1};\ p \ge p'\};\ p' \in \mathcal{Q}\}$ converges to $x_i \in M_o(A_i)$, say, and since $\mathcal{Q}$ is a chain, $x_i \in P(A_i)$. Thus $x = x_1 + x_2 \in P(A_1) + P(A_2)$. If $A_n \downarrow \emptyset$, $P(A_n) \longrightarrow \{0\}$ is easily established, so is the extremality of P(A) in $M_o(A)$ for every $A \in \mathcal{A}_o$. From Zorn's lemma we infer the existence of a maximal $P_o \in \mathcal{M}$.
2. Given $x' \in E'$, $A \in \mathcal{A}_o$, define $m(x',A)$ as the set of all maximum points of x' in $P_o(A)$, i.e. $m(x',A) := \{x \in P_o(A);\ (x';x) = S(x';P_o(A))\}$, then $m(x',A)$ is an extremal subset of $M_o(A)$, and, if $x' \in E'$ is fixed, $m(x',\cdot)$ is additive, and $A_n \downarrow \emptyset$ implies $m(x',A_n) \longrightarrow \{0\}$. Since P_o is maximal, $P_o = m(x',\cdot)$ for any $x' \in E'$, hence there exists a vector valued premeasure m_o such that $P_o(A) = \{m_o(A)\}$ for every $A \in \mathcal{A}_o$. □

1.3 Proposition: The Monotone Extension Theorem implies the Extension Theorem.

Proof: If $M_o'(A) := M_o(A) - m_o(A)$, where m_o is chosen according to 1.2, M_o' is a monotone premeasure. Under the assumption that the Monotone Extension Theorem holds, there exists a uniquely determined monotone measure $M': s(\mathcal{A}_o) \longrightarrow C(E)$ that extends M_o', and obviously, $A \longmapsto S(x';M'(A))$ is a finite realvalued measure for every $x' \in E'$, hence |18|, Theorem 1, implies that there exists a weak vector valued measure $\hat{m}$ defined on $s(\mathcal{A}_o)$ with values in E (i.e. $A \mapsto (x';\hat{m}(A))$ is a finite signed measure for every $x' \in E'$) such that $\hat{m}(A) \in M'(A)$ for every $A \in s(\mathcal{A}_o)$, and $\hat{m}$ is not identical 0. $\hat{m}$ is a strong vector valued measure, and $\{\hat{m}(A);\ A \in s(\mathcal{A}_o)\}$ is contained in a weakly compact subset $K_1 \subset E$, and since $K_2 := (M_o - \hat{m})(X)$ is compact, $K := K_1 + K_2$ is weakly compact. Given $A \in \mathcal{A}_o$, $m_o(A) \in (M_o - \hat{m})(A) + \hat{m}(A) \subset K$, hence the Theorem on Extension in |13|, p. 176, implies the existence of an extension $m | s(\mathcal{A}_o)$ of $m_o | \mathcal{A}_o$. Define $M := M' + m$, $M | s(\mathcal{A}_o)$ is the looked-for extension of $M_o | \mathcal{A}_o$, and is obviously uniquely determined. □

Thus only the monotone version of 1.1 has to be proven. One might suspect that this can be done by means of the Hopf-Kolmogoroff Theorem (|10|, (10.36)) in the following way: let $m_o(x')(A) := S(x';M_o(A))(x' \in E')$, then $m_o(x')$ extends uniquely to a finite measure $m(x')$ on $s(\mathcal{A}_o)$, and $m(x')(A)$ is the support function of the compact, and convex set $M(A) := \{x \in E; \forall x' \in E': (x';x) \leq m(x')(A)\}$. Hence M has to be shown a measure in the strong sense considered here, but this is rather cumbersome in the general case, however easily done if the premeasure we are concerned with is continuous with respect to a finite realvalued premeasure, as is shown after the proof of the Monotone Extension Theorem. Clearly, M is a weak set valued measure, but the class of those measures properly includes the class of (strong) set valued measures, as the following example shows.

<u>1.4 Example:</u> Let $E := \mathbb{R}^{\mathbb{N}}$, and $d((x_n),(y_n)) := \sum_{n \in \mathbb{N}} 2^{-n} |x_n - y_n| (1 + |x_n - y_n|)^{-1}$, hence E is a Fréchet space, and E' consists of those elements of E all of which but a finite number are zero (|7|, IV.2.28, IV.13.46). If $X := \mathbb{N}$, $A \subset X$, define $M(A) := \{(x_n); \forall n \in \mathbb{N}: 0 \leq x_n \leq 2^n 1_A(n)\}$, 1_A denoting the indicator function of A. Obviously, M(A) is compact and convex, and M is additive. Given $x' := (x'_n) \in E'$, $|S(x';M(A))| \leq \sum_{i \in \mathbb{N}} |x'_i| 2^i 1_A(i)$; this is a finite sum, and since M is monotone, $S(x';M(\cdot))$ is a finite positive measure on the power set of X. However, M fails to be a set valued measure, since if $A_n := \{k; k \geq n\}$, the Hausdorff distance between $M(A_n)$ and $\{0\}$ is at least $\frac{1}{2}$, thus $A_n \downarrow \emptyset$ does not imply $M(A_n) \longrightarrow \{0\}$.

Since $\mathcal{A}_o$ is a sublattice of the power set of X, and C(E) is a complete Abelian uniform semigroup with cancellation law, when endowed with Hausdorff's metric, the Monotone Extension Theorem is deduced from a general extension theorem for modular, and σ-smooth functionals, which are defined on a sublattice of a topological σ-lattice, and have their values in a complete Abelian uniform semigroup with a uniform cancellation law. This theorem, which seems to be interesting in its own right, since both the Hopf-Kolmogoroff Theorem, and the Daniell-Stone Theorem may be deduced from it, is proved in § 2; in § 3 the discussion of set valued measures is continued.

§ 2

If H is a complete Abelian separated uniform semigroup with identity, L is a distributive lattice, $f: L \to H$ is said to be <u>modular</u> (or a <u>valuation</u>) iff $\forall a,b \in L: f(a \vee b) + f(a \wedge b) = f(a) + f(b)$ holds; f is said to be <u>smooth</u> iff, given a monotone sequence $(a_n) \subset L$, $(f(a_n))$ converges in H, and $f(\lim_n a_n) = \lim_n f(a_n)$ holds, whenever $\lim_n a_n \in L$ (where $\lim_n a_n := \sup (a_n)$, if $(a_n) \uparrow$, i.e. if (a_n) increases, and $\lim_n a_n := \inf (a_n)$, if $(a_n) \downarrow$, i.e. if (a_n) decreases, respectively). L is assumed to be a sublattice of a topological σ-lattice W, i.e. $\wedge$ and $\vee$, respectively, are continuous with respect to order convergence for sequences. This is equivalent to $\sup (b \wedge a_n) = b \wedge \sup(a_n)$, and $\inf (b \vee a_n) = b \vee \inf (a_n)$, whenever $(a_n) \subset W$, $b \in W$ (|3|, p. 248). According to |19|, Satz I.3, and |8|, 14 D, respectively, every Boolean σ-algebra, and every bounded subset of a Dedekind σ-complete Riesz space is a topological σ-lattice.

A subset A of W is said to be a <u>s-system</u> (<u>i-system</u>) iff every sequence in A has its supremum (infimum) in A; given $F \subset W$, F^s (F^i) denotes the smallest s-system (i-system) containing F. If $A = A^s = A^i$, A is called a <u>Borel-system</u>, B(F) denotes the smallest Borel-system containing F. A routine argument (cp. |10|, (21.6)) shows that B(L) is the smallest subset of W, that contains L, and that is closed under monotone convergence.

In H, a uniform cancellation law is assumed to hold. If $*a\ (c,d) := (c+a, d+a)$, define for $V \subset H^2$ $V^* := \bigcup\{*a^{-1}(V);\ a \in H\}$, hence $V \subset V^* = V^{**}$, and $(V_1 + V_2)^* \subset V_1^* + V_2^*$ hold. Since addition in H is uniformly continuous (|17|, p. 2), $G^* := \{V^*;\ V \in G\}$ is a base for a uniformity, provided G is a base for the given uniformity. The uniform cancellation law then is said to hold in H iff $G_s^* := \{V^*;$ V is open and symmetric} generates the given uniformity. Since H is separated, the cancellation law holds in H; in every topological group the uniform cancellation law holds, hence the results obtained here generalize those of |6|, § 2.

2.1 Definition: Let $A \subset W$, $g: A \to H$. $g \mid A$ is said to be order convex iff $a \le b \le c$, and $(g(a), g(c)) \in V$ together imply $(g(b), g(c)) \in V$, whenever $a, b, c \in A$, $V \in G_s^*$. Order convexity is Yoon's monotony for group-valued measures, adapted to the situation considered here. $f \mid L$ is assumed to be smooth, order convex, and modular. Since $L^s = \{\sup\ (a_n);\ L \supset (a_n)\uparrow\}$, and $L^i = \{\inf\ (a_n);\ L \supset (a_n)\downarrow\}$ hold (W is a topological σ-lattice), $f^s : \sup\ (a_n) \longmapsto \lim_n f(a_n)$, and $f^i : \inf\ (a_n) \longmapsto \lim_n f(a_n)$ define modular and order convex mappings $f^s \mid L^s$, and $f^i \mid L^i$, respectively; if $L^s \supset (a_n)\uparrow$, $f^s(\sup\ (a_n)) = \lim_n f^s(a_n)$ holds, and analoguously $L^i \supset (a_n)\downarrow$ implies $f^i(\inf\ (a_n)) = \lim_n f^i(a_n)$. These properties are however not yet sufficient for an extension of f to B(L).

2.2 Definition: $f|L$ is said to be σ-smooth iff

(a) $f^s \mid L^s \cap L^i = f^i \mid L^s \cap L^i$,

(b) $f' \mid L^s \cup L^i$ is order convex, where $f'(a) := f^s(a)$, if $a \in L^s$, $:= f^i(a)$ if $a \in L^i$.

Simple examples (|6|, 2.2) show that (a) does not necessarily holds, if $f \mid L$ is modular, order convex, and smooth. If W is a Boolean σ-algebra, and L is a Boolean subalgebra, $f \mid L$ is σ-smooth, provided $f \mid L$ is smooth, modular, and order convex. For the remainder of § 2, $f \mid L$ is assumed to be σ-smooth.

Following Kats (|12|), $(\underline{a}, \bar{a}) \in L^s \times L^i$ is said to be a V-boundary for $a \in W$ iff $\underline{a} \le a \le \bar{a}$, and $(f'(\underline{a}), f'(\bar{a})) \in V$ $(V \in G_s^*)$.

From a technical point of view, the following lemma will be of interest for the construction of an extension.

2.3 Lemma: Let $(a_n) \subset W$ be a monotone sequence such that there exists for every a_n a V-boundary for any $V \in G_s^*$. Then there exists a V-boundary for $a := \lim_n a_n$ $\forall V \in G_s^*$.

Proof:

1. Given $(V_n) \subset G_s^*$, define inductively $|V_1| := V_1$, $|V_1; \ldots; V_{n+1}| := (|V_1; \ldots; V_n| + V_{n+1})^*$. Since G_s^* is a base for the uniformity, and since addition is uniformly continuous, there exists for every entourage V a sequence $(V_n) \subset G_s^*$ such that $\forall n \in \mathbb{N}$: $|V_1; \ldots; V_n| \subset V$.

2. Since the cases "$(a_n)\uparrow$", and "$(a_n)\downarrow$" can be treated in complete analogy, (a_n) is assumed to be increasing. Given $V \in G_s^*$, there exists an entourage U such that $\bar{U} \subset V$. If $(V_n) \subset G_s^*$ is chosen for U according to 1., let $(\underline{a}_n, \bar{a}_n)$ be a V_n-boundary for a_n. Putting $\bar{c}_n := \sup\{\bar{a}_i;\ 1 \le i \le n\}$, $(\underline{a}_n, \bar{c}_n)$ is a $|V_1;\ldots;V_n|$-boundary for a_n. If $n = 1$, this is evident; assume, $(f'(\underline{a}_n), f'(\bar{c}_n))$ is in $U_n := |V_1;\ldots;V_n|$. Since

$\underline{a}_n \le \bar{a}_{n+1} \wedge \bar{c}_n \le \bar{c}_n$, $U_n \in G_s^*$, $(f'(\bar{c}_n), f'(\bar{a}_{n+1} \wedge \bar{c}_n)) \in U_n$, this implies

$(f'(\underline{a}_{n+1}) + f'(\bar{c}_n \wedge \bar{a}_{n+1}), f'(\bar{c}_{n+1}) + f'(\bar{c}_n \wedge \bar{a}_{n+1}))$

$= (f'(\underline{a}_{n+1}) + f'(\bar{c}_n \wedge \bar{a}_{n+1}), f'(\bar{c}_n) + f'(\bar{a}_{n+1})) \in U_n + V_{n+1}$,

thus $(f'(\underline{a}_{n+1}), f'(\bar{c}_{n+1})) \in (U_n + V_{n+1})^* = |V_1;\ldots;V_{n+1}|$.

Since $(\underline{a}_n, \bar{c}_n)$ is a $|V_1;\ldots;V_n|$-boundary for a_n, $(\underline{c}_n, \bar{c}_n)$ is, too, where

$\underline{c}_n := \sup\{\underline{a}_i;\ 1 \le i \le n\}$, since $f|L$ is σ-smooth. Define $\bar{a} := \sup(\bar{c}_n)$, $\underline{a} := \underline{c}_n$ for

a suitable chosen n, then $(\underline{a}, \bar{a})$ is a V-boundary for a, since $(f'(\underline{c}_n))$, and

$(f'(\bar{c}_n))$, respectively, converge in H. □

This auxiliary result yields the extension of $f \mid L$.

<u>2.4 Theorem:</u> If $f \mid L$ is modular, and σ-smooth, there exists a uniquely determined extension $\hat{f} \mid B(L)$ such that $\hat{f}$ is modular, smooth, and order convex.

<u>Proof:</u> To begin with, $f \mid L$ is extended to a subset $C(L,f)$ of W that contains L; this extension is shown to be smooth, and order convex. Since $C(L,f)$ turns out to be closed under monotone convergence, from this the assertion is deduced.

1. Following the construction in |12|, define

   $C(L,f) := \{a \in W;$ there exists $h_a \in H$ such that there exists for every $V \in G_s^*$ a

   V-boundary $(\underline{a}, \bar{a})$ with $(h_a, f'(\underline{a})) \in V\}$,

   then evidently $L^s \cup L^i \subset C(L,f)$ holds. Given $a \in C(L,f)$, h_a is uniquely determined; for, if h_a, and h'_a have the required properties, and if W' is an arbitrary entourage, there exists $V \in G_s^*$ such that $V^6 \subset W'$. Let $(\underline{a}, \bar{a})$, and $(\underline{a}', \bar{a}')$ be V-boundaries for a such that $(h_a, f'(\underline{a})) \in V$, and $(h'_a, f'(\underline{a}')) \in V$, respectively, then, since $\underline{a} \le \underline{a} \vee \underline{a}' \le a \le \bar{a} \vee \bar{a}' \le \bar{a}$ holds, $(f'(a \vee a'), h_a) \in V^3$.

   Analogously, $(f'(\underline{a} \vee \underline{a}'), h'_a) \in V^3$ holds, hence $(h_a, h'_a) \in W$, thus $h_a = h'_a$. Defining $f''(a) := h_a$, $f'' \mid C(L,f)$ is univalued, and evidently an extension of $f \mid L$.

2. If (a_n) is a monotone sequence in $C(L,f)$, its limit a is in $C(L,f)$, and $f''(a_n)$ converges to $f''(a)$. This is shown for increasing sequences, a completely symmetric argument proves it for decreasing ones. If $V \in G_s^*$ is arbitrary, there exists a V-boundary $(\underline{a}_n, \bar{a}_n)$ for a_n. Because of part 2 in the proof of 2.3, we may, and do assume that $(\underline{a}_n)\uparrow$. Since $(f'(\underline{a}_n))$ converges in H, $(f''(a_n), f''(a_m)) \in V^3$ for almost all $n,m \in \mathbb{N}$; since H is complete, $h_a := \lim_n f''(a_n)$

exists. If $(\underline{a},\bar{a})$ is a V-boundary for a (which exists by 2.3), we have $(f'(\underline{a}),h_a) \in V$: there exists $V_o \in G_s^*$ such that $(f'(\underline{a}),f'(\bar{a})) \in V_o$, and $\bar{V}_o \subset V$; if $V_1 \in G_s^*$ is arbitrary, $V_2 \in G_s^*$ with $\bar{V}_2 \subset V_1$, there is a V_2-boundary $(\underline{c}_n,\bar{c}_n)$ for a_n such that $(\bar{c}_n) \uparrow$. Defining $\hat{c} := \sup (c_n)$, $(f'(\hat{c} \wedge \bar{a}),h_a) \in V_1$, and $(f'(\underline{a}),f'(\hat{c} \wedge \bar{a}) \in V_o$ hold, since $f \mid L$ is σ-smooth. From this, we get $(h_a,f'(\underline{a})) \in V_1 \circ V_o$. V_1 is chosen arbitrarily, thus $(h_a,f'(\underline{a})) \in \bigcap\{V_1 \circ V_o;$ $V_1 \in G_s^*\} = \bar{V}_o \subset V$.

3. $f'' \mid C(L,f)$ is order convex, for if $a \le b \le c$, $V \in G_s^*$, and $(f''(a),f''(c)) \in V$, there is $V_o \in G_s^*$ such that $(f''(a),f''(c)) \in V_o \subset \bar{V}_o \subset V$, and $(f''(b),f''(c)) \in V_1^3 \circ V_o \circ V_1^3$ is shown easily for an arbitrary $V_1 \in G_s^*$, hence $(f''(b),f''(c)) \in V$.

4. Since $L \subset C(L,f)$, and since $C(L,f)$ is closed with respect to monotone convergen[ce] $B(L) \subset C(L,f)$ holds. Let $f\hat{}$ be the restriction of f'' to $B(L)$, then evidently $f\hat{}$ is a smooth, and order convex extension of $f \mid L$.

   a) $f\hat{} \mid B(L)$ is modular: given $M \subset B(L)$, define

   $R(M) := \{a \in B(L);\ f\hat{}(a \wedge b) + f\hat{}(a \vee b) = f\hat{}(a) + f\hat{}(b)\ \forall b \in M\}$.

   Thus $L \subset R(L)$, and the latter set is closed with respect to monotone convergenc[e] since W is a topological σ-lattice, and since $f\hat{} \mid B(L)$ is smooth, hence $R(L)$ equals $B(L)$. This implies $L \subset R(B(L))$, and $R(B(L)) = B(L)$ is proved analogously.

   b) $f\hat{} \mid B(L)$ is uniquely determined, since any smooth extension of $f \mid L$ coincides with $f\hat{}$ on a subset of $B(L)$ that contains L, and that is closed under monotone convergence, hence coincides with $f\hat{}$ on $B(L)$. □

$C(L,f)$ will not be used further, it has been introduced in order to avoid those transfinite arguments, which were used for a proof of |6|, 2.6, an analogon of 2.4 for the group valued case.

We now turn to set valued measures in order to prove the Monotone Extension Theorem by means of 2.4.

§ 3

Fix an algebra $\mathcal{A}_o$ of subsets of a set X, and a monotone premeasure $M_o: \mathcal{A}_o \longrightarrow C($[...] According to |4|, Theorem II-12, the uniformity induced by the Hausdorff metric i[s] generated by $\{v(U);\ U$ is a neighbourhood of 0 in E$\}$, where $(A,B) \in v(U)$ iff $A \subset B + U$ and $B \subset A + U$ $(A,B \in C(E))$.

$C(E)$ is a separated Abelian complete semigroup with $\{0\}$ as identity (|4|, Theorem II-14), and the uniform cancellation law holds in $C(E)$, since, if $A,B,C \in C(E)$, and if U is a closed convex neighbourhood of 0 in E, $A + B \subset C + B + U$ implies $A \subset C + U$ (|14|, Lemma 2).

Since $\mathcal{A}_o$ is a sublattice of the topological σ-lattice of all subsets of X, 2.4 can be applied, provided $M_o \mid \mathcal{A}_o$ is shown to be modular, and σ-smooth. Since additivity, and modularity are equivalent, this can be deduced from 3.1. For this, let $\mathcal{B}$ be a σ-algebra on X, and $\mathcal{M} := \mathcal{M}(X,\mathcal{B})$ be the Banach space of all finite signed measures on $(X,\mathcal{B})$, where the norm is the total variation. $\mathcal{M}'$ is the topological dual of $\mathcal{M}$, and τ is the topology on $\mathcal{M}$ induced by $\mathcal{M}'$. $\mathcal{E}'$ denotes the Borel sets on E' with respect to the weak * topology. If (A_n) is a summable sequence of subsets of E (i.e. $\forall (x_n \in A_n)_{n \in \mathbb{N}}$: (x_n) is summable), denote $\{\sum_{n \in \mathbb{N}} x_n;\ \forall n \in \mathbb{N}: x_n \in A_n\}$ by $\widetilde{\sum_{n \in \mathbb{N}}} A_n$.

<u>3.1 Theorem:</u> Let $M: \mathcal{B} \longrightarrow C(E)$ be monotone, then the following properties are equivalent:

(a) M is a measure,

(b) M is additive, and continuous with respect to monotone convergence,

(c) M is additive, and $B_n \downarrow \emptyset$ implies $M(B_n) \longrightarrow \{0\}$,

(d) defining $S_M(x')(B) := S(x';M(B))$, S_M is a finite transition kernel (|2|, § 56) from $(E', \mathcal{E}')$ to $(X, \mathcal{B})$ such that $\{S_M(x'); x' \in K\}$ is relatively compact in $\mathcal{M}$ with respect to τ, whenever $K \subset E'$ is equicontinuous.

<u>Proof:</u> Evidently, (b) and (c) are equivalent. If $(A_n) \subset C(E)$ is summable, for every $x' \in E'$ $S(x'; \widetilde{\sum_{n \in \mathbb{N}}} A_n) = \sum_{n \in \mathbb{N}} S(x';A_n)$ holds (cp. |16|).

1. Let M be a measure, then M is additive, since $M(\emptyset) = \{0\}$.
If $B_n \downarrow \emptyset$, $B_n = \bigcup_{m \geq n} (B_m - B_{m+1})$ thus $S(x';M(B_n)) \to 0$ for every $x' \in E'$. Since $(M(B_n))$ decreases, $M(B_n) \to \bigcap_{n \in \mathbb{N}} M(B_n)$, as is shown with the Hausdorff metric. Assume, there is $y \neq 0$ such that y is contained in every $M(B_n)$, then there exists $x' \in E'$ such that $S(x';M(B_n)) \geq 1$ holds for all $n \in \mathbb{N}$, which is a contradiction. Hence (a) implies (c).

2. Assume, (c) holds. If $(B_n) \subset \mathcal{B}$ is a disjoint sequence, $(M(B_n))$ is summable, since M is monotone, and since $M(\bigcup_{m \geq n} B_m) \longrightarrow \{0\}$ as $n \to \infty$. Furthermore, the equality $M(\bigcup_{n \in \mathbb{N}} B_n) = \widetilde{\sum_{n \in \mathbb{N}}} M(B_n)$ holds, since
$\widetilde{\sum_{n \in \mathbb{N}}} M(B_n) \subset M(\bigcup_{n \in \mathbb{N}} B_n) = \lim_m M(\bigcup_{1 \leq i \leq m} B_i) = \{x; \forall m \in \mathbb{N} \; \exists y_m \in M(\bigcup_{1 \leq i \leq m} B_i): y_m \longrightarrow x\}$,
and from this the following equalities are deduced for any $x' \in E'$:
$S(x';M(\bigcup_{n \in \mathbb{N}} B_n)) = \sum_{n \in \mathbb{N}} S(x';M(B_n)) = S(x';\widetilde{\sum_{n \in \mathbb{N}}} M(B_n))$.
The geometric version of the Hahn-Banach Theorem yields that $A = B$, whenever $S(x';A) = S(x';B)$ holds for every $x' \in E'$, provided $A,B \in C(E)$. Thus the desired equality holds, and (a) is proved, too.

3. Assume, (a) holds.
(i) Given $x' \in E'$, $S_M(x')$ is a non-negative measure (cp. |16|, Lemma 2.1), and is finite, since $M(X)$ is compact. Fix $B \in \mathcal{B}$, then $x' \longmapsto S_M(x')(B)$ is lower semicontinuous with respect to the weak* topology (|4|, Theorem II-16); since the Borel sets are the initial σ-algebra with respect to the semicontinuous functions, $x' \longmapsto S_M(x')(B)$ is $\mathcal{E}'$-measurable. Hence S_M is a finite transition kernel from $(E', \mathcal{E}')$ to $(X, \mathcal{B})$.
(ii) If $K \subset E'$ is equicontinuous, $p_K: (A,B) \longmapsto \sup\{|S(x';A) - S(x';B)|; x' \in K\}$ is continuous, and $\{p_K; K \subset E'$ is equicontinuous$\}$ generates the uniformity on $C(E)$ (|4|, Theorem II-18). Thus, if $B_n \downarrow \emptyset$, $S_M(x')(B_n) \to 0$ uniformly in $x' \in K$. Since K is equicontinuous, $\{S_M(x'); x' \in K\}$ is bounded in $\mathcal{M}$ (|15|, Corollary for III. 4.1), thus |7|, IV. 9.1 together with the Eberlein-Šmulian Theorem implies the relative compactness of the latter set. Hence (d) holds.

4. The implication (d) $\Rightarrow$ (c) is easily deduced from |7|, IV. 9.1, since $\{p_K; K \subset E'$ equicontinuous$\}$ generates the uniformity on $C(E)$. □

<u>3.2 Corollary:</u> Let $M_o: \mathcal{A}_o \longrightarrow C(E)$ be monotone. Then M_o is a premeasure iff M_o is additive, and continuous with respect to monotone convergence. □

Now 1.1 can be proved in its monotone version.

Proof of the Monotone Extension Theorem: Given a neighbourhood U of 0, $(A,B) \in v(U)$, $A \subset C \subset B$ implies $(C,B) \in v(U)$, thus M_o is order convex. Because of 3.2, and the remark after 2.2, M_o is σ-smooth, thus the theorem follows from 2.4, and the implication (b) $\Rightarrow$ (a) in 3.1, since $B(\mathcal{A}_o)$ equals $s(\mathcal{A}_o)$. □

On closing this note, it is remarked that the situation becomes far more simple if $M_o \mid \mathcal{A}_o$ is continuous with respect to a realvalued finite nonnegative premeasure $\hat{m}_o \mid \mathcal{A}_o$, i.e. if given a neighbourhood U of 0 in E, there is $\varepsilon > 0$ such that $M_o(A) \subset U$ whenever $\hat{m}_o(A) < \varepsilon$ $(A \in \mathcal{A}_o)$.

From this starting point, there are two possibilities to prove the monotone version of 1.1. The first is to note that, given a disjoint sequence $(A_n) \subset \mathcal{A}_o$, $M_o(A_n)$ converges to $\{0\}$, thus $M_o \mid \mathcal{A}_o$ is s-bounded. Hence Sion's Extension Theorem (|17|, Theorem 6.1) can be applied. The second is, as mentioned to extend for any $x' \in E'$ $m_o(x') := S(x';M_o(\cdot))$ to the finite measure $m(x') \mid s(\mathcal{A}_o)$ then $m = S_M$, where $M(A) := \{x \in E;\ \forall x' \in E': (x';x) \le m(x')(A)\}$ $(A \in s(\mathcal{A}_o))$, hence $M: s(\mathcal{A}_o) \longrightarrow C(E)$. Evidently, S_M is a finite transition kernel from $(E', \mathcal{E}')$ to $(X, s(\mathcal{A}_o))$. If $\hat{m} \mid s(\mathcal{A}_o)$ is the extension of $\hat{m}_o \mid \mathcal{A}_o$, $\hat{m}$ dominates the bounded set $A_K := \{S_M(x');\ x' \in K\}$, whenever $K \subset E'$ is equicontinuous. Hence, A_K is relatively compact in $\mathcal{M}(X, s(\mathcal{A}_o))$ with respect to τ (|7|, IV. 9.2), and 3.1 implies that M is a monotone measure, thus the looked-for extension.

Acknowledgement: The author is indebted to Prof. B. Fuchssteiner for some helpful discussions.

References

|1| Artstein, Z.: Set Valued Measures. Trans. AMS 165 (1972) 103 - 125.

|2| Bauer, H. (1968): Wahrscheinlichkeitstheorie und Grundzüge der Maßtheorie. Verlag Walter de Gruyter, Berlin.

|3| Birkhoff, G. (1967): Lattice Theory. AMS Coll. Publ., vol. XXV, Providence.

|4| Castaing, C., Valadier, M. (1977): Convex Analysis and Measurable Multifunctions. Lecture Notes in Mathematics 580, Springer Verlag, Berlin.

|5| Costé, A., Pallu de la Barrière, R.: Un théorème de Radon-Nikodym pour les multimesures à valeurs convexes fermées localement compactes sans droite. C.R. Acad. Sci. Paris 280 (1975) 255 - 258.

|6| Doberkat, E.-E.: Extending σ-Smooth Group-Valued Functionals on Distributive Lattices. Preprint (1977)

|7| Dunford, N., Schwartz, J.T. (1958): Linear Operators I. Interscience Publishers, New York

|8| Fremlin, D.H. (1974): Topological Riesz Spaces and Measure Theory. University Press, Cambridge.

|9| Godet-Thobie, C.: Sélection de multimeasures. Application à un théorème de Radon-Nikodym multivoque. C.R. Acad. Sci. Paris, 279 (1974) 603 - 606.

|10| Hewitt, E., Stromberg, K. (1969): Real and Abstract Analysis. Springer Verlag, Berlin.

|11| Hildenbrand, W. (1974): Core and Equilibria of a Large Economy. Princeton University Press, Princeton.

|12| Kats, M.P.: On the Continuation of Vector Measures. Siberian Math. J. 13 (1972) 802 - 806.

|13| Kluvánek, I.: The Extension and Closure of Vector Measures. In: Tucker, D.H., Maynard, H.B. (Eds.): Vector and Operator Valued Measures and Applications. (Academic Press, New York, 1973) 175 - 190.

|14| Rådström, H.: An Embedding Theorem for Spaces of Convex Sets. Proc. AMS 3 (1952) 165 - 169.

|15| Schaefer, H.H. (1971): Topological Vector Spaces. Springer Verlag, New York.

|16| Schmeidler, D.: Convexity and Compactness of Countably Additive Correspondences. In: Kuhn, H.W., Szegö, G.P. (Eds.): Differential Games and Related Topics. (North Holland, Amsterdam, 1971) 235 - 242

|17| Sion, M. (1973): A Theory of Semigroup Valued Measures. Lecture Notes in Mathematics 355, Springer Verlag, Berlin.

|18| Tolstonogov, A.A.: On the Theorems of Radon-Nikodým and A.A. Ljapunov for a Multivalued Measure. Soviet Math. Dokl., 16 (1975) 1588 - 1592.

|19| Vladimiroff, D.A. (1972): Boolesche Algebren. Akademie-Verlag, Berlin.

|20| Yoon, J.: On the Atoms of Group-Valued Measures. J. Korean Math. Soc. 11 (1974) 71 - 76 (MR 50, # 10202).

GAME THEORY AND RELATED TOPICS
O. Moeschlin, D. Pallaschke (eds.)

SOME PROPERTIES OF DISTRIBUTIONS OF CORRESPONDENCES

W. Eberl, jun.
Department of Statistics
University of Dortmund

In this paper we are concerned with the concept of the distribution of a correspondence. Using the weak resp. the strong inverse of a correspondence we get the weak resp. the strong distribution, for which we summarize some properties and give a representation as the upper resp. lower envelope of the distributions of its measurable selections. It turns out that the weak resp. the strong distributions enjoy dual properties, which should be helpful in further investigations on these set-functions.

I INTRODUCTION, PRELIMINARIES

In the mathematical economics correspondences play a more and more important role. In this context in a variety of papers concepts like the measurability or e.g. the integral of a correspondence were considered for recent years.
But there had not been introduced any reasonable concept for the distribution of a correspondence up to 1974, when HART and KOHLBERG suggested such one in |1|. They used for their definition the weak inverse of the correspondence and supposed the underlying measure space to be complete, whereas the range space was required to be a Polish space with its Borel subsets.
In this paper we take up Hart-Kohlberg's concept with a slight modification to get rid of the two mentioned assumptions. Of course, the completeness of the underlying measure space is unessential. The second requirement, a topological one, will be substituted by a weaker "non-topological" one, which guarantees for the range space just the structure needed.

We shall refer to the Projection Theorem and to the Measurable Choice Theorem. Since there are several versions of these fundamental theorems in the literature, we first formulate the versions which we shall use later on.

Formulating the Projection Theorem we need the concept of a Blackwell space. To introduce this, we first define a measurable space $(D,\mathcal{D})$ to be semicompact, if there exists a subsystem $\mathcal{D}_o$ of $\mathcal{D}$ containing $\emptyset$ and D, which is closed against finite intersections, enjoys the finite intersection property for countable sequences, such that $\mathcal{D} \subset S(\mathcal{D}_o)$ ($S(\mathcal{D}_o)$: the system of all $\mathcal{D}_o$-analytic subsets of D). Now, a Blackwell space is a measurable space $(T,\mathcal{C})$ such that there exist a semicompact measurable space $(D,\mathcal{C})$ and a $\mathcal{D}$-$\mathcal{C}$-measurable function $f:D \rightarrow T$ of D onto T. Examples for semicompact measurable spaces are all σ-compact Hausdorff spaces with its Borel subsets. Any euclidian space, any non-countable Borel subset of any Polish space with its Borel subsets is a Blackwell space. Thus all range spaces, which occur usually in economic applications of the Projection Theorem, are Blackwell spaces.

Projection Theorem

Let (Ω,A) be a measurable space, let $(T,\mathcal{C})$ be a Blackwell space. Then the following implication holds:

$$G \in S(A \times \mathcal{C}) \implies Pr_{\Omega}(G) \in S(A).$$

If $(T,\mathcal{C})$ is semicompact, the theorem is covered by a theorem of MARCZEWSKI and RYLL-NARDZEWSKI |4|. Then, the general case is easily deduced from the previous one, see |3|.

For the Measurable Choice Theorem we need the concept of a standard measurable space. This is a measurable space which is isomorphic to a countable product of copies of $\{0,1\}$ with all subsets of $\{0,1\}$ measurable. In fact, any standard measurable space is a Blackwell space; any euclidian space, any non-countable Borel subset of any Polish space with its Borel subsets is a standard measurable space.

(Ω,A) and $(T,\mathcal{C})$ being measurable spaces, we call a correspondence $\varphi:\Omega \rightarrow \mathcal{P}_o(T)$ analytical, if

$$G_{\varphi} := \{(\omega,t) \mid t \in \varphi(\omega)\} \in S(A \times \mathcal{C}).$$

Measurable Choice Theorem

Let (Ω,A,ν) be a σ-finite measure space, let $(T,\mathcal{C})$ be a standard measurable space and let $\varphi:\Omega \rightarrow \mathcal{P}_o(T)$ be an analytical correspondence. Then

$$M_{\varphi} := \{f:\Omega \rightarrow T \mid f \text{ measurable}, f(\omega) \in \varphi(\omega) \quad \nu\text{-a.e.}\} \neq \emptyset.$$

II WEAK DISTRIBUTIONS

Let (Ω,A) be a measurable space, (T,C) a Blackwell space and $\varphi:\Omega \to P_0(T)$ an analytical correspondence. Then, in virtue of the Projection Theorem, the weak inverse $\varphi^{-1}(C)$ is A-analytical for all C-analytical sets C:

$$\varphi^{-1}(C) := \{\omega\in\Omega \mid \varphi(\omega)\cap C \neq \emptyset\} = Pr_\Omega(G_\varphi \cap (\Omega\times C)) \in S(A) \qquad (C\in S(A)).$$

Now, because of the universal measurability of analytical sets, see e.g. |5|, we can introduce the weak distribution of φ in the following way.

Definition

Let (Ω,A,ν) be a σ-finite measure space, (T,C) a Blackwell space and $\varphi:\Omega \to P_0(T)$ an analytical correspondence. Then, the weak distribution $D_\varphi:S(C) \to \overline{\mathbb{R}}_+$ of φ let be defined by

$$D_\varphi(C) := (\overline{\nu}\circ\varphi^{-1})(C) = \overline{\nu}(\varphi^{-1}(C)) \qquad (C\in S(C)),$$

where $(\Omega,A_\nu,\overline{\nu})$ denotes the completion of (Ω,A,ν).

Remark: The above concept coincides with the usual one for the distribution of a universally measurable resp. measurable function, if $|\varphi(\omega)| = 1$ $(\omega\in\Omega)$.

In our first theorem concerning weak distributions we fix a correspondence φ and summarize some properties of D_φ as a (nonnegative, possibly infinite) set-function.

Theorem 1

Let (Ω,A,ν) be a σ-finite measure space, (T,C) a Blackwell space and $\varphi:\Omega \to P_0(T)$ an analytical correspondence. Then we have:

(a) D_φ is concave, i.e., it holds $D_\varphi(\emptyset) = 0$ and

$$D_\varphi(C_1) + D_\varphi(C_2) \geq D_\varphi(C_1\cup C_2) + D_\varphi(C_1\cap C_2) \qquad (C_1,C_2 \in S(C));$$

(b) D_φ is σ-subadditive:

$$D_\varphi(\bigcup C_n) \leq \sum D_\varphi(C_n) \qquad (C_n \in S(C));$$

(c) D_φ is lower σ-continuous:

$$C_n \in S(C),\ C_n \uparrow C \implies D_\varphi(C_n) \uparrow D_\varphi(C);$$

(d) $$D_\varphi(\underline{\lim}\, C_n) \leq \underline{\lim}\, D_\varphi(C_n) \qquad (C_n \in S(C))$$

("lower half of sequential continuity").

Proof: The definition of φ^{-1} implies

$$\varphi^{-1}(\bigcup C_n) = \bigcup \varphi^{-1}(C_n)$$

and

$$\varphi^{-1}(\bigcap C_n) \subset \bigcap \varphi^{-1}(C_n)$$

for any $C_n \in S(C)$ $(n \in \mathbb{N})$; from these relations we can readily deduce the statements of the theorem.

In the next theorem we give a representation of the weak distribution of a correspondence as the upper envelope of the distributions of its measurable selections.

Theorem 2

Let (Ω, A, ν) be a σ-finite measure space, (T, C) a standard measurable space and $\varphi: \Omega \to P_0(T)$ an analytical correspondence. Then

$$D_\varphi(C) = \max_{f \in M_\varphi} \nu_f(C) \qquad (C \in S(C))$$

with $\nu_f = \overline{\nu} \circ f^{-1}$.

Proof: Obviously we have $D_\varphi(C) \geq \nu_f(C)$ $(C \in S(C),\ f \in M_\varphi)$; thus

(i) $$D_\varphi(C) \geq \sup_{f \in M_\varphi} \nu_f(C) \qquad (C \in S(C)).$$

Now, fix $C \in S(C)$ and assume w.l.o.g. $D_\varphi(C) > 0$. Then we have

$$\Omega_0 := \varphi^{-1}(C) \in A_\nu .$$

Consider the (σ-finite) measure space (Ω_0, A_0, ν_0) with $A_0 := A_\nu | \Omega_0$, $\nu_0 := \overline{\nu} | A_0$ and define the correspondence $\psi: \Omega_0 \to P_0(T)$ by

$$\psi(\omega) := \varphi(\omega) \cap C \neq \emptyset \qquad (\omega \in \Omega_0).$$

Because of the relation

$$G_\psi = G_\varphi \cap (\Omega_0 \times C)$$

ψ is analytical. We have $\psi^{-1}(C) = \varphi^{-1}(C) = \Omega_0$; thus

(ii) $$D_\varphi(C) = D_\psi(C) = \nu_0(\Omega_0).$$

According to the Measurable Choice Theorem there exists a function $f_1 \in M_\psi$; i.e., let $f_1: \Omega_0 \to T$ be A_0-C-measurable with

$$f_1(\omega) \in \psi(\omega) = \varphi(\omega) \cap C \quad \nu_0\text{-a.e.}$$

Further, choose $f_2 \in M_\varphi$, i.e., let $f_2: \Omega \to T$ be A-C-measurable with

$$f_2(\omega) \in \varphi(\omega) \quad \nu\text{-a.e.}$$

Now, define $f:\Omega \to T$ A-C-measurable such that

$$f(\omega) = \begin{cases} f_1(\omega) & (\omega\in\Omega_0) \\ f_2(\omega) & (\omega\in\Omega-\Omega_0) \end{cases} \quad \nu\text{-a.e.}$$

In virtue of $f_1^{-1}(C) = \Omega_0$ ν_0-a.e. we then get

$$f^{-1}(C) \supset \Omega_0 \quad \nu\text{-a.e.}$$

and therefore $f \in M_\varphi$. Finally, we conclude

$$D_\varphi(C) \leq \bar{\nu}(f^{-1}(C)) = \nu_f(C),$$

which proves the theorem.

Remark: If in the above theorem $(T,C) = (\mathbb{R}^n, \mathcal{B}^n)$ and φ is integrably bounded (i.e., $\exists g \in L^1(\nu)$: $|t| \leq g(\omega)$ $(t\in\varphi(\omega))$ ν-a.e., see |1|, |2|), then $L_\varphi := M_\varphi \cap L^1(\nu) = M_\varphi$ and the above theorem says

$$D_\varphi(C) = \max_{f\in L_\varphi} \nu_f(C) \qquad (C\in C).$$

Concerning sequences of correspondences we have the following conclusions.

Theorem 3

Let (Ω,A,ν) be a σ-finite measure space, (T,C) a Blackwell space and $\varphi_n:\Omega \to P_0(T)$ $(n\in\mathbb{N}_0)$ analytical correspondences. Then:

(a) $\varphi_n \uparrow \varphi_0 \implies D_{\varphi_0} = \sup D_{\varphi_n}$;

(b) $\varphi_n \downarrow \varphi_0 \implies D_{\varphi_0} \leq \inf D_{\varphi_n}$;

(c) $\underline{\lim}\, \varphi_n \neq \emptyset \implies D_{\underline{\lim}\, \varphi_n} \leq \underline{\lim}\, D_{\varphi_n}$;

(d) $\varphi_1 \cap \varphi_2 \neq \emptyset \implies D_{\varphi_1} + D_{\varphi_2} \geq D_{\varphi_1\cup\varphi_2} + D_{\varphi_1\cap\varphi_2}$

Proof: The first two statements follow immediately from the relations

$$(\bigcup\varphi_n)^{-1} = \bigcup\varphi_n^{-1}$$

resp.

$$(\bigcap\varphi_n)^{-1} \subset \bigcap\varphi_n^{-1} ,$$

which can be verified readily.
The third conclusion being a consequence of the first two ones, the last follows in a similar way like the concavity of D_φ in Theorem 1.

In our last theorem on weak distributions we state a certain associative law for the weak distribution of the composition of correspondences.

Theorem 4

Let (Ω,A,ν) be a σ-finite measure space, (T,C), $(U,\mathcal{D})$ Blackwell spaces and $\varphi:\Omega \to P_o(T)$ resp. $\psi:T \to P_o(U)$ analytical correspondences. Further, define $\psi\circ\varphi:\Omega \to P_o(U)$ by

$$\psi\circ\varphi(\omega) := \bigcup_{t\in\varphi(\omega)} \psi(t) \qquad (\omega\in\Omega).$$

Then $\psi\circ\varphi$ is an analytical correspondence with the following weak distribution:

$$D_{\psi\circ\varphi} = D_\varphi \circ \psi^{-1}.$$

Proof: Obviously $\psi\circ\varphi$ is a correspondence, i.e., $\psi\circ\varphi \neq \emptyset$. Since the set

$$\{(\omega,t,u)\,|\,t\in\varphi(\omega), u\in\psi(t)\} = (G_\varphi \times U) \cap (\Omega\times G_\psi)$$

is an element of $S(A\times C\times\mathcal{D})$, we conclude by means of the Projection Theorem

$$G_{\psi\circ\varphi} = Pr_{\Omega\times U}[(G_\varphi\times U) \cap (\Omega\times G_\psi)] \in S(A\times\mathcal{D}),$$

i.e., $\psi\circ\varphi$ is analytical.
From the validity of $(\psi\circ\varphi)^{-1} = \varphi^{-1}\circ\psi^{-1}$ we deduce

$$D_{\psi\circ\varphi}(D) = \overline{\nu}((\varphi^{-1}\circ\psi^{-1})(D)) = D_\varphi(\psi^{-1}(D)) = (D_\varphi\circ\psi^{-1})(D) \qquad (D\in S(\mathcal{D})),$$

which proves the theorem.

Remarks

(1) The given definition of $\psi\circ\varphi$ coincides with that in |2|.

(2) Recalling the definition $D_\varphi = \overline{\nu}\circ\varphi^{-1}$ we can write the statement of the above theorem in the more suggestive form

$$\overline{\nu}\circ(\psi\circ\varphi)^{-1} = (\overline{\nu}\circ\varphi^{-1})\circ\psi^{-1}.$$

III STRONG DISTRIBUTIONS

Let (Ω,A) be a measurable space, (T, C) a Blackwell space and $\varphi:\Omega \to P_o(T)$ an analytical correspondence. Then, we define the strong inverse $\varphi^*(C)$ in the usual way:

$$\varphi^*(C) := \{\omega\in\Omega\,|\,\varphi(\omega) \subset C\} \qquad (C\in P(T)).$$

Then

$$\varphi^*(C) = \Omega - \varphi^{-1}(T - C) \qquad (C\in P(T)).$$

Therefore, if $S^c(\mathcal{A})$ resp. $S^c(\mathcal{C})$ denote the system of all complements of $\mathcal{A}$-resp. $\mathcal{C}$-analytic sets, we have $\varphi^*(C) \in S^c(\mathcal{A})$ for all $C \in S^c(\mathcal{C})$, by the Projection Theorem. Thus we are lead to the following definition.

Definition

Let $(\Omega,\mathcal{A})$ be a measurable space, $(T,\mathcal{C})$ a Blackwell space and $\varphi:\Omega \to P_0(T)$ an analytical correspondence. Then, the strong distribution $D^*_\varphi:S^c(\mathcal{C}) \to \overline{\mathbb{R}}_+$ of φ let be defined by

$$D^*_\varphi(C) := (\overline{\nu}\circ\varphi^*)(C) = \overline{\nu}(\varphi^*(C)) \qquad (C \in S^c(\mathcal{C})).$$

Remark: On the common domain $\mathcal{C}$ of D_φ and D^*_φ it holds

$$D_\varphi(C) + D^*_\varphi(T-C) = \nu(\Omega) \qquad (C\in\mathcal{C}).$$

The last remark indicates that strong distributions enjoy dual properties of weak distributions, at least on their common domain $\mathcal{C}$. In fact, one can prove the dual theorems of the previous section for strong distributions on their whole domain $S^c(\mathcal{C})$. As the corresponding proofs essentially are much the same as for weak distributions, we omit them.

Theorem 5

Let $(\Omega,\mathcal{A},\nu)$ be a σ-finite measure space, $(T,\mathcal{C})$ a Blackwell space and $\varphi:\Omega \to P_0(T)$ an analytical correspondence. Then we have:

(a) D^*_φ is convex, i.e., it holds $D_\varphi(\emptyset) = 0$ and

$$D^*_\varphi(C_1) + D^*_\varphi(C_2) \leq D_\varphi(C_1 \cup C_2) + D_\varphi(C_1 \cap C_2) \qquad (C_1, C_2 \in S^c(\mathcal{C}));$$

(b) D_φ is σ-superadditive:

$$D^*_\varphi(\textstyle\sum C_n) \geq \sum D^*_\varphi(C_n) \qquad (C_n \in S^c(\mathcal{C}) \text{ pairwise disjoint});$$

(c) D^*_φ is upper-σ-continuous:

$$\left.\begin{array}{l} C_n \in S^c(\mathcal{C}),\ C_n \downarrow C \\ D^*_\varphi(C_n) < \infty \end{array}\right\} \Longrightarrow D^*_\varphi(C_n) \downarrow D_\varphi(C);$$

(d) $$\overline{\lim}\, D^*_\varphi(C_n) \leq D^*_\varphi(\overline{\lim}\, C_n) \qquad (C_n \in S^c(\mathcal{C}))$$

("upper half of sequential continuity").

Theorem 6

Let (Ω,A,ν) be a σ-finite measure space, $(T,\mathcal{C})$ a standard measurable space and $\varphi:\Omega \to \mathcal{P}_o(T)$ an analytical correspondence. Then

$$D^*_\varphi(C) = \min_{f\in M_\varphi} \nu_f(C) \qquad (C \in S^c(A)).$$

Theorem 7

Let (Ω,A,ν) be a σ-finite measure space, $(T,\mathcal{C})$ a Blackwell space and $\varphi_n:\Omega \to \mathcal{P}_o(T)$ $(n\in \mathbb{N}^o)$ analytical correspondences. Then

(a) $\varphi_n \uparrow \varphi_o \implies D^*_{\varphi_o} = \inf D^*_{\varphi_n}$;

(b) $\varphi_n \downarrow \varphi_o \implies D^*_{\varphi_o} \geqq \sup D^*_{\varphi_n}$;

(c) $\underline{\lim}\, \varphi_n \neq \emptyset \implies \overline{\lim}\, D^*_{\varphi_n} \leqq D^*_{\underline{\lim}\, \varphi_n}$;

(d) $\varphi_1 \cap \varphi_2 \neq \emptyset \implies D^*_{\varphi_1} + D^*_{\varphi_2} \leqq D^*_{\varphi_1\cup\varphi_2} + D^*_{\varphi_1\cap\varphi_2}$.

Theorem 8

Under the assumptions and with the notation of Theorem 4 it holds

$$D^*_{\psi\circ\varphi} = D^*_\varphi \circ \psi^*.$$

REFERENCES

|1| S. Hart, E. Kohlberg: Equally distributed correspondences, Journal of Mathematical Economics 1 (1974)

|2| W. Hildenbrand: Core and Equilibria of a large Economy, Princeton University Press (Princeton and London, 1974)

|3| J. Hoffmann-Jørgensen: The Theory of analytic Spaces, Various Publication Series No. 10, Aarhus University (1970)

|4| E. Marczewski, C. Ryll-Mardzewski: Projections in abstract Sets, Fund. Math. 40, (1953) 160-164

|5| S. Saks: Theory of the Integral, New York Wafer Publishing Company (1937)

GAME THEORY AND RELATED TOPICS
O. Moeschlin, D. Pallaschke (eds.)

A STATISTICAL CHARACTERIZATION OF LOCALIZABLE MEASURE SPACES

D. Plachky

Institute of Mathematical Statistics
University of Münster

F.R. Diepenbrock

Siemens A.G.,
Bereich Daten- und Informationssysteme
8000 München

If (X, Σ, μ) denotes a measure space, where Σ is a σ-ring of subsets of the set X, a locally Σ-measurable function $\varphi: X \longmapsto [0,1]$ is usually called test function in mathematical statistics, because φ can be viewed as the probability for a certain decision in favour of a hypothesis. The hypothesis, which is usually described by a set of probability measures on Σ, is often not dominated by a σ-finite measure on Σ, but in spite of this the underlying probability measures have a density $f \in L_1(X, \Sigma, \mu)$ with respect to a measure μ on Σ, which is not σ-finite. Consider for example the set of all discrete probability measures on the Borel subsets of the real line. Now the power of a test function φ is given by $\int \varphi f \, d\mu$ and it is therefore interesting to know (for example in connection with questions concerning the existence of optimal test functions), whether the set of all test functions defined on X are compact with respect to the weakest topology, such that all mappings defined by $\varphi \longmapsto \int \varphi f \, d\mu$, $f \in L_1(X, \Sigma, \mu)$, are continuous. This is clearly equivalent with the fact, that every net $\{\varphi_\alpha\}$ of test functions has a subnet $\{\varphi_\beta\}$, such that $\int \varphi_\beta f \, d\mu \longrightarrow \int \varphi f \, d\mu$, $f \in L_1(X, \Sigma, \mu)$, holds for some test function φ. The main result of this paper is that this property of the underlying measure space (X, Σ, μ) is equivalent to the condition on (X, Σ, μ) to be localizable in the sense of Kelley and Namioka ([1], p. 129). This means that every cross section $\{g_A \mid A \in \Sigma_o\}$, i. e. $g_A = g_B \mu$ - a. e. holds on $A \cap B$ for any $A, B \in \Sigma_o$, where Σ_o denotes the set of all $A \in \Sigma$ with $\mu(A) < \infty$ and g_A is defined on X and is Σ-measurable for any $A \in \Sigma_o$, can be pieced together into a single locally measurable function g,

i. e. $gI_A = g_A$ holds μ - a. e. on A for any $A \in \Sigma_o$. Here I_A denotes the indicator function of $A \subset X$. It should be pointed out that the main result is proved elementary without using Segal's functional analytical characterization of (X, Σ, μ) to be localizable, i. e. the dual of $L_1(X, \Sigma, \mu)$ is equal to the space $L_\infty(X, \Sigma, \mu)$ of all locally measurable and almost everywhere locally bounded functions on X.

MAIN RESULT

If $\{\varphi_A | A \in \Sigma_o\}$ denotes a cross section of test functions it is clear from Segal's result about localizable measure spaces that the net $\{\varphi_A I_A | A \in \Sigma_o\}$, where Σ_o is directed by inclusion, has a subnet $\{\varphi_{A_\alpha} I_{A_\alpha}\}$ with the property $\int \varphi_{A_\alpha} I_{A_\alpha} f d\mu \longrightarrow \int \varphi f\, d\mu$, $f \in L_1(X, \Sigma, \mu)$, for some test function φ. The following techniques of proof shows that φ pieces the cross section $\{\varphi_A | A \in \Sigma_o\}$ together. This is one part of the following

THEOREM. The measure space (X, Σ, μ) is localizable if and only if for every cross section $\{\varphi_A | A \in \Sigma_o\}$ of test functions the corresponding net $\{\varphi_A I_A | A \in \Sigma_o\}$ has the property $\int \varphi_A I_A f\, d\mu \longrightarrow \int \varphi f\, d\mu$ for all $f \in L_1(X, \Sigma, \mu)$ and for some test function φ. In this case $\varphi_A I_A = \varphi I_A \mu$ - a. e. holds for all $A \in \Sigma_o$.

Proof. Let $\{g_A | A \in \Sigma_o\}$ denotes a cross section and $\varphi_A = \frac{1}{2} + \frac{1}{\pi} \arctan g_A$ the test function corresponding to g_A, $A \in \Sigma_o$. If the cross section $\{\varphi_A | A \in \Sigma_o\}$ is pieced together by a test function φ, then $g = \tan \pi(\varphi - 1/2)$ does the same with respect to $\{g_A | A \in \Sigma_o\}$. Let us assume instead of $\int \varphi_A I_A f\, d\mu \longrightarrow \int \varphi f\, d\mu$ only that $\int \varphi_{A_\alpha} I_{A_\alpha} f\, d\mu \longrightarrow \int \varphi f\, d\mu$ holds for all $f \in L_1(X, \Sigma, \mu)$ and some test function φ, where $\{\varphi_{A_\alpha} I_{A_\alpha}\}$ is some subnet of the net $\{\varphi_A I_A | A \in \Sigma_o\}$. Now the following assertion can be disproved: there exists $A_o \in \Sigma_o$ such that $\varphi_{A_o} I_{A_o} = \varphi I_{A_o} \mu$ - a. e. does not hold. For this purpose it is enough to disprove the assertion: $\mu(\{x \in X | \varphi_{A_o}(x) I_{A_o}(x) < \varphi(x) I_{A_o}(x)\}) > 0$ because of symmetry.

Since $\{x \in X \mid \varphi_{A_o}(x) I_{A_o}(x) < \varphi(x) I_{A_o}(x)\} \in \Sigma_o$, the last assertion implies the existence of a real number $a > 0$, such that $\mu(B_o) > 0$ with $B_o = \{x \in X \mid a + \varphi_{A_o}(x) I_{A_o}(x) < \varphi(x) I_{A_o}(x)\}$. Furthermore from the definition of subnet follows the existence of A_{α_o}, such that $B_o \subset A_{\alpha_o}$ holds, which implies for all A_α with $A_{\alpha_o} \subset A_\alpha$ the following chain of inequalities:

$$\int I_{B_o} \varphi \, d\mu = \int I_{A_{\alpha_o}} \varphi I_{B_o} \, d\mu \geq \int (I_{A_{\alpha_o}} \varphi_{A_{\alpha_o}} + a) I_{B_o} \, d\mu =$$

$$\int I_{A_{\alpha_o}} \varphi_{A_{\alpha_o}} I_{B_o} \, d\mu + a \, \mu(B_o) = \int I_{A_{\alpha_o}} \varphi_{A_\alpha} I_{B_o} \, d\mu + a \, \mu(B_o) =$$

$$\int I_{A_\alpha} \varphi_{A_\alpha} I_{B_o} \, d\mu + a \, \mu(B_o).$$

This is a contradiction to the assumption $\int \varphi_{A_\alpha} I_{A_\alpha} f \, d\mu \longrightarrow \int \varphi f \, d\mu$ for all $f \in L_1(X, \Sigma, \mu)$, if one chooses $f = I_{B_o}$. But this finishes the proof, because in the case, where the measure space (X, Σ, μ) is localizable, one has for a cross section $\{\varphi_A \mid A \in \Sigma_o\}$ a test function φ with $\varphi_A I_A = \varphi I_A \mu$ - a. e. for all $A \in \Sigma_o$ and $\int \varphi I_A f \, d\mu \longrightarrow \int \varphi f \, d\mu$ for any $f \in L_1(X, \Sigma, \mu)$ on account of the σ-finiteness of $\{x \in X \mid f(x) \neq 0\}$, which together implies $\int \varphi_A I_A f \, d\mu \longrightarrow \int \varphi f \, d\mu$ for all $f \in L_1(X, \Sigma, \mu)$.

The method of proof yields immediately that a compactness assumption implies the localizability of the underlying measure space. Segal's characterization of localizable measure spaces together with the theorem of Alaoglu yields the other direction of the following

THEOREM. The measure space (X, Σ, μ) is localizable if and only if the unit ball of $L_\infty(X, \Sigma, \mu)$ is compact with respect to the weakest topology on $L_\infty(X, \Sigma, \mu)$ such that all mappings $g \longmapsto \int fg \, d\mu$, $f \in L_1(X, \Sigma, \mu)$, $g \in L_\infty(X, \Sigma, \mu)$ are continuous.

Proof. A non elementary but short proof of this theorem may be based on a result of Dixmier (see [3]). According to this result a Banach space B that has a locally convex Hausdorff topology τ such that the unit ball of B is τ-compact is the dual B'* of the

Banach space B' consisting of all linear functionals on B whose restrictions to the unit ball of B are τ-continuous. For B = $L_\infty(X, \Sigma, \mu)$ and τ being the weakest topology such that the mappings $g \longmapsto \int fg\, d\mu$, $f \in L_1(X, \Sigma, \mu)$, $g \in L_\infty(X, \Sigma, \mu)$, are contiuous it follows, that $L_1(X, \Sigma, \mu)$ can be isometrically imbedded in B' and by a minimax theorem (see [2]), that the image of $L_1(X, \Sigma, \mu)$ is dense in B' and therefore $B' = L_1(X, \Sigma, \mu)$ holds. According to the theorem of Dixmier one has $B'^* = L_\infty(X, \Sigma, \mu)$ and hence Segal's result yields that (X, Σ, μ) is localizable.

REMARKS

A non elementary proof of the second theorem can be given also with the help of Šmulians weak compactness theorem (see [1], 16.6 resp. 18.8). Furthermore it should be pointed out that the compactness assumption in both theorems proved above implies also localizability of the underlying measure space, where piecing together a cross section into a single measurable (in stead of a locally measurable) function is possible. In this case the space $L_\infty(X, \Sigma, \mu)$ is equal to the set of all almost locally bounded and measurable functions, because this modified version of localizability implies that every locally measurable function must be equal up to a zero set to a measurable function. That this version is indeed a modification of localizability in the sense of Kelley and Namioka shows the example, where $X = [0, 1]$, $\Sigma = \{A \subset X \mid A \text{ or } A^c \text{ is countable}\}$ and μ is equal to the counting measure. Finally it should be pointed out, that localizability in the sense of Kelley and Namioka can also be characterized topologically in the following sense: For every net $\{I_{A_\alpha} \mid A_\alpha \in \Sigma_o\}$ directed by inclusion there exists a locally measurable set A_o such that $\int I_{A_\alpha} f\, d\mu \longrightarrow \int I_{A_o} f\, d\mu$ holds for all $L_1(X, \Sigma, \mu)$.

REFERENCES

[1] Kelley, J. L. and Namioka, J. (1965): Linear Topological Spaces. Van Nostrand, New York.

[2] König, H.: Über das von Neumannsche Minimaxtheorem. Arch. Math. XIX (1968), 482 - 487.

[3] Ng, K.-F.: On a theorem of Dixmier. Math. Scand. 29(1971),279f.

GAME THEORY AND RELATED TOPICS
O. Moeschlin, D. Pallaschke (eds.)

NEW APPROACH TO SYSTEMS WITH TRANSFORMED ARGUMENT

Danuta Przeworska-Rolewicz
Institute of Mathematics
Polish Academy of Sciences
Warszawa

A good generalization of differential operator, difference operators and some partial differential operators (for instance: Laplace operator, ware operator, even with variable coefficients) are right invertible operators. The theory of right invertible operators in linear spaces (in general, without any topology) is now well developed (cf. References). We shall shown here some aspect of this theory connected with notions of shifts and periodicity, Several applications to functional-differential systems and their generalizations could be obtained by this method. Also for some problems of controllability, observability and optimization of systems described by right invertible operators this approach is very useful.

Let X be a linear space over $\mathbb{C}$ (this assumption is not essential, we can consider an arbitrary algebraically closed field). Let $L(X)$ denote the set of all linear operators A whose domains $\mathfrak{D}_A$ are linear subspaces of X and with ranges in X. Write: $L_o(X) = \{A \in L(X) : \mathfrak{D}_A = X\}$. Let $R(X)$ $L(X)$ denote the set of all right invertible operators. Suppose that $D \in R(X)$. By $\mathfrak{R}_D$ we denote the set of all right inverses of . Without loss of generality, we can assume that $\mathfrak{R}_D \subset L_o(X)$. The set $\ker D$ plays the role of <u>space of constants</u> for D. An arbitrary projection $F \in L_o(X)$ <u>onto</u> $\ker D$ is an <u>initial operator</u> for D corresponding to an $R \in \mathfrak{R}_D$. By these definitions we have:

(1) $\quad DR = I, \quad F^2 = F, \quad FX = \ker D, \quad F = I - RD \quad \text{on} \quad \mathfrak{D}_D.$

An operator $S \in L_o(X)$ is said to be <u>D-invariant</u> if $SD = DS$ on $\mathfrak{D}_D$. An operator $A \in L_o(X)$ is said to be <u>S-periodic</u> if S is D-invariant and $AS = SA$.

Suppose that N is an arbitrary positive integer, $S \in L_o(x)$ is D-invariant and $X_{S,N} = \{x \in X : S^N x = x\}$. If $\dim X_{S,N} > 0$ then every $x \in X_{S,N}$ is called an S^N-periodic element. Since on the space $X_{S,N}$ the operator S is an involution of order N then this space is a direct sum of N eigenspaces of S corresponding to the eigenvalued being N-th roots of unity. This permit to apply even in this general case methods developed in [5].

Suppose now that D,R,F are defined, as above and that either $A(\mathbb{R}) = \mathbb{R}_+$ or $A(\mathbb{R}) = \mathbb{R}$. We say that $S_{A(\mathbb{R})} = \{S_h\}_{h\in A(\mathbb{R})} \subset L_o(X)$ is a family of R-shifts if the following conditions are satisfied:

$$(2) \qquad S_o = I; \quad S_h R^k F = \sum_{j=0}^{k} \frac{(-1)^{k-j}}{(k-j)!} h^{k-j} R^j F \quad \text{for all} \quad h \in A(\mathbb{R}), \quad k \in \{0\} \cup \mathbb{N}.$$

One can prove (cf. [8]) that

1) R-shifts preserve all constants;
2) If $A(\mathbb{R}) = \mathbb{R}_+$ $(= \mathbb{R})$ then $S_{A(\mathbb{R})}$ is a commutative semigroup (an Abelian group) with respect to the superposition of operators as a structure operation;
3) R-shifts are uniquely determined on the set $P(R) = \mathrm{lin}\{R^k z : z \in \ker D,\ k \in \{0\} \cup \mathbb{N}\}$ of D-polynomials;
4) If X is a complete linear topological space, $\overline{P(R)} = X$, S_h are continuous for all $h \in A(\mathbb{R})$ then the family $S_{A(\mathbb{R})}$ is uniquely determined on X;
5) If X is a complete locally convex linear topological space, the operator D is closed, $\overline{P(R)} = X$ and $S_{A(\mathbb{R})}$ is a strongly continuous semigroup (group) of R-shifts then D is the infinitesimal generator for $S_{A(\mathbb{R})}$, $\overline{\mathfrak{D}_D} = X$ and S_h are D-invariant for all $h \in A(\mathbb{R})$.
6) Write for a fixed $h \in A(\mathbb{R})$:

$$(3) \qquad D_1 = D - \frac{1}{h} FS_h; \quad R_1 = (I + FS_h + \frac{1}{h} RFS_h)R; \quad X^{(1)}_{S_h,N} = X_{S_h,N} \cap \mathfrak{D}_D.$$

Then $D_1 : X^{(1)}_{S_h,N} \longrightarrow X_{S_h,N}$; $R_1 : X_{S_h,N} \longrightarrow X^{(1)}_{S_h,N}$ and $D_1 R_1 = I$, $R_1 D_1 = I$. This implies the existence of periodic solutions of equations with D and an R-shift.

EXAMPLE 1. Suppose that $X = C(\mathbb{R})$, $D = \frac{d}{dt}$, $R = \int_0^t$, $(Fx)(t) = x(0)$, $(S_h x)(t) = x(t-h)$ for $x \in X$, $h \in \mathbb{R}$. Then $X_{S_h,1}$ is the space of all ω-periodic functions where $\omega = -h$. We have: $(D_1 x)(t) = x^1(t) + \frac{1}{\omega}x(0)$, $(FS_h Rx)(t) = \int_0^\omega x(s)ds$ and $(R_1 x)(t) = \int_0^t x(s)ds + (\frac{t}{\omega} + 1)\int_0^\omega x(s)ds$. We therefore have obtained in this way classical tools to prove the existence of periodic solutions.

EXAMPLE 2. Consider the equation: $x'(t) = x(g(t))$, where $g: \mathbb{R} \longrightarrow \mathbb{R}$, $g \in C^1(\mathbb{R})$ and $g' > 0$. Write: $(Ax)(t) = x(g(t))$, $\hat{D} = DA^{-1}$, $\hat{R} = AR$, $\gamma_h(t) = g(g^{-1}(t) - h)$, $(\hat{S}_h x)(t) = x(\gamma_h(t))$ for all $h \in \mathbb{R}$, $x \in X$. The function γ_h is a general solution of so-called translation equation. Then $\hat{S}_h$ are $\hat{R}$-shifts.

For some applications the notion of R-shifts is not sufficient. For that reason we shall introduce now another class of shifts. We assume, in addition, that R is a Volterra operator, i.e. the operator $I-\lambda R$ is invertible for all $\lambda \in \mathbb{C}$. We shall write berifly: $R \in V(X)$. If $R \in \mathcal{R}_D \cap V(X)$, one can define exponential operators as follows: $e_\lambda = (I - \lambda R)^{-1}$ for all $\lambda \in \mathbb{C}$.

We say that $S_{A(\mathbb{R})}$ is a family of D-shifts if

(4) $\quad S_0 = I; \quad S_h e_\lambda F = e^{-\lambda h} e_\lambda F \quad$ for all $\lambda \in \mathbb{C}$, $h \in A(\mathbb{R})$.

Now, if we defined the set of all exponential elements as follows: $E(R) = \text{lin}\{e_\lambda(z) : z \in \ker D, \lambda \in \mathbb{C}\}$ and if we put in Properties 1), 2), 3), 4), 5) D-shifts instead of R-shifts and the set $E(R)$ instead of $P(R)$ we obtain similar statements for D-shifts.

EXAMPLE 3. $X = C[0,T]$, $D = t\frac{d}{dt}$, $R = \int_0^t$ for an $a > 0$, $(Fx)(t) = x(a)$, $(S_h x)(t) = x(rt)$, where $h = -\ln r$. Then $R \in V(X)$, exponential elements are of the form Ct^λ $(\lambda \in \mathbb{C})$ and $\{S_h\}_{h \in \mathbb{R}}$ is a family of D-shifts.

One can show that there are R-shifts which are not D-shifts, and conversely, and examples of spaces without R-shifts and D-dhifts. If X is a Banach space and R is quasi-nilpotent then the notions of R-shifts and D-shifts coincide.

Having already defined both these kinds of shifts one can prove existence of exponential-periodic and polynomial-exponential-periodic solutions of some problems and different kind of perturbation theorems.

References

[1] Naguyen dinh Quyet: Controllability and observability of linear systems described by the right invertible operators in linear spaces, Preprint N$^{\underline{o}}$ 113, Institute of Mathematics, Polish Academy of Sciences, October 1977.

[2] ——: On linear systems described by right invertible operator acting in a linear space, Control and Cybernetics, 2,7(1978), 33-45.

[3] ——: A minimal problem of a quadratic functional for the system described by right invertible operators in Hilbert space, Control and Cybernetics, 3,7(1978), 27-36.

[4] ——: On the stability and observability of D-R systems in Banach spaces, Demonstratio Math. (to appear).

[5] D.Przeworska-Rolewicz: Equations with transformed argument. An algebraic approach, Elsevier and PWN - Polish Scientific Publishers, Amsterdam-Warszawa, 1973.

[6] ——: Algebraic theory of right invertible operators, Studia Math. 48(1973), 129-144.

[7] ——: Introduction to Algebraic Analysis and its applications, (Polish), WNT, Warszawa (to appear in 1979).

[8] ——: Shifts and periodicity for right invertible operators, Preprint N$^{\underline{o}}$ 157, Institute of Mathematics, Polish Academy of Sciences, October, 1978.

[9] S.Staniaszek: Necessary and sufficient conditions for optimization in non-linear systems with right invertible operators Working paper, System Research Institute, Polish Academy of Sciences, Warszawa, (to appear).

[10] M.Tasche: Funktionalanalytische Methoden in der Operatorenrechnung, Nova Acta Leopoldina, Band 49, 231(1978).

[11] H.von Trotha: Structure properties of D-R spaces, Preprint N$^{\underline{o}}$ 102, Institute of Mathematics, Polish Academy of Sciences, May, 1977. Dissertationes Math. (to appear).

GAME THEORY AND RELATED TOPICS
O. Moeschlin, D. Pallaschke (eds.)

A NOTE ON PERIODIC SOLUTIONS OF DIFFERENTIAL-DIFFERENCE EQUATIONS WITH APPLICATIONS TO ECONOMICS

Hartmut von Trotha

GMD Birlinghoven 5205 St. Augustin 1
West Germany

0. Introduction

Many problems in economics can be described properly by delay differential systems. We consider the following linear autonomous delay differential system

$$(1)\quad B\,x'(t) = L(x_t) + y(t) = \sum_{k=o}^{m} A_k\, x(t-k) + y(t),\ t \in \mathbb{R}$$

with B and each A_k real (constant) matrices and $y(t)$ an n-vector-valued function to be specified later. In the sequel we shall restrict our attention to N-periodic functions $y(t)$ only, with N a common multiple of $1,\ldots,m$. We then shall look for

a) N-periodic solutions of (1) in an explicit form
b) Solutions of (1), subject to the initial condition

$$(2)\quad x(t) = g(t) \quad o \le t \le m$$

with $g(t)$ a n-vector-valued function (real or complex).

Clearly [1],[3],[5], the stability of system (1), as well as the existence of periodic solutions is determined by the associated characteristic function

$$(3)\quad H(s) = Bs - \sum_{k=o}^{m} A_k (e^{-s})^k,\ s \in \mathbb{C},$$

so we shall have to make some assumptions about (3) to solve problem a.
If the periodic solutions are known, then problem b can be dealt with in the following way:

First determine a periodic solution x_N of (1) and then solve the homogeneous initial value problem

$$(4)\quad Bx'(t) - L(x_t) = o;\ x(t) = g(t) - x_N(t);\ o \le t \le m.$$

Now, if x_o is a solution of (4), then obviously $x(t) = x_N(t) + x_o(t)$ is the desired solution.

To simplify matters we shall only discuss the cases $B = o$ and B invertible (hence we can put $B = I$). $B = o$ describes a pure delay system as, for instance, a linear econometric model, whereas, for example, a dynamic Leontieff model requires B to be invertible ($B = I$).

1. Notations and Basic Results

We use the following notation :

Let $F(I)$, $I \subset \mathbb{R}$ an interval, be the space of n-vector-valued complex functions of a real variable t. We denote by $C(I)$ the subspace of continuous functions and by F_w and C_w the subspaces of w-periodic and continuous w-periodic functions of $F(\mathbb{R})$ with $w \in \mathbb{R}$. Finally, we shall mean by Q_w either the space F_w or C_w.

To begin with, we cite some results which can be found in [7].
Clearly, the shift operator $Sx(t) = x(t-1)$ is an involution of order N on Q_N, hence $Q_N = E_1 \oplus .. \oplus E_N$ with $E_j = \ker(S-e^{hj})$, $h = 2\pi i/N$. The corresponding projection operators P_j, $1 \le j \le N$ are given by

$$(5) \quad P_j x(t) = \frac{1}{N} \sum_{k=o}^{N-1} (e^{-hj})^k x(t-k) \overset{\text{def}}{=} x_{(j)}(t), \quad x \in Q_N.$$

and $E_j = P_j Q_N$. The elements of E_j can then be characterized by

$$(6) \quad x(t) \in E_j \text{ iff } x(t) = e^{-hjt} f(t),$$

where $f(t) \in Q_1 \subset Q_N$ i.e. f is of period 1.
If not otherwise stated all what follows will refer to Q_N.

1.1 Case $B = I$, $Q_N = C_N$

It is well-known that the delay differential system (see e.g. [1])

$$(7) \quad x'(t) = L(x_t) + y(t) = \sum_{k=o}^{m} A_k x(t-k) + y(t)$$

subject to the initial conditions (2) with $g(t) \in C[o,m]$ has a unique solution $x(t) \in C[o,\infty)$ if $y(t) \in C[o,\infty)$.[1)] Moreover, a sufficient asymptotic stability of system (7) is given if all roots of the characteristic equation

$$(8) \quad \det H(s) = o$$

have negative real parts only.
Now, if $y(t) \in C_N$, it will be seen that a similar condition, namely

$$(9) \quad \det(H(h\cdot\ell) - 2\pi ik) \neq o; \; k \in \mathbb{Z}$$
$$h = 2\pi i/N; \; -N \le \ell \le -1$$

is sufficient for the existence of a <u>unique</u> solution $x(t) \in C_N^1$ (in fact, conditio (9) is necessary too, see [5]).
To prove the last assertion, we observe first that by a result of Przeworska-Rolewicz [7], system (7) <u>with $y \in C_N$</u> has an <u>N-periodic</u> solution iff the N ordinary differential equations

$$(10) \quad x_j'(t) = \sum_{k=o}^{m} A_k (e^{hj})^k x_j(t) + y_{(j)}(t), \quad y_{(j)}(t) \in E_j, \; 1 \le j \le N$$

have <u>N-periodic</u> solutions. In this case a solution of (7) is of the form

$$(11) \quad x = \sum_{j=1}^{N} x_j, \text{ with each } x_j \text{ a solution of (10).}$$

Moreover, (7) has a unique N-periodic solution iff equations (10) have unique

1) and the solution is real if g and y are real

solutions in E_j, $1 \le j \le N$. By linearity we then can infer the existence of a unique real solution $x(t) \in C_N^1$, provided $y(t) \in C_N$ is a real function. Indeed, it is rather easy to see that statements (i) - (iii) are equivalent

(i) $x_j'(t) - \sum_{k=o}^{m} A_k (e^{hj})^k x_j(t) = o$ has a nontrivial solution in E_j.

(ii) $f'(t) = A_j f(t)$ has a nontrivial solution $f(t) \in C_1$, where

$$(12)\quad A_j = hj + \sum_{k=o}^{m} A_k (e^{hj})^k;\quad 1 \le j \le N,\ h = 2\pi i/N.$$

(iii) $\det(\exp A_j - I) = o$, $1 \le j \le N$.

Simple spectral considerations show that (iii) is equivalent to $\det(A_j - 2\pi ik) = o$; $k \in \mathbb{Z}$, $1 \le j \le N$. Condition (9) then implies $\det(\exp A_j - I) \ne o$ and thus (7) has only the trivial solution in C_N. But from this follows, see [7], [5], that (7) has exactly one solution $x(t) \in C_N^1$ for $y(t) \in C_N$.

A. General solution

Integrating systems (10), we see that

$$(13)\quad x_j(t) = \exp(\bar{A}_j t)(d_j + \int_o^t \exp(-\bar{A}_j s)\, y_{(j)}(s) ds);\ d_j \in \mathbb{C}^n$$

with

$$(14)\quad \bar{A}_j = \sum_{k=o}^{m} A_j (e^{hj})^k = A_j - hj,\quad 1 \le j \le N \text{ and } A_j \text{ as in (12)},$$

is a solution of (10), but not necessarily a N-periodic one. The existence of a (unique) N-periodic solution of (7) now entails a d_j such that $x_j \in E_j$ i.e. $x_j = x_{(j)}$, $1 \le j \le N$. To determine d_j we use (6) and have together with (12)

$$x_j(t) e^{hjt} = \exp(A_j t)(d_j + \int_o^t \exp(-\bar{A}_j s)\, y_{(j)}(s) ds) \in C_1.$$

By periodicity and putting $t = o$, one gets through condition (9)

$$(15)\quad d_j = (I - \exp A_j)^{-1} \int_o^t \exp(-\bar{A}_j s)\, y_{(j)}(s) ds,\quad 1 \le j \le N.$$

Thus, according to (11), $x = \sum_{j=1}^{N} x_j$ is the unique solution of (7) where each x_j is given by (13) together with (15) and (5).

B. Fourier-series solution

Let $y(t) \in C_N$ be of bounded variation on [o,N].

By (5) and (6) $y_{(j)}(t) = e^{-hjt} f_j(t)$, $1 \le j \le N$, and

$$f_j(t) = \frac{1}{N} \sum_{k=o}^{N-1} e^{h \cdot j(t-k)} y(t-k)$$

with $f(t) \in C_1$ and of bounded variation on [o,1]. Hence

$$(16) \quad f_j(t) = \sum_{k=-\infty}^{+\infty} a_{jk} e^{2\pi ikt}, \; a_{jk} = \int_o^1 f_j(t)\, e^{-2\pi ikt} dt \in \mathbb{C}^n, k \in \mathbb{Z}$$

and the integration of systems (10) can be carried out because of the uniform convergence of (16), see [9]. This leads to (unique) solutions x_j of (10) in E_j, $1 \le j \le N$, of the form

$$x_j(t) = \Big(\sum_{k=-\infty}^{+\infty} b_{jk} e^{2\pi ikt}\Big) e^{-hjt} \text{ with } b_{jk} = (2\pi ik - A_j)^{-1} a_{jk}, \; k \in \mathbb{Z}.$$

Therefore, analogous to case A.

$$x(t) = \sum_{j=1}^{N} \Big(\sum_{k=-\infty}^{+\infty} b_{jk}\, e^{2\pi ikt}\Big) e^{-hjt}$$

is the solution of (7) for $y(t) \in C_N$ and of bounded variation.

1.2 Case $B = o$, $\mathcal{Q}_N = F_N$

We proceed very much as in case $B = I$ and consider the delay system

$$(17) \quad L(x_t) = \sum_{k=o}^{m} A_k x(t-k) = y(t).$$

We make the further assumptions

$$(18) \quad \text{(i)} \;\; \det A_o \neq o, \;\; \text{(ii)} \;\; \det\Big(\sum_{k=o}^{m} A_k\Big) \neq o.$$

Then, due to (18), point (i), system (17), subject to the initial condition (2) with $g(t) \in F[o,m]$, has a unique solution $x(t) \in F[o,\infty)$ if $y(t) \in F[o,\infty)$[1]. (18) point (ii) entails (canonical form of λ-matrices [4]) that $\det H(s)$ has only a finite number of zeros. A sufficient (and necessary) condition for the asymptotic stability of (17) is then given [6],[8] if

(19) All roots of the characteristic equation have positive real parts only.

What remains is to show that by (19), system (17) has a unique solution $x(t) \in F_N$ for $y(t) \in F_N$, which is real if y is real.

---

[1] The solution is real if g and y are real.

In fact, as in the case $B = I$, system (17), with $y \in F_N$ has a solution in F_N iff the N matrix equations

$$(20)\quad \bar{A}_j x_j(t) = \sum_{k=o}^{m} A_k (e^{hj})^k x_j(t) = y_{(j)}(t) \in E_j$$

have solutions $x_j(t)$ in F_N and this solution is given by (11).
Now, $\bar{A}_j = H(-hj)$, $1 \le j \le N$, hence by condition (19), $\bar{A}_j$ is invertible. From (6) we can infer that any nxn-matrix maps E_j into E_j. Consequently, using (5)

$$(21)\quad x_j(t) = \frac{1}{N} \sum_{k=o}^{N-1} (e^{-hj})^k (\bar{A}_j)^{-1} y(t-k)$$

is the unique solution of (20) (in E_j). Thus $x(t) = \sum_{j=1}^{N} x_j(t)$ is the unique N-periodic solution of system (17) for $y \in F_N$.

2. References

[1] Bellman-Cooke, *Differential-Difference Equations*, Academic Press, New York 1963

[2] Deistler, Z-Transform and Identification of Linear Econometric Models with Autocorrelated Errors, *Metrica*, Band 22, 1975.

[3] Driver, *Ordinary and Delay Differential Equations*, Springer-Verlag, New York 1977.

[4] Gantmacher, *Matrizenrechnung 1*, VEB Deutscher Verlag der Wissenschaften, Berlin 1958.

[5] Hale, *Theory of Functional Differential Equations*, Springer-Verlag, New York 1977.

[6] Miller, *Linear Difference Equations*, W.A. Benjamin, Inc., New York, Amsterdam 1968.

[7] Przeworska-Rolewicz, *Equations with Transformed Argument*, Elsevier, Amsterdam 1973.

[8] Rubio, *The Theory of Linear Systems*, Academic Press, New York and London 1971

[9] Schönhage, *Approximationstheorie*, de Gruyter, Berlin, New York 1971

GAME THEORY AND RELATED TOPICS
O. Moeschlin, D. Pallaschke (eds.)

ELEMENTARY PROOF OF THE MEASURABLE UTILITY THEOREM

Andrzej Wieczorek
Institute of Computer Science
Polish Academy of Sciences
00-901 Warszawa, P.O. Box 22

A relation $\precsim$ in a set P is a *linear pre-order* iff it is reflexive, transitive and $x \precsim y$ or $y \precsim x$ holds for every $x,y \in P$. We write "$x \sim y$" for "$x \precsim y$ and $y \precsim x$" and "$x \prec y$" for "$x \precsim y$ and not $y \precsim x$". The equivalence class of an $x \in P$ with respect to $\sim$ is denoted by $[x]$.

An *open gap* is an ordered pair of the form $([x],[y])$ such that $x \prec y$ and there exists no $z \in P$ for which $x \prec z \prec y$. In this case $[x]$ is called the *lower extreme* and $[y]$ is the *upper extreme* of the gap.
If P is a topological space, a linear pre-order $\precsim$ in P is said to be *continuous* iff sets $\{y \in P | x \precsim y\}$ and $\{y \in P | y \precsim x\}$ are closed for every $x \in P$.

A real function u on P is called *monotone* (resp. it is a *representation* of $\precsim$) iff for every $x,y \in P$, $x \precsim y$ implies (resp. is equivalent to) $u(x) \leq u(y)$.

Throughout the paper correspondences $F: K \to 2^L$ are identified with their graphs $\{(k,l) \in K \times L | l \in F(k)\}$. Consequently, for a set $F \subset K \times L$ and a $k \in K$, the symbol $F(k)$ will denote $\{l \in L | (k,l) \in F\}$.

"Measurability" of a set will always mean the membership of a σ-field in question; similarly, "measurability" of a function means that counter-images of measurable sets belong to this σ-field; in both cases there is no measure involved.
If a metric space is regarded as a measurable space, its measurable subsets are, by definition, Borel sets.

We are now in a position to formulate the main result of the paper:

<u>Measurable Utility Theorem:</u>
Let $(T,\mathcal{T})$ *be a measurable space and let* X *be a Polish (i.e. metric separable complete) space. Let* Φ[1] *and* Ψ *be measurable subsets of* $T \times X$ *and* $T \times X^2$, *respectively, such that for every* $t \in T$, $\Psi(t)$ *is a graph of a continuous linear pre-order* $\precsim_t$ *in* $\Phi(t)$.

---

[1] *It has been noticed by Gisela Brentzel that measurability of* Φ *is a consequence of the other assumptions.*

Then for every probabilistic measure τ *on* T *there exists a measurable real function* u *on* Φ *such that* $u(t,\cdot)$ *is, for* τ*-almost every* $t \in T$, *a continuous representation of* $\precsim_t$.

The first formulation of the Theorem and a proof in a particular case: $\Phi(t)$ *is connected for every* t, are due to Aumann [1969].
Aumann was also interested in the general case. He noticed in [1969] without a proof that the Theorem would be true in its general form within a system of axioms of set theory in which every set both PCA and CPCA is Lebesgue measurable, provided such a system was known to be noncontradictory.

Another contribution has been made by Wesley [1976] who succeeded to avoid the assumption of connectedness of $\Phi(t)$ (dealing however with a particular case: $(T,\mathcal{T})$ *is the unit interval and* τ *is the Lebesgue measure*). Unfortunately, his proof requires very strong methods of mathematical logic (like generic models, forcing etc.) which are usually avoided, if it is possible.

The present paper contains an elementary proof of the Measurable Utility Theorem, the formulation of which is more general than both cases considered by Aumann and Wesley.

Many theorems on numerical representation of pre-ordering relations have been known for a long time. The Measurable Utility Theorem is a natural "measurable" extension of theorems of this group; similarly, proofs of the Theorem may be related to their proofs. We formulate below the most popular among those theorems, due to Debreu [1954]:

Utility Representation Theorem: *A continuous linear pre-order* $\precsim$ *in a topological space with a countable base has a continuous numerical representation* u.

A standard method to prove the Utility Representation Theorem is the following: Choose a countable dense subset S of the space in which the pre-order is defined; by means of a "dyadic" mapping define a representation on S and finally extend it continuously to all the space. The so constructed function fails to be a representation if there are open gaps. However, in this case, it suffices to enlarge S at the beginning by one element from every extreme of every open gap and to proceed as before.

The proof of the Measurable Utility Theorem given by Aumann [1969] rests upon this idea. One has to select, in a measurable manner, a countable dense subset of $\Phi(t)$ for almost every t (this is done by means of Castaing's [1967] Selection

Theorem quoted in the sequel as 9.b) and to adopt the procedure mentioned above taking care of measurability. In Aumann's case $\Phi(t)$ is connected for every t and therefore there are no open gaps at all. In our general case we must also measurably select points from the extremes of every open gap in $\Phi(t)$ for almost every t: this is the main difficulty to overcome in this paper. This is done by means of a theorem of Lusin [1930], quoted in the sequel as 8.

There is also an alternative method to prove the Utility Representation Theorem found by Mehta [1977] and related to Nachbin's [1965] theory of normal preorders. The method makes use of a countable open base in the space in question rather than of its countable dense subset as in the method mentioned above. It is also possible to find a proof of the Measurable Utility Theorem related to this idea. The author will publish such a proof in a separate paper (Wieczorek [1980]).

1. PRELIMINARIES

Throughout the paper we use the common notation: $\mathbb{N}$ = positive integers; $\mathbb{R}$ = reals ; $\mathbb{I}$ = unit interval [0,1] ; $\mathrm{Proj}_A C$ = projection of C to A ; χ_A = indicator function to A.

The facts and definitions listed below will be used in the proof:

A subset of a Polish space is called *analytic* iff it is an image, under a measurable function, of a Borel subset of a Polish space.

1. *The intersection and the union of two analytic sets is analytic.*
2. *A subset of a Polish space which is a measurable image of an analytic set is analytic.*
3. *The counter-image of an analytic set under a measurable function defined over an analytic set is analytic.*
4. *A subset of a Polish space which is a one-to-one measurable image of a Borel set is Borel.*
5. *A measurable function defined on a subset of a Polish space is extendable, in a measurable manner, to any Borel set containing its domain.*
6. *If* f *and* g *are measurable real functions defined over the same Borel set then* $\{x \mid f(x) = g(x)\}$ *is Borel.*
7. *If* $A \subset \mathbb{I} \times \mathbb{R}$ *is analytic then so is* $\{t \in \mathbb{I} \mid A(t) \text{ is uncountable}\}$.
8. *If* $A \subset \mathbb{I} \times \mathbb{R}$ *is analytic and for every* $t \in \mathbb{I}$, $A(t)$ *is at most countable*

then there exist measurable functions $f_i:\Pi \to \mathbb{R}$, $i\in\mathbb{N}$, *such that for every* $t\in\Pi$, $A(t)\subset\{f_i(t)\,|\,i\in\mathbb{N}\}$.

In 9 - 12 below D is a Polish space and in 10 - 13 $(M,\mathcal{M})$ is a measurable space:

9. *Let* $A\subset\Pi\times D$ *be analytic and let* μ *be a probabilistic measure on Borel subsets of* Π. *Then there exist Borel sets* C_1, C_2 *of equal measure* μ *such that* $C_1\subset\mathrm{Proj}_\Pi A\subset C_2$ *and*
   a. *a measurable function* $f:C_1\to D$ *such that* $f(t)\in A(t)$ *for all* $t\in C_1$
   b. *a sequence* (f_i) *of measurable functions from* C_1 *to* D *such that for every* $t\in C_1$, $\{f_i(t)\,|\,i\in\mathbb{N}\}$ *is a dense subset of* $A(t)$.
10. *Let* μ *be a probabilistic measure on* $\mathcal{M}$. *For every set* $B\subset M$ *which is a projection of a measurable subset of* $M\times D$ *(in particular, if* $(M,\mathcal{M})$ *is the unit interval with Borel sets, for every analytic set* $B\subset\Pi$ *) there exist sets* $C_1, C_2\in\mathcal{M}$ *of equal measure* μ *such that* $C_1\subset B\subset C_2$.
11. *The graph of a measurable function* $f:M\to D$ *is a measurable set.*
12. *For every measurable set* $A\subset M\times D$ *there exists a countably generated* σ*-field* $\mathcal{M}'\subset\mathcal{M}$ *such that* A *is measurable in the product of* $(M,\mathcal{M}')$ *and* D.

A measurable space $(N,\mathcal{N})$ is a *subspace* of $(M,\mathcal{M})$ iff $N\subset M$ and $\mathcal{N}=\mathcal{M}|N := \{A\cap N\,|\,A\in\mathcal{M}\}$ (without requiring $N\in\mathcal{M}$).

13. *If* $\mathcal{M}$ *is countably generated and contains all singletons* $\{t\}$, $t\in M$, *then* $(M,\mathcal{M})$ *is isomorphic with a subspace of* Π.
14. *Given a topological space with a countable base, equipped with a continuous linear pre-order, there are at most countably many open gaps.*

Most of the propositions formulated above are either generally known or can be found in Kuratowski [1966]. As for the rest, 8 is due to Lusin [1930, pp. 243 and 247]; proofs of 9.b and 12 are in Aumann [1969].

2. CONSTRUCTION FOR A SINGLE RELATION

Let P be a topological space equipped with a continuous linear pre-order $\precsim$. With an arrangement $b=(b_i)$ of a countable dense subset of P we associate a function $\delta:\mathbb{N}\to\mathbb{R}$ defined inductively for i=1,2,.. as follows (we take here $\sup\emptyset=0$ and $\inf\emptyset=1$):

$$\delta(i) = \tfrac{1}{2}[\inf\{\delta(j)\,|\,j<i,\ b_i\precsim b_j\} + \sup\{\delta(j)\,|\,j<i,\ b_j\precsim b_i\}].$$

Then a function $d:P\to\mathbb{R}$ is defined by

$$d(x) = \sup\{\delta(i)\,|\,b_i\precsim x\}.$$

This construction (or an equivalent) is actually the major step in the usual procedure of constructing a continuous representation.
The following proposition can be obtained by techniques of Debreu [1954] and Jaffray [1975].

PROPOSITION 1

a. d *is continuous and monotone;*

b. *the restriction of* d *to* $|b| = \{b_i \mid i \in \mathbb{N}\}$ *is a representation of* $\precsim$ *restricted to* $|b|$;

c. *for every rational* r *of the form* $m/2^n$, $m,n \in \mathbb{N}$, *not in* $d(|b|)$ *but in* $\text{conv}(d(|b|))$, *there exist* $i,j \in \mathbb{N}$ *such that* $[d(b_i) < r < d(b_j)$ *and there exists no* $k \in \mathbb{N}$ *with* $d(b_i) < d(b_k) < d(b_j)]$.

We say that the sequence b *discriminates an extreme* $[x]$ in P iff $[x]$ contains no element of b. It *discriminates an open gap* iff it discriminates either its upper extreme or its lower extreme.

By means of Proposition 1 we shall prove:

PROPOSITION 2

For every real number ξ *one of the cases* a-d *holds:*

a. $d^{-1}(\xi)$ *is empty;*

b. $d^{-1}(\xi)$ *is equal to an equivalence class* $[x]$;

c. $d^{-1}(\xi) = [x] \cup [z]$, *where* $([x],[z])$ *is a discriminated open gap;*

d. $d^{-1}(\xi) = [x] \cup [y] \cup [z]$, *where* $([x],[y])$ *and* $([y],[z])$ *are discriminated open gaps; in this case* $[y]$ *is not discriminated.*

PROPOSITION 3

If an open gap $([x],[z])$ *is discriminated then* $d(x) = d(z)$.

For the proof of Proposition 2 suppose that $d^{-1}(\xi)$ is nonempty. It is clear, by the definition of d, that whenever $d^{-1}(\xi)$ contains an x, it contains the whole equivalence class $[x]$.
Suppose that $d^{-1}(\xi)$ is a union of exactly two equivalence classes, say $[x]$ and $[z]$ and let $x \prec z$. If the set $\{u \in P \mid x \prec u \prec z\}$ were nonempty, it would contain (being open) an element b_k of the dense set $|b|$; in this case however, by monotonicity of d, $d(x) \leq d(b_k) \leq d(z)$; since $d(x) = d(z)$, $d^{-1}(\xi) \supset [x] \cup [z] \cup [b_k]$, a contradiction. Thus $\{u \in P \mid x \prec u \prec z\}$ is empty which means that $([x],[z])$ is an

open gap. If it were not discriminated, there would exist $b_i, b_j \in |b|$ such that $b_i \sim x$ and $b_j \sim z$. Since $b_i \prec b_j$, by Proposition 1.b $d(b_i) < d(b_j)$ and consequently also $d(x) < d(z)$, a contradiction.
Suppose that $d^{-1}(\xi)$ contains more than three equivalence classes, say $[x]$, $[y]$, $[z]$ and $[v]$ with $x \prec y \prec z \prec v$. In this case the set $\{u \in P | x \prec u \prec z\}$ is open and nonempty, hence it contains some b_i. The set $\{u \in P | b_i \prec u \prec v\}$ is also open and nonempty (it contains z), thus it contains some b_j. Since $b_i \prec b_j$, we have $d(x) \leq d(b_i) < d(b_j) \leq d(v)$ which contradicts $d(x) = d(v)$. The case is impossible.
Finally, suppose that $d^{-1}(\xi)$ is a union of exactly three equivalence classes, say $[x]$, $[y]$ and $[z]$ with $x \prec y \prec z$. Then, as before, both sets $\{u \in P | x \prec u \prec y\}$ and $\{u \in P | y \prec u \prec z\}$ must be empty and hence $([x],[y])$ and $([y],[z])$ are open gaps indeed. Similarly, both of them are discriminated. Since $\{u \in P | x \prec u \prec z\} = [y]$ is open, there exists $b_i \in [y]$, i.e. $[y]$ is not discriminated.

In Proposition 3, first suppose that $[z]$ is discriminated: then $d(x) = d(z)$ immediately by the definition of d. Now suppose that $[x]$ is discriminated while $[z]$ is not and $d(x) < d(z)$. Obviously $0 \leq d(x)$, $d(z) \leq 1$; choose $r = m/2^n$, $m, n \in \mathbb{N}$, such that $d(x) < r < d(z)$. Then by Proposition 1.c there exists b_i with $d(b_i) < r$ and such that $d(b_i) < d(b_j) < r$ holds for no $j \in \mathbb{N}$. Choose $b_k \in [z]$. Since $d(b_i) < d(b_k)$, by Proposition 1.b, $b_i \prec b_k$ and hence $b_i \precsim x$, i.e. $d(b_i) \leq d(x)$; actually $b_i \prec x$ because $[x]$ is discriminated. The set $\{u \in P | b_i \prec u \prec b_k\}$ being open and nonempty contains some b_j; clearly $b_i \prec b_j \precsim x$. Thus, by Proposition 1.a and 1.b, $d(b_i) < d(b_j) \leq d(x) < r$, contradicting the way i has been chosen.

From Propositions 1.a and 2 we immediately obtain a corollary:

COROLLARY 1

If there is no open gap discriminated by b *then* d *is a representation of* $\precsim$.

Moreover, from Propositions 2 and 3:

COROLLARY 2

d *assumes different values on representatives of different discriminated upper extremes. The same holds for discriminated lower extremes. Generally, the word "discriminated" cannot be dropped.*

For P metric it will be convenient to have an alternative formula for d.
Let for $i, j \in \mathbb{N}$, K_{ij} denote an open ball in P of radius j^{-1} around b_i and let $L_{ij} = K_{ij} \setminus \bigcup_{k<i} K_{kj}$. Notice that for every j the family $\{L_{ij} | i \in \mathbb{N}\}$ is a partition of P.

For every j a function $q_j : P \to \mathbb{R}$ is defined by

$$q_j(x) = \sum_{i=1}^{\infty} \delta(i)\, \chi_{L_{ij}}(x).$$

The continuity of d implies

PROPOSITION 4

For every $x \in P$, $d(x) = \limsup_j q_j(x)$.

3. MEASURABILITY

The measurability of functions of certain type will be required twice in the proof. Therefore this property is formulated below in more general terms.

In the notation of the Theorem and under its assumptions we assume that $T' \in \mathcal{T}$ is a set of measure 1 and that $\mathfrak{f} = (f_i)$ is a sequence of measurable functions from T' to X such that for every $t \in T'$, $\{f_i(t) \mid i \in \mathbb{N}\}$ is a dense subset of $\Phi(t)$. In this case, for every $t \in T'$ and $i,j \in \mathbb{N}$, the symbols

$$\delta^t \ , \ d^t \ , \ q_j^t \ , \ K_{ij}^t \text{ and } L_{ij}^t$$

have the same meaning,respectively, as the symbols

$$\delta \ , \ d \ , \ q_j \ , \ K_{ij} \text{ and } L_{ij} \ ,$$

in Section 2, after replacement of the terms P, $\precsim$, $\mathfrak{f}$ in Section 2 by $\Phi(t)$, $\precsim_t$ and $\mathfrak{f}(t)$, respectively, where $\mathfrak{f}(t)$ denotes the sequence $(f_i(t))_{i \in \mathbb{N}}$.

A function $V_{\mathfrak{f}}$ from $\Phi' = \Phi \cap (T' \times X)$ to the reals is defined by

$$V_{\mathfrak{f}}(t,x) = d^t(x).$$

LEMMA 1

There exists a set $T'' \in \mathcal{T}$, $T'' \subset T'$, *of measure* 1 *such that* $V_{\mathfrak{f}}$ *is measurable on* $\Phi'' = \Phi \cap (T'' \times X)$.

For the proof, notice that by 10 and 11, for every i,j there is a set $N_{ij} \in \mathcal{T}$ of measure 0 such that $\{t \in T' \setminus N_{ij} \mid f_i(t) \precsim_t f_j(t)\} \in \mathcal{T}$. Write $T'' = T' \setminus \bigcup_{i,j} N_{ij}$. Then for every i the function $\delta^t(i)$ is measurable in t on T''.

Let, for every i,j

$$K_{ij}^* = \{(t,x) \in \Phi'' \mid \mathrm{dist}(x, f_i(t)) < j^{-1}\}$$

and

$$L_{ij}^* = K_{ij}^* \setminus \bigcup_{k<i} K_{kj}^*.$$

All sets K^*_{ij} and L^*_{ij} are measurable. Moreover, for every j the family $\{L^*_{ij} | i \in \mathbb{N}\}$ is a partition of Φ''.

For every j define a function $Q_j : \Phi'' \to \mathbb{R}$ by

$$Q_j(t,x) = \sum_{i=1}^{\infty} \delta^t(i)\ \chi_{L^*_{ij}}(t,x)$$

and a function $D: \Phi'' \to \mathbb{R}$ by

$$D(t,x) = \limsup_j Q_j(t,x).$$

We see that all the functions Q_j and D are measurable. Moreover, for every $(t,x) \in \Phi''$ and every j, $Q_j(t,x) = q^t_j(x)$.

Hence, by Proposition 4, $D(t,x) = V_{\mathfrak{f}}(t,x)$.

4. PROOF OF THE THEOREM

Suppose the following lemma is true:

LEMMA 2

Under the assumptions of the Theorem *there exist: a set* $T' \in \mathcal{T}$ *such that* $\tau(T') = 1$ *and a sequence* $\mathfrak{f} = (f_i)$ *of measurable functions from* T' *to* X *such that for every* $t \in T'$, $\{f_i(t)\}$ *is a dense subset of* $\Phi(t)$ *and there is no open gap in* $\Phi(t)$ *discriminated by the sequence* $\mathfrak{f}(t) = (f_i(t))_{i \in \mathbb{N}}$.

Then, by Lemma 1, we find a set $T'' \in \mathcal{T}$, $T'' \subset T'$ with $\tau(T'') = 1$ for which the function $V_{\mathfrak{f}}$, as defined in Section 3, is measurable on $\Phi'' = \Phi \cap (T'' \times X)$. By Corollary 1 for every $t \in T'$, $V_{\mathfrak{f}}(t,\cdot)$ is a continuous representation of $\precsim_t$.

Any measurable function $u: \Phi \to \mathbb{R}$ which coincides with $V_{\mathfrak{f}}$ on Φ'' satisfies the hypothesis of the Theorem.

Thus we must only prove Lemma 2.

a. This part of the proof is taken from Aumann [1969]. We repeat the construction in order to introduce the notation.

By means of 12 the general case is easily reduced to the case: $\mathcal{T}$ is countably generated. Identifying points in every atom we may additionally assume that $\mathcal{T}$ contains all singletons $\{t\}$, for $t \in T$.

By 13 we may assume that $(T, \mathcal{T})$ is a subspace of the unit interval $\mathbb{I}$.

Define a measure σ on Borel subsets of $\mathbb{I}$ letting $\sigma(B) = \tau(B \cap T)$.

By definition, there exist Borel sets $\tilde{\Phi}$ and $\tilde{\Psi}$ in $\mathbb{I} \times X$ and $\mathbb{I} \times X^2$, respectively, such that $\Phi = \tilde{\Phi} \cap (T \times X)$ and $\Psi = \tilde{\Psi} \cap (T \times X^2)$.

By 9.b, there exists a Borel set $S_1 \subset \mathrm{Proj}_{\Pi}\tilde{\Phi}$ with $\sigma(S_1)=1$ for which there exist measurable functions $h_i : S_1 \to X$, $i \in \mathbb{N}$, such that for every $t \in S_1$, $\{h_i(t)\}$ is a dense subset of $\tilde{\Phi}(t)$.
Write $T_1 = T \cap S_1$ and notice that $\tau(T_1)=1$.
For every i let $\tilde{h}_i$ be the restriction of h_i to T_1 and let h denote the sequence $(\tilde{h}_i)$. By Lemma 1 the function

$$V_h : \Phi \cap (T_1 \times X) \to \mathbb{R}$$

is measurable on $\Phi_1 = \Phi \cap (T_2 \times X)$ for some $T_2 \in \mathcal{T}$ with $\tau(T_2)=1$.

b. The restriction $V_{h|\Phi_1}$ being measurable has, by 5, a measurable extension W to $\Phi_2 = \tilde{\Phi} \cap (S_1 \times X)$.
Write

$$\Xi = \{(t,x,y) \in S_1 \times X^2 \mid W(t,x) = W(t,y),\ (t,x,y) \in \tilde{\Psi} \text{ and } (t,y,x) \notin \tilde{\Psi}\}.$$

By 6, Ξ is a Borel set.
Notice that for every $t \in T_2$

$$\Xi(t) = \{(x,y) \in \Phi(t) \times \Phi(t) \mid V_h(t,x) = V_h(t,y) \text{ and } x \prec_t y\}.$$

Denote

$$\Xi^* = \{(t,y) \in S_1 \times X \mid \exists x\ (t,x,y) \in \Xi\}$$

and

$$\Xi_* = \{(t,x) \in S_1 \times X \mid \exists y\ (t,x,y) \in \Xi\}.$$

Both sets Ξ^* and Ξ_*, as projections of Ξ, are analytic.

In view of Propositions 2 and 3, for every $t \in T_2$, $\Xi^*(t)$ (resp. $\Xi_*(t)$) is equal to the union of all upper (resp. lower) extremes of all discriminated open gaps in $\Phi(t)$; thus every discriminated extreme in $\Phi(t)$ is a subset of either $\Xi^*(t)$ or $\Xi_*(t)$.

c. Comparing the last remark with Corollary 2 we shall in this step eliminate from $\Xi^*(t)$ and $\Xi_*(t)$, measurably, those extremes which are not discriminated.

For every $i \in \mathbb{N}$ let

$$\Lambda_i = \{(t,x,y) \in S_1 \times X^2 \mid (t,x,y) \in \tilde{\Psi},\ (t,y,x) \in \tilde{\Psi} \text{ and } y = h_i(t)\}.$$

By 11, all Λ_i are Borel sets.
Let, for every i,

$$L_i = \{(t,x) \in S_1 \times X \mid \exists y\ (t,x,y) \in \Lambda_i\}.$$

As a one-to-one measurable image of Λ_i (actually a projection), every set L_i is, by 4, Borel.

Notice that for every $t \in T_2$

$$L_i(t) = \{x \in \Phi(t) \mid x \sim_t h_i(t)\}.$$

Write $L = \bigcup_i L_i$ and $\Theta^* = \Xi^* \backslash L$, $\Theta_* = \Xi_* \backslash L$.
Obviously both sets Θ^* and Θ_* are analytic.

Thus we can deduce that for every $t \in T_2$, $\Theta^*(t)$ is equal to the union of all discriminated upper extremes while $\Theta_*(t)$ is equal to the union of all discriminated lower extremes.

d. Our next objective is to find measurable partitions of Θ^* and Θ_*, every member of which selects, for almost every $t \in T_2$, at most one discriminated extreme in $\Phi(t)$.

Let $\theta : \Phi_2 \to \mathbb{I} \times \mathbb{R}$ be defined by

$$\theta(t,x) = (t, W(t,x)).$$

Since θ is measurable, $K = \theta(\Theta^* \cup \Theta_*)$ is analytic.
From 14 and Corollary 2 it follows that for every $t \in T_2$, $K(t)$ is at most countable.
By 7 the set $U = \{t \in S_1 \mid K(t) \text{ is countable}\}$ has an analytic complement. Since $U \supset T_2$ and $\tau(T_2) = 1$, thus every Borel set containing U is of σ-measure 1. Therefore, by 10, we can find a Borel set $S_2 \subset U$ such that $\sigma(S_2) = 1$.
Consider the set $K_1 = K \cap (S_2 \times \mathbb{R})$. It is also analytic, therefore by 8, there exist measurable functions $\gamma_i : \mathbb{I} \to \mathbb{R}$, $i \in \mathbb{N}$, such that $K_1(t) \subset \{\gamma_i(t) \mid i \in \mathbb{N}\}$ for every $t \in S_2$.
By 11, each graph $Gr\gamma_i$ is a Borel set, thus each set

$$G^i = \Theta^* \cap \theta^{-i}(Gr\gamma_i) \quad \text{and} \quad G_i = \Theta_* \cap \theta^{-1}(Gr\gamma_i)$$

is analytic.

Obviously by Corollary 2, for every $t \in T_3 = S_2 \cap T$, every discriminated upper extreme in $\Phi(t)$ is of the form $G^i(t)$ for some i while every discriminated lower extreme is of the form $G_i(t)$ for some i.

e. It is now natural to find measurable selections from the sets G^i and G_i.
From 9.a we find, for every i, Borel sets N^i and M^i such that $\sigma(N^i) = 0$ and $N^i \cup M^i \supset \mathrm{Proj}_{\mathbb{I}}\, G^i$ and a measurable function $g^i : M^i \to X$ such that $g^i(t) \in G^i(t)$ for every $t \in M^i$.
Analogously, for every i there are Borel sets N_i and M_i such that $\sigma(N_i) = 0$ and $N_i \cup M_i \supset \mathrm{Proj}_{\mathbb{I}}\, G_i$ and a measurable function $g_i : M_i \to X$ such that $g_i(t) \in G_i(t)$ for every $t \in M_i$.
For the set $S_3 = S_2 \backslash \bigcup_i (N^i \cup N_i)$ we have $\sigma(S_3) = 1$; hence we can define $T' = S_3 \cap T$ with $\tau(T') = 1$.

f. Concluding the construction we shall define a sequence $\mathfrak{f} = (f_i)$ of measurable functions from T' to X: for every i

$$f_{3i-2} \quad \text{is the restriction of } h_i \text{ to } T';$$

$$f_{3i-1}(t) = \begin{cases} g^i(t) & \text{if} \quad t \in T' \cap M^i, \\ h_1(t) & \text{if} \quad t \in T' \setminus M^i; \end{cases}$$

$$f_{3i}(t) = \begin{cases} g_i(t) & \text{if} \quad t \in T' \cap M_i, \\ h_1(t) & \text{if} \quad t \in T' \setminus M_i. \end{cases}$$

Since for every $t \in T'$, $\{h_i(t)\}$ is a dense subset of $\Phi(t)$, so is $\{f_i(t)\}$. By the construction, for every $t \in T'$, $\{g^i(t)\}$ and $\{g_i(t)\}$ have nonempty intersections with every discriminated upper and lower, respectively, extreme in $\Phi(t)$. Thus $\mathfrak{f}$ satisfies the hypothesis of Lemma 2.

Acknowledgements: I wish to thank Janusz Kaniewski for drawing my attention to the theorem of Lusin and to Tadeusz Bromek and Marian Srebrny whose suggestions allowed to improve an earlier version of the paper.

References

Aumann, R.J., 1969, Measurable utility and the measurable choice theorem, in: La Decision (Editions du Centre National de la Recherche Scientifique), 15 - 26

Castaing, C., 1967, Sur les multi-applications mesurables, Thèse (Caen)

Debreu, G., 1954, Representation of a preference ordering by a numerical function, in: R.M. Thrall, C.H. Combs and R.L. Davis, eds., Decision Processes, (Wiley, New York)

Jaffray, J., 1975, Existence of a continuous utility function: an elementary proof, Econometrica 43, 981 - 983

Kuratowski, K., 1966, Topology I (Academic Press - PWN, New York - Warszawa)

Lusin, N., 1930, Leçon sur les ensembles analytiques et leurs applications, (Gauthier - Villars, Paris)

Mehta, G., 1977, Topological ordered spaces and utility functions, International Economic Review 18, 779 - 782

Nachbin, L., 1965, Topology and Order (Van Nostrand, Princeton)

Wesley, R., 1976, Borel preference orders in markets with a continuum of traders, Journal of Mathematical Economics 3, 155 - 165

Wieczorek, A., 1980, Measurable utility theorem, to appear

PART IV
MATHEMATICAL ECONOMICS

GAME THEORY AND RELATED TOPICS
O. Moeschlin, D. Pallaschke (eds.)

AGGREGATION OF DEMAND IN CASE OF NONCONVEX PREFERENCES

E. Dierker, H. Dierker, and W. Trockel
Department of Economics, SFB 21
University of Bonn

1. Introduction

Economists commonly believe that the consistency of individual decisions in a purely competitive economy is achieved by the price mechanism. However, the consistency of individual decisions cannot be expected if the aggregate decision is not uniquely determined by the price system. If consumers' preferences exhibit nonconvexities, individual demand need not be uniquely determined by the price system. A first question then is: When will aggregate demand of the consumption sector be uniquely determined? Because of the upper hemi-continuity of the mean demand correspondence this amounts to the question of when mean demand will be a continuous function.

But even a continuous aggregate demand function is not suitable to explain the consistency of individual demand decisions if it has extremely steep slopes. If aggregate demand is too sensitive with respect to prices, very small price variations lead to a considerable deviation from equilibrium. Furthermore, one would like to be able to show that equilibria are generically regular. So a next question is: When will aggregate demand be a differentiable function? Our approach in this paper has been motivated by this question.

The problem of smoothing demand by aggregation was posed by Debreu (1972) and W. Hildenbrand (1974). The continuity and differentiability of aggregate demand has been studied by Sondermann (1975, 1976, 1978) and by Araujo and Mas-Colell (1978). A major difference between their work and ours is that they stipulate a finite-dimensional manifold structure on the space of preferences considered, an assumption which we want to avoid. The manifold structure is used to formulate the notion of dispersed preferences. The conditions in the papers just mentioned imply that mean demand is a continuous function, but they do not yield differentiability everywhere. For the study of continuity of mean demand without the use of derivatives, see Mas-Colell and Neuefeind (1977), Neuefeind (1978), Yamazaki (1978), and W. Hildenbrand (1978).

In our approach to the problem of differentiability of aggregate demand we concentrate on smooth preferences. Since the space of consumers' characteristics consists of two components, namely the space of preferences, which has little structure, and the wealth space, which has the structure of the real line, we decided to carry out the aggregation in two steps, first with respect to wealth, then with respect to preferences. In this paper we deal with the first step only and show that aggregation with respect to wealth already brings about a considerable smoothing effect.

Section 2 introduces the model. In section 3 we show that for 1 large subset of utility functions of a certain class and for all price systems the mean demand of all consumers whose tastes are represented by a given utility function in that subset is a uniquely determined bundle of commodities. This together with the well known upper hemi-continuity of the mean demand correspondence implies that the mean demand of the consumption sector is a continuous function. This result is essentially a consequence of the multijet transversality theorem (Golubitsky and Guillemin (1975)).

In section 4 we point out that for a fixed preference relation, aggregation with respect to a continuous income distribution leads to a continuously differentiable demand function except for prices in a closed null set. The prices in this null set correspond to three types of difficulties: vanishing Gaussian curvature of indifference surfaces, critical jumps, and multiple jumps.

Points with vanishing Gaussian curvature can be considered as points where a catastrophe occurs. In case of a cusp catastrophe we show that vanishing Gaussian curvature does not destroy differentiability of mean demand of a fixed preference as long as no other disturbance occurs simultaneously.

2. The Model

Let us consider the consumption sector of an economy with $\ell \geq 2$ commodities The commodity space is $\mathbb{R}^\ell$. We consider prices in

$$S = \{p \in \mathbb{R}^\ell \mid p \gg o,\ ||p|| = 1\}.$$

$||\cdot||$ denotes the Euclidean norm, $p \gg o$ means $p_h > o$ for all $h \in \{1,\ldots,\ell\}$. Every consumer has the consumption set

$$X = \{\ x \in \mathbb{R}^\ell \mid x \gg o\}.$$

Let $\mathcal{U}$ denote the set of C^∞ utility functions $u : X \to \mathbb{R}$ satisfying assumptions U1), U2), U3) below.

U1) $Du\ (x) \gg o$ for all $x \in X$ (monotonicity).

The following boundary assumption keeps demand away from the boundary of X.

U2) $cl\ (u^{-1}(u(x))) \subset X$ for all $x \in X$.

Let $g(x) = Du(x)\ ||Du(x)||^{-1}$. The third assumption makes the sets $g^{-1}\ (p), p \in S$, smooth one-dimensional manifolds.

U3) $g : X \to S$ is a submersion.

This assumption rules out the simultaneous vanishing of two or more principal curvatures of the indifference hypersurfaces, it does not preclude, however, the Gaussian curvature of these surfaces to become zero.

Let $\mathcal{P}$ denote the set of preference relations $\lesssim$ representable by utility functions in $\mathcal{U}$, i.e. there is $u \in \mathcal{U}$ such that $x \lesssim y \iff u(x) \leq u(y)$.

The *wealth* of a consumer is a number $w \in\ ]o, \infty[$. For an agent, described by his wealth $w \in\ ]o, \infty[$ and his preference $\lesssim\ \in \mathcal{P}$, the demand at price system $p \in S$ is

$$\varphi(\lesssim, p, w) = \{x \in X \mid px \leq w,\ x \prec y \Rightarrow py > w\}.$$

The *mean demand* of all consumers is the integral of the demand correspondence φ with respect to a measure on the space $\mathcal{P} \times\ ]o, \infty[$ of consumers' characteristics. The *integral of a correspondence* is defined as follows: Let $(\Omega, \mathcal{A}, \nu)$ be a measure space, $\psi : \Omega \to \mathbb{R}^m$ be a correspondence. The integral of ψ with respect to ν is the set

$\int \psi \, d\nu := \{\int s \, d\nu \in \mathbb{R}^m | s \in L^1 (\Omega, \mathcal{A}, \nu), s(\omega) \in \psi(\omega) \ \nu\text{-a.e. in } \Omega\}$.

For details see Hildenbrand (1974).

3. Continuous mean demand

Let $\delta_{\lesssim}$ be the wealth distribution for the fixed preference $\lesssim \in \mathcal{P}$. Let $\Phi_{\lesssim}(p) = \int_0^\infty \varphi_{\lesssim}(p,w)\, \delta_{\lesssim}(dw)$ denote the mean demand at price system p with respect to the measure $\delta_{\lesssim}$ on $(]0, \infty[, \mathcal{B}(]0, \infty[))$ given the preference $\lesssim \in \mathcal{P}$ where $\mathcal{B}$ denotes the Borel σ-algebra. Clearly, $\Phi_{\lesssim}(p)$ is a singleton if and only if $\varphi_{\lesssim}(p,w)$ is a singleton for $\delta_{\lesssim}$ - almost every wealth w. To prove this last property for any measure $\delta_{\lesssim}$ which is absolutely continuous with respect to Lebesgue measure λ, it suffices to prove it for λ on $]0, \infty[$. Note that the single - valuedness of $\Phi_{\lesssim}$ at every $p \in S$ implies that $\Phi_{\lesssim}: S \to \mathbb{R}^{\ell}$ is a continuous function.

Theorem 1: *In the space $\mathcal{U}$ of utility functions endowed with the C^∞ Whitney topology there is a residual subset $\mathcal{U}_{res}$ such that each element in $\mathcal{U}_{res}$ represents a preference relation $\lesssim$ for which $\varphi_{\lesssim}(p,w)$ is a singleton λ- almost everywhere on $]0, \infty[$.*

We sketch the proof of the theorem [1]. First we show that

$$\lambda(\{w = g(x)x | g(x) = p, \mathcal{K}(x) = 0\}) = 0,$$

where $\mathcal{K}(x)$ denotes the Gaussian curvature of $u^{-1}(u(x))$ at x. The rank condition U3) makes $g^{-1}(p)$ a one - dimensional differentiable manifold. The manifold $g^{-1}(p)$ is tangent to the budget hyperplane through $x \in g^{-1}(p)$ with normal vector p, if the Gaussian curvature of the indifference surface $u^{-1}(u(x))$ vanishes at x. Hence x is a critical point of the mapping $x \mapsto g(x)x$ defined on $g^{-1}(p)$. The set of critical values has Lebesgue measure zero.

It suffices to show that for any u in a residual subset $\mathcal{U}_{res}$ of $\mathcal{U}$ and for any $p \in S$ there is, outside that null set, only a set of isolated points $w \in]0,\infty[$ for which demand fails to be single - valued. Take any $u \in \mathcal{U}$ and suppose that the demand set at (p,w) does not contain any point with vanishing Gaussian curvature. The demand set at (p,w) is contained in the intersection of the manifold $g^{-1}(p)$ with the budget hypersurface $B_{p,w} = \{x \in X | px = w\}$ corresponding to (p,w). This intersection is transversal because of the nonvanishing Gaussian curvature. Therefore, and because of boundary assumption U2), $B_{p,w} \cap g^{-1}(p)$ consists of finitely many points. For fixed p any point in $B_{p,w} \cap g^{-1}(p)$ can be traced locally if w

1) For a detailed proof see Dierker, Dierker, Trockel (1978a).

varies. This means that there are $\varepsilon > o$ and r smooth functions $h_i :]w - \varepsilon, w + \varepsilon[\to X$, $i=1, \ldots ,r$, $r = \# (B_{p,w} \cap g^{-1}(p))$, such that for any $w' \in]w - \varepsilon, w + \varepsilon[$ the set $B_{p,w'} \cap g^{-1}(p)$ equals $\{h_1(w'),\ldots, h_r(w')\}$.

Now suppose $x_1 = h_1(w)$ and $x_2 = h_2(w)$ are demanded at (p,w). Then, in particular, $u(x_1) = u(x_2)$, and $Du(x_1)$ is proportional to $Du(x_2)$. If $Du(x_1)$ exceeds $Du(x_2)$, then a slight increase of wealth from w to w' prevents $h_2(w')$ from belonging to the demand set at (p,w'), because $u(h_1(w')) > u(h_2(w'))$. Similarly, a slight decrease of wealth from w to w" prevents $h_1(w'')$ from belonging to the demand set at (p,w'').

However, the case $Du(x_1) = Du(x_2)$ cannot be excluded, not even in the case of only two commodities. Therefore, assume now $Du(x_1) = Du(x_2)$. Then one is led to consider the second order variation of u at x_i along h_i. If the second order increase of u at x_1 along h_1 exceeds that of u at x_2 along h_2, then a similar reasoning as in the first order case shows that a slight variation of wealth prevents one of the commodity bundles from belonging to the demand set. If the first and the second order increase of utility at x_1 and x_2 along h_1 and h_2, respectively, happen to coincide, apply a similar argument to the third order increase of utility, and so on.

The condition that all utility increases up to the order k coincide becomes more and more restrictive for growing k. It turns out that there is a residual set of utility functions for which it is impossible that all utility increases up to the order ℓ at x_1 and x_2 coincide. However, taking derivatives for any orde into account, one can let the exceptional set of utility functions shrink much more.

4. Towards Differentiability

In this section we want to show that, for a fixed preference ordering $\lesssim$, aggregation with respect to wealth leads to a mean demand $\Phi_{\lesssim} : S \to \mathbb{R}^{\ell}$ which is almost everywhere C^1. The preference ordering $\lesssim$ is assumed to be represented by the utility function $u : X \to \mathbb{R}$ which satisfies U1) to U3). We need two additiona assumptions on u.

The first of these assumptions concerns the behavior of u in a neighborhoo of a point $\tilde{x}$ whose associated indifference surface has vanishing Gaussian curva ture $\mathcal{K}$ at $\tilde{x}$. Let $\mathcal{K}(\tilde{x}) = o$, $\tilde{p} = g(\tilde{x})$, $\tilde{w} = \tilde{p}\tilde{x}$. Consider the family of budget hyp surfaces $B_{p,w} = \{x \in X | px = w\}$. The function $u|B_{\tilde{p},\tilde{w}}$ has a degenerate critical point at $\tilde{x}$. The family $u|B_{p,w}$, $(p,w) \in S \times]o, \infty[$, can be regarded as an unfoldin of the germ of $u|B_{\tilde{p},\tilde{w}}$ at $\tilde{x}$. We require this unfolding to be stable or, equivalently, to be versal (cf. Bröcker (1975)). To be more specific, define $U: \mathbb{R}_+^{\ell-1} \times S \times \mathbb{R}_+ \to \mathbb{R}$ by

$$U(x_1, \ldots, x_{\ell-1}, p, w) = u(x_1, \ldots, x_{\ell-1}, x_\ell(p,w)),$$

where $x_\ell(p,w) = (w - \sum_{h=1}^{\ell-1} p_h x_h)(1 - \sum_{h=1}^{\ell-1} p_h^2)^{-\frac{1}{2}}$.

We assume:

U4) The unfolding U of $U(\cdot, \tilde{p}, \tilde{w})$ is stable.

Furthermore we postulate:

U5) Let $(x,y,z) \in X \times X \times X$, $x \neq y, x \neq z, y \neq z$. Suppose $u(x) = u(y) = u(z)$, $\alpha\, Du(x) = \beta\, Du(y) = \gamma\, Du(z) = p$ for $\alpha, \beta, \gamma \in \mathbb{R}$, $px = py = pz$, and $x-y = \lambda(y-z)$. Then $\lambda \neq \frac{\alpha-\beta}{\beta-\gamma}$.

This assumption implies that triples (x, y, z) of pairwise distinct, collinear commodity bundles which are demanded simultaneously are isolated. Counting equations and unknowns makes it plausible that U5) is fulfilled in "most" cases.

Furthermore, we need the following assumption on the wealth-distribution of agents with preference ordering $\lesssim$:

M) The probability measure $\delta_{\lesssim}$ on $(]0, \infty[, \mathcal{B}(]0, \infty[))$ has a continuous density, $h_{\lesssim}$, with respect to Lebesgue measure λ, and the support of $\delta_{\lesssim}$ is contained in a compact interval $[\underline{w}, \bar{w}] \subset]0, \infty[$.

The mean demand $\Phi_{\lesssim} : S \to \mathbb{R}^\ell$ of all agents with preference ordering $\lesssim$ is defined by

$$\Phi_{\lesssim}(p) = \int_{\underline{w}}^{\bar{w}} \varphi_{\lesssim}(p,w)\, h_{\lesssim}(w)\, \lambda(dw).$$

<u>Theorem 2</u>: *Let $\lesssim$ be represented by the utility function u satisfying* U1) *to* U5) *and let the wealth-distribution $\delta_{\lesssim}$ satisfy* M). *Then there is a closed null set $N_{\lesssim}$ in the price space S such that the restriction of the mean demand $\Phi_{\lesssim}$ to $S \setminus N_{\lesssim}$ is a C^1 function.*

We sketch the idea behind the theorem.[2)]

There are three phenomena which may prevent $\Phi_{\lesssim}$ from being C^1 : vanishing Gaussian curvature, critical jumps, and multiple jumps.

To study the first of these phenomena, let X_0 be the set of points $x \in X$ such that $\mathcal{K}(x) = 0$, $\underline{w} \leq g(x)\, x \leq \bar{w}$, and x is a maximum of $u|B_{g(x),g(x)x}$.

The latter requirement is fulfilled whenever x is in the demand set $\varphi_{\lesssim}(g(x), g(x)x)$. Points in X_0 correspond to catastrophes of corank 1, codimension ≥ 2. Hence the set $g(X_0)$ is null.

Two different commodity bundles x and y are demanded simultaneously at price

2) For a detailed proof see Dierker, Dierker, Trockel (1978b).

system p only if $g(x) = g(y) = p$, $u(x) = u(y)$, and $px = py$. This system of equations is used to define an $(\ell - 1)$- dimensional differentiable manifold of triples (p, x, y), which are called *jumps*. A jump is *critical* if it is a critical point of $(p, x, y) \mapsto p$. According to Sard's theorem critical jumps give rise to a null set of prices.

Mean demand $\Phi_{\lesssim}$ need not be differentiable at a price system p associated with a noncritical jump if more than two commodity bundles are demanded at p. Thus one is led to consider the following system of equations:

$$g(x) = g(y) = g(z) = p,\ u(x) = u(y) = u(z),\ px = py = pz.$$

Its solutions are points (p, x, y, z) which are called *multiple jumps*. Multiple jumps form an $(\ell - 2)$- dimensional manifold and thus give rise to a null set of prices.

Let $N_{\lesssim}$ be the union of the three null sets corresponding to vanishing Gaussian curvature, critical jumps, and multiple jumps, respectively. Due to the compactness of $[\underline{w}, \bar{w}]$ and the boundary assumption U2) the set $N_{\lesssim}$ is closed. Let $\bar{p} \in S \backslash N_{\lesssim}$. For p near $\bar{p}$, individual demand $\varphi_{\lesssim}(p,w)$ is C^1 for all but a finite number of w's that correspond to jumps which are neither critical nor multiple. It follows that mean demand $\Phi_{\lesssim}$ is C^1 at $\bar{p}$.

One would like to find reasonable assumptions yielding a mean demand which is C^1 everywhere on S. First observe that the vanishing of Gaussian curvature does not necessarily destroy differentiability. For the case of the cusp catastrophe we have:

Theorem 3: *Let* $\lesssim$ *be represented by the utility function* u *satisfying* U1), U2), U3) *and let* $\delta_{\lesssim}$ *satisfy* M). *Suppose* $\varphi_{\lesssim}(\tilde{p},\tilde{w}) = \{\tilde{x}\}$ *and* $\mathcal{K}(\tilde{x}) = 0$. *Let* $g(\tilde{x}) = \tilde{p}$, $\tilde{w} = \tilde{p}\tilde{x}$. *Furthermore, assume that the unfolding* U *associated with* $u|B_{\tilde{p},\tilde{w}}$ *is equivalent to* f *defined by* $f(y_1,\ldots, y_{\ell-1}, v_1, v_2,\ldots, v_\ell) = -(y_1^4 + v_2 y_1^2 + v_1 y_1 + y_2^2 + \ldots + y_{\ell-1}^2)$.
Then there is $\varepsilon > 0$ *such that*
$\int_{\tilde{w}-\varepsilon}^{\tilde{w}+\varepsilon} \varphi_{\lesssim}(\cdot, w)\, h_{\lesssim}(w)\lambda(dw)$ *is* C^1 *at* $\tilde{p}$.

For a proof see Dierker, Dierker, Trockel (1978b.)

5. Concluding Remarks

To deal with the lack of smoothness caused by critical and by multiple jumps one has also to aggregate with respect to preferences. At this point one needs assumptions expressing that preferences are sufficiently dispersed. As Lebesgue measure is not available on the space of preferences it is not clear what precisely dispersion of preferences is supposed to mean. Multiple jumps involving no critical jumps lead to kinks of the mean demand function $\Phi_{\lesssim}$. To get differentiability by aggregation with respect to preferences one would require that, for every $p \in S$, the set of preferences exhibiting such a jump is null. Critical jumps, however , require a more sophisticated analysis. Moreover, situations where several disturbing phenomena, vanishing Gaussian curvature, critical, and multiple jumps, occur simultaneously have to be studied. It is not known, at present, which conditions must be satisfied so that the mean demand of an economy becomes everywhere C^1.

References

Araujo, A., and A. Mas-Colell (1978):"Notes on the smoothing of aggregate demand", Journal of Mathematical Economics, 5, 113-127.

Bröcker, T. (1975): Differentiable Germs and Catastrophes, Cambridge University Press, Cambridge.

Debreu, G. (1972): "Smooth preferences", Econometrica, 40, 603-615.

Dierker, E., H. Dierker, and W. Trockel (1978a): "Continuous mean demand functions derived from nonconvex preferences", Working Paper IP-257, University of California, Berkeley, CA.

Dierker, E., H. Dierker, and W. Trockel (1978b): "Smoothing demand by aggregation with respect to wealth", Discussion Paper No. 35, SFB 21, Projektgruppe "Theoretische Modelle", University of Bonn.

Golubitsky, M. and V. Guillemin, (1973): Stable Mappings and Their Singularities, Springer, Berlin.

Hildenbrand, W. (1974): Core and Equilibria of a Large Economy, Princeton University Press, Princeton, New Jersey.

Hildenbrand, W. (1978): "On the uniqueness of mean demand for dispersed preferences", Discussion Paper No. 39, SFB 21, Projektgruppe "Theoretische Modelle", University of Bonn.

Mas-Colell, A. and W. Neuefeind (1977): "Some generic properties of aggregate excess demand and an application", Econometrica, 45, 591-599.

Neuefeind, W. (1978): "A note on the denseness of differentiable aggregate demand," Journal of Mathematical Economics, 5, 129-131.

Sondermann, D. (1975): "Smoothing demand by aggregation", Journal of Mathematical Economics, 2, 201-223.

Sondermann, D. (1976): "On a measure theoretical problem in mathematical economics", in: Springer Lecture Notes in Mathematics, 541.

Sondermann, D. (1978): "Uniqueness of mean maximizers and continuity of aggregate demand", Working Paper IP-263, University of California, Berkeley, CA.

Yamazaki, A. (1978): "Continuously dispersed preferences, regular preference-endowment distribution and mean demand function", Department of Economics, University of Illinois at Chicago Circle, No. 78-21.

GAME THEORY AND RELATED TOPICS,
O. Moeschlin, D. Pallaschke (eds.)

GENERALIZED NASH EQUILIBRIUM POINTS FOR THE THEORY OF OLIGOPOLY

Wolfgang Eichhorn
Institut für Wirtschaftstheorie und
Operations Research
Universität Karlsruhe
D-7500 Karlsruhe

INTRODUCTION

Although the notion of the Cournot [1838]-Nash [1951] equilibrium point plays an important role in noncooperative game theory, it fails to explain the "rigidity" of prices as often observed in oligopoly situations. The fact that oligopolists, even in the noncooperative case, very often hold their product prices constant for a relatively long period of time makes no sense within the context of the Cournot-Nash equilibrium point for the following reasons: Why should an oligopolist in the case of noncooperation fix his prices, when

(i) there does not exist any Cournot-Nash equilibrium point in the underlying "game", or when

(ii) there exist one or more Cournot-Nash equilibrium points whose locations depend heavily on the lasting impact of the exogenous parameters ?

In this note we shall introduce into oligopoly theory some equilibrium concepts that can be considered as economically motivated generalizations of the Cournot-Nash concept. One of our definitions leads us to a set S of equilibrium points which contains, on the one hand, the set of Cournot-Nash points, and, on the other hand, has the following property: for every oligopoly situation given by finite or compact strategy sets and discrete or continuous pay-off functions, respectively, the set S is not empty. Hence, in a theory based on this definition there is no situation like (i). Moreover, the dependence of S on exogenous parameters is much weaker than that of the set of the Cournot-Nash equilibrium points. Hence, our set S of generalized Cournot-Nash points yields an explanation of price rigidity in oligopoly.

DEFINITIONS AND RESULTS

We consider an oligopolistic noncooperative market, where competition among n firms (competitors, oligopolists) takes place. Let Σ^ν be the set of strategies σ^ν of firm ν ($\nu = 1,2,\ldots,n$) and let $k_\nu \geqq 1$ be the number of criteria according to which competitor ν measures his success or failure. Such criteria may be profit or loss, sales volume, market share, and so on. We assume that there are (vector-valued) functions

$$(1) \quad \underline{F}^\nu : \Sigma^1 \times \Sigma^2 \times \ldots \times \Sigma^n \to \mathbb{R}^{k_\nu} \qquad \begin{cases} \nu = 1,2,\ldots,n; \\ \mathbb{R} \text{ the real line} \end{cases}$$

which relate to the strategy vector or strategy point

$$\underline{\sigma} := (\sigma^1,\sigma^2,\ldots,\sigma^n) \in \Sigma^1 \times \Sigma^2 \times \ldots \times \Sigma^n =: \underline{\Sigma}$$

chosen by the n firms the <u>success</u> <u>vector</u>

$$\underline{F}^\nu(\underline{\sigma}) = (F_1^\nu(\underline{\sigma}), F_2^\nu(\underline{\sigma}),\ldots,F_{k_\nu}^\nu(\underline{\sigma})) \qquad (\nu = 1,2,\ldots,n).$$

Let us call the functions $\underline{F}^\nu$ the <u>success</u> <u>functions</u> of the oligopolists.

Clearly, a real-world oligopoly is <u>not</u> a game in which player ν knows the values of his pay-off function $F^\nu : \underline{\Sigma} \to \mathbb{R}$ exactly for <u>every</u> $\underline{\sigma} \in \underline{\Sigma}$. In practice firm ν knows its success vector at most for the <u>present</u> $\underline{\sigma} \in \underline{\Sigma}$, provided that this $\underline{\sigma}$ has been valid for a sufficiently long period of time. Let us consider such a strategy point $\underline{\sigma} = (\sigma^1,\sigma^2,\ldots,\sigma^n)$. When will competitor ν change his strategy $\sigma^\nu \in \Sigma^\nu$? Assuming that he has learned enough from Pareto he will replace σ^ν with $\bar{\sigma}^\nu \in \Sigma^\nu$ if and only if the success vector that he <u>expects</u> from $\bar{\sigma}^\nu$, say $\underline{E}_{\underline{\sigma}}^\nu(\bar{\sigma}^\nu)$, is greater than or equal to $\underline{F}^\nu(\underline{\sigma})$, where at least one component of $\underline{E}_{\underline{\sigma}}^\nu(\bar{\sigma}^\nu)$ is greater than the corresponding component of $\underline{F}^\nu(\underline{\sigma})$:

$$\underline{E}_{\underline{\sigma}}^\nu(\bar{\sigma}^\nu) \geq \underline{F}^\nu(\underline{\sigma}) \qquad (\nu = 1,2,\ldots,n).$$

We point out here that competitor ν bases a strategy change on his <u>more or less subjective expectation</u> about his success after the change has taken place. Since, on the one hand, risk aversion often is an enterpreneurial principle and since, on the other hand, competitor ν

(i) can estimate the reaction of his competitors only vaguely and

(ii) does not know much about the value of $\underline{F}^\nu$ for any $\underline{\sigma}^*$ different from the present $\underline{\sigma}$

in most strategic situations $\underline{\sigma}$ we will have

$$(2) \qquad \underline{E}_{\underline{\sigma}}^\nu(\bar{\sigma}^\nu) \ngeq \underline{F}^\nu(\underline{\sigma}) \qquad (\nu = 1,2,\ldots,n).$$

We assume that the oligopolists do not change their strategies σ^ν if (2) holds. In all these cases we have the phenomenon of <u>strategy</u> <u>rigidity</u> or, if the strategies are price strategies, of <u>price</u> <u>rigidity.</u>

<u>Definition 1.</u> In the oligopoly described above we call a strategy point $\underline{\sigma} \in \underline{\Sigma}$ an <u>equilibrium</u> <u>point</u>, if (2) holds for every $\bar{\sigma}^\nu \neq \sigma^\nu$ $(\sigma^\nu \in \Sigma^\nu, \bar{\sigma} \in \Sigma^\nu)$. (Note that (2) is required for $\nu = 1,2,\ldots,n$).

Remark 1. The notion of the Cournot-Nash equilibrium point is a special case of Definition 1.

Proof. Take $k_1 = k_2 = \ldots = k_n = 1$, whence $\underline{F}^\nu(\underline{\sigma}) = F^\nu(\underline{\sigma})$, and $\underline{E}^\nu_{\underline{\sigma}}(\bar{\sigma}^\nu) = F^\nu(\sigma^1,\ldots,\sigma^{\nu-1},\bar{\sigma}^\nu,\sigma^{\nu+1},\ldots,\sigma^n)$ in order to obtain, from (2),

$$F^\nu(\sigma^1,\ldots,\sigma^{\nu-1},\bar{\sigma}^\nu,\sigma^{\nu+1},\ldots,\sigma^n) \leqq F^\nu(\underline{\sigma}) \qquad (\nu = 1,2,\ldots,n),$$

which are the defining inequalities for Cournot-Nash equilibrium points.

Remark 2. The success expectations

$$\text{(3)} \qquad \underline{E}^\nu_{\underline{\sigma}} \ : \ \Sigma^\nu \setminus \{\sigma^\nu\} \to \mathbb{R}^{k_\nu} \qquad (\nu = 1,2,\ldots,n)$$

of the oligopolists characterize, in a certain sense, their degree of risk aversion. Whenever the components of $\underline{E}^\nu_{\underline{\sigma}}(\bar{\sigma}^\nu)$ are sufficiently small for all $\bar{\sigma}^\nu \neq \sigma^\nu$ and for all $\nu = 1,2,\ldots,n$, then, obviously, every strategy point $\underline{\sigma}$ is an equilibrium point according to Definition 1. Note that the set S of equilibrium points belonging to (3), where we assume $\underline{E}^\nu_{\underline{\sigma}}$ to be defined for every $\underline{\sigma} \in \underline{\Sigma}$, contains the set $\hat{S}$ of equilibrium points belonging to another system $\hat{\underline{E}}^\nu_{\underline{\sigma}}$ of success expectations, if

$$\underline{E}^\nu_{\underline{\sigma}}(\bar{\sigma}^\nu) \leqq \hat{\underline{E}}^\nu_{\underline{\sigma}}(\bar{\sigma}^\nu) \text{ for every } \underline{\sigma} \in \underline{\Sigma},\ \bar{\sigma}^\nu \neq \sigma^\nu, \qquad \nu = 1,2,\ldots,n.$$

From now on, we assume that the success functions (1) are scalar-valued, that is, $k_1 = k_2 = \ldots = k_n = 1$. If the oligopolists know their success functions F^ν, then their success expectations $E^\nu_{\underline{\sigma}}$ will satisfy

$$\underset{\sigma^1,\ldots,\sigma^{\nu-1},\sigma^{\nu+1},\ldots,\sigma^n}{\text{infimum}} F(\sigma^1,\ldots,\sigma^{\nu-1},\bar{\sigma}^\nu,\sigma^{\nu+1},\ldots,\sigma^n) \leqq E^\nu_{\underline{\sigma}}(\bar{\sigma}^\nu)$$
$$\leqq \underset{\sigma^1,\ldots,\sigma^{\nu-1},\sigma^{\nu+1},\ldots,\sigma^n}{\text{supremum}} F^\nu(\sigma^1,\ldots,\sigma^{\nu-1},\bar{\sigma}^\nu,\sigma^{\nu+1},\ldots,\sigma^n)$$

for every $\underline{\sigma} \in \underline{\Sigma}$, $\sigma^\nu \neq \bar{\sigma}^\nu$, $\nu = 1,2,\ldots,n$.

Note that success expectations $E^\nu_{\underline{\sigma}}(\bar{\sigma}^\nu)$ satisfying these inequalities may be interpreted as expectation values in the mathematical sense by assigning appropriate subjective probabilities to the reactions of the competitors.

Definition 2. We call a strategy point $\underline{\sigma} \in \underline{\Sigma}$ a minimum risk equilibrium point, if

$$(4) \qquad \inf_{\sigma^1,\ldots,\sigma^{\nu-1},\sigma^{\nu+1},\ldots,\sigma^n} F^\nu(\sigma^1,\ldots,\sigma^{\nu-1},\bar{\sigma}^\nu,\sigma^{\nu+1},\ldots,\sigma^n) \leqq F^\nu(\underline{\sigma})$$

for all $\bar{\sigma}^\nu \neq \sigma^\nu$, $\nu = 1,2,\ldots,n$, and a <u>maximum risk equilibrium point</u>, if (4) holds with supremum instead of infimum.

Obviously, in any given oligopoly with scalar-valued success functions the set of all minimum risk equilibrium points contains the set of all Cournot-Nash equilibrium points, which, on the other hand, contains the set of all maximum risk equilibrium points.

It is well known that many oligopolies (games) based on scalar-valued success functions (pay-off functions) do not have any Cournot-Nash points. It is interesting to know that our notion of a minimum risk equilibrium point does not have this "drawback".

<u>Theorem.</u> Every noncooperative oligopoly (or game) given by

(a) arbitrary success functions (or pay-off functions) on arbitrary finite strategy sets or by

(b) arbitrary continuous success functions on arbitrary compact strategy sets

has at least one minimum risk equilibrium point.

<u>Proof.</u> For every $\bar{\sigma}^\nu \in \Sigma^\nu$ let us determine

$$(5) \qquad \min_{\sigma^1,\ldots,\sigma^{\nu-1},\sigma^{\nu+1},\ldots,\sigma^n} F^\nu(\sigma^1,\ldots,\sigma^{\nu-1},\bar{\sigma}^\nu,\sigma^{\nu+1},\ldots,\sigma^\nu),$$

where $\sigma^1 \varepsilon\, \Sigma^1,\ldots,\sigma^{\nu-1} \varepsilon\, \Sigma^{\nu-1},\ \sigma^{\nu+1} \varepsilon\, \Sigma^{\nu+1},\ldots,\sigma^n \varepsilon\, \Sigma^n$.
Note that each of the assumptions (a) or (b) yields the existence of these minima and of the other extrema to follow. Now, choose $\tilde{\sigma}^\nu \varepsilon \Sigma^\nu$, for which the maximum of these minima will be obtained, and consider the set of strategy points

$$\{(\sigma^1,\ldots,\sigma^{\nu-1},\tilde{\sigma}^\nu,\sigma^{\nu+1},\ldots,\sigma^n) | \sigma^1 \varepsilon \Sigma^1,\ldots,\sigma^{\nu-1} \varepsilon \Sigma^{\nu-1},\sigma^{\nu+1} \varepsilon \Sigma^{\nu+1},\ldots,\sigma^n \varepsilon \Sigma^n$$

All these points are <u>status quo points of oligopolist</u> ν in the sense that he will not change his strategy $\tilde{\sigma}^\nu$ as long as his risk aversion is as extreme as expressed by (4). Since for every $\nu = 1,2,\ldots,n$ at least one $\tilde{\sigma}^\nu$ maximizing the minima (5) exists, strategy point $\underline{\tilde{\sigma}} = (\tilde{\sigma}^1, \tilde{\sigma}^2, \ldots, \tilde{\sigma}^n)$ is a minimum risk equilibrium point of our oligopoly (or game). Our theorem is proved.

From now on, let us assume that the extrema of the success functions F^ν under consideration always exist.

Remark 3: Obviously, for any fixed $\nu \in \{1,2,\ldots,n\}$, the set S^ν of the status quo points of oligpolist ν is

$$(6) \qquad \{\underline{\sigma} \in \underline{\Sigma} \mid F^\nu(\underline{\sigma}) \geqq \max_{\sigma^\nu} \min_{\sigma^1,\ldots,\sigma^{\nu-1},\sigma^{\nu+1},\ldots,\sigma^n} F^\nu(\underline{\sigma})\},$$

and the set S of the minimum risk equilibrium points of the oligopoly can be obtained by intersecting $S^1, S^2, \ldots, S^n$:

$$S = S^1 \cap S^2 \cap \ldots \cap S^n.$$

Is it possible that S is equal to the set $\underline{\Sigma}$ of all strategy points? Yes, if $S^1 = S^2 = \ldots = S^n = \underline{\Sigma}$, that is, if each of the n sets (6) is equal to $\underline{\Sigma}$. We see immediately:

Remark 4: If for every fixed $\nu \in \{1,2,\ldots,n\}$ and for all $\bar{\sigma}^\nu \in \Sigma^\nu$ the minimum in (5) is equal to a constant α^ν, then $S = \underline{\Sigma}$, that is, every strategy point $\underline{\sigma} \in \underline{\Sigma}$ is a minimum risk equilibrium point.

Remark 5: The result of our theorem is a strict one in the following sense: If we replace $\leqq$ in (4) with $<$ (or $\geqq$ in (6) with $>$) then the theorem is no more true.

Proof. Consider a duopoly, where the success functions F^1 and F^2 are given by

(7)

F^1	σ^2_1	σ^2_2
σ^1_1	1	2
σ^1_2	3	1

F^2	σ^2_1	σ^2_2
σ^1_1	4	2
σ^1_2	2	3

Here all strategy points $\underline{\sigma} \in \underline{\Sigma}$ are minimum risk equilibrium points as is to be expected by Remark 4, but no $\underline{\sigma} \in \underline{\Sigma}$ satisfies (4) for both $\nu = 1$ and $\nu = 2$ when $\leqq$ is replaced with $<$.

Remark 6: Examples like (7) suggest to require strict inequality signs in (4) or (6). If, for instance, $\underline{\sigma} = (\sigma_1^1, \sigma_1^2)$, then the first duopolist can go on to his second strategy without any risk and with the chance of improving his success. But note that in real-worl oligopolies there are costs of going on to another strategy. This suggests to allow for the very inequality signs in (4) and (6), or even to modify (4) or (6) by explicitly taking into account the costs of changing the present strategy into another .

Remark 7: Since in every noncooperative oligopoly the set of the minimum risk equilibrium points contains the set of the Cournot-Nash points, the following is true: When the values $F^{\nu}(\underline{\sigma})$ of the success functions are changed by the impact of exogenous parameters, then strategy points always lose the property of being Cournot-Nash points earlier (or at most at the same time) than they lose the property of being minimum risk equilibrium points. From (6) it follows that the class of changes of the values of the success functions $F^1, F^2, \ldots, F^n$, against which a minimum risk equilibrium point $\underline{\sigma}$ is invariant, is the greater the greater the differences

$$F^{\nu}(\underline{\sigma}) - \max_{\sigma^{\nu}} \min_{\sigma^1, \ldots, \sigma^{\nu-1}, \sigma^{\nu+1}, \ldots, \sigma^n} F^{\nu}(\underline{\sigma}) \qquad (\nu = 1, 2, \ldots, n$$

are.

REFERENCES

A. Cournot, Recherches sur les Principes Mathématiques de la Théorie des Richesses, Paris 1838. German translation by W. Waffenschmidt, Untersuchungen über die mathematischen Grundlagen des Reichtums, Jena 1924.

J.F. Nash, Noncooperative Games, Annals of Mathematics 54 (1951), 286-295.

GAME THEORY AND RELATED TOPICS
O. Moeschlin, D. Pallaschke (eds.)

PRODUCTION AND DISTRIBUTION

B. Fuchssteiner and A. Schröder

Department of Mathematics
Gesamthochschule Paderborn

We treat in this paper a supply-demand model with infinite commodity set. The main feature of the model is that the producer as well as the consumer do have alternatives concerning their production and consumption respectively. This generalizes a situation considered by Heinz König in his lectures on Mathematical Economy (for finite commodity sets). The principal tools needed in the investigation are disintegration - type arguments which were developed in the decomposition-theory of linear functionals.

THE PROBLEM

We consider a commodity set X and, for the moment, it is assumed that $X = \{1,\dots,n\}$ is finite. Furthermore we consider one producer and one consumer whose production and consumption are measured by the functions α and ν respectively. We look for conditions which guarantee that production can satisfy consumption. An immediate guess for such a condition is that α has to exceed ν. But in real life α and ν are rather functions on $\mathcal{P}'(X) = \{Y \subset X \mid Y \neq \phi\}$ instead of functions on X, since the consumer allows alternatives and the producer usually has at his disposal capacities which he uses according to the market situation. So, for solving our problem, we have to establish a production plan p and a distribution plan v such that

(1) p observes the limits given by the production capacities α (p is then said to be *possible*),

(2) v is *satisfactory*, i.e. v satisfies the demand ν.

(3) we do not distribute more than we produce (p and v are then called *compatible*).

To make this more precise we assume that the producer consists of subunits $U(Y)$, $Y \in \mathcal{P}'(X)$, where $U(Y)$ is the collection of all factories where the commodities $i \in Y$ can be produced equivalently but where production cannot switch to commodities outside of Y. Then the production capacities $\alpha : \mathcal{P}'(X) \to \mathbb{R}_+$ are given by the numbers $\alpha(Y)$, measuring the maximal output of the subunit $U(Y)$. On the consumers side the situation is quite similar. His demand is given by a function $\nu : \mathcal{P}'(X) \to \mathbb{R}_+$, where $\nu(Y)$ measures that fraction of his total demand which can be satisfied by assignment of an arbitrary combination of commodities out of Y (in total amount $\nu(Y)$, of course).

Then the problem is to find a *production - distribution plan* (p,v), where the production plan p as well as the distribution plan v are mappings from $X \times \mathcal{P}'(X)$ to $\mathbb{R}_+$. The quantity $p(i,Y)$ measures the amount of i being produced by the subunit $U(Y)$. Similarily, $v(i,Y)$ measures the amount of i being assigned to the demand $\nu(Y)$.

The requirements (1) - (3) immediatly lead to the following definitions:

The production plan p is said to be *possible* if for all $Y \in P'(X)$ we do have

$$(4)^* \quad \sum_{i \in Y} p(i,Y) \leq \alpha(Y).$$

The distribution plan v is defined to be *satisfactory* if

$$(5)^* \quad \nu(Y) \leq \sum_{i \in Y} v(i,Y) \quad \text{for all} \quad Y \in P'(X) \; .$$

And finally, the plans p and v are said to be *compatible* if for every commodity $i \in X$ production exceeds distribution, i.e.

$$(6)^* \quad \sum_{\{Y | i \in Y\}} v(i,Y) \leq \sum_{\{Y | i \in Y\}} p(i,Y) \; .$$

If we put formally $p(i,Y) = v(i,Y) = 0$ whenever $i \notin Y$ then we can rewrite (4) - (6) in the following form:

$$(4) \quad \sum_{i \in X} p(i,Y) \leq \alpha(Y)$$

$$(5) \quad \nu(Y) \leq \sum_{i \in X} v(i,Y)$$

$$(6) \quad \sum_{Y \in P'(X)} v(i,Y) \leq \sum_{Y \in P'(X)} p(i,Y) \; .$$

Now, our (preliminary) problem is:

Are there plans, which are possible, satisfactory and compatible ?

A simple necessary condition for the existence of these plans is easily found. For this purpose we consider an arbitrary subset $Y \subset X$. Then obviously the total amount of that part of the demand which can only be satisfied by assignment of goods out of Y has to be less than or equal to the amount the producer can produce when he concentrates all his efforts on the production of goods belonging to Y, that is, whenever he has the alternative to produce something in Y or in $X \setminus Y$ then he chooses the production of the good in Y. In formulas this condition reads as follows:

$$(7) \quad \sum_{\{Z \in P'(X) | Z \subset Y\}} \nu(Z) \leq \sum_{\{Z \in P'(X) | Z \cap Y \neq \phi\}} \alpha(Z) \quad \text{for all} \quad Y \in P'(X) \; .$$

At this stage we have to admit that we are not interested in the problem we have stated so far, but rather in its generalization to infinite commodity sets X. In the first moment this sounds like an intended contribution to "general abstract nonsense", which is not so. For example, already in the simple case when one is interested in the dynamical behaviour of supply-demand models one has to take into account that the assignment of a commodity today is different from its assignment in two years from now.

So in this case the "mathematical" commodity set has to be the product $X \times \mathbb{R}$

($\mathbb{R}$ being the time scale), which is clearly infinite. For this reason, in the next chapter our problem will be reformulated (and solved) in a measure-theoretic setup.

1 Remarks : (i) Our problem originates from a problem considered by Heinz König in his beautiful lectures on "Mathematische Wirtschaftstheorie" [6]. There he gives a complete description of the case when X is a finite set and when the production plan is already known, or equivalently, when the producer has no alternatives (i.e. α is a function $X \to \mathbb{R}_+$). König's solution of this problem already incorporates the classical "Heiratssatz".

(ii) For finite X our problem can be completely solved by a (somewhat sophisticated) application of network methods. The best way to proceed is to apply Gales theorem ([5] or [1,p.38]).

(iii) In this paper we do neither deal with algorithms nor with integer-valued plans. But all this can be done in the framework of methods which is developed in the next chapter. Furthermore these methods allow the treatment of more sophisticated models. For example when further restrictions (raw-material constraints) are imposed on the production.

(iv) Finally we should mention that the case of several producers and consumers can be investigated by summing up their production-capacity functions α and their demand functions ν, and that further applications (time-tables, labor-market models, etc.) are possible.

THE PRODUCTION - DISTRIBUTION THEOREM

As commodity set we consider now a (possibly infinite) metric space X. By $\mathcal{K}$ we denote the family of its nonempty compact subsets. We endow $\mathcal{K}$ with the Hausdorff-metric [9] :

$$h(K_1,K_2) = \max(\sup_{x\in K_1} d(x,K_2), \sup_{x\in K_2} d(x,K_1)), \quad K_1,K_2 \in \mathcal{K},$$

where $d(\cdot,\cdot)$ stands for the metric of X.

A $[0,+\infty]$- valued countably additive measure on a topological space will be called *tight* if it is a measure with respect to the σ-algebra generated by the compact sets and if it is finite on every compact set and if it is inner-regular with respect to the family of compact sets. By Σ_X and $\Sigma_{\mathcal{K}}$ we denote the σ-algebras generated by the compact subsets of X and $\mathcal{K}$ respectively.

Now, assume that there are given two tight measures α and ν on $\mathcal{K}$. We call these measures *production-capacity* and *demand-measure*. Throughout this paper we shall assume that there is no *superabundance*. That means that there is no compact set of commodities which can be produced in an infinite amount. In formulas this reads as follows:

$$(8) \quad \alpha\{Z \in \mathcal{K} \mid Z \cap K \neq \phi\} < \infty \quad \text{for all} \quad K \in \mathcal{K}.$$

The measures α and ν are said to fulfill the *balance condition* if:

$$(9) \quad \nu\{Z \in \mathcal{K} \mid Z \subset K\} \leq \alpha\{Z \in \mathcal{K} \mid Z \cap K \neq \phi\} \quad \text{for all} \quad K \in \mathcal{K}.$$

Our aim is to find a reasonable production plan and a suitable distribution plan. By this we mean tight measures p, v on $X \times \mathcal{K}$ such that:

(10) p and v are *plans*, i.e. $p(A \times B) = v(A \times B) = 0$ whenever $A \in \Sigma_X$ and $B \in \Sigma_K$ are such that $A \cap K = \phi$ for all $K \in B$,

(11) p is *possible*, i.e. $\alpha(B) \geq p(X \times B)$ for all $B \in \Sigma_K$,

(12) v is *satisfactory*, i.e. $\nu(B) \leq v(X \times B)$ for all $B \in \Sigma_K$,

(13) p and v are *compatible*, i.e.

$$v(A \times K) \leq p(A \times K) \quad \text{for all} \quad A \in \Sigma_X.$$

These are the suitable generalizations of the properties given by (4) to (6). Of course, because of the regularity condition it suffices to check (10) to (13) for compact A and B.

Main Theorem: *There are plans, which are possible, satisfactory and compatible if and only if the balance condition does hold.*

THE PRINCIPAL TOOLS

In this chapter we gather the principal tools which will be used in the analysis of our supply-demand model.

I. $F = (F,\leq)$ denotes a preordered convex cone, i.e. $\leq$ is reflexive and transitive on F such that

$$f_i \leq g_i,\ 0 \leq \lambda_i \in \mathbb{R}\ (i = 1,2) \Rightarrow \lambda_1 f_1 + \lambda_2 f_2 \leq \lambda_1 g_1 + \lambda_2 g_2 \ .$$

Functionals are maps $\mu : F \to \mathbb{R} = \bar{\mathbb{R}} \cup \{-\infty\}$, where $0 \cdot (-\infty)$, is defined to be 0 and where the other operations are extended to $\bar{\mathbb{R}}$ in the obvious way. In the set of functionals we consider the pointwise order on F. Since no confusion can arise this order relation is also denoted by $\leq$. Linear (sublinear, superlinear) means positive homogeneous (i.e. $\mu(\lambda f) = \lambda \mu(f) \quad \forall\ \lambda \geq 0,\ f \in F$) and additive (subadditive, superadditive). A functional μ is called *order-preserving* if $f \geq g \Rightarrow \mu(f) \geq \mu(g)$.

Sandwich Theorem ([2]): *Let* π *be a sublinear and order-preserving functional and let* $\delta \leq \pi$ *be superlinear. Then there is a linear order-preserving* μ *with* $\delta \leq \mu \leq \pi$.

II. If T is a regular topological space then we denote by $UC_0^+(T)$ the convex cone of $\mathbb{R}_+$ - valued uppersemicontinuous functions with compact support. We endow $UC_0^+(T)$ with the T- pointwise order.

2. Lemma: If $\mu : UC_0^+(T) \to \mathbb{R}_+$ is linear and order-preserving then there is a tight measure m on T such that

$$\mu(f) = \int_T f\, dm \quad \text{for all} \quad f \in UC_0^+(T) \ .$$

For the proof we restrict μ to the subcone $UC_0^+(K)$, where K is some compact subset of T. Then we extend this restriction linearily and order-preservingly to the cone $\Phi = U C_0^+(K) + C(K)$. This is a subcone of $U C_0^+(K) - U C_0^+(K)$, hence the extension must be unique. Then by the Riesz representation theorem we find a measure m_K on K representing our linear functional on Φ, hence on $U C_0^+(K)$. Taking the limit of all those m_K we obtain our measure m. □

III. We consider two regular spaces T_1, T_2 and the σ-algebras Σ_1, Σ_2 generated by the compact subsets of T_1 and T_2. Let $(A,B) \to \tau(A,B)$ be a $\mathbb{R}_+ \cup \{+\infty\}$-valued map on $\Sigma_1 \times \Sigma_2$ such that $\tau(A,B) = \sup\{\tau(K_1 \cap A, K_2 \cap B) \mid K_i \text{ compact} \subset T_i\}$ $\forall (A,B) \in \Sigma_1 \times \Sigma_2$ and having in addition the property that if one variable is put equal to a fixed subset of a compact set then the map is a tight measure in the other variable.

<u>3 Lemma</u>: There is a unique tight measure m on $T_1 \times T_2$ such that $m(A \times B) = \tau(A,B)$ $\forall (A,B) \in \Sigma_1 \times \Sigma_2$.

For the proof one considers the ring P of all finite disjoint unions of measurable rectangles [6,p 149]. Then one defines for finite disjoint unions $m(\bigcup A_i \times B_i) = \Sigma\, \tau(A_i, B_i)$ and gets an unambiguously defined finitely additive map on P since τ is separately additive. Then for an arbitrary rectangle $K = K_1 \times K_2$ the restriction of m to $P_K = \{S \in P \mid S \subset K\}$ extends uniquely to a Borel measure m_K on K (Riesz representation theorem). Again taking the limit one finds the desired m. □

IV. The following pre-disintegration theorem (a special case can be found in [10]) is an interesting application of the sandwich theorem.

<u>Situation:</u> Let (Ω,Σ,m) be a measure space and let $\omega \to \pi_\omega$ be a map from Ω into the set of $\mathbb{R}_+$- valued order-preserving sublinear functionals on the convex cone $F = (F, \leq)$ such that $\omega \to \pi_\omega(f)$ is in $L^1(m)$ for all $f \in F$. Furthermore we consider a superlinear $\mathbb{R}_+$- valued functional δ on F with

$$\delta(f) \leq \int_\Omega \pi_\omega(f)\, dm(\omega) \quad \text{for all } f \in F.$$

<u>4 Theorem</u>: *There is a map* $\tau : \Sigma \times F \to \mathbb{R}_+$ *having the following properties:*

(14) $f \to \tau(A,f)$ *is for every* $A \in \Sigma$ *linear and order-preserving on* F,

(15) $A \to \tau(A,f)$ *is for every* $f \in F$ *a positive measure on* (Ω,Σ),

(16) $\tau(A,f) \leq \int_A \pi_\omega(f)\, dm(\omega)$ *for all* $A \in \Sigma$, $f \in F$.

(17) $\delta(f) \leq \tau(\Omega,f)$ *for all* $f \in F$.

<u>Proof</u> (compare [3, proof of the sum theorem]): We consider the cone Φ of all simple and measurable functions $\varphi : \Omega \to F$, where simple means that $\varphi(\Omega)$ is a finite subset of F and where measurable means that $\varphi^{-1}(\{f\}) \in \Sigma$ $\forall f \in F$. We endow Φ with the preorder

$$\varphi_1 \leq \varphi_2 \Leftrightarrow \varphi_1(\omega) \leq \varphi_2(\omega) \quad \forall \omega \in \Omega .$$

Then we define an order-preserving sublinear π and a superlinear $q \le \pi$ by :

$$\pi(\varphi) = \int_\Omega \pi_\omega(\varphi(\omega))\, d\,m\,(\omega)$$

$$q(\varphi) = \begin{cases} \delta(f) & \text{if } \varphi \text{ is constant and equal to } f \\ -\infty & \text{otherwise.} \end{cases}$$

According to the sandwich theorem and by virtue of Zorn's lemma there is a maximal order-preserving and linear $\mu : \Phi \to \overline{\mathbb{R}}$ with $q \le \mu \le \pi$. Now, the desired τ is given by

$$\tau(A,f) = \mu(1_A \cdot f) ,$$

where $1_A(\omega) = \{1$ if $\omega \in A$ and 0 otherwise$\}$ is the characteristic function of A. All the properties are easily checked, the only difficulty arises in proving that $A \to \tau(A,f)$ is countably additive. For this purpose it is sufficient to show that

$$(18) \quad \tau(\bigcup_{n\in\mathbb{N}} A_n, f) = \liminf_{m\to\infty} \sum_{n=1}^{m} \tau(A_n,f)$$

for every sequence $A_n \in \Sigma$ with $A_n \cap A_m = \phi$ if $n \neq m$. Given such a sequence we consider the superlinear $\rho \le \pi$ defined by

$$\rho(\varphi) = \mu(1_Y \varphi) + \liminf_{m\to\infty} \sum_{n=1}^{m} \mu(1_{A_m} \varphi) ,$$

where $Y = \Omega \setminus \bigcup_{n\in\mathbb{N}} A_n$. We claim that $\rho \ge \mu$. Then this proves (18) because the sandwich theorem gives us an order-preserving linear $\bar{\mu}$ with $\rho \le \bar{\mu} \le \pi$. Since μ is maximal $\bar{\mu}$ must be equal to μ, hence $\rho = \mu$ and (18) is proved. Now, all what remains is the proof of the claim. If $Z_m = \bigcup_{n\ge m} A_n$ then we have

$$\rho(\varphi) + \limsup_{m\to\infty} \mu(1_{Z_m} \cdot \varphi) \ge \mu(\varphi) .$$

And we obtain $\rho \ge \mu$ from

$$\limsup_{m\to\infty} \mu\,(1_{Z_m} \cdot \varphi) \le \limsup_{m\to\infty} \pi\,(1_{Z_m} \varphi) =$$

$$= \limsup_{m\to\infty} \int_{Z_m} \pi_\omega(\varphi(\omega))\, d\,m\,(\omega) ,$$

where the last term ≤ 0 since $Z_{m+1} \subset Z_m$ and $\bigcap_{m\in\mathbb{N}} Z_m = \phi$. □

V. Combination of the results from II to IV leads to a powerful tool.

Situation: Assume that we are given:
Two regular topological spaces T and Ω,

- a tight measure m on Ω ,
- a superlinear $\delta : U\,C_0^+(T) \to \mathbb{R}_+$

- and a map $\omega \to \pi_\omega$ from Ω into the set of $\mathbb{R}_+$- valued order-preserving sublinear functionals on $UC_0^+(T)$ such that $\omega \to \pi_\omega(f)$ is always in $L^1(m)$ and fulfills the following inequality:

$$(19) \quad \delta(f) \le \int_\Omega \pi_\omega(f)\, d\,m(\omega) \quad \text{for all} \quad f \in UC_0^+(T) .$$

<u>5 Theorem</u>: *There is a tight measure* s *on* $\Omega \times T$ *such that*

$$(20) \quad s(K_1 \times K_2) \le \int_{K_1} \pi_\omega\,(1_{K_2})\, d\,m(\omega)$$

$$(21) \quad \delta(1_{K_2}) \le s(\Omega \times K_2)$$

for all compact $K_1 \subset \Omega$, $K_2 \subset T$.

BACK TO THE MAIN THEOREM

Now, let us return to our commodity set X and the tight measures α and ν given on the metric space $\mathcal{K}$ consisting of the nonempty compact subsets of X.

For $K \in \mathcal{K}$ we consider on the convex cone $UC_0^+(X)$ the sublinear functional

$$\pi_K(f) = \sup\{f(x) \mid x \in K\}$$

and the superlinear functional

$$\delta_K(f) = \inf\{f(x) \mid x \in K\} .$$

The functions $K \to \pi_K(f)$ and $K \to \delta_K(f)$ are $\Sigma_{\mathcal{K}}$- measurable since for all $\varepsilon \in \mathbb{R}$
$\{K \mid \pi_K(f) < \varepsilon\} = \{K \in \mathcal{K} \mid K \subset f^{-1}([0,\varepsilon[)\}$ is open and
$\{K \mid \delta_K(f) \ge \varepsilon\} = \{K \in \mathcal{K} \mid K \subset f^{-1}([\varepsilon,\infty[)\}$
is either compact ($\varepsilon > 0$) or equal to $\mathcal{K}$ ($\varepsilon \le 0$).
Integration with respect to α and ν yields the sublinear

$$\pi(f) = \int_{K \in \mathcal{K}} \pi_K(f)\, d\,\alpha(K)$$

and the superlinear

$$\delta(f) = \int_{K \in \mathcal{K}} \delta_K(f)\, d\,\nu(K) .$$

<u>6 Lemma:</u> The following are equivalent:

(i) $\delta \le \pi$

(ii) α and ν fulfill the balance condition.

<u>Proof:</u> (i) $\Rightarrow$ (ii): For $K \in \mathcal{K}$ take f to be 1_K (the characteristic function of K). Then

$\delta(f) = \nu\{Y \in \mathcal{K} \mid Y \subset K\}$ and $\pi(f) = \alpha\{Y \in \mathcal{K} \mid Y \cap K \neq \phi\}$.

Hence $\delta(f) \le \pi(f)$ gives the balance condition for K.

(ii) $\Rightarrow$ (i). It suffices to prove $\delta(f) \le \pi(f)$ for step functions. Thus we assume $f = \sum_{i=1}^{n} \lambda_i 1_{K_i}$, $K_i \in \mathcal{K}$, $\lambda_i \geq 0$. Without loss of generality we may assume that

$$(22) \quad K_1 \subset K_2 \subset K_3 \ldots \subset K_n$$

This assumption implies

$$\pi_K(f) = \sum_{i=1}^{n} \lambda_i 1_{T_i}(K)$$

$$\delta_K(f) = \sum_{i=1}^{n} \lambda_i 1_{S_i}(K) ,$$

where 1_{T_i} and 1_{S_i} are the characteristic functions of the sets $T_i = \{Y \in \mathcal{K} \mid Y \cap K_i \neq \phi\}$ and $S_i = \{Y \in \mathcal{K} \mid Y \subset K_i\}$. Now, we obtain from (ii) the desired inequality:

$$\pi(f) = \sum_{i=1}^{n} \lambda_i \alpha(T_i) \geq \sum_{i=1}^{n} \lambda_i \nu(S_i) = \delta(f) . \qquad \square$$

For the final proof of our main theorem we need another lemma of a similar flavour. Let us define for $x \in X$ via

$$\rho_x(\varphi) = \sup\{\varphi(K) \mid x \in K\}$$

a sublinear functional on $UC_0^+(\mathcal{K})$. For fixed φ the function $x \to \rho_x(\varphi)$ is certainly Σ_X- measurable, since the set

$$(23) \quad \{x \in X \mid \rho_x(\varphi) \geq \varepsilon\} = \bigcup \{K \mid \varphi(K) \geq \varepsilon\}$$

is compact whenever $\varepsilon > 0$. Now, let m be a tight measure on X.

<u>7 Lemma</u>: The following are equivalent:

(i) $\nu\{Y \in \mathcal{K} \mid Y \subset K\} \leq m(K) \quad \forall K \in \mathcal{K}$

(ii) $\int_{\mathcal{K}} \varphi(K)\, d\nu(K) \leq \int_X \rho_x(\varphi) dm(x) \quad \forall \varphi \in UC_0^+(\mathcal{K})$.

<u>Proof</u> (compare [8,p. 26]):

(ii) ⇒ (i): For arbitrary K we consider φ to be the characteristic function of the compact set $\mathcal{K}(K) = \{Y \in \mathcal{K} \mid Y \subset K\}$. Then according to (23) $x \to \rho_x(\varphi)$ is equal to the characteristic function of K. So, inequality (i) is a special case of (ii).

(i) ⇒ (ii): Again it suffices to prove (ii) for step-functions. Without loss of generality we can write a step function $\varphi \in UC_0^+(\mathcal{K})$ in the form

$$\varphi = \sum_{i=1}^{n} \lambda_i 1_{K_i} ,$$

where K_i are compact subsets of K with $K_1 \subset K_2 \subset K_3 \ldots \subset K_n$. Then the function $x \to \rho_x(\varphi)$ is equal to $\sum_{i=1}^{n} \lambda_i 1_{K_i}$, where $K_i = \bigcup\{Y \mid Y \in K_i\}$. From this and (i) we obtain

$$\int_X \rho_x(\varphi) d\, m(x) \geq \sum_{i=1}^{n} \lambda_i m(K_i) \geq \sum_{i=1}^{n} \lambda_i \nu\{ Y \in K \mid Y \subset K_i\} \geq \sum_{i=1}^{n} \lambda_i \nu(K_i)$$

$$\geq \int_K \varphi(K) \, d\nu(K). \qquad \square$$

<u>Proof of the Main Theorem</u>: The fact that the existence of a production-distribution plan implies the balance condition is very easy and therefore left to the reader as an exercise.

Now, let us assume that the balance condition holds for α and ν. Then we have $\delta \leq \pi$ (lemma 6). And theorem 5 gives us a tight measure p on $X \times K$ such that we have for compact $K_1 \subset X$ and compact $\tilde{K} \subset K$

$$(24)^* \quad p(K_1 \times \tilde{K}) \leq \int_{K \in \tilde{K}} \pi_K(1_{K_1}) \, d\alpha(K)$$

$$(25)^* \quad p(K_1 \times K) \geq \int_{K \in K} \delta_K (1_{K_1}) \, d\nu(K) .$$

Using the definition of δ_K and π_K we may rewrite these inequalities in the following form:

$$(24) \quad p(K_1 \times \tilde{K}) \leq \alpha\{Y \in \tilde{K} \mid Y \cap K_1 \neq \phi\}$$

$$(25) \quad p(K_1 \times K) \geq \nu\{Y \in K \mid Y \subset K_1\}.$$

The inequality (24) already implies that p is a plan in the sense of (10) and that this plan is possible (11).

Now, we define a measure m on X by

$$m(A) = p(A \times K) .$$

This is a tight measure because it is finite on compact sets which is a consequence of (24) and the fact that superabundance was not allowed. From (25) we get

$$(26) \quad \nu\{Y \in K \mid Y \subset K\} \leq m(K) \quad \text{for all } K \in K .$$

Hence we know (lemma 7) that

$$(27) \quad \int_K \varphi(K)\, d\nu(K) \le \int_X \rho_x(\varphi)\, dm(x) \quad \text{for all} \quad \varphi \in UC_o^+(K) .$$

Since the left-hand side is linear we may apply theorem 5 for a second time. This gives us a tight measure v on $X \times K$ with

$$(28)^* \quad v(K_1 \times \tilde{K}) \le \int_{K_1} \rho_x(1_{\tilde{K}})\, dm(x)$$

$$(29) \quad v(X \times \tilde{K}) \ge \int_K 1_{\tilde{K}}\, d\nu = \nu(\tilde{K}) ,$$

for compact $K_1 \subset X$ and compact $\tilde{K} \subset K$.

Using the definition of ρ_x the inequality $(28)^*$ may be rewritten in the following form

$$(28) \quad v(K_1 \times \tilde{K}) \le m(\bigcup\{K_1 \cap Y | Y \in \tilde{K}\}).$$

This shows that v is a plan in the sense of (10). Another consequence (by regularity) of (28) is

$$(30) \quad v(K_1 \times K) \le m(K_1) = p(K_1 \times K) ,$$

which implies that p and v are compatible (13). And finally, using (29) we come to the conclusion that v is satisfactory. □

REFERENCES

|1| L.R. Ford and D.R. Fulkerson (1962), Flows in Networks, Princeton

|2| B. Fuchssteiner (1974), Sandwich Theorems and Lattice Semigroups, J. functional Analysis 16, 1-14

|3| B. Fuchssteiner (1977), Decomposition Theorems, Manuscripta Mathematica 22, 151-164

|4| B. Fuchssteiner (1977), When does the Riesz representation theorem hold? , Arch.Math. 28, 173-181

|5| D. Gale (1957), A Theorem on Flows in Networks, Pacific J.Math. 7, 1073-1082

|6| P.R. Halmos (1974), Measure Theory, Springer-Verlag, New York - Heidelberg-Berlin

|7| K. Jacobs (1969), Der Heiratssatz, Selecta Mathematica I, 103-141, Springer-Verlag, New York - Heidelberg - Berlin

|8| H. König, M. Neumann (1976) Mathematische Wirtschaftstheorie, Vorlesungsausarbeitung, Saarbrücken

|9| E. Michael (1951), Topologies on Spaces of Subsets, Trans.Amer.Math.Soc. 71 151-182

|10| M. Neumann (1977), On the Strassen Disintegration Theorem, Arch.Math. 29, 413-420

GAME THEORY AND RELATED TOPICS
O. Moeschlin, D. Pallaschke (eds.)

EQUILIBRIA UNDER PRICE RIGIDITIES AND EXTERNALITIES*

Joseph Greenberg and Heinz Müller
Department of Economics
V.P.I. & S.U., U.S.A.
and
Cambridge University, England

The existence of equilibria under price rigidities and externalities is proved by using results on abstract economies. This approach allows for preferences to be incomplete, nontransitive, dependent on prices, on the rationing scheme and on the actions of other individuals. Moreover, the family of admissible rationing schemes is extended. An example shows that if externalities occur, such a Drèze equilibrium may Pareto dominate the Walras equilibria. Due to its generality the result also applies to temporary equilibrium theory.

I. Introduction

In [2], Drèze considers an economy where a Walrasian equilibrium may fail to exist because prices are not sufficiently flexible. These price rigidities may be given by constraints on prices either in nominal terms or relative to some index. In such markets the imbalance between supply and demand is typically absorbed by some kind of quantity rationing. As noted by Drèze, the rather "fashionable" research area of microeconomic foundations to macroeconomics is closely related to that work: "The phenomenon of price rigidity, i.e., the persistence of prices at which supply and demand are not equal, is frequently observed, and plays an important role in some macroeconomic models. Downwards wage rigidities in the presence of underemployment, with or without minimum wage laws, is the foremost example. Rent controls, price controls aimed at curbing inflationary pressures, usury laws, price uniformity over time or space, provide other examples." ([2], p. 301).

We therefore feel that it might be important to generalize Drèze's result as well as to provide a simpler and a shorter proof than the one given in [2] (thereby answering Drèze's own explicit invitation ([2], pp. 319-320)). The technique of our proof rests upon an application of the abstract economies approach, originally introduced by Debreu [1]. More specifically, we use the result in [6] which is an extension of the Shafer-Sonnenschein Theorem [8]. This approach allows us to make the following extensions on Drèze's work:

(i) *Externalities in preferences*: In this case we show by means of an example that a Drèze equilibrium may Pareto dominate the Walras one. This fact may partly explain and perhaps justify incentives for price controls if externalities occur.

---

*We are greatly indebted to Jacques Drèze for encouragement and helpful comments. This work was written when the authors were visiting C.O.R.E., Belgium.

(ii) *Dependence of preferences on both the prices and the rationing scheme*: Thus, expectations can be easily embedded in our model. Indeed, some results in temporary equilibrium ([3], [4], [5]), where preferences are given by a Bernoulli index, depending, in general, on prices and quantity constraints, become corollaries of our theorem. This generalization is obtained at the cost of an additional assumption, which is automatically satisfied, when preferences are independent of the prices. Such an assumption is also made in [3], [4], [5].

(iii) *Interdependent rationing scheme*: Generally the rationing in market j may depend on the actions an individual chooses in other markets. Moreover, different individuals may be rationed differently. This extension generalizes the results in [2], [4], [5].

(iv) *Preferences need be neither transitive nor complete*.

We present the model in Section II. In Section III the formulation and the proof of the existence theorem are given. The remarks in Section IV conclude the paper. First, the example mentioned in (i) is presented. Then, two counterexamples show that the two assumptions strengthening those of [2] are necessary - one of them due to our more general framework, the other to correct an oversight in [2]. Remark 4 elaborates on (iii) above.

II. The Model

An economy E consists of m commodities, given by the set $M = \{1,\dots,m\}$; n individuals, given by the set $N = \{1,\dots,n\}$; a nonempty index set[1] $J \subseteq M$; and two vectors $\bar{p} \varepsilon [R_+ \cup \{\infty\}]^m$, $\underline{p} \varepsilon R^m_+$, such that[2]:

(1) $$\underline{p} \leqq \bar{p} \text{ and } \sum_J \underline{p}_j < 1 \leq \sum_J \bar{p}_j .$$

The *admissible price set* Δ for E is given by:

$$\Delta = \{p \varepsilon R^m_+ | \underline{p} \leqq p \leqq \bar{p} \text{ and } \sum_J p_j = 1\}.$$

Note that Δ is not necessarily compact, and by (1) $\Delta \neq \emptyset$.[3]

For each $i \varepsilon N$, $\tilde{X}^i \subseteq R^m_+$ represents i's *consumption set* and $w^i \varepsilon R^m_+$ is his *initial endowment*. A *rationing scheme* is a pair of vectors $(L,\ell) \varepsilon R^m_+ x R^m_-$. Each $i \varepsilon N$ has a *preference correspondence* $\tilde{D}^i$ which depends on the consumption of all individuals, on the prices and on the rationing scheme, i.e., $\tilde{D}^i$: $\tilde{X} x \Delta x R^m_+ x R^m_- \to R^m_+$, where $\tilde{X} = \prod_{i \varepsilon N} \tilde{X}^i$. Thus, $\tilde{D}^i (x,p,L,\ell) \cap \tilde{X}^i$ is interpreted to be the set

---

[1] J is the set of commodities that enter the price index, $\sum_J p_j$. When J consists of a single element, the corresponding commodity serves as the numeraire.

[2] Let $a, b \varepsilon R^S$, then $a \geqq b$ means $a_i \geq b_i$ $\forall i$, $a > b$ denotes $a_i > b_i$ $\forall i$, and $a \geq b$ implies $a \geqq b$ but $a \neq b$.

[3] If $\sum_J \bar{p}_j = 1$ then $p, q \varepsilon \Delta$ imply that $p_j = q_j = \bar{p}_j$ $\forall j \varepsilon J$. Hence, no loss of generality results from the condition: $\sum_J \underline{p}_j < 1$.

of all $\bar{x}^i \varepsilon \tilde{X}^i$ which consumer i *prefers* to x^i, when all other individuals $h \neq i$ consume x^h, the price system is p and the rationing scheme is (L,ℓ). Such a representation allows for noncomplete and nontransitive preferences, as well as for externalities. Given a rationing scheme (L,ℓ) and prices $p\varepsilon\Delta$, the *budget set* of individual i is:

$$B^i(p,L,\ell) = \{x^i \varepsilon \tilde{X}^i | p\, x^i \leq pw^i \text{ and } \ell \leqq x^i - w^i \leqq L\}$$

We denote: $z_j^i = x_j^i - w_j^i$, $z_j = \sum_{i\varepsilon N} z_j^i$ and $z = (z_1,\ldots,z_m)$. An *equilibrium under price rigidities and rationing*, or a *Drèze equilibrium*, for E, consists of $x\varepsilon\tilde{X}$, $p\varepsilon\Delta$, $(L,\ell)\varepsilon R_+^m x R_-^m$ such that:

(E.1) $z = 0$

(E.2) $\forall i\varepsilon N$, $x^i \varepsilon B^i(p,L,\ell)$ and $\tilde{D}^i(x,p,L,\ell) \cap B^i(p,L,\ell) = \emptyset$

(E.3) $\forall j\varepsilon M$, $L_j = z_j^i$ for some $i\varepsilon N$ implies $z_j^h > \ell_j$ $\forall h\varepsilon N$,
$\ell_j = z_j^i$ for some $i\varepsilon N$ implies $z_j^h < L_j$ $\forall h\varepsilon N$

(E.4) $\forall j\varepsilon M$, $p_j < \bar{p}_j$ implies $L_j > z_j^i$ $\forall i\varepsilon N$
$p_j > \underline{p}_j$ implies $\ell_j < z_j^i$ $\forall i\varepsilon N$

(E.5) $L \neq 0$ and $\ell \neq 0$.

(E.1) is stated in equality form because quantity rationing may be used to eliminate excess supply. (E.3) guarantees that only one (the short) side of the market may be rationed. (E.4) states that no rationing is allowed unless price rigidities are binding and (E.5) eliminates the trivial rationing scheme.

III. The Main Theorem and Its Proof:

Assumptions:

For each agent $i\varepsilon N$:

(A.1) $\tilde{X}^i$ is a closed convex subset of R_+^m with $\tilde{X}^i \supseteq \tilde{X}^i + R_+^m$.

(A.2) $w^i \varepsilon \text{Int}.\tilde{X}^i$.

(A.3) $\tilde{D}^i$ has an open graph relative to its domain, is convex valued and is irreflexive, i.e., $x^i \notin \tilde{D}^i(x,p,L,\ell)$.

(A.4) Strict desirability in all $j\varepsilon J$:
$\forall (x,p,L,\ell)\varepsilon \tilde{X} x \Delta x R_+^m x R_-^m$ and $\forall \bar{x}^i \varepsilon \tilde{X}^i$
if $\bar{x}^i \geq x^i$, $\bar{x}_h^i = x_h^i$ $\forall h \notin J$ then $\bar{x}^i \varepsilon \tilde{D}^i(x,p,L,\ell)$.

(A.5) Boundary condition: Let
$(x^k, p^k, L^k, \ell^k)\varepsilon \tilde{X} x \Delta x R_+^m x R_-^m$ with $||p^k|| \to \infty$,
$x^k \to x\varepsilon\tilde{X}$, $(L^k, \ell^k) \to (L,\ell)\varepsilon R_+^m x R_-^m$, and let
$\bar{x}^{i,k} \varepsilon \tilde{X}^i$, $\bar{x}^{i,k} \to \bar{x}^i$ with $\bar{x}^i \geq x^i$, $\bar{x}_h^i = x_h^i$ $\forall h \notin J$.
Then, $\exists K$ s.t. $\bar{x}^{i,K} \varepsilon \tilde{D}^i(x^K, p^K, L^K, \ell^K)$.

Remarks

(1) Assumptions (A.1) - (A.3) are standard and require no comments.

(2) Remark 2 in Section IV shows that (A.4) cannot be replaced by weak desirability (not even for all $j\varepsilon M$).

(3) Remark 3 in Section IV shows that (A.5) is indispensable. Note that whenever either D^i is independent of p or that Δ is compact, (A.5) follows from (A.3) and (A.4).

(4) In a temporary equilibrium model using dynamic programming one derives a Bernoulli index $v(x,p,L,\ell)$, representing the individual's utility function. This Bernoulli index is jointly continuous in all arguments, provided the expectations are continuous relative to the topology of weak convergence and intertemporal preferences satisfy some regularity properties ([3], [5]). Define $h_k(.) \equiv v(.,p^k,L^k,\ell^k)$, then the "tightness" assumption in [3] guarantees that a subsequence $h_{k'}(.)$ converges continuously to some $\bar{h}(.)$. Moreover, specific assumptions in [3] guarantee that $\bar{h}$ is strictly monotonic. Therefore (A.5) follows from traditional assumptions in temporary equilibrium theory. The connection between temporary equilibrium and price rigidities is studied in [4].

(5) If $\forall i \varepsilon N$, the domain of $\tilde{D}^i$ is $\tilde{X} x (R^m_+ \setminus \{0\}) x R^m_+ x R^m_-$, (i.e., Δ is replaced by $R^m_+ \setminus \{0\}$), and $\forall q \varepsilon R^m_+$

$$\tilde{D}^i(x,q,L,\ell) = \tilde{D}^i(x, \frac{q}{||q||},L,\ell)$$

(i.e., preferences depend only on *relative* prices), then (A.3) and (A.4) imply (A.5).

Theorem: *For every economy E that satisfies* (A.1) - (A.5), *there exists a Drèze equilibrium.*

The proof of the theorem is divided into three parts: First we consider an economy E' which satisfies (A.1) - (A.4) and

(A.6) $\forall j \varepsilon M$, $0 \leq \underline{p}_j < \bar{p}_j < \infty$, and $\sum_J \underline{p}_j < 1 < \sum_J \bar{p}_j$.

We then convert E' into an abstract economy Γ with n+3 players. Lemma 1 shows that the result in [6] can be applied to establish the existence of a quasi-equilibrium for Γ. In Lemma 2 we verify that this quasi-equilibrium is a Drèze equilibrium for E'. Finally, by taking a sequence of economies that satisfy (A.1) - (A.4) and (A.6) and which tend to E, the theorem is established.

Denote:

$$\alpha = 2 \sum_{j \varepsilon M} \sum_{i \varepsilon N} w^i_j,$$

$$C = \{x \varepsilon R^m_+ | x_j \leq 2\alpha,\ j=1,\dots,m\}.$$

Note that by (A.1) and (A.2), for any $i \varepsilon N$ and for any $j=1,\dots,m$,

$$z^i_j = \alpha \text{ and } z^h + w^h \varepsilon \tilde{X}^h\ \forall h \varepsilon N, \text{ imply that } z_j > 0,$$

$$z^i_j = -\alpha \text{ implies that } z^i + w^i \notin \tilde{X}^i.$$

Let $\Gamma = (X^i, A^i, D^i)_{i=0}^{n+2}$ be an abstract economy, generated from E', where:

For $i \varepsilon N$ $\quad X^i = \tilde{X}^i \cap C$

$X^\circ = \Delta$ (which, by (A.6) is compact),

$X^{n+1} = \{L \varepsilon R^m_+ | 0 \leq L_j \leq \alpha\}$

$X^{n+2} = \{\ell \varepsilon R^m_- | -\alpha \leq \ell_j \leq 0\}$.

$A^i(p,L,\ell) = B^i(p,L,\ell) \cap C$, $\forall i \varepsilon N$

$A^i = X^i \quad i = 0, n+1, n+2.$

For $i\varepsilon N$, $D^i = \tilde{D}^i \cap C$

$D^\circ(p, x, L, \ell) = \{q\varepsilon\Delta | qz > pz\}$,

$D^{n+1}(p,x,L,\ell) = \{\tilde{L}\varepsilon X^{n+1} | \tilde{L}(z+p-\bar{p}) < L(z+p-\bar{p})\}$,

and $D^{n+2}(p,x,L,\ell) = \{\tilde{\ell}\varepsilon X^{n+2} | \tilde{\ell}(z+p-\underline{p}) < \ell(z+p-\underline{p})\}$.

Moreover, let $\Psi = (\Psi^\circ,\ldots,\Psi^{n+2})$ be the vector of n+3 functions,

Ψ^i: $\Delta x R^m_+ x R^m_- \to R_+$, where:

For $i\varepsilon N$, $\Psi^i(p,L,\ell) = -p\ell$, and

$\Psi^\circ = \Psi^{n+1} = \Psi^{n+2} \equiv K > 0$.

Lemma 1: Γ *has a* Ψ *quasi equilibrium. That is, there exists* (p,x,L,ℓ) *such that*:

$$(p,x,L,\ell)\varepsilon \prod_{i=0}^{n+2} A^i(p,L,\ell), \text{ and } \forall i \textit{ with } \Psi^i(p,L,\ell) > 0$$

(E) $$A^i(p,L,\ell) \cap D^i(x,p,L,\ell) = \emptyset.$$

Proof: Under (A.1) - (A.4) and (A.6), all the conditions in [6] will be satisfied if we prove that $\forall i\varepsilon N$, $A^i(p,L,\ell)$ is u.h.c. and that it is also l.h.c. whenever $\Psi^i(p,L,\ell) > 0$.

Denote, for $i\varepsilon N$,

$G^i(p) = \{x^i \varepsilon \text{ Int } X^i | px^i < pw^i\}$

$\bar{G}^i(p) = \{x^i \varepsilon X^i | px^i \leq pw^i\}$, and

$F^i(L,\ell) = \{x^i \varepsilon R^m_+ | \ell \leqq x^i - w^i \leqq L\}$.

(i) u.h.c.: Since $A^i(p,L,\ell) = \bar{G}^i(p) \cap F^i(L,\ell)$, it is u.h.c. as a nonempty intersection of u.h.c. closed-valued correspondences ([7], pp. 23-24).

(ii) l.h.c.: By (A.1) and (A.2), $G^i(p)$ is l.h.c., open and convex-valued at every $p\varepsilon\Delta$. Moreover, $G^i(p) \cap F^i(L,\ell) \neq \emptyset$ whenever $\Psi^i(p,L,\ell) > 0$. Therefore, $G^i(p) \cap F^i(L,\ell)$ is l.h.c. being the intersection of a l.h.c. open and convex-valued correspondence with a continuous correspondence ([7], p. 35). To conclude that $A^i(p,L,\ell)$ is l.h.c. whenever $\Psi^i(p,L,\ell) > 0$, note that

$A^i(p,L,\ell) = \text{cl } \{G^i(p) \cap F^i(L,\ell)\}$ ([7], p, 26). Q.E.D.

Lemma 2: *Let* (p,x,L,ℓ) *be a* Ψ *quasi equilibrium for* Γ. Then, (p,x,L,ℓ) *is a Drèze equilibrium for* E'.

Proof: Denote:

$\bar{R} = \{j\varepsilon M | z_j > 0\}$, $\underline{R} = \{j\varepsilon M | z_j < 0\}$.

(i) $j\varepsilon\bar{R} \Rightarrow p_j < \bar{p}_j$: Otherwise, by (E) for i = n+1, $L_j = 0$ implying $z^i_j \leq 0$ $\forall i\varepsilon N$.

(ii) $j\varepsilon\underline{R} \Rightarrow p_j > \underline{p}_j$: Otherwise, by (E) for i = n+2, $\ell_j = 0$ implying $z^i_j \geq 0$ $\forall i\varepsilon N$.

(iii) $\bar{R}\subset J$: Otherwise, by (i) $\exists \hat{p}\varepsilon\Delta$ with $\hat{p}z > pz$ which contradicts (E) for i=0.

(iv) $\underline{R} \subset J$: Similar to the argument in (iii), otherwise, (ii) contradicts (E) for i=0.

(v) $\bar{R} = \emptyset$ or $\underline{R} = \emptyset$: By (iii) and (iv), $\bar{R} \cup \underline{R} \subset J$. If $\bar{R} \neq \emptyset$ and $\underline{R} \neq \emptyset$, using (i) and (ii) we get a contradiction to (E) for i=0.

(vi) $\bar{R} = \emptyset$: Otherwise, by (v), $z_j \geq 0$ $\forall j \in M$, and for some $k \in J$, $z_k > 0$. (A.6) implies that $\exists \hat{p} \in \Delta$ with $\hat{p}_k > 0$. Hence $\hat{p}z > 0$. But by Walras Law, $pz \leq 0$, and hence, $\hat{p}z > pz$, contradicting (E) for i=0.

(vii) $\underline{R} = \emptyset$: Otherwise, by (v), $z_j \leq 0$ $\forall j \in M$, and $\exists k \in M$ with $z_k < 0$, which implies by (E) for i=n+1, that $L_k = \alpha$. By (ii) $p_k > \underline{p}_k \geq 0$, thus $pz < 0$. Hence, $\exists i \in N$ with $pz^i < 0$. Moreover, $p\ell \leq pz^i < 0$ implying $\psi^i(p,L,\ell) > 0$. By (E), (A.1) and (A.4) $z_k^i = L_k = \alpha$. Hence, by the definition of α, $z_k > 0$, contradiction.

Thus, by (vi) and (vii) z=0, i.e., (E.1) is satisfied. (E.1) together with (E) for i = n+1, n+2, yield:

(viii) $\bar{p}_k > p_k \Rightarrow L_k = \alpha$ and $p_j > \underline{p}_j \Rightarrow \ell_j = -\alpha$.
Thus, (E.4) is satisfied. Moreover, since by (A.6) $\underline{p} < \bar{p}$, $\forall j \in M$ either $L_j = \alpha$ or $\ell_j = -\alpha$, establishing (E.3). By (A.6) and (viii), we have:

(ix) $\exists k, j \in J$ with $L_k = \alpha$ and $\ell_j = -\alpha$,
which verifies (E.5). Also, (A.6) and (viii) imply $\psi^i(p,L,\ell) > 0$ $\forall i \in N$. Using (E) for $i \in N$, we derive (E.2).
By (A.4), any maximal point in A^i with respect to D^i, is also a maximal point in B^i w.r.t. $\tilde{D}^i$.

Q.E.D.

So far we have shown that for every economy E' that satisfies (A.1) - (A.4) and (A.6) there exists a Drèze equilibrium. We shall now replace (A.6) by (A.5) and conclude the proof of the theorem.

Let E be an economy that satisfies (A.1) - (A.5), and let $(\underline{p}^k, \bar{p}^k)$ be a sequence of prices that tend to $(\underline{p}, \bar{p})$, such that for every k=1,2,..., $(\underline{p}^k, \bar{p}^k)$ satisfies (A.6). Define $\{E^k\}$ to be the sequence of economies that is obtained from E by altering only the price bounds, i.e., in E^k, $(\underline{p}^k, \bar{p}^k)$ replaces $(\underline{p}, \bar{p})$. Also, if either $\underline{p} \not< \bar{p}$, or $\sum_J \bar{p}_j = 1$, the preference correspondences $\tilde{D}^i$ are extended so that (A.1) - (A.4) are satisfied by E^k for all k. Denote by $e^k = (x^k, p^k, L^k, \ell^k)$ a Drèze equilibrium for E^k. (Such an e^k exists by Lemma 2 above). We shall show that

$$e = (x,p,L,\ell) = \lim e^k, \text{ is a Drèze equilibrium for } E.$$

<u>Claim 1</u>: $\{p^k\}$ *is bounded.*

<u>Proof</u>: Suppose to the contrary that $||p^k|| \to \infty$[1]. Then, $\pi^k = \dfrac{p^k}{||p^k||} \to \pi$, with

[1]Whenever needed, take the appropriate subsequence.

$||\pi|| = 1$, $(x^k, L^k, \ell^k) \to (x,L,\ell)$. (Recall that X is compact and that $||L^k|| \leq n\alpha$ and $||\ell^k|| \leq n\alpha$).

Since $p \in R^m_+$, $\pi_j > 0 \Rightarrow p^k_j > \underline{p}_j$ for all k large enough. By (viii), $\ell^k_j = -\alpha$ hence $\ell_j = -\alpha$. Thus, $\forall i \in N$ $\Psi^i(p^k, L^k, \ell^k) > 0$ and $\tilde{\Psi}(\pi, L, \ell) \equiv -\pi\ell > 0$. For all $j \in J$, $\pi_j = 0$, and therefore $\bar{x}^i \in A^i(\pi, L, \ell)$[1], where

$$\bar{x}^i_j = \begin{cases} x^i_j & j \notin J \\ w^i_j + L_j & j \in J. \end{cases}$$

Since $z^k = 0$ $\forall k$, z=0, and by (ix), $\exists j \in J$ with $\bar{x}^i_j > x^i_j$. Recall that $-\pi\ell > 0$ and using the l.h.c. of A^i at (π,L,ℓ) there exists $\bar{x}^{i,k} \to \bar{x}^i$ with $\bar{x}^{i,k} \in A^i(p^k,L^k,\ell^k)$. By (A.5) $\exists K$ s.t. $\bar{x}^{i,K} \in A^i(p^K,L^K,\ell^K) \cap D^i(-x^K,p^K,L^K,\ell)$, contradicting the fact that e^K is an equilibrium for E^K.

Q.E.D.

To conclude the proof of the theorem, we shall show that $e = (x,p,L,\ell) = \lim e^k$ is a Drèze equilibrium for E. (E.1) and (E.3) - (E.5) follow immediately from (viii) and (ix). Since $\sum_J \underline{p}_j < 1$, by (viii) we have $\Psi^i(p,L,\ell) > 0$. The continuity of A^i together with (A.3) imply (E.2).

Q.E.D.

IV. Remarks

1. The following example, though perhaps not surprising, shows that in the presence of externalities, a Drèze equilibrium (that involves a binding rationing scheme) may Pareto dominate the Walras equilibria:

The economy is given by: m=n=2, J=M, $\tilde{X}^i = R^2_+$, $i \in N$, $w^1 = (6,0)$, $w^2 = (0,4)$, and the preferences are represented by the following utility functions[2]:

$$u^1(x^1,x^2) = 4x^1_1\, x^1_2\, (x^2_2 + 0.25),\quad u^2(x^2) = x^2_1 + x^2_2.$$

(Thus u^2 depends only on the consumption bundle of individual 2). The unique Walras equilibrium is given by:

$$p = (1,1),\ x^1 = (3,3) \text{ and } x^2 = (3,1),$$

giving rise to the utility levels: $u^* = (45,4)$.

Let $\Delta = \{p \in R^2_+ | p_1 + p_2 = 1,\ 0 \leqq p \leq (\frac{1}{3},\frac{2}{3})\}$. (Thus, $\Delta = \{(\frac{1}{3},\frac{2}{3})\}$). Hence,

$p = (\frac{1}{3},\frac{2}{3})$, $x^1 = (3,\frac{3}{2})$, $x^2 = (3,\frac{5}{2})$, $L = (3,100)$ and $\ell = (-100,-100)$ is a Drèze equilibrium for this economy, giving rise to the utility levels:

$\bar{u} = (49.5,\ 5.5) > u^*$.

---

[1] A^i is invariant under price normalization; its continuity properties hold also for the point (x,π,L,ℓ). [See the proof of Lemma 1].

[2] Recall that x^i_j denotes the jth coordinate of the consumption bundle $x^i \in \tilde{X}^i$.

2. The following example verifies that (A.4) cannot be replaced by weak monotonicity over all $j\varepsilon M$;

$n=1$, $m=3$, $J=\{2,3\}$, $\tilde{X} = R^3_+$, $w = (1,1,1)$, $u(x) = x_1 + x_3$, and $\Delta = \{p\varepsilon R^3_+ | p_2 + p_3 = 1,\ 0 \leqq p \leqq (\infty, 0.6, 0.6)\}$.

Suppose that (x,p,L,ℓ) is a Drèze equilibrium for this economy. Then $p\varepsilon\Delta \Rightarrow p_2 > 0$ and therefore $\ell_2 < x_2 - w_2$. Moreover, $p_1 < \infty$ implies $L_1 > x_1 - w_1$, and by (E.1) $x=2$. Therefore, $\exists\delta > 0$ such that

$$\hat{x} = x + (\delta, -\frac{\delta p_1}{p_2}, 0)\varepsilon A(p,L,\ell) \cap D(x,p,L,\ell).$$

Hence there exists no Drèze equilibrium in this case.

3. (A.5) is indispensable. Indeed, let

$n=1$, $m=2$, $J=\{1\}$, $\tilde{X}=R^2_+$, $w=(1,1)$, $u(x,p) = \frac{x_1}{1+p_2} + x_2$, and $\Delta=\{p\varepsilon R^2_+ | p_1=1,\ 0 \leqq p \leqq (2,\infty)\}$.

Suppose that (x,p,L,ℓ) is a Drèze equilibrium. Then, by (E.1), $x=w$. Using (E.4) we have $\ell_1 < 0 < L_1$ and $L_2 > 0$. Hence, $\hat{x}\varepsilon A(p,L,\ell) \cap D(p,x,L,\ell)$, where

$$\hat{x} = (1-\varepsilon,\ 1 + \frac{\varepsilon}{p_2}),\ 0 < \varepsilon < 1,$$

contradicting (E.2). To see that (A.5) is not satisfied by $u(x,p)$, choose $x^k \equiv x = (1,1)$, $p^k = (1,k)$, $\bar{x}^k = (2, 1 - \frac{2}{k+1}) \to \bar{x} = (2,1) \geq x$. But, for all k, $1 = u(\bar{x}^k, p^k) < u(x^k, p^k) = 1 + \frac{1}{k+1}$.

4. More generally, the rationing in market j may depend on the actions an individual chooses in other markets. For example, *total* amount of energy consumed by an individual may be restricted while leaving the consumer the freedom of choice as to which sources of energy he will use. (A similar flavour is present in the Diet problem). Formally, an individual faces a set of "admissible excess demands". Moreover, these sets may vary among individuals. The following formulation allows for such possibilities: $\forall i\varepsilon N$, let $F^i: [0,1]^m \times [-1,0]^m \to R^m$ be a correspondence whose section $F^i(L,\ell)$ denotes the set of admissible excess demands by individual i. (In our model above, $F^i(L,\ell) \equiv \{z^i | \ell \leqq z^i \leqq L\}$). The new budget set is

$$\tilde{B}^i(p,L,\ell) = \{x^i\varepsilon X^i | px^i \leqq pw^i,\ x^i - w^i\varepsilon F^i(L,\ell)\}$$

Assume:

R.1 F^i is continuous and convex compact valued

R.2 $0\varepsilon F^i(L,\ell)\ \forall(L,\ell)$

R.3 $L_j = 0 \Rightarrow z_j \leqq 0\ \forall z\varepsilon F^i(L,\ell)$

$\ell_j = 0 \Rightarrow z_j \geqq 0\ \forall z\varepsilon F^i(L,\ell)$

R.4 $\quad L_j = 1 \Rightarrow$ If $z \varepsilon F^i (L,\ell)$ then $\check{z}^j \varepsilon F^i(L,\ell)$

$$\text{with } \check{z}^j_h = \begin{cases} z_h & \text{if } h \neq j \\ \alpha & \text{if } h = j \end{cases}$$

$\ell_j = -1 \Rightarrow \exists\, z \varepsilon F^i(L,\ell)$ s.t. $z_j < 0$.

Then by a straightforward modification of our proof we can show:

Theorem: *Under* (A.1) - (A.5) *there exists* (x,p,L,ℓ) s.t.

$\tilde{E}.1 \quad z = \sum_{i \varepsilon N}(x^i - w^i) = 0$

$\tilde{E}.2 \quad \forall i \varepsilon N,\ x^i \varepsilon \tilde{B}^i(p,L,\ell)$ and $\tilde{D}^i(x,p,L,\ell) \cap \tilde{B}^i(p,L,\ell) = \emptyset$

$\tilde{E}.3 \quad \forall j \varepsilon M,\ L_j < 1 \Rightarrow \ell_j = -1$ and $\ell_j > -1 \Rightarrow L_j = 1$

$\tilde{E}.4 \quad \bar{p}_j > p_j \Rightarrow L_j = 1$

$\quad\quad p_j > \underline{p}_j \Rightarrow \ell_j = -1$

$\tilde{E}.5 \quad L \neq 0,\ \ell \neq 0.$

Using the representation in [5], p. 560, rationing by priorities turns out to be a special case of this remark.

5. Since preferences may be price dependent, production could be easily introduced. Indeed, for firm j we add an agent n+2+j with

$$X_{n+2+j} = A_{n+2+j} = Y_j$$

and

$$D_j(y_j,p,\ldots) = \{\bar{y}_j \varepsilon Y_j \,|\, p\bar{y}_j > py_j\}.$$

References

[1] Debreu, G. (1952) "A Social Equilibrium Existence Theorem", *Proceedings of the National Academy of Sciences of the U.S.A.* 38.

[2] Drèze, J. (1975), "Existence of an Exchange Equilibrium Under Price Rigidities", *International Economic Review*, 301-320.

[3] Grandmont, J.M. (1974), "On the Short-run Equilibrium in a Monetary Economy", in Drèze (ed.), *Allocation Under Uncertainty, Equilibrium and Optimality*, John Wiley, New York.

[4] Grandmont, J.M., Laroque, G. (1976), "On Temporary Keynesian Equilibria", *Review of Economic Studies 43*, 53-67.

[5] Grandmont, J.M. (1977), "Temporary General Equilibrium Theory", *Econometrica 45*, 535-572.

[6] Greenberg, J. (1977), "Quasi-equilibrium in Abstract Economies Without Ordered Preferences", *Journal of Mathematical Economics*, 163-165.

[7] Hildenbrand, W. (1974), *Core and Equilibria of a Large Economy*, Princeton University Press.

[8] Shafer, W., and H. Sonnenschein, (1975), "Equilibrium in Abstract Economies Without Ordered Preferences", *Journal of Mathematical Economics 2*, 345-348.

GAME THEORY AND RELATED TOPICS
O. Moeschlin, D. Pallaschke (eds.)

A CONTINUOUS MAX-FLOW PROBLEM

K. Jacobs
Institute of Mathematics
University of Erlangen - Nürnberg

The purpose of this paper is to draw the attention of linear programming, and in particular max-flow, specialists to a measure-theoretical version of the max-flow problem. A piecewise differentiable version of it has been treated successfully by Beckmann 1 2.

Let Q be a source and S a sink in - say - three-space $\mathbb{R}^3$: $Q, S \in \mathbb{R}^3$, $Q \neq S$. Consider all Lip_1 paths joining Q with S. They form a compact metric space Ω under uniform topology. A measure μ on Ω is also called a <u>flow</u> from Q to S. This is in accordance with the well-known fact that in finite network theory every flow is a sum of one-path flows (see e.g. Jacobs [1]).

Imagine an animal feeding on oxygen along an $\omega \in \Omega$. Its oxygen consumption can be described by arc length measure $P(\omega,.)$ distributed over ω. This defines a measure kernel P from Ω to the compact set $X \subseteq \mathbb{R}^3$ covered by all $\omega \in \Omega$ together. We may use P in order to

a) transport a given measure μ from Ω to X:
$(\mu P)(f) = \mu(d\omega)P(\omega,f)$

b) transport a given bounded measurable function f on X back to Ω :
$(Pf)(\omega) = P(\omega, dx)f(x)$.

A finite or infinite measure $c \geqslant 0$ on X will be interpreted as a <u>capacity condition</u> in X. We may think of $c(E)$ = oxygen available in E.

We will be interested in flows μ <u>obeying</u> a given c in the sense that the animals in a flow consume no more oxygen than is available : $\mu P \leqslant c$. To <u>maximize the strength</u> $\mu(\Omega) = \mu(1)$ of the flow within this condition is a classical optimization problem in, however, infinitely many dimensions (see Lempio [1]).

The two sections of this discussion paper present some basic facts about the problem thus described, dealing, however, with a situation with $X \subsetneq R^2$. I am calling for further results, and, in particular, answer to the questions put forth in the text.

I am aware of some connection of the problem discussed here with former papers of Appell [1] and Friedrich [1].

The Deutsche Forschungsgemeinschaft generously supported an attempt to elaborate a continuous analogon of the Ford-Fulkerson algorithm, which, I regret to say, has so far been not very successful.

§ 1. A special kernel and its continuity properties.

Consider the mutilated square

$$X = \{(x,y) \mid |x| + |y| \leq 1, \quad -\tfrac{1}{2} \leq x \leq \tfrac{1}{2}\} \leq \mathbb{R}^2$$

and define by

$$\Omega = \{\omega \mid \omega\colon \left[-\tfrac{1}{2}, \tfrac{1}{2}\right] \longrightarrow \mathbb{R},\ |\omega(0)| \leq \tfrac{1}{2} \geq |\omega(1)|,$$

$$\left[-\tfrac{1}{2} \leq s, t \leq \tfrac{1}{2} \Longrightarrow |\omega(s) - \omega(t)| \leq |s-t|\right]\}$$

the set of all Lip_1 functions on $\left[-\frac{1}{2}, \frac{1}{2}\right]$ whose graph lies in X. Ω is compact metric for uniform convergence. For every $\omega \in \Omega$ the curve $t \longrightarrow (t, \omega(t))$ is rectifiable. Denote by $P(\omega,.)$ the arc length measure in Ω, living on the graph of ω: for every $f \in C(X, \mathbb{R})$ we have

$$P(\omega,f) = \int_{-\frac{1}{2}}^{\frac{1}{2}} f(t,\omega(t)) \sqrt{1 + \omega'^2(t)}\, dt,$$

where ω' is the Lebesgue-a.e. defined derivative of the function ω. We may clearly consider P as a measure kernel from Ω to X. By

$$f \longrightarrow P(.,f) = \int P(.,dx) f(x)$$

P induces a positive linear mapping - again denoted by P - from $C(X, \mathbb{R})$, and even from the space $\mathrm{mble}^b(X, \mathbb{R})$ of all bounded measurable functions on X, to the space $\mathbb{R}^\Omega$ of all real functions on Ω. In fact, $P(.,f)$ is lower semicontinuous for every $0 \leq f \quad C(X, \mathbb{R}) : \omega_n \longrightarrow \omega$ implies, by easy geometric analysis, that the arc length of the graph of ω over any subinterval of $\left[-\frac{1}{2}, \frac{1}{2}\right]$ is at most the lim inf over the corresponding arc lengths of the ω_n; Approximation of $\omega \equiv 0$ by zigzag curves shows that <

may actually occur; this yields $P(\omega,f) \leq \underline{\lim}_n \inf P(\omega_n,f)$ for every $0 \leq f \in C(X, \mathbb{R})$. Thus $P(.,f)$ is measurable on Ω for every $f \in C(X, \mathbb{R})$; therefore $\int \mu(d\omega)P(\omega,f)$ makes sense and defines a positive linear form $\mu P = m$ on $C(X, \mathbb{R})$, i.e. a finite measure m on X, given a finite measure μ on Ω. How about continuity properties of $\mu \longrightarrow \mu P$ for weak convergence of the μ? Putting μ_n as point mass one on a zigzag line approximating $\omega \equiv 0$ we see that $\mu_n \longrightarrow \mu$ (weakly) is compatible with $\lim\inf (\mu_n P)(f) > (\mu P)(f)$, $\forall 0 \leq f \in C(X, \mathbb{R})$. Thus we will expect lower semicontinuity of $\mu \longrightarrow \mu P$. In fact, fix some $0 \leq f \in C(X, \mathbb{R})$ and consider the lower semicontinuous φ defined on Ω by $\varphi(\omega) = P(\omega,f)$. Fix a positive measure μ on Ω and choose any $\varepsilon > 0$ and find some $\psi \in C(\Omega, \mathbb{R})$ such that $0 \leq \psi \leq \varphi$ and $\mu(\psi) \geq \mu(\varphi) - \varepsilon$. Let now $0 \leq \mu_n \longrightarrow \mu$ weakly. Then

$$\lim_n\inf (\mu_n P)(f) = \lim_n\inf \mu_n(Pf)$$

$$= \lim_n\inf \mu_n(\varphi) \geq \underline{\lim}_n \mu_n(\psi) = \mu(\psi) \geq \mu(\varphi) - \varepsilon$$

$$= (\mu P)(f) - \varepsilon$$ follows. This proves the desired result.

The mapping $\mu \longrightarrow \mu P$ is not one-to-one at all: define

a) μ_a by equidistribution of mass one over a one-parameter set of zigzag paths with slopes ± 1, where every path in the set is a translate mod 1 of all others in horizontal direction.

b) and define μ_b by equidistribution of mass $\sqrt{2}$ over all constant paths covering the same strip in X as those envisaged in a)

and $\mu_a P = \mu_b P$ follows.

The mapping $f \longrightarrow Pf$ is one-to-one, however, even for bounded measurable functions f. The proof is easy for continuous functions f. In fact let $U \subseteq X$ be open such that $f(x) \geq \alpha > 0$ for $x \in U$.

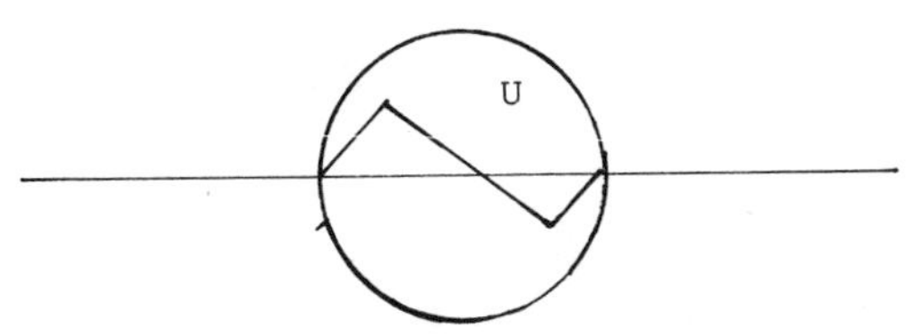

Consider a constant path ω passing through U and a zigzag modification ω' of it within U. If d is the length of the piece of ω which we have modified, then ω' is by $d\sqrt{2}$ longer than ω. We conclude $P(\omega',f) - P(\omega,f) \geq \alpha\, d \sqrt{2} > 0$. Repeating a similar argument where $f < 0$, we see that $P(\omega, f) = 0 \Longrightarrow f = 0$ for every $f \in C(X, \mathbb{R})$, hence $f \longrightarrow Pf$ is one-to-one on continuous functions on X. Actually our method of proof gives us a means of identifying the values $f(x)$ by inspecting $P(\omega,f)$ for paths ω varying only in a neighborhood of x.

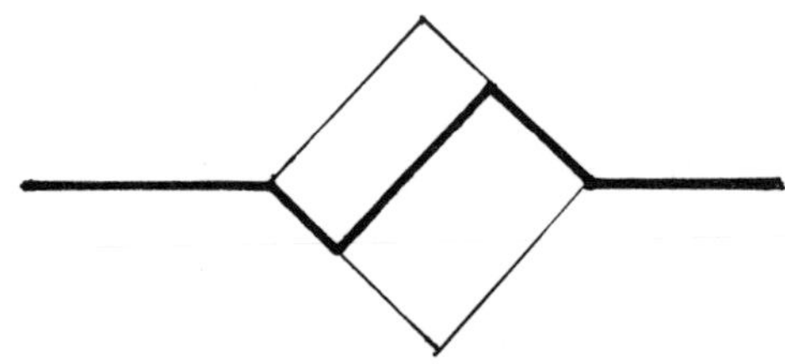

By inspection of zigzag paths as designed here, plus an application of Fubini's theorem. we easily reconstruct local integral mean values of a bounded measurable f on X from values $P(\omega,f)$.

It would be of general interest to know what the class $\{\mu P \mid \mu \geq 0$ a finite measure on $\Omega\}$ is like. Obviously μP assigns the value 0 to say that $N \times [0,1] \leq X$ where N is a one-dimensional Lebesgue nullset, no matter what μ is like. But I don't know what else of common nullsets of all μP resp. all $P(\omega,.)$ we have. It would be desirable to have kernels Q from X to Ω which invert $\mu \longrightarrow \mu P$ at least for large classes of measures μ.

§ 2. The optimization (max flow) problem. Existence of solutions.

For the max-flow problem which we are going to formulate, we will call every finite measure $\mu > 0$ in Ω a <u>(Lipschitz) flow</u> in X, and every finite measure $c \geq 0$ in X a <u>capacity restriction</u> in X.

We will say that μ obeys the capacity restriction c if $\mu P \leq c$ holds. We will call μ(with $\mu P \leq c$) a maximal flow for the capacity restriction c if the total variation $\|\cdot\|$ of measures in Ω fulfils

$$\|\mu\| \geq \|\nu\|$$

for every flow ν obeying the capacity restriction c.

Our max-flow problem now runs as follows: For any given capacity restriction c in X find a maximal flow obeying c. As $\mu \geq 0$ implies $\mu = \mu(1)$, i.e. weak continuity of μ, the existence of solutions follows from

Theorem 2.1. For every capacity restriction c the set

$$K_c = \{\mu | \mu \text{ is a Lipschitz-flow in } X \text{ and obeys the capacity restriction } c\}$$

is weakly compact.

Proof. Clearly it suffices to consider a sequence $\mu, \mu_1, \mu_2, \ldots$ of flows such that $\mu_n(\varphi) \longrightarrow \mu(\varphi)$ $(\varphi \in C(\Omega, \mathbb{R}))$, $\mu_1 P, \mu_2 P, \ldots \leq c$, and to prove $\mu P \leq c$. Let in fact $0 \leq f \in C(X, \mathbb{R})$. We have $(\mu_n P)(f) \leq c(f)$ $(n = 1,2,\ldots)$ and we have to prove the same inequality for μ. For any given $\varepsilon > 0$ find a continuous $0 \leq \varphi \leq P(.,f)$ such that $\mu(\varphi) \geq (\mu P)(f) - \varepsilon$. This is possible as $P(.,f)$ is lower semicontinuous. Clearly $\mu_n(\varphi) \leq \mu_n(P(.,f)) \leq c(f)$ $(n = 1,2,\ldots)$ and thus $\mu(\varphi) \leq c(f)$ follows. As $\varepsilon > 0$ was arbitrary, we obtain the desired inequality.

Varying the proof slightly, we even get

Theorem 2.2. Let $c, c_1, c_2, \ldots$ be capacity restrictions such that $c_n \longrightarrow c$ (weakly).
Let $\mu, \mu_1, \mu_2, \ldots$ be Lipschitz flows such that μ_n obeys the capacity restriction c_n $(n = 1,2,\ldots)$ and $\mu_n \longrightarrow \mu$ (weakly). Then μ obeys the capacity restriction c. In particular the strength of a maximal flow for c depends weakly lower semicontinuously on c.

If we construct c by putting positive multiples of linear Lebesgue measure on suitable segments in X, we arrive at a max-flow problem which can be solved by purely combinatorial methos: Ford-Fulkerson's max-flow-min-cut theorem and its proof by a marking algorithm. It would be desirable to get a measure-theoretical analogon to the concept of a cut and the marking algorithm such as to be able to construct arbitrarily good approximations to maximal flows.

Example 2.3. Let $\omega_o(t) \equiv 0$ and c defined by putting $\varrho(t)\,\lambda(dt)$ on ω_o, where λ is linear Lebesgue measure and ϱ is a bounded measurable function on $[0,1]$, with $\inf \{\varrho(t) \mid 0 \leq t \leq 1\} = 0$ but $\int \varrho \, d\lambda = 1$. Clearly $\mu P \leq c$ implies $\mu \leq \alpha\, \delta_\omega$ here where δ_ω is point mass one at $\omega \in \Omega$, and in our case only $\alpha = 0$ is possible. Now what might be a substitute for a cut in the sense of Ford-Fulkerson, i.w. a segment in ω_o bearing capacity 0. It can only be a nested sequence of segments, constructed, say, in the following way. Decompose $[0,1] =$

$$= \left[0, \frac{1}{2^n}\right] \cup \left[\frac{1}{2^n}, \frac{2}{2^n}\right] \cup \dots \cup \left[\frac{2^n-1}{2^n}, 1\right] \quad \text{and}$$

determine the leftmost of these intervals in which ϱ has infimum 0. Doing this for $n = 1,2,\dots$ we get a decreasing sequence $J_1 \geq J_2 \geq \dots$ of compact dyadic intervals with exactly one point t_o in common. ϱ has infimum 0 in every neighborhood of t_o. Here the point t_o might be taken as a "cut": no flow can pass by t_o. If we replace this c by a general measure c living on ω_o, we find that any $\mu P \leq c$ must minorize the absolutely continuous part of e with respect to linear Lebesgue measure, and again a point t_o may serve as a "cut".

I have the general impression that germs of Lipschitz paths might serve as a substitute for cuts. This would necessitate a disintegration of any given c into measure living on Lipschitz paths. I don't see such a theorem at present. A continuous marking algorithm would have to deal with such germs in an appropriate way, I guess.

I conclude with another example of maximal flow zero.

Example 2.4. For every closed rectangle $R = [a,b] \times [c,d] \subseteq \mathbb{R}^2$ let Z_R be the decomposition (mod boundary points) of R resulting by tripartitioning

$$[a,b] = \left[a, a + \frac{b-a}{3}\right] \cup \left[a + \frac{b-a}{3}, a + 2\frac{b-a}{3}\right] \cup \left[a + 2\frac{b-a}{3}, b\right]$$

and tripartitioning c,d analogously. Let R_1, R_2, R_3, R_4 be the four corner rectangles and R_o the central one in this partition and R' their union. Let now $\varrho_o(x,y) = 1 \quad ((x,y) \in X)$ and

consider $R_o = \left[-\frac{1}{2}, \frac{1}{2}\right] \times \left[-\frac{1}{2}, \frac{1}{2}\right] \subseteq X$.
Let

$$\rho_1(x,y) = \begin{cases} \rho_o(x,y) & \text{in } R \setminus R_o \\ \frac{1}{2}\,\rho_o(x,y) & \text{in } R'_o \\ 1 & \text{else} \end{cases}$$

Apply the same formula in each of the nine rectangles of Z_{R_o} in order to get ρ_2. Continue by induction. As step 1 we had n = 5 "sinking rectangles", at step 2 we have $n_2 = 9 \cdot 5$ of them, at step 3 we have $n_3 = 9^2 \cdot 5$ of them etc. No $\omega \in \Omega$ can avoid having a piece of positive arc length in common with rectangles that have "sunk" arbitrarily often. Clearly c, defined as being absolutely continuous with density $\rho = \lim \rho_k$ with respect to 2-dimensional Lebesgue measure has positive value in every relatively open subset of X, but a flow obeying the capacity restriction c can only be zero. Question: what should one consider as a "cut" here?

REFERENCES

[1] Appell, P., Le problème géométrique des deblais et remblais, Mem.Sci.Math. 27, Paris 1928

[1] Beckmann, M., A continuous model of transportation, Econometrica 20 (1952), 643-660

[2] Beckmann, M., Ein kontinuierliches max-flow-min-cut Theorem, preprint May 1978

[1] Friedrich, V., Stetige Transportoptimierung, ihre Beziehung zur Theorie der hölderstetigen Funktionen und einige ihrer Anwendungen, Berlin (VEB Deutscher Verlag der Wissenschaften) 1972

[1] Göpfert, A., Mathematische Optimierung in allgemeinen Vektorräumen, Leipzig (BSG B.G. Teubner) 1973

[1] Jacobs, K., Der Heiratssatz, Selecta Mathematica I Berlin-Heidelberg-New York (Springerverlag) 1969

[1] Lempio, F., Lineare Optimierung in unendlichdimensionalen Vektorräumen, Computing 8 (1972) 284-290, Springer-Verlag

GAME THEORY AND RELATED TOPICS
O. Moeschlin, D. Pallaschke (eds.)

SOCIO - COMBINATORICS

K. Jacobs

Institute of Mathematics
University of Erlangen - Nürnberg

By socio-combinatorics I understand the totality of all those results of a purely combinatorial character which allow of a sociological interpretation. The present paper presents three results of this character:

1) Volker Strehl's nice little theorem on unreliable coalitions,
2) the friendship theorem of Erdös-Rényi-Sós,
3) Arrow's paradox along with the recent counter-result of Peleg [1],

There are other results in socio-combinatorics which I have left aside here deliberately, such as

4) the marriage theorem (and the harem theorem),
5) Ford-Fulkerson's max-flow-min-cut theorem,
6) Kuhn's minimax theorem for tree games along with recent extensions by Seinsche [1],
7) the theory of NIM (see Berge [1]),
8) the combinatorics of lattice gases (Peierls -Kasteleyn and others) that would apply to interaction economics (see Föllmer [1]).

My presentation of Strehl's unpublished result is based on oral communications of Volker Strehl.

The friendship theorem, originally a purely graph theoretical result of hungarian brand, soon found its present interpretation, and Longyear-Parsons [1] have given a self-contained proof. I prefer, however, to present the original proof of Erdös-Rényi-Sós [1] resp. Wilf [1] here, simply because it brings the reader into contact with a field in combinatorics which he should see once in his lif anyhow: finite projective planes and the eigenvalue theory of their incidence matrices (see Baer [1], Ball [1], Hall [1], Ryser [1], Pickert [1], [2] , Dembowski [1], Hughes-Piper [1]). In both these cases combinatorics was there first, and the sociological interpretation was given only afterwards. It is quite different in Arrow's case. The renowned economist (Nobel prize 1972) first presented a proof of this statement on the formal impossibility of formal democracy. The mathematicians then worked out his idea to the effect of a theorem on dictatorship. I present the proof given by Kirman-Sondermann [1] here ("the ruling families form an ultrafilter"). I am sure Arrow found his statement as regrettable as any good democrat does, but a result of Gibbard [1] and Satterthwaite [1] seemed even to corroborate dictatorship. The gentle massaging of the theory by Fishburn [1], Sen [1], Dutta-Pattanaik [1] and many others finally culminated in the result of Peleg [1] who constructed decision schemes in which coalitions of a certain size can abolish any king or dictator. Our paper concludes with a description of Peleg's result.

§1. Unreliable coalitions.

Let N be a finite nonempty set with n elements. Call N the parliament and any nonempty subset C of N a coalition. Let $P'(N) = P(N) \setminus \{\emptyset\}$ be the set of all coalitions.

Definition 1.1. A map

$$d : P'(N) \longrightarrow \{-1, +1\}$$

is called a voting function if

$$d(C \cup D) \in \{d(C), d(D)\} \quad (C, D \in P'(N),\ C \cap D = \emptyset)$$

i.e.

$$C, D \in P'(N),\ C \cap D = \emptyset,\ d(C) = d(D)$$
$$\Longrightarrow d(C \cup D) = d(C) = d(D).$$

This is certainly a plausible definition. The following one is equally plausible.

Definition 1.2. A coalition $E \in P'(N)$ is called unreliable for the voting function d if

1) $C, D \in P'(N),\ C \cap D = \emptyset,\ C \cup D = E$
$\Longrightarrow d(C) \neq d(D)$

2) $i \in N \setminus E \Longrightarrow d(E \cup \{i\}) = d(\{i\})$.

For every $C \subseteq N$ let 1_C be the indicator function of C, i.e. $1_C(i) = 1$ for $i \in C$, and $1_C(i) = 0$ else. A subset P of $P'(N)$ is said to span $P'(N)$ if every 1_C with $C \in P'(N)$ is a real linear combination of some $1_{D_1}, \ldots, 1_{D_r}$ with $D_1, \ldots, D_r \in P$.

It would be really amusing if the unreliable coalitions would span $P'(N)$. Unfortunately it is unknown whether they do it in general. But they do it under an additional hypothesis on d, and this is just the nice little theorem of Volker Strehl.

<u>Definition 1.3.</u> A voting function d is said to be <u>linear</u> if there is a real function $w : N \longrightarrow \mathbb{R}$ such that

$$d(C) = 1 \iff \sum_{i \in C} w(i) > 0$$

$$d(C) = -1 \iff \sum_{i \in C} w(i) < 0$$

holds for every $C \in P'(N)$.

<u>Theorem 1.4.</u> For $n \leq 4$, every voting function on $N = \{1,\dots,n\}$ is linear. For $n = 5$ there is a non-linear voting function on $\{1,\dots,5\}$.

<u>Proof.</u> 1) $n = 1$ and $n = 2$ are trivial. Consider now $n = 3$, $N = \{1,2,3\}$ and let d be a voting function. If $d(\{1\}) = d(\{2\}) = d(\{3\})$, then d is linear with e.g. $w(i) = d(\{i\})$. The remaining cases are easily reduced to the case

$$d(\{1\}) = d(\{2\}) = 1, \ d(\{3\}) = -1$$

We will put $w(1) = 1$ at any rate. We are constrained to $w(2) > 0$, $w(3) < 0$. This makes $d(\{1,2\}) = d(\{1\}) = d(\{2\}) = 1$, $d(\{3\}) = -1$ sure, as it should. Thus we have only to deal with $\{1,3\}$, $\{2,3\}$ and $\{1,2,3\}$. If $d(\{1,3\}) = 1$, then $d(\{2\}) = 1$ yields $d(\{1,2,3\}) = 1$. In this case we choose $w(3) = -\frac{1}{2}$ and choose $w(2) > 0$ such that $d(\{2,3\}) = 1$ iff $w(2) + w(3) > 0$, which is easy. The case $d(\{1,3\}) = -1$ is handled similarly. We leave case $n = 4$ to the reader.

2) Consider the case $n = 5$ and define (we dispense with waved brackets)

$$d(C) = \begin{cases} 1 \text{ for } C = 1,\ 2,\ 12,\ 13,\ 14,\ 15,\ 23,\ 24,\ 25, \\ \qquad\qquad 123,\ 124,\ 125,\ 134,\ 135,\ 245,\ 1234, \\ \qquad\qquad 1235,\ 1245,\ 12345 \\ -1 \text{ for all other } C \in P'(\{1,2,3,4,5\}) \end{cases}$$

This is in fact a voting function (exercise). Assume now that d is linear, and $w : N \longrightarrow R$ is chosen accordingly.

It is easily seen that

$$w(1) + w(3) + w(5) > 0$$
$$w(2) + w(3) + w(5) < 0,$$

hence $w(1) > w(2)$. Similarly

$$w(1) + w(4) + w(5) < 0$$
$$w(2) + w(4) + w(5) > 0$$

implies $w(1) < w(2)$, a contradiction. Thus d cannot be linear.

Exercise 1.5. Construct non-linear voting functions for any $n > 5$.

Theorem 1.6. (V. Strehl, unpublished).
If d is a linear voting function, then the unreliable coalitions span $P'(N)$.

Proof. Let $C, D, E \in P'(N)$, $C \neq D$. We call $\{C, D\}$ a derivation of E if one of the following two conditions holds:

A) $E = C \cup D$, $C \cap D = \emptyset$, $d(C) = d(D)$.

b) $E \cap C = \emptyset$, $D = E \cup C$, $d(C) \neq d(D)$ (and hence $d(E)=d(D)$)

The first point is: E is an unreliable coalition iff it has no derivation.

The second point is: If d is linear and $w : N \longrightarrow \mathbb{R}$ is defined accordingly, if we, furthermore, put $w(C) = \sum_{i \in C} w(i)$ $(C \in P'(N))$, then for any derivation $\{C, D\}$ of some E, we have $|w(C)|<|w(E)|>|w(D)|$.

The third point is: we can write

$$E = C \cup D, \; C \cap D = \emptyset.$$

as a linear relation

$$1_E = 1_C + 1_D,$$

and similarly

$$E \cap C = \emptyset, \; D = E \cup C$$

as

$$1_E = 1_D - 1_C.$$

It is now obvious how to prove the theorem: for every $E \in P'(N)$ we build a binary "tree" of derivations. The w-values strictly decrease at every derivation, hence every such tree has "tips" allowing no more derivation, i.e. representing unreliable coalitions $C_1,\ldots,C_r$. The linear relations belonging to the derivation fit together into a linear combination $1_E = \alpha_1 1_{C_1} + \ldots + \alpha_r 1_{C_1}$ with even integer-valued $\alpha_1,\ldots,\alpha_r$.

§ 2. The friendship theorem

appeared for the first time in Erdös-Rényi-Sós [1] 1968 and received its present name probably by Higman on a symposium in Oxford 1969. Wilf [1] gave a comparatively simple proof involving finite projective geometry, and Longyear-Parsons [1] produced a more lengthy but completely elementary derivation. I prefer to give the reader a chance to see more of mathematics, and am following Erdös-Rényi-Sós [1] and Wilf [1] here.

Friendship is a binary symmetric relation. We may assume that it is not reflexive. In general it is not transitive. Thus let $N = \{P, Q, \ldots\}$ be a finite set of $n \geq 2$ "persons" and assume such a friendship relation $\sim$ be given there:

$$P \sim Q \Longrightarrow Q \sim P$$
$$P \sim Q \Longrightarrow P \neq Q$$

is the full list of our general requirements.

We may represent N as a set in the plane and link P and Q by an unoriented edge $\overline{PQ} = \overline{QP}$ if $P \sim Q$. This yields our <u>friendship graph</u>, as we call it.

Now we impose a rather peculiar condition, the

Common Friend Condition CFC:

$$P, Q \in N, \; P \neq Q \Longrightarrow P \sim F \sim Q \quad \text{for exactly one } F \in N$$

i.e. any two different persons have exactly one friend in common.

It is easy to draw friendship graphs with this property, namely, <u>star-shaped</u> graphs like:

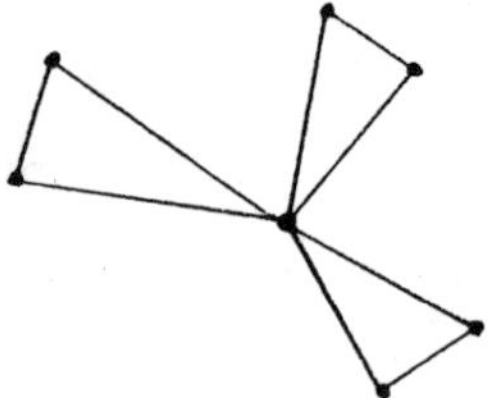

To be precise, $(N,\sim)$ yields a star-shaped graph iff n is odd, one person is the friend of everybody else (some people call such a person a "politician"), and the remaining even number of persons is grouped in couples of friends.

<u>Theorem 2.1 (Friendship Theorem)</u>
If $\sim$ is a friendship relation in N fulfilling CFC, then the corresponding friendship graph is star-shaped.

The proof consists of a long chain of conclusions from CFC and the assumption that the friendship graph is not star-shaped, ending up with a contradiction. Our main business will be the establishment of a projective plane in this imaginary world. Namely, we will define the sets of the form

$$L(P) = \{Q \mid Q \in N,\ Q \sim P\} \qquad (P \in N)$$

as "lines" and establish the following statements:

> Axiom I: If $Q, R \in N$, $Q \neq R$, then there is exactly one line $L(P)$ containing both Q and R.

This is in fact true: we have to take P as the unique common friend of Q and R.

> Axiom II: If $L(P)$, $L(Q)$ are lines, $L(P) \neq L(Q)$, then their intersection $L(P) \cap L(Q)$ consists of exactly one point R.

This is trivial again: We have to take R as the unique common friend of P and Q.

Thus our situation has already a basic projective structure. But for a projective plane we need still one more axiom, namely

> Axiom III: There are four pairwise distinct P, Q, R, S in N such that no three of these are on one line.

The establishment of axiom III is the crucial point of our proof. It requires the assumption that the friendship graph in question is not star-shaped.

Now let us check our forces. CFC implies "no quadrangle", i.e. there are no four pairwise distinct $P, Q, R, S \in N$ such that $P \sim Q \sim R \sim S \sim P$. In fact such a quadrangle

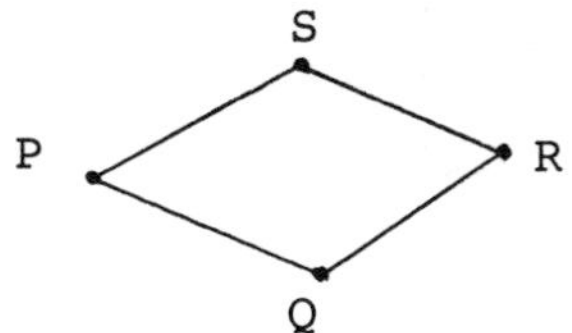

would imply that e.g. P, R have two different common friends: both Q and S.

Consequently: for any two distinct $P, Q \in N$ with $P \sim Q$ there is exactly one R with $P \sim Q \sim R \sim P$ (a "triangle"). A second triangle containing the edge $\overline{PQ}$ would immediate lead to a forbidden quadrangle.

For any P let $v(P)$ be the number of friends of P. We prove now the following

<u>Lemma 2.2.</u> For every $P \in N$, $v(P)$ is an even number ≥ 2 and P is a vertex of exactly $\frac{v(P)}{2}$ triangles.

<u>Proof.</u> Consider $L(P)$ and map this set into itself by $f_P : L(P) \longrightarrow L(P)$, where $f_P Q$ is the unique common friend of P and Q. This mapping f_P is injective: if

$f_P Q = S = f_P R$ would happen with $Q, R \in L(P)$, $Q \neq R$, we would immediately get a quadrangle, which is forbidden. As $L(P)$ is finite, f_P simply permutes the points of $L(P)$, and it is clear that

1) $f_P \circ f_P$ = identity

2) f_P has no fixed points.

Thus $L(P)$ splits into two-element sets $\{Q, R\}$ such that $P \sim Q \sim R \sim P$. Clearly this proves the lemma.

We would, of course, be through with our proof of the Friendship Theorem if we would find a $P \in N$ with $v(P) = n-1$.

Let now $P \sim Q \sim R \sim P$ a triangle. We form the sets

$$L'(P) = L(P) \setminus \{Q,R\}$$
$$L'(Q) = L(Q) \setminus \{R,P\}$$
$$L'(R) = L(R) \setminus \{P,Q\}$$
$$W = N \setminus [L'(P) \cup L'(Q) \cup L'(R) \cup \{P,Q,R\}].$$

We may sketch this situation as follows:

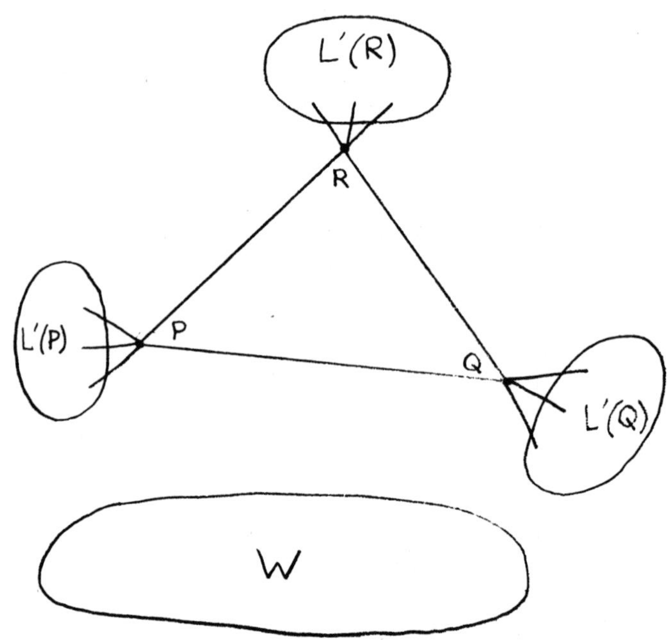

Since there is no quadrangle, there is no edge joining a point in $L'(P)$ with a point in $L'(Q)$ etc., and for similar reasons, $L'(P)$, $L'(Q)$, $L'(R)$ are pairwise disjoint.

Lemma 2.3. If there is a P with $v(P) = 2$, then there is a Q with $v(Q) = n - 1$.

Proof. If $v(P) = 2$, then $L(P) = \{Q,R\}$, say, and we have $Q \sim R$, $L'(P) = \emptyset$. If there is any $S \in W$, then we look for the unique common friend F of P and S. As $F \in L(P)$, we have $F = Q$ or $F = R$, hence $S \in L'(Q)$, or $S \in L'(R)$, a contradiction. Thus W is empty. Similarly, $L'(Q)$ and $L'(R)$ cannot be both nonempty, because otherwise we would choose F as the common friend of some $U \in L'(Q)$ and some $V \in L'(R)$. An easy discussion leads to $F = P$, hence to a quadrangle, a contradiction. Thus $N = \{P, Q, R\} \cup L'(Q)$, say, and $v(Q) = n-1$ follows.
We are thus finished with our proof of the friendship theorem as soon as we have found some P with $v(P) = 2$. Thus we may henceforth assume $v(P) > 2$ for all $P \in N$, and thus $v(P) \geq 4$.
We now set out to derive Axiom III from the assumption that our friendship graph is not star-shaped. Don't forget that we may and shall assume $v(P) \geq 4$ $(P \in N)$ as well.

Choose any $P \in N$ and three distinct points T, U, V on $L(P)$. The line joining, say, P and T, contains at least one more point $R \neq P$, not on $L(P)$. Let Q be the common friend of P and R. There is a point $S \in \{T,U,V\}$ which is not Q nor the common friend T of P and Q. The situation may be pictured as follows

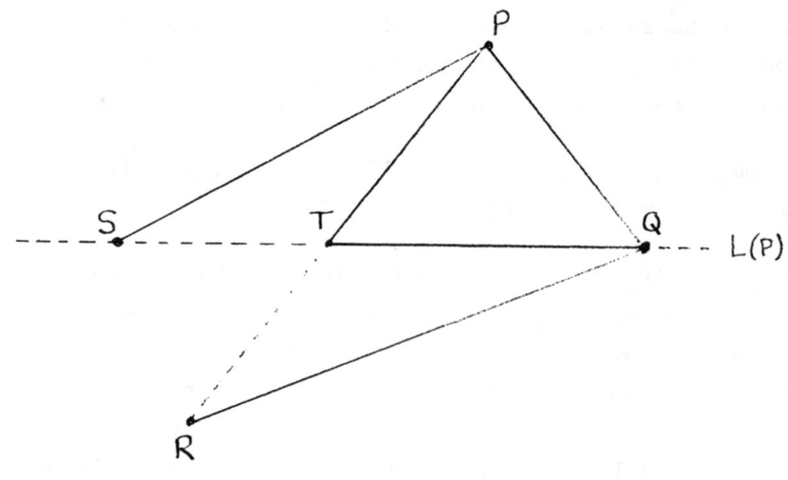

We have $S \sim P \sim Q \sim R$. Assume now that there is an $X \in N$ with three of P, Q, R, S on $L(X)$. We check through all four cases and get a contradiction each time.

Case 1: $P, Q, R \in L(X)$. Then $X \sim P \sim Q \sim R \sim X$ is a quadrangle.

Case 2: $Q, R, S \in L(X)$. Then $X = P$, contradicting the fact that $R \in L(P)$.

Case 3: $R, S, P \in L(X)$. Then $X = Q$, but S was chosen not to be on $L(Q)$

Case 4: $S, P, Q \in L(X)$. Then $X = T$ and $S \sim T \sim Q \sim P \sim S$ is a quadrangle.

We have now proved Axioms I - III, and hence established N as a projective plane in which the $L(P)$ are the lines. Moreover it is clear that the mapping $P \longrightarrow L(P)$ is bijective and preserves incidences. Now there is a general theorem on finite projective planes saying that for such a mapping there is necessarily some P which lies on the corresponding line: we get $P \in L(P)$, i.e. $P \sim P$, a contradiction. The Friendship Theorem is proved.

The theorem on projective planes is e.g. to be found in Hughes-Piper [1], lemma 12.3. The proof is not hard: square the incidence matrix of the plane and prove that this square has the eigenvalues $q^2, q-1,\dots,q-1$, hence the incidence matrix has $\pm q, \pm\sqrt{q-1},\dots, \pm\sqrt{q-1}$. Here $q + 1$ is the number of points on a line. If we have s times the plus and r times minus sign here, we get $r + s = n - 1$, hence $q + 1 + s\sqrt{q} = r\sqrt{q}$ (as the trace of the incidence matrix is 0). This leads to a contradiction rather easily.

§ 3. Arrow's paradox and its solution.

1. The problem. Simple cases.

Let N be a set of $n \geq 2$ "agents" and A a set of $a \geq 2$ "alternatives". A linear ordering $<$ on A is also called a preference (order) on A. Clearly there are $a!$ such preferences. In many papers quasi-orderings that need not be linear are considered, but the solid nucleus of sweeping results becomes clear already in the linear case, and we will stick to linear orderings throughout this paper.

A mapping $p : N \longrightarrow \text{perm}(A)$ = the set of all preferences on A is called a preference pattern. The set of all such patterns is obviously denoted by $\text{perm}(A)^N$ and has $(a!)^n$ elements.

<u>Definition 3.1.</u> A mapping

$$d : \text{perm}(A)^N \longrightarrow \text{perm}(A)$$

is called a social decision function (SDF) if the following holds.

1) (<u>Unanimity Rule</u>) Let $(<_i)_{i \in N}$ be a preference pattern and $< = d((<_i)_{i \in N})$. Then

$$a, b \in A, \ a \underset{i}{<} b \ (i \in N) \Longrightarrow a < b$$

2) (<u>Independence Rule</u>) Let $(<_i)_{i \in N}$ and $(\prec_i)_{i \in N}$ be two preference patterns, $< = d((<_i)_{i \in N})$, $\prec = d((\prec_i)_{i \in N})$. Let $a, b \in A$ and assume

$$\{i \mid a \underset{i}{<} b\} = \{i \mid a \underset{i}{\prec} b\}.$$

Then

$$a < b \Longleftrightarrow a \prec b.$$

<u>Definition 3.2.</u> Let $d : \mathrm{perm}(A)^N \longrightarrow \mathrm{perm}(A)$ be a SDF. An agent $i \in N$ is called a <u>dictator</u> for d if $<_i = d((<_j)_{j \in N})$ for every preference pattern $(<_j)_{j \in N}$. The SDF is called <u>dictatorial</u> if it has a dictator.

It is obvious that there can be at most one dictator for any given SDF.

If $a = 2$, we can easily design non-dictatorial SDF's. If $n > 2$, we simply follow the majority rule, with an arbitrarily appointed chairman whose vote decides in the case of a tie, in the case of an even n. If we would follow this device also for $n = 2$, we would get a dictator, but here we have another method. Unanimity plus a constant decision in the two non-unanimous cases.

<u>2. Two agents, three alternatives.</u>

Let $N = \{1,2\}$, $A = \{a,b,c\}$. It should be clear which preference pattern is meant by the compound symbol

$$\begin{array}{cc} a & a \\ b & c \\ c & b \\ \hline 1 & 2 \end{array}$$

Clearly we have to say here that agent 1 prefers b to c etc.

<u>Definition 3.3.</u> 1 <u>beats</u> 2 at $x, y \in A$, $x \neq y$, for a given SDF d if the following holds: for any preference pattern $x <_1 y$, $y <_2 x$ implies $x < y$, where $< = d((<_1, <_2))$.

<u>Lemma 3.4. (Exchange lemma).</u> If 1 beats 2 at x,y for d, then 1 beats 2 at any couple of distinct alternatives

for d.

<u>Proof.</u> Assume that 1 beats 2 at a,b. Look at

b	a
a	c
c	b
—	—
1	2

and consider the resulting $<$. It must put $a < b$ by beating and $c < a$ by unanimity. Hence $c < a < b$ results. But this means that 1 beats 2 at c,b; we see this in this particular pattern, but transfer the result to any pattern with $c \underset{1}{<} b$, $b \underset{2}{<} c$, by independence. Thus we see that we may "exchange" a by c. Similarly we prove that we may "exchange" b by c. A suitable chain of "exchanges" links every couple x,y to a,b. The lemma is proved.

<u>Proposition 3.5.</u> (Arrow's paradox for $n = 2$, $a = 3$). If $n = 2$, $a = 3$, then every SDF is dictatorial.

<u>Proof.</u> Let $x, y \in A$, $x \neq y$. Let $A = \{x,y,z\}$ and consider the pattern

z	z
y	x
x	y
—	—
1	2

and look at the $<$ resulting from a fixed SDF d and this pattern. If $x < y$, then 1 beats 2 at x,y, hence at any couple, and 1 is a dictator by beating plus unanimity. If $y < x$, then 2 beats 1 at x,y, and 2 is a dictator.

<u>3. The general case.</u>

We now pass over to $n \geq 2$, $a \geq 3$. We picture preference patterns in an obvious way by arrays like

x	b		a
y	y		t
.	.	...	.
.	.		
.	.		
b	z		x
a	x		y
1	2		n

Definition 3.6. Let $\emptyset \neq F \subseteq N \neq F$ and $x, y \in A$, $x \neq y$. We say that F beats $N \setminus F$ at x,y for a given SDF d if the following holds:
let $(<_i)_{i \in N}$ be any preference pattern such that
$i \in F \Longrightarrow x <_i y$ and $i \in N \setminus F \Longrightarrow y <_i x$; let $<$ be the preference resulting from that pattern via d; then $x < y$.

Remark. Independence implies that $x < y$, for one such pattern implies $x < y$ for any such pattern. Hence our definition is well-formed.

Lemma 3.7. (Exchange Lemma). If $\emptyset \neq F \subsetneq N \neq F$ and F beats $N \setminus F$ at x,y for d, then it beats $N \setminus F$ at any couple of distinct alternatives for d. In this case F is called a ruling family for d. N is called a ruling family for any d, by definition.

The proof is nearly obvious: adapt the proof of lemma 3.4.

Clearly a dictator is nothing but a one-element ruling family, provided we can prove that any set containing a ruling family is a ruling family again. This is, however, one of the statements implicit in

Theorem 3.8. (Kirman-Sonderman [1]). Let $n \geq 2$, $a \geq 3$ and d any SDF. Then the ruling families for d form an ultrafilter.

<u>Proof.</u> 1) Let $\emptyset \neq F \subseteq G \subseteq N \neq G \neq F$ and assume that F is a ruling family for d. Choose three distinct alternatives $a, b, c \in A$. We picture a preference pattern (or rather those features of it that count in the present situation; independence proves that this is sufficient) in the following obvious way:

$a <_i c <_i b$	$N \setminus G$
$a <_j b <_j c$	$G \setminus F$
$b <_k a <_k c$	F

N

We check the order of a, b, c in the preference $<$ resulting by d. Clearly $b < a$ as F is a ruling family. Clearly $a < c$ by unanimity. Thus G beats $N \setminus G$ at b, c, hence is a ruling family.

2) Let $\emptyset \neq F \subseteq N \neq F$. Then F or $N \setminus F$ is a ruling family. In fact, let us consider a pattern

$b <_i a$	$N \setminus F$
$a <_j b$	F

N

and check, for the $<$ resulting by d, the order of a, b. If $a < b$, then F rules, if $b < a$, then $N \setminus F$ rules.

3) Let $\emptyset \neq F \subseteq N \neq F$, $\emptyset \neq G \subseteq N \neq G$ and assume that F and G are ruling families. Clearly $F \cap G = \emptyset$ is impossible since otherwise F and $N \setminus F$ would both be ruling families. We prove that $F \cap G$ is a ruling family. For this, we consider a pattern

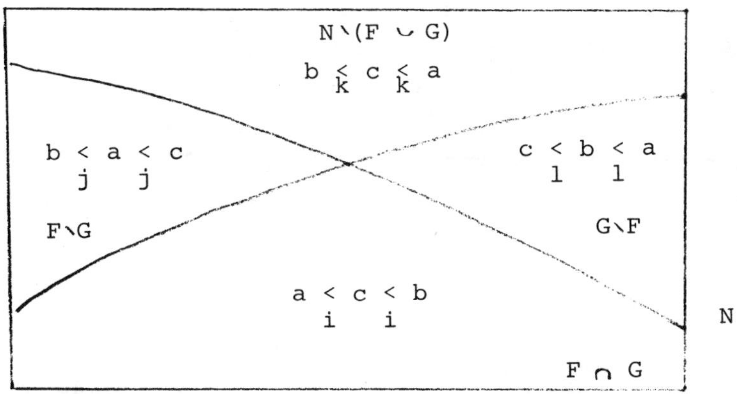

Let $<$ be the preference resulting by d and let us check the position of a, b, c for $<$. $a < c$ follows because F is a ruling family. $c < b$ follows because G is a ruling family. Thus $F \cap G$ beats its complement at a, b, hence is a ruling family. The theorem is proved.

As a corollary we obtain

<u>Theorem 3.9. (Arrow's Paradoxon).</u> If $n \geq 2$, $a \geq 3$, then every SDF d is dictatorial.

<u>Proof.</u> The ruling families for d form an ultrafilter in the finite set N, hence there is one $i \in N$ such that $\{i\}$ is a ruling family, i.e. i is a dictator for d.

4. Peleg's solution

Arrow's paradox (theorem 3.9) poses, of course, the problem of proving the existence of non-dictatorial decision schemes and to give explicit constructions. There is an abundance of attempts to solve this problem, see Kelly [1], Fishburn [3].

Of all the solutions the one preferred by B. Peleg [1] seems to me most appealing. I will state his result here, but give no proof. Let us begin with some discussion. A first obvious modification of the Arrow setup replaces the requirement that a SDF yields a linear order on the set A of alternatives by the weaker requirement that it yields only a top preference.

Definition 3.10. A mapping

$$t : \text{perm}(A)^N \longrightarrow A$$

is called a social choice function (SCF). A SCF t is called anonymous if

$$t((<_{\tau(i)})_{i\in N}) = t((<)_{i\,i\in N})$$

for every permutation τ of N. Let t be any SCF.

1) A coalition $\emptyset \neq F \subseteq N$ is said to be winning (for t) if $(<)_{i\,i\in N} \in \text{perm}(A)^N$, $a \in A$, $b <_j a$ $(j \in F, b \in A)$

$$t((<)_{i\,i\in N}) = a.$$

2) t is said to have the property (W_h) if every $F \subseteq N$ with $|F| = h$ is winning.

3) An $i \in N$ is called a dictator for t if $\{i\}$ is winning for t. t is called dictatorial if there is a dictator for t.

Clearly, a t with property (W_h) for some $h < n$ cannot be dictatorial.

Let t be a SCF and $(<)_{j\,j\in N} \in \text{perm}(A)^N$. With these data we may define a non-cooperative n-person game $(t, (<)_{j\,j\in N})$ with perm(A) = set of all (pure) strategies for agent i.

Given a n-tuple $(<'_j)_{j\in N}$ of strategies, agent i considers $t((<'_j)_{j\in N}) \in A$ as his payoff, and he considers his payoff as better or worse according to $<_i$. It should now be clear what an <u>equilibrium</u> in this game is: it is some $(<'_j)_{j\in N}$ such that no agent can improve his payoff by just changing his own strategy while all other agents keep their strategies fixed.

<u>Definition 3.11</u>. A SCF t is called <u>non-manipulable</u> if every $(<_j)_{j\in N}$ is itself an equilibrium point for the game $(t,(<_j)_{j\in N})$. Otherwise, t is called <u>manipulable</u>.

Non-manipulability means that nobody has a reason to change his preference order after the choice has been made according to t.

Here we may mention a result of Gibbard [1] and Satterthwaite [1]: If t is non-manipulable with a range that equals A , then t is dictatorial. Consequently, every anonymous t with property (W_n) is manipulable. This looks rather dissatisfactorial. Peleg [1] proposes the

<u>Definition 3.12.</u> A SCF t is called <u>exactly consistent</u> if there is a mapping

$$m : \text{perm}(A)^N \longrightarrow \text{perm}(A)^N$$

such that

1) for each $(<_j)_{j\in N}$, $m((<_j)_{j\in N})$ is an equilibrium point of $(t,(<_j)_{j\in N})$.

2) $t \circ m = t$.

Now, what does exact consistency mean? It means that individuals might like to change their preferences after t has been applied. But there is a mechanism m which allows them to make their changes in such a fashion that a renewed application of t yields the same result as before, and now nobody has a reason anymore to change his preferences, if he uses his old preference order for judgement. Sophisticated enough, isn't it? But Peleg [1] puts in even more sophistication. Let us return to our game $(t, (<_i)_{i \in N})$.

<u>Definition 3.13.</u> Let t be a SCF. $(<'_i)_{i \in N}$ is called a <u>strong equilibrium</u> point for the game $(t, (<_i)_{i \in N})$ if for every $\emptyset \neq F \subseteq N$ and for every $(<''_i)_{i \in N}$ that coincides with $(<'_i)_{i \in N}$ in $N \setminus F$ there is at least one $j \in F$ whose payoff has not really improved under the replacement of $(<'_i)_{i \in N}$ by $(<''_i)_{i \in N}$. t is said to be <u>exactly and strongly consistent</u> if there is an m as in definition 3.12. such that $m((<_i)_{i \in N})$ is always even a strong equilibrium point of $(t,(<_i)_{i \in N})$.

<u>Definition 3.14.</u> Let t be a SCF.

1) $\emptyset \neq F \subseteq N$ is said to be <u>strongly winning</u> for t if $(<_i)_{i \in N} \in \text{perm}(A)^N$, $a, b \in A$, $b <_j a$ $(j \in F)$

$$\Longrightarrow \quad t((<_i)_{i \in N}) \neq b$$

2) t is said to have <u>property (W_h^*)</u> if every F with $|F| = h$ is strongly winning for t.

Clearly a t with property (W_h^*) for some $h < n$ cannot be dictatorial.

The basic result of Peleg [1] states the following ($[x]$ denotes the maximal integer $\leq x$).

Theorem 3.15. For $n, a \geq 2$ let $k(n,a) = \left[\frac{n(a-1)}{a}\right] + 1$. Assume $n \geq a - 1$. Then there is an exactly and strongly consistent anonymous t with property (W_h^*) where

$$n = 0 \bmod a \Longrightarrow h = k(n,a)$$

$$n = -1 \bmod a \Longrightarrow h = k(n,a)$$

and $h = k(n,a) + 1$ in all other cases.

Anonymity wards off dictatorship.
(W_h^*) wards off veto players, but it does the job only for $h < n$. Now, if n is small and a is large, then $\frac{n(a-1)}{a}$ is nearly n, hence $k(n, a) = n$, and the theorem yields (W_n^*) at most, which is of no use. But if n is large enough as compared to a, then $\frac{n(a-1)}{a}$ is considerably smaller than n, and the theorem yields property (W_h^*) with some $h < n$, thus warding off veto players . In short: If the population is large and the number of alternatives is small, every would-be veto player can be kept in check if only a certain number $h < n$ of people resolves to do so.

How is this compatible with Arrow's paradox (theorem 3.9)? One of the basic points is, of course the fact that a SCF doesn't determine a full preference order $<$ on A but only the topmost element of it. One would be tempted to construct a SDF from a sequence of non-dictatorial SCF's such that the first SCF decides the top alternative, the second for the alternative next to the top etc. But even if we would agree to allow a dictator for each step (and a different one for every step) in such a procedure, the independence condition of Arrow would quickly fail. In fact consider the preference patterns $(n = 2, a = 3)$

a	c		b	c
b	a	and	c	a
c	b		a	b
1	2		1	2

Assume that agent 1 dictates the top preference, and agent 2 the next positions. Then

$$b < c < a \quad \text{resp.} \quad a < c < b$$

result. Clearly independence is violated here for b, c. This is in agreement with the widespread opinion that independence is the crucial reason for dictatorship.

I conclude with a sketch of Peleg's [1] construction of his SCF t. Choose a pattern $(<_i)_{i\in N}$ and consider, for each $a \in A$, the set $\{i \mid a <_i b \ (a \neq b \in A)\} = C(a)$. For every $\emptyset \neq F \subseteq C(a)$ we say that F blocks a. Peleg then eliminates one a with $C(a) \neq \emptyset$ from A and also deletes a suitable $\emptyset \neq F \subseteq C(a)$. With the remaining setup he repeats this construction until only one alternative is left. This alternative is taken as the value of $t((<_i)_{i\in N})$. The fine properties requested for t in theorem 3.15 are achieved by choosing sort of an optimal variant of the above elimination procedure.

REFERENCES

[1] Arrow, K.J.:
Social choice and individual values.
2nd ed., New York (Wiley) 1963

[1] Baer, R.:
Polarities in finite projective planes.
Bull.AMS 52, 77-93 (1946)

[1] Ball, R.W.:
Dualities of finite projective planes.
Duke Math.J. 15, 929-940 (1948)

[1] Berge, C.:
Théorie des graphes et ses applications.
Paris (Dunod) 1958.

[1] Brown, D.J.:
Aggregation of preferences.
22 pp, preprint, Institute of Business and Econ. Research,
Univ. of Calif., Berkeley 1973

[1] Burger, E.:
Einführung in die Theorie der Spiele.
2.Aufl.Berlin (de Gruyter) 1966.

[1] Dembowski, P.:
Finite Geometries.
Berlin-Heidelberg-New York (Springerverlag) 1968

[1] Dutta, B., and P. Pattanaik:
On nicely consistent voting systems,
Econometrica 46 (1978), 163-170.

[1] Erdös, P., A. Rényi and V.T. Sós:
On a problem of graph theory.
Stud.Sci.Math.Hung. 1, 215-235 (1966)

[1] Fishburn, P.C.:
Arrow's impossibility theorem: concise proof and infinite voters.
J. Econ.Theory 2, 103-106 (1970).

[2] Fishburn, P.C.:
Utility for Decision Making, New York (Wiley) 1970

[3] Fishburn, P.C.:
The theory of social choice,
Princeton (UP) 1973

[1] Föllmer, H.:
Random economies with many interacting agents.
J.Math.Econ. 1 (1974), 51-62

[1] Gibbard, A.:
Manipulation of voting schemes: a general result.
Econometrica 41 (1973), 587-601.

[1] Hall, M.:
Combinatorial theory.
X+310 pp, Waltham(Mass.)-Toronto-London (Blaisdell) 1967.

[1] Hughes, D.R. and P.C. Piper:
Projective planes.
Berlin-Heidelberg-New York.

[1] Kelly, J.S.:
Arrow impossibility theorems,
New York (Academic Press) 1978.

[1] Kirman, A.P. and D. Sondermann:
Arrow's theorem,many agents and invisible dictators
J. Econ.Theory 5, 267-277 (1972).

[1] Longyear, J.Q., and T.D. Parsons:
The friendship theorem.
Indag.Math. 34, 257-262 (1972).

[1] Peleg, B.:
Consistent voting systems.
Econometrica 46, 153-161 (1978).

[1] Pickert, G.:
Projective Ebenen.
Berlin-Göttingen-Heidelberg (Springerverlag) 1955

[2] Pickert, G.:
Einführung in die endliche Geometrie.
Stuttgart (Klett) 1974.

[1] Ryser, H.J.:
Combinatorial mathematics.
MAA & Wiley 1963.

[1] Satterthwaite, M.:
Strategy-proofness and Arrow's conditions: Existence and correspondence theorems for voting procedures and social welfare functions.
J.Ec.Th. 10, 187-217 (1975).

[1] Seinsche, D.:
Unverschränkte Graphen und strategische Spiele in kombinatorischer Form.
Z.Wahrsch.verw.Geb. 27, 297-334 (1973).

[1] Sen, A.K.:
Collective choice and social welfare.
San Francisco (Golden-Day) 1970.

[1] Skala, H.L.:
A variation of the friendship theorem.
SIAM J.Appl.Math. 23, 214-220 (1972).

[1] Steiner, H.-G.:
Exemples of exercises in mathematization: an extension of the theory of voting bodies.
Educ.Stud.Math. 1, 289-299 (1969).

[1] Wilf, H. S.:
The friendship theorem.
Com.Math.Appl.ed.Welsh, 307-309, New York (Acad. Press) 1971.

GAME THEORY AND RELATED TOPICS
O. Moeschlin, D. Pallaschke (eds.)

ATOMLESS ECONOMIES WITH COUNTABLY MANY AGENTS

Christoph Klein

GMD Birlinghoven 5205 St. Augustin 1
West Germany

INTRODUCTION. Following R.J. Aumann [1], one usually takes the unit interval equipped with the Lebesgue measure $([0,1], \mathcal{B}[0,1], \lambda)$ for the space of agents of a large economy with pure competition. The Borel field yields the allowed coalitions. The points symbolize the agents. Large economies with pure competition are a kind of limit of finite economies (cf. [3] chapter 2.1, example 3 and prop. 2). It would be natural, if large economies had countably many agents. But card $[0,1] = \aleph_1 = \text{card } 2^{\mathbb{N}}$. Moreover all coalitions should be allowed (cf. [3] page 126). As there are non-Lebesgue-measurable sets, the extension of λ onto $\mathbb{P}([0,1])$ (:= the set of all subsets of [0,1]) is not possible. Here we introduce the concept of a space of agents, which tries to avoid these points of criticism. We mention, that there is no σ-additive, atomless and non-trivial measure on $\mathbb{N}$.

COUNTABLE ECONOMIES D. Pallaschke [5] defined a large economy with countably many traders as a map $\mathcal{E} : \mathbb{N} \to K_0 \subset \mathcal{P} \times \mathbb{R}^n_+$, where K_0 is a compact subset of the agents characteristic space (cf. [3] chapter 2.1), together with a finitely-additive non-negative and normalized measure $\nu_{\mathbb{N}}$ on $\mathbb{N}$. Since $\nu_{\mathbb{N}}(A)$ is defined for every $A \subset \mathbb{N}$, every coalition is allowed.

$\mathcal{E}$ yields canonically a continuous map $\tilde{\mathcal{E}} : \beta\,\mathbb{N} \to K_0, \tilde{\mathcal{E}}|_{\mathbb{N}} \equiv \mathcal{E}$ where $\beta\,\mathbb{N}$ is the Stone Cech compactification of $\mathbb{N}$. $\nu_{\mathbb{N}}$ extends to a normed and non-negative Radon measure ν on $\beta\,\mathbb{N}$ (cf. [5] or [6] 18.7) satisfying $\nu_{\mathbb{N}}(A) = \nu(\bar{A}^{\beta\mathbb{N}})\ \forall\ A \subset \mathbb{N}$. Therefore $\tilde{\mathcal{E}} : (\beta\mathbb{N}, \mathcal{B}(\beta\mathbb{N}), \nu) \to K_0$ is an exchange economy in the sense of W. Hildenbrand ([3] chapter 2.1, definition 1).

Since $\dim \beta\mathbb{N} = 0$ and the open and closed sets of $\beta\mathbb{N}$ are $\{\bar{A}^{\beta\mathbb{N}} | A \subset \mathbb{N}\}$ (cf.[8] 3.16), the linear hull of $\{\chi(\bar{A}^{\beta\mathbb{N}}) \mid A \subset \mathbb{N}\}$ is dense in $C(\beta\mathbb{N})$ (cf. [6] 8.2.2). $C(\beta\mathbb{N})$ denotes the space of continuous real functions on $\beta\mathbb{N}$ endowed with the supremum norme. $\chi(\bar{A}^{\beta\mathbb{N}})$ denotes the characteristic function of $\bar{A}^{\beta\mathbb{N}}$. $C(\beta\mathbb{N})$ is dense in $L_1(\nu)$ with respect to the L_1 norm. Therefore the linear hull of $\{\chi(\bar{A}^{\beta\mathbb{N}}) \mid A \subset \mathbb{N}\}$ is dense in $L_1(\nu)$. The analogous result is true for $\mathbb{R}^n$- valued functions. Let $e_j \in \mathbb{R}^n$ be the canonical j-th unit vector of $\mathbb{R}^n$, $1 \le j \le n$.
If the core $C(\tilde{\mathcal{E}})$ of $\tilde{\mathcal{E}}$ equals the set of Walras equilibrium allocations $W(\tilde{\mathcal{E}})$, then a sequence in lin.hull $\{\chi(\bar{A}^{\beta\mathbb{N}})e_i | i \le n, A \subset \mathbb{N}\}$ approximates a Walras equilibrium allocation if and only if it approximates an element of the core. Moreover if $f \in l^\infty$ and $\tilde{f}$ is its continuous extension onto $\beta\,\mathbb{N}$, we have $\int_{\beta\mathbb{N}} \tilde{f}\, d\nu = \int_{\mathbb{N}} f\, d\nu_{\mathbb{N}}$ (cf.[6] 18.7.) and the same remains true for a bounded $f : \mathbb{N} \to \mathbb{R}^n$. So we identify a coali-

tion $A \subset \mathbb{N}$ with its closure $\bar{A}^{\beta\mathbb{N}} \subset \beta\mathbb{N}$.

Let $B := \{f \mid f : \mathbb{N} \to \mathbb{R}^n$, f is bounded with respect to the sup norm$\}$. We define the core of $\mathscr{E}$ by $C(\mathscr{E}) := \{\{f_n\}_{n\in\mathbb{N}} \subset B \mid \tilde{f}_n \to g \in C(\tilde{\mathscr{E}})$ with respect to the $L_1(\nu)$ norm$\}$ and $W(\mathscr{E})$ analogously. We suppose that $\tilde{\mathscr{E}}$ fulfills the assumptions of theorem 1 in [3] chapter 2.1. Then $\mathscr{E}$ is an economy with countably many traders, with pure competition i.e. $\nu_{\mathbb{N}}(\{n\}) = o\ \forall\ n\in\mathbb{N}$, in which every coalition is allowed and where $C(\mathscr{E}) = W(\mathscr{E})$.

So the mentioned points of criticism do not apply here. But there are other questions, especially: How to characterize $\nu_{\mathbb{N}}$ such that ν is atomless? How to have a concrete picture of ν, how to calculate ν? What meaning do the points of $\beta\mathbb{N}\setminus\mathbb{N}$ have?

<u>ATOMLESS MEASURES ON $\beta\mathbb{N}$</u>. We characterize those $\nu_{\mathbb{N}}$ which extend to atomless ν :

<u>Lemma 1</u> Suppose X to be a compact topological space with dim X = 0 (i.e. X has a basis of open and closed-i.e. clopen - sets) and let μ be a Radon measure on X. If $M \subset X$ is a Borel set and $\varepsilon > 0$, then there is a clopen set $C_{M,\varepsilon} \subset X$, such that $|\mu|([C_{M,\varepsilon}\setminus M] \cup [M\setminus C_{M,\varepsilon}]) < \varepsilon$. $|\mu|$ denotes the total variation of μ.

P r o o f: μ is regular. Therefore there is a closed set $A_{M,\varepsilon}$, an open set $O_{M,\varepsilon}$ such that $O_{M,\varepsilon} \supset M \supset A_{M,\varepsilon}$ and $|\mu|(O_{M,\varepsilon}\setminus A_{M,\varepsilon}) < \varepsilon$. Let $x \in A_{M,\varepsilon}$. There is a clopen set $U(x)$ with $x \in U(x) \subset O_{M,\varepsilon}$ because dim X = 0. A finite covering of $A_{M,\varepsilon}$ by such sets yields $C_{M,\varepsilon}$.

We recall that $\{\bar{A}^{\beta\mathbb{N}} \mid A \subset \mathbb{N}\}$ are the clopen sets of $\beta\mathbb{N}$ (cf. [8] 3.16) and that $\nu_{\mathbb{N}}(A) := \nu(\bar{A}^{\beta\mathbb{N}})$. $|\nu_{\mathbb{N}}|$ denotes the total variation of $\nu_{\mathbb{N}}$ (cf. [6] 17.2.1). We obtain the following result:

<u>Proposition 2</u> Let ν be a Radon measure on $\beta\mathbb{N}$. ν is atomless iff for every $\varepsilon > 0$ there is a disjoint decomposition $\{A_{i,\varepsilon}\}_{i=1}^{n(\varepsilon)}$ of $\mathbb{N}$ such that $|\nu_{\mathbb{N}}|(A_{i,\varepsilon}) < \varepsilon$, $i=1,\ldots n(\varepsilon)$.

P r o o f: It suffices to prove the only if part. Since ν is a Radon measure, $|\nu|$ is one also. ν is atomless iff $|\nu|$ is. As $|\nu|$ is atomless, $\{|\nu|(E) \mid E \in \mathscr{B}(F)\} = [0, |\nu|(F)]$ where $F \in \mathscr{B}(\beta\mathbb{N})$ is given arbitrarily (cf. [6] 26.5.12). Using Lemma 1, one constructs the decomposition.

<u>MEASURES AND COMPACTIFICATIONS.</u> We want to explain a procedure which will generate atomless Radon measures on $\beta\mathbb{N}$. The concretisation of these measures generally is equivalent to the concretisation of a free ultrafilter on $\mathbb{N}$ (cf. [4]). Therefore we will give concrete atomless Radon measures on smaller compactifications of $\mathbb{N}$ than $\beta\mathbb{N}$ and then we raise these up onto $\beta\mathbb{N}$.

Let K be a compactification of the discrete topological space $\mathbb{N}$ and let dim K = 0.

The map $U \to \{n \in \mathbb{N} | n \in U\}$ is a boolean isomorphism from the algebra of clopen subsets of K onto a subalgebra of $\mathbb{P}(\mathbb{N})$. This subalgebra of $\mathbb{P}(\mathbb{N})$ contains $\mathbb{E}(\mathbb{N})$, the ideal of finite subsets of $\mathbb{N}$. On the other hand, let $\emptyset \neq \mathcal{A} \subset \mathbb{P}(\mathbb{N})$ be a set, then we denote by $<\mathcal{A}>$ the generated subalgebra in $\mathbb{P}(\mathbb{N})$ i.e. the generated field of sets in $\mathbb{N}$. $\tau(<\mathcal{A}>)$ is defined to be the Stone space of $<\mathcal{A} \cup \mathbb{E}(\mathbb{N})>$ seen as space of ultrafilters. $\tau(<\mathcal{A}>)$ is a compactification of $\mathbb{N}$ because $\{\{n\}|n \in \mathbb{N}\}$ is dense in $<\mathcal{A} \cup \mathbb{E}(\mathbb{N})>$. The continuous surjection $\phi : \beta\mathbb{N} \longrightarrow\!\!> \tau(<\mathcal{A}>)$, $\phi|_{\mathbb{N}} \equiv id$ is given by $x \longmapsto \mathfrak{T} \;\forall\; x \in \mathbb{F}_{\mathfrak{T}} := \cap \{\overline{F}^{\beta\mathbb{N}} | F \in \mathfrak{T}\}$, where $\mathfrak{T}$ is an ultrafilter in $<\mathcal{A} \cup \mathbb{E}(\mathbb{N})>$. Hence there is a bijection (more precisely an anti-isomorphism cf. [6] 10.1.2) between the zero-dimensional compactifications of $\mathbb{N}$ and subalgebras of $\mathbb{P}(\mathbb{N})$ which contain $\mathbb{E}(\mathbb{N})$.

Let Komp be a compactification of $\mathbb{N}$, dim Komp = 0 and SKomp the above mentioned subalgebra of $\mathbb{P}(\mathbb{N})$. As dim Komp = 0 and $\mathbb{N}$ is dense in Komp, we have the isometric isomorphism $C(\text{Komp}) \cong \overline{\text{linear hull } \{\chi(A) | A \subset \text{Komp}, A \text{ clopen}\}}^{C(\text{Komp})} \cong \overline{\text{linear hull } \{\chi(A) | A \in \text{SKomp}\}}^{\ell\infty}$ given by $C(\text{Komp}) \ni f \mapsto f|_{\mathbb{N}}$ (cf. [6] 8.2.2 and 16.2).

We denote by $J : \mathbb{P}(\mathbb{N}) \longrightarrow\!\!> \mathbb{P}(\mathbb{N}) / \mathbb{E}(\mathbb{N}) =: \mathbb{P}_0(\mathbb{N})$ the boolean quotient map. $\mathbb{A} := \{A_i\}_{i \in I} \subset \mathbb{P}(\mathbb{N})$ is called an essentially independent set (short EIS), if card $I \geq \aleph_0$ and $\{JA_i\}_{i \in I}$ is an independent set in $\mathbb{P}_0(\mathbb{N})$. Examples of EIS's are won canonically by surjections of $\mathbb{N}$ onto dense subsets of the Cantor-$\aleph_1$-space $\{-1,+1\}^S$, card $S = \aleph_1$ (cf. [7] § 14 E).

Let ν_0 be the canonical product measure on $\{-1,+1\}^I \cong \tau(<\mathbb{A}>) \setminus \mathbb{N}$ = Stone space of $<\mathbb{A} \cup \mathbb{E}(\mathbb{N})>/ \mathbb{E}(\mathbb{N})$ defined by $\nu_0(A_i) = 0.5 \;\forall\; i \in I$ (cf. [6] 16.5.4). ν_0 generates an atomless Radon measure ν on $\tau(<\mathbb{A}>)$ with $\nu(\{n|n \in \mathbb{N}\}) = \{o\}$. We will lift ν onto $\beta\mathbb{N}$:

Let X, Y be compact, topological spaces. Let $\phi : X \longrightarrow\!\!> Y$ be a continuous surjection and let ν be a Radon measure on Y. There is a Radon measure μ on X with $|\nu|(Y) = \|\nu\| = \|\mu\| = |\mu|(X)$ and $\nu = \mu \circ \phi^{-1}$ because of the Hahn-Banach theorem and the isometric injection $C(Y) \hookrightarrow C(X)$.

<u>Proposition 3</u> If ν is atomless, then μ is also.
The proof is easy (cf. [6] 19.7.8 B,c). Now we have an universal procedure to generate atomless Radon measures on $\beta\mathbb{N}$. We will look at a special example.

<u>THE REPLICA ECONOMY.</u> Every real economy has finitely many agents. Therefore the whole set of agents as well as subgroups of agents could be seen as replica subsets of $\mathbb{N}$ i.e. their characteristic functions considered as binary numbers are periodic. The knowledge of the cardinality or of properties of groups of agents are gotten by random-sampling (the sampling size can be the whole set of agents) i.e.

one concludes properties of the set of all agents by replica-set-building with the random sample. Notice that this interpretation is correct because the whole set of agents is a replica set. Hence the interpretation works also, if there is no $n \in \mathbb{N}$ such that $n \cdot$ (sample size) = cardinality of the set of all agents. The sampling size is the exactness of the measuring or sampling.

Therefore the idealized space of agents which we propose here is $(\mathbb{N}, R, d)$. $\mathbb{N}$ is the set of agents. The boolean algebra R of all replica subsets of $\mathbb{N}$ symbolizes the set of all allowed coalitions, i.e. measuring with every precision is allowed. The density measure d on $\mathbb{N}$ yields the influence of a coalition. d is defined by $d(A) = \lim_{n \to \infty} n^{-1} \Sigma_{i=1}^{n} \chi(A)(i)$. We will deduce that d generates an atomless Radon measure on Stone (R).

Let $\{p_i\}_{i \in \mathbb{N}}$ be the strictly monotonly increasing sequence of all primes. If $n \in \mathbb{N} = \{1,2,\dots\}$, $j \in \{0,\dots,(p_i)^n - 1\}$, let $M(p_i,n,\mathbb{N},j) := (p_i)^n\mathbb{N} - j = \{(p_i)^n k - j \mid k \in \mathbb{N}\}$. Let $\hat{\mathcal{A}}(p_i)$ be the subalgebra of $\mathbb{P}(\mathbb{N})$ which is generated by $\{M(p_i,n,\mathbb{N},j) \mid n=1,\dots;\ j \in \{0,\dots,(p_i)^n - 1\}\}$.

<u>Lemma 4</u> The subalgebras $\hat{\mathcal{A}}(p_i)$ are independent in $\mathbb{P}_o(\mathbb{N})$ (with respect to J).

P r o o f: $M(p_i,n,\mathbb{N},j) = \{s \in \mathbb{N} \mid s = -j \bmod (p_i)^n\}$. By a well known theorem on simultaneous congruences (cf. [2] theorem 121), we get the proof.

Hence $\tilde{R} := <\{\hat{\mathcal{A}}(p_i) \mid i \in \mathbb{N}\}> = \prod_{i=1}^{\infty} \hat{\mathcal{A}}(p_i)$ is just the boolean product. $\tau(\tilde{R})$ is a compactification of $\mathbb{N}$ and $\tau(\tilde{R}) \setminus \mathbb{N} \cong$ Stone space of $(<\{\hat{\mathcal{A}}(p_i) \mid i \in \mathbb{N}\} \cup \mathbb{E}(\mathbb{N})\}>/\mathbb{E}(\mathbb{N})) \cong$ Stone space of $\prod_{i=1}^{\infty} \hat{\mathcal{A}}(p_i)$ is homeomorphic to the Stone space of the boolean product. It is $R = <\tilde{R} \cup \mathbb{E}(\mathbb{N})>$: If $M \subset \mathbb{N}$ and $\chi(M)$ is periodic as a binary number, then $M \in <\tilde{R} \cup \mathbb{E}(\mathbb{N})>$, because $M \in <\mathbb{E}(\mathbb{N}) \cup \hat{\mathcal{A}}(p_{i_1}) \cup \dots \cup \hat{\mathcal{A}}(p_{i_k})>$ where m is the period of M and $p_{i_1}^{n1} \dots p_{i_k}^{n_k} = m$ is the decomposition of m by primes. Therefore $\tau(R) = \tau(\tilde{R})$ = Stone space of R.

The density d is a measure on every boolean algebra $\hat{\mathcal{A}}(p_i), i \in \mathbb{N}$ (cf. [6] 17.1.1). Therefore we obtain canonically the product measure on $\tilde{R}$. This product measure is just the density:

<u>Lemma 5</u> The subalgebras $\hat{\mathcal{A}}(p_i)$ are d-independent, i.e. $\prod_{k=1}^{m} d(A_{r_k}) = d(\bigcap_{k=1}^{m} A_{r_k})$ if the r_k are pairwise distinct, $A_{r_k} \in \hat{\mathcal{A}}(p_{r_k})$, $m \in \mathbb{N}$.

P r o o f: Notice that an element of $\hat{\mathcal{A}}(p_k)$ is a disjoint union of sets $M(p_k,n,\mathbb{N},j)$ modulo $\mathbb{E}(\mathbb{N})$ with suitable n. The proof is easy for such sets (cf. [2] theorem 121).

The measure d on $\tilde{R}$ extends to R because $d(\mathbb{E}(\mathbb{N})) = \{o\}$. Therefore d induces a Radon measure μ on $\tau(R)$, the Stone space of R. μ is atomless and carrier $\mu = \tau(R) \setminus \mathbb{N}$.

Moreover $\mu(\bar{A}^{\tau(R)}) = d(A)\ \forall A \in R$. It is not difficult to generate other atomless Radon measures on $\tau(R)$ because of the product structure of $R/\mathbb{E}(\mathbb{N})$.

Now $\tau(R)$ is zero dimensional. Hence the linear hull of $\{\chi(\bar{A}^{\tau(R)}) \mid A \in R\}$ is dense in $C(\tau(R))$ (cf. [6] 8.2.2). If $n \in \mathbb{N}$, let $d_n \in C(\tau(R))^*$ be defined by $d_n(f) = n^{-1}\sum_{i=1}^{n} f(i)$. As $d_n(\chi(\bar{A}^{\tau(R)})) \to d(A) = \mu(\bar{A}^{\tau(R)})\ \forall A \in R$, we have $d_n \to \mu$ in the weak * topology of $C(\tau(R))^*$. Here we identified μ with the generated element of $C(\tau(R))^*$.

The pair $\{\tau(R),\mu\}$ is universal in the following sense: Let K be a compactification of $\mathbb{N}$ such that dim K = 0 and $A \subset K$ is clopen only if $A \cap \mathbb{N}$ is a replica set and let ν be a Radon measure on K with $\nu(S) = d(S \cap \mathbb{N})$ if $S \subset K$ is clopen. Then the unique continuous surjection $\psi : \tau(R) \twoheadrightarrow K$, $\psi|_{\mathbb{N}} \equiv \mathrm{id}$ fulfills $\nu = \mu \cdot \psi^{-1}$ (cf. [6]).

We have shown that there is an atomless measure space $(\tau(R), \mathcal{B}(\tau(R)), \mu)$ related to the idealized space of agents $(\mathbb{N},R,d)$. We call a map $\mathcal{E} : (\mathbb{N},R,d) \to K_0 \subset \mathcal{P} \times \mathbb{R}^n_+$ a replica economy with pure competition, if there is a continuous extension $\tilde{\mathcal{E}}$ i.e. a continuous map $\tilde{\mathcal{E}} : \tau(R) \to K_0$, $\tilde{\mathcal{E}}|_{\mathbb{N}} \equiv \mathcal{E}$. Then in analogy to D. Pallaschke, one can apply R.J. Aumann's result on core and Walras equilibrium.

We now want to give an interpretation of the points of $\beta\mathbb{N} \setminus \mathbb{N}$ respectively $\tau(R) \setminus \mathbb{N}$.

MEASUREMENT OF GROUPS. One examines or analyses the existence, the cardinality and the properties of subgroups of the set of all agents by compiling statistics or with public opinion polls. There is, for every subgroup G, a public opinion poll or a statistic (theoretically possible or thinkable not necessarily practically possible), using which one can measure G. In the mathematical language : If there is a subgroup G, then there is a method of measurement (i.e. a map) $\phi_G : \mathbb{N} \to \{0,1\}$ with $G = \{n \mid \phi_G(n) = 1\}$. $\mathbb{N}$ is the set of agents. From the viewpoint of the agent, that means that he has a property or a sign such that one can determine the value of ϕ_G.

If there is a family $\{\phi_i \mid i \in I\}$ of methods of measurement, then there are $2^{\mathrm{card}\, I}$ theoretically posible combinations of the properties which an agent can have. If card $I = \aleph_0$, then $2^{\mathrm{card}\, I} = \aleph_1$ = card [0,1]. We have only card $\mathbb{N} = \aleph_0$ many agents which means that there are at most $\aleph_0$ many real combinations of the properties. Suppose that there is always a method of measuring which distinguishes a pair of two nonidentical agents. Then we can inject the real combinations of properties $\{(\phi_i(n))_{i \in I} \mid n \in \mathbb{N}\}$ into the set of all theoretically possible combinations of properties $\{(a_i)_{i \in I} \mid a_i \in \{0,1\}\} = 2^I =: \mathcal{G}(I)$. We will call $\mathcal{G}(I)$ the group space. If ϕ_{i_0} is a method of measuring, then $G_{i_0} = \{n \in \mathbb{N} \mid \phi_{i_0}(n) = 1\}$ is the real group of agents and $\{(a_i)_{i\in I} \mid a_i \in \{0,1\}$ and $a_{i_0} = 1\}$ is the subset of $\mathcal{G}(I)$ which we call the property-group $\bar{G}_{i_0}$. The property-groups generate a boolean algebra of subsets of $\mathcal{G}(I)$. These subsets are just the clopen subsets of $\mathcal{G}(I)$ where $\mathcal{G}(I)$ is topologized as usual i.e. as the Cantor-I-space.

Until now we treated the properties or the methods of measuring as independent because we constructed the $2^{\text{card } I}$ theoretically possible combinations of properties i.e. the group-space $\mathcal{G}(I)$. But one can measure the cardinality of groups (if these are finite) or the weight of groups i.e. $\lim_{m\to\infty} \frac{1}{m}$ card $\{n \mid \phi(n)=1$ and $n\leq m\}$ (if these are replica sets) as well as the cardinality or the weight of finite intersections of groups. Doing this, one obtains canonically a Radon measure ν on $\mathcal{G}(I)$ which fulfills $\nu(\bar{G})$ = weight (G). With respect to ν, the property-groups can be independent or dependent.

The pair {carrier ν , ν} is therefore a space of agents of a large economy, in which only the measured groups (including the generated subalgebra) have nonzero influence (although there exist on carrier ν more Borel sets with positive measure But these are not clopen). The space C(carrier ν) has to be constructed by the linear hull of $\{\chi(\bar{A}) \mid A$ is a measurable intersection of groups$\}$.

During the description of $\tau(R)$, we had methods of measurement of the kind : $\phi : \mathbb{N} \to \{1,..,p_i^k\}$. It is possible to reduce each one to finitely many methods of the normal kind : $\mathbb{N} \to \{0,1\}$. Then we can imbed $(\tau(R) \setminus \mathbb{N}, \nu)$ as (carrier $\tilde{\nu}$,$\tilde{\nu}$) in a space 2^I.

I want to emphasize that the last part is a motivation for the building of τ (R). It is not a detailed description how to approximate an atomless space of agents by real finite economies.

I thank A. Klodt, who proved a first version of Lemma 5.

References

[1] R.J. Aumann, Markets with a continuum of traders, Econometrica 32, (1964)
[2] G.H. Hardy and E.M Wright, An introduction to the theory of numbers, Oxfort 1960
[3] W. Hildenbrand, Core and equilibria of a large economy, Princeton 1974
[4] Ch. Klein, Eine Characterisierung atomloser Radonmaße auf β ℕ und ihre Anwendung auf große Ökonomien, Dissertation Bonn 1978
[5] D. Pallaschke, Markets with uncountably many traders, Journal of Applied Math. and Comp.,4,201-212(1978)
[6] Z. Semadeni, Banach spaces of continuous functions I, Warszawa 1971
[7] R. Sikorski, Boolean algebras, Berlin-Heidelberg-New York 1969
[8] R.C.Walker, The Stone Čech compactification, Berlin-Heidelberg-New York 1974

GAME THEORY AND RELATED TOPICS
O. Moeschlin, D. Pallaschke (eds.)

MANY AGENTS IN A VON NEUMANN MODEL

Jerzy Łoś and Maria W. Łoś

Institute of Computer Science
Polish Academy of Sciences, Warsaw

The main feature of a von Neumann model, like any other model with "constant return to scale", is that it can not serve to investigate problems concerning quantities of production and of prices, instead it is adjusted to problems dealing with their proportions, in particular with the rate of growth and the rate of profit. This is manifested by the von Neumann equilibrium which is a non-cooperative equilibrium of a quotient game, homogeneous both in production and prices (see e.g. Łoś 1976). As presented by von Neumann (1937) and then applied to more general models it is a game of two players only : production and prices, thus representing a very centralized economy without possibilities of expressing diverse interests of economic agents, like production sectors, groups of people or cooperating countries.

In this paper we show that it is possible to extend the game leading to the von Neumann equilibrium to a many-persons game, covering in this way many instances of conflicts or cooperations of agents following their own interests. This extension does not violate the usual idea behind the von Neumann equilibrium, but even reinforce it by showing that the usual equilibrium is a special case of the new one. In order to stress the novelty of the many-persons game underlying the von Neumann model we resign in this paper from all known generalizations, like negative entries in matrices, cone ordered spaces and extended (four matrix) models. We start with recalling the two-persons game for a non-negative two-matrix model in order to make the generalization more evident. We end with two examples of applications to economic theory, one is this of cooperation of two economies (countries) through a common foreign market, the other is that of labour and consumption, which leads directly into the capital theory. Albeit the first one is a more straitforward development of the usual closed von Neumann model and of the Morgenstern-Thompson model of an open economy, the second one seems to have more impact on the economic theory.

1. The simple von Neumann model.

A simple von Neumann model is a pair of non-negative matrices A, B, both of the same dimension, say $n \times m$. The matrix A is called the input matrix, the matrix B is called the output matrix and when a row n-vector $x = (x_1,\ldots,x_n)$ is applied to these matrices we get row m-vectors $xA = (y_1,\ldots,y_m) = y$ and $xB = (y_1',\ldots,y_m') = y'$ called input and output bundles (because we consider them as commodity bundles). The vectors $x = (x_1,\ldots,x_n)$ applied to A and B from the left are called intensity vectors. A pair consisting of the i-th row in A and the i-th row in B is called a (linear) process and the i-th coordinate x_i of the intensity vector x is the intensity applied to the i-th process.

A column m-vector $p = \begin{pmatrix} p_1 \\ \vdots \\ p_m \end{pmatrix} = \langle p_1,\ldots,p_m \rangle$ can be applied to A and B from the right producing column n-vectors Ap and Bp called cost and revenue vectors. The vector p is called the price vector and p_j is the price of the j-th commodity. In $Ap = \langle q_1,\ldots,q_n \rangle = q$, the coordinate q_i is the cost per unit of intensity of the i-th process and in $Bp = \langle q_1',\ldots,q_n' \rangle = q'$, the number q_i' is the revenue obtained from running the i-th process at intensity one. Of course, the numbers xAp and xBp are the total cost and the total revenue at intensities x and prices p.

The system is supposed to work in time what means that there is given a producti period (time interval) such that if the input xA is inserted in to the system at the beginning of the production period then, after remaining in the system throu out the whole interval, it produces the output xB emerging from the system at th end of the period. It follows that, at given prices p, the total cost xAp is incurred at the beginning of the production period, while the revenue xBp is colle ted at the end of the period, thus the amount xAp is invested for the whole leng of the production period.

As intensity vectors x and price vectors p only non-negative vectors are allowec and it is the only restriction on x´s and p´s. It follows that it makes no sense to ask for the largest nett production xB - xA or for maximum profit at given pr ces xBp - xAp, because applying instead of x, the vector ζ x with $\zeta > 0$ large enough we can make both quantities as large as we want (provided they are positive). Instead it makes sense to ask about the different rates, for instance thi of growth and that of profit. These rates are proportions between xA and xB, respectively xAp and xBp, and consequently they do not depend on multiplication of x (or p) by positive numbers.

2. The von Neumann game.

To denote that a vector z is non-negative in all coordinates we write $z \geq 0$. Consequently, for two vectors z and z´ (of the same dimensions) $z \leq z'$ denotes that $z' - z \geq 0$ or that every coordinate of z´ is not less than the correspondi coordinate of z.

Let $z \geq 0$ and $z' \geq 0$ be two non-negative vectors (of any dimension, row or colun We define the "rate function" as

$$^{z}/\!\!/_{z'} = \sup \{\lambda \mid \lambda z' \leq z\}.$$

It follows at once that if $z = \alpha$ and $z' = \beta$ are non-negative numbers, then

$$(2.1) \qquad ^{\alpha}/\!\!/_{\beta} = \begin{cases} ^{\alpha}/_{\beta} & \text{, iff } \beta > 0 \text{ ,} \\ +\infty & \text{, iff } \beta = 0 \text{ ,} \end{cases}$$

and in general

$$(2.2) \qquad ^{z}/\!\!/_{z'} = \min \{ ^{z_i}/\!\!/_{z'_i} \mid \text{all coordinates } i\} =$$

$$= \min \{ ^{z_i}/_{z'_i} \mid \text{all coordinates } i\text{, such that } z'_i > 0\}.$$

It follows also that if, for instance, $z \geq 0$ and $z' \geq 0$ are row vectors and $v \geq 0$ is a column vector of the same dimension, then

$$(2.3) \qquad ^{z}/\!\!/_{z'} \leq \, ^{zv}/\!\!/_{z'v} \text{ ,}$$

and, moreover, that

$$(2.4) \qquad ^{z}/\!\!/_{z'} = \, ^{zv}/\!\!/_{z'v} \text{ may be always reached with a suitable } v \geq 0.$$

The number $^{z}/\!\!/_{z'} - 1$ is called the rate of z over z´.

With help of the rate function we define for a von Neumann model a game of two player in which the first one has the set of strategies

$\{x \mid 0 \leq x \neq 0\}$ and the loss-function

$$f(x,p) = \, ^{xAp}/\!\!/_{xBp}.$$

The second player has strategies in $\{ p | 0 \le p \ne 0 \}$ and the loss-function

$$g(x,p) = {}^{xBp}/\!\!/_{xAp}.$$

This is a non-zero-sum game, played by both players with the intention to minimize the corresponding loss-functions.

With the interpretation given in Sect. 1, we may say that the first player - called production - tries to choose intensities minimizing the rate of cost $f(x,p) - 1$ and the second player - prices - tries to minimize the rate of profit $g(x,p) - 1$.

Following the well known definition, a non-cooperative equilibrium (Nash equilibrium) of this game consists of two vectors (strategies) $o \le \bar{x} \ne 0$ and $0 \le \bar{p} \ne 0$ such that

$$f(\bar{x},\bar{p}) \le f(x,\bar{p}) \ , \quad \text{for all} \quad 0 \le x \ne 0 \ ,$$

$$g(\bar{x},\bar{p}) \le g(\bar{x},p) \ , \quad \text{for all} \quad 0 \le p \ne 0 \ .$$

Since, following (2.3), for all $x \ge 0$ and $p \ge 0$:

$${}^{Ap}/\!\!/_{Bp} \le {}^{xAp}/\!\!/_{xBp} = f(x,p)$$

and

$${}^{xB}/\!\!/_{xA} \le {}^{xBp}/\!\!/_{xAp} = g(x,p) \ ,$$

we see that each player choosing his strategy bounds from below the loss-function of the other. From (2.4) it follows that

(2.5) Two strategies $\bar{x}$, $\bar{p}$ are in equilibrium if and only if

$${}^{A\bar{p}}/\!\!/_{B\bar{p}} = {}^{\bar{x}A\bar{p}}/\!\!/_{\bar{x}B\bar{p}} \quad \text{and} \quad {}^{\bar{x}B}/\!\!/_{\bar{x}A} = {}^{\bar{x}B\bar{p}}/\!\!/_{\bar{x}A\bar{p}}$$

A simple checking shows that the following is true.

<u>Theorem 1.</u> Two strategies $\bar{x}$, $\bar{p}$ are in equilibrium and the value $f(x,p) = \sigma$, $g(x,p) = \lambda$ are finite and positive if and only if $\lambda = \sigma^{-1}$ and the following inequalities hold:

$$\lambda \ \bar{x}A \le \bar{x}B, \quad B\bar{p} \le \lambda \ A\bar{p}, \quad \bar{x}B\bar{p} > 0.$$

It follows that at finite equilibrium $xAp > 0$ and $\lambda xAp = xBp$.

It is also easy to prove that:

<u>Theorem 2.</u> If the model has both KMT properties:
(KMT1) no row of A is zero,
(KMT2) no column of B is zero,

then each equilibrium is finite, i.e. the values of f and g are finite and positive.

It is known that a von Neumann model with both KMT properties has equilibria and that the number of λ´s which can be the values of g at equilibrium (equilibrium level) is finite, bounded by both n and m (Kemeny 1956, Thompson 1956).

3. Partitioned von Neumann models.

The von Neumann model, as presented in the preceeding Section, represents a centralized economy with just one player in control of production and just one market on which prices are formed. We are not going to discuss how the prices on the market are formed, i.e. what is the nature of the price controlling player. Is it an "invisible hand" or even a government appointed authority aiming to decrease the rate of profit. We note, however, that with two or more markets for different commodities we should have as many players controlling prices of different commodity groups.

The partitioned von Neumann model is a simple model with processes (rows) and commodities (columns) partitioned to allow for independent production decisions for different groups of processes and independent price forming on markets for different groups of commodities.

Formally such a model consists of two matrices A and B partitioned in blocks A_{ij} and B_{ij} of the same dimensions, let us say $n_i \times m_j$, $\Sigma\, n_i = n$ and $\Sigma\, m_j = m$. We have then the following picture :

$$A = \begin{pmatrix} A_{11} & \cdots & A_{1r} \\ \vdots & & \vdots \\ A_{s1} & & A_{sr} \end{pmatrix} = \begin{pmatrix} A_{1*} \\ \vdots \\ A_{s*} \end{pmatrix} = (A_{*1},\dots,A_{*r}),$$

$$B = \begin{pmatrix} B_{11} & \cdots & B_{1r} \\ \vdots & & \vdots \\ B_{s1} & & B_{sr} \end{pmatrix} = \begin{pmatrix} B_{1*} \\ \vdots \\ B_{s*} \end{pmatrix} = (B_{*1},\dots,B_{*r}),$$

where $A_{i*} = (A_{i1},\dots,A_{ir})$, $A_{*j} = \begin{pmatrix} A_{1j} \\ \vdots \\ A_{sj} \end{pmatrix} = < A_{1j},\dots,A_{sj} >$,

and similarly for B_{i*} and B_{*j}. The intensity vectors x and the price vectors p split accordingly into $x = (x^1,\dots,x^s)$ and $p = < p^1,\dots,p^r >$, with x^i being n_i-vectors (rows) and p^j being m_j-vectors (columns).

The game connected with the partitioned model is a $s + r$ -players game with $\{x^i \mid 0 \leq x^i \neq 0\}$, $i = 1,\dots,s$, being sets of strategies for the first s players which have the loss functions

$$f_i(x,p) = f_i(x^i,p) = x^iA_{i*}p \,/\!/\, x^iB_{i*}p \quad , \; i = 1,\dots,s \; ,$$

and with sets of strategies $\{p^j \mid 0 \leq p^j \neq 0\}$, $j = 1,\dots,r$ and the loss-functions

$$g_j(x,p) = g_j(x,p^j) = xB_{*j}p^j \,/\!/\, xA_{*j}p^j \quad , \; j = 1,\dots,r$$

for the other r players.

There are two kind of players, s production controlling players and r price controlling players. The loss-function of each player depends on his own strategy and all strategies of players in the other group. It does not depend immediately on strategies of other players in the same group.

The equilibrium of this game is a $r + s$ -tuple of strategies $\bar{x} = (\bar{x}^1,\dots,\bar{x}^s)$, $\bar{p} = < \bar{p}^1,\dots,\bar{p}^r >$ such that

$$f_i(\bar{x}^i,\bar{p}) \leq f_i(x^i,\bar{p}), \text{ for all } 0 \leq x^i \neq 0,\; i = 1,\dots,s \; ,$$

$$g_j(\bar{x},\bar{p}^j) \le g_j(\bar{x},\bar{p}^j) \text{ , for all } 0 \le p^j \ne 0,\ j = 1,\dots,r.$$

Theorem 3. Suppose $\bar{x}$ and $\bar{p}$ are in equilibrium.

If $f_i(\bar{x}^i,\bar{p}) = \sigma_i$ is finite and positive, then for $\mu_i = \sigma_i^{-1}$ we have
(3.1.i) $B_{i*}\bar{p} \le \mu_i A_{i*}\bar{p}$, (3.2.i) $\bar{x}^i B_{i*}\bar{p} = \mu_i \bar{x}^i A_{i*}\bar{p}$,

(3.3.i) $\bar{x}^i B_{i*}\bar{p} > 0$.

If $g_j(\bar{x},\bar{p}^j) = \lambda_j$ is finite and positive, then we have (3.4.j) $\lambda_j \bar{x} A_{*j} \le \bar{x} B_{*j}$,

(3.5.j) $\lambda_j \bar{x} A_{*j}\bar{p}^j = \bar{x} B_{*j}\bar{p}^j$, (3.6.j) $\bar{x} A_{*j}\bar{p}^j > 0$.

If $\bar{x}$ and $\bar{p}$ are s- and r-tuples of strategies such that all (3.1.i), (3.2.i), (3.3.i), $i = 1,\dots,s$ and all (3.4.j), (3.5.j), (3.6.j), $j = 1,\dots,r$ are satisfied with finite and positive numbers $\mu_1,\dots,\mu_s$, $\lambda_1,\dots,\lambda_r$, then $\bar{x}$ and $\bar{p}$ is an

equilibrium with values $f_i(\bar{x}^i,\bar{p}) = \mu_i^{-1}$ and $g_j(\bar{x},\bar{p}^j) = \lambda_j$.

We get the proof by a straightforward application of (2.3) and (2.4).

An easy checking shows that the following is true.

Theorem 4. If one of the matrices B_{ij}, $j = 1,\dots,r$, has the KMT2 property, i.e. no column of B_{ij} is zero, then at every equilibrium $\bar{x}$, $\bar{p}$, the value $f_i(\bar{x},\bar{p})$ is finite (but can be 0). If one of the matrices A_{ij}, $i = 1,\dots,s$, has the KMT1 property, i.e. no row of A_{ij} is zero, then at every equilibrium $\bar{x}$, $\bar{p}$, the value $g_j(\bar{x},\bar{p})$ is finite (but can be 0).

The relation between an equilibrium of the simple model A,B and the partitioned model $A = (A_{ij})$, $B = (B_{ij})$ is as follows.

Theorem 5. If $\bar{x} = (\bar{x}^1,\dots,\bar{x}^s)$, $\bar{p} = \langle \bar{p}^1,\dots,\bar{p}^r \rangle$ is an equilibrium of the partitioned model with all $f_i(\bar{x},\bar{p}) = \mu^{-1}$ equal, finite and positive and all $g_j(\bar{x},\bar{p}) = \lambda$ equal, finite and positive, then $\bar{x}$ and $\bar{p}$ is an equilibrium of the simple model and $\lambda = \mu$.

The inverse implication does not hold, albeit it is true that with $s = r = 1$ both equilibria coincide.

4. Two economies connected by a market.

Let (A_1,B_1) and (A_2,B_2) be two von Neumann models describing two economies, we shall say two countries. We form a partitioned model.

$$A = \left(\begin{array}{cc|c} A_1 & 0 & 0 \\ C_1 & E_1 & 0 \\ \hline 0 & 0 & A_2 \\ 0 & E_2 & C_2 \end{array}\right) , \qquad B = \left(\begin{array}{cc|c} B_1 & 0 & 0 \\ F_1 & D_1 & 0 \\ \hline 0 & 0 & B_2 \\ 0 & D_2 & F_2 \end{array}\right) ,$$

with matrices C_i, D_i, E_i, F_i, $i = 1,2$, of any appropriate dimensions. In the corresponding game there are two players controlling intensities called the first and the second country and three players controlling prices called markets : two inner markets of the countries and one foreign market (in the middle). The production processes of each country, these in A_i, B_i, are extended by trade processes in C_i, D_i, E_i, F_i, which connect the country with the foreign market. Therefore, each country controls intensity vectors (x^i,t^i), with x^i containing intensities of production processes and t^i containing intensities of trade processes. The markets control prices p^1, p, p^2 respectively.
The game of such model has the following loss-functions :

$$f_i(x^i,t^i;p^i,p) = x^iA_ip^i + t^iC_ip^i + t^iE_ip \Big/\!\!\Big/ x^iB_ip^i + t^iF_ip^i + t^iD_ip\ ,$$

$i = 1,2$, for the two countries,

$$g_i(x^i,t^i;p^i) = x^iB_ip^i + t^iF_ip^i \Big/\!\!\Big/ x^iA_ip^i + t^iC_ip^i\ ,\ i = 1,2,$$

for the inner markets, and

$$g(t^1,t^2,p) = t^1D_1p + t^2D_2p \Big/\!\!\Big/ t^1E_1p + t^2E_2p\ ,$$

for the foreign market.

An equilibrium (x^1,t^1), (x^2,t^2), p^1, p^2, p (all five vectors nonnegative and non-zero) with finite and positive values $f_i = \mu_i^{-1}$, $g_i = \lambda_i$, $g = \lambda$, is uniquely determined by the following conditions :

(4.1.i) $\lambda_i(x^iA_i + t^iC_i) \le x^iB_i + t^iF_i$,

(4.2.i) $\lambda_i(x^iA_i + t^iC_i)p^i = (x^iB_i + t^iF_i)p^i > 0$,

(4.3.i) $\begin{cases} B_ip^i \le \mu_iA_ip^i\ , \\ F_ip^i + D_ip \le \mu_i(C_ip^i + E_ip)\ , \end{cases}$

(4.4.i) $(x^iB_i + t^iF_i)p^i + t^iD_ip = \mu_i\ (x^iA_i + t^iC_i)p^i + t^iE_ip > 0$,

(4.5) $\lambda(t^1E_1 + t^2E_2) \le t^1D_1 + t^2D_2$,

(4.6) $\lambda(t^1E_1 + t^2E_2)p = (t^1D_1 + t^2D_2)p > 0$.

The interpretation of our model depends, of course, on the blockmatrices, especially on the properties of matrices C_i and F_i containing the input and the output of trade processes within the country. A reasonable assumption seems to be that by operating these processes the country can not make any additional profit, what may be expressed for instance as follows. If $x^i \ge 0$, $t^i \ge 0$, $p^i \ge 0$, $\lambda > 0$, $\lambda\ (x^iA_i + t^iC_i) \le x^iB_i + t^iF_i$, $B_ip^i \le \lambda A_ip^i$ and $F_ip^i \le \lambda C_ip^i$, then $t^iF_ip^i = 0$ (compare the definition of "auxiliary only" processes in Łoś, 1976, p.82).

The simplest form of C_i, D_i, E_i, F_i is as follows. We assume that the same commodities are involved in both countries and in the foreign market and we accept

$C_i = \begin{pmatrix} I \\ 0 \end{pmatrix} = D_i$, $E_i = \begin{pmatrix} 0 \\ I \end{pmatrix} = F_i$, with I being the (square) identity matrix. With these matrices the country can ship everything to the foreign market and order everything from the foreign market and transportation is free. A more realistic situation we get by accepting as before $D_i = \begin{pmatrix} I \\ 0 \end{pmatrix}$, but changing C_i to $\begin{pmatrix} I + T_i \\ 0 \end{pmatrix}$. The matrix T_i expresses the input for transporation. There are many other possibilities to define the trade processes. The most interesting seems to be those in which the matrices are not non-negative. By assuming here, for the sake of simplicity, that all matrices are non-negative we have excluded the discussion of those possibilities.

How does work our model? Of course t^iC_i is the exported bundle and t^iF_i the imported bundle of commodities. Those bundles together with input x^iA_i and output x^iB_i satisfy the balance inequality in physical units (4.1.i) with the factor of growth λ_i and inside the country at the same factor the money market is cleared (4.2.i).

The factor of interest μ_i, which is the ration of revenues to costs of the country can be different than λ_i, as it is to be seen from (4.3.i) and (4.4.i). If t^iC_i is the bundle of commodities which the country exports (or partially spend for transportation), then t^iD_i is this bundle which arrives at the foreign market. The cost of the former bundle at internal prices $t^iC_ip^i$ is added to the total costs, while the later bundle sold at the foreign market brings the revenue t^iD_ip, what is accounted for on the revenue side of balance equation (4.4.i). Analogously, we have $t^iF_ip^i$ added to the total revenues and t^iE_ip to the total costs for the import activities.

The conditions (4.5) and (4.6) express the balance on the foreign market in physical units and at prices p.

It should be emphasized that at equilibrium all λ_1, λ_2, λ, μ_1, μ_2 may be different. In such a case the equilibrium is not a stationary policy : the policy can not be repeated, because one country growths at a different rate than the other, thus - if foreign exchange is essential - it delivers in the second step of planning inappropriate amounts of commodities. It is nothing peculiar in it. An equilibrium expresses the balance of interests of the parties concentrated only on the rates of cost and profit. This may not lead to a long-run equilibrium.

It can be proved (with some assumptions on the model) that the equilibrium of the simple model A, B at the maximal rate of growth is indeed an equilibrium of the partitioned model with all rates equal,thus a stationary policy. But the partitioned model will have a lot of other equilibria (continuum of them) with different rates and these equilibria seems to be the most interesting for the economic theory.

5. The von Neumann model with labour and consumption.

Let A, B be a simple von Neumann model, $l = \langle l_1,\dots,l_n \rangle \gg 0$ a column n-vector with positive coordinates and $c = (c_1,\dots,c_m) \geq 0$ a row m-vector with non-negative coordinates, not all equal zero : $c \neq 0$. By the model with labour input coefficients l and a consumption basket c, we understand the partitioned von Neumann model of dimension $(n + 1) \times (m + 1)$:

$$A' = \left(\begin{array}{c|c} A & l \\ \hline c & 0 \end{array}\right), \qquad B' = \left(\begin{array}{c|c} B & 0 \\ \hline 0 & 0 \end{array}\right).$$

We have in this model a homogeneous labour and consumption only in multiplicities of the consumption basket c.

The game related to this model is performed between four players, the first one (production) has row vectors $0 \neq x \geq 0$ as his strategies, the second (population) chooses the consumption level $\alpha > 0$, the third (market) sets prices $0 \neq p \geq 0$ and the last (labour) decides about the wage rate $w > 0$. The loss-functions are defined as follows :

$$f_1 = {}^{xAp + xlw}/\!\!/_{xBp} \quad \text{(production : x)} ,$$

$$f_2 = {}^{\alpha cp}/\!\!/_{0} = +\infty \quad \text{(population : } \alpha\text{)} ,$$

$$g_1 = {}^{xBp}/\!\!/_{xAp + \alpha cp} \quad \text{(market : p)} ,$$

$$g_2 = {}^{0}/\!\!/_{xlw} = 0 \quad \text{(labour : w)} .$$

Of course neither population nor labour has any great interest in the game, since their loss-functions are constant. This is because consumption is only added to input, population is to be feed, and labour increases only costs, workers are to be paid, hence they influence only the outcome of the other players. It seems, therefore, that a cooperative solution would be more appropriate here, but let us look at the non-cooperative equilibria of the game.

At an equilibrium only f_1 and g_1 may be finite and positive. If A and B satisfy both KMT conditions (see Section 2), then at every equilibrium both f_1 and g_1 are finite and positive and therefore each equilibrium x, α, p, w is characterized by the existence of two numbers $\lambda > 0$, $\mu > 0$ which satisfy the following conditions :

(5.1) $\quad \lambda(xA + \alpha c) \leq xB$, $\qquad$ (5.2) $\quad Bp \leq \mu(Ap + lw)$,

(5.3) $\quad \lambda(xA + \alpha c)p = xBp = \mu x(Ap + lw) > 0$.

An equilibrium is what is also called a steady state of an economy with the growth rate $\lambda - 1$ and the interest rate $\mu - 1$. Indeed, (5.1) says that with x and α the economy reproduces the input at least with the rate $\lambda - 1$, (5.2) says that p and w bound the profit rate of every process from above with $\mu - 1$ and (5.3) says that actually both rates are achieved.

If A and B satisfy the KMT conditions, then there exist equilibria with any l and c. But here the problem of existence does not seem to be the most important one. More important seems to be the problem which λ , μ may serve as factors of growth and interest at equilibrium. Father on, given such allowable λ and μ which x, α, p, w can form an equilibrium at those λ and μ. Especially interesting is, how large can be α for x bounded by the inequality $xl \leq 1$; and for which processes i, we can find an equilibrium with $x_i > 0$. The first one is the problem of optimal consumption level (with the total supply of labour equal 1), the second one is the problem of techniques used at given rates. All these problems belong to the capital theory and have been extensively studied for Leontief model exhibiting the so called pradoxes of capital theory : reswitching of techniques, non-monotonicity of consumption and others. For more general models, like those considered here (joint production), it is possible to reconstruct the known paradoxes and to construct new ones. It may be shown, for instance, that for some models equilibria can exist at λ_1, μ and at λ_2, μ , but for some λ ´s in between : $\lambda_1 < \lambda < \lambda_2$, no equilibrium exists at λ , μ. Detailed description of such phenomena will be given in a separate publication of the second author.

References

J.G. Kemeny, 1956, Game theoretic solution of an economic problem, Progress Report No. 2, Darmouth Mathematics Project.

J. Łoś, 1976, Extended von Neumann models and game theory, in Computing Equilibria: How and Why, eds. J. Łoś and M.W. Łoś, North-Holland Publ. Co. Amsterdam, PWN, Warszawa, pp. 141-157.

J. Łoś, 1956, Von Neumann models of open economies, in : Warsaw Fall Seminars in Mathematical Economics 1975, eds. M.W. Łoś, J. Łoś and A. Wieczorek, Springer Verlag, Berlin, pp. 67-95.

J. von Neumann, 1937, Über ein ökonomisches Gleichungssystem und eine Verallgemeinerung des Brouwerschen Fixpunksatzes, Ergebnisse eines mathematischen Kolloquium 8, pp. 73-83.

G.Thompson, 1956, On the solution of a game theoretic problem, in : Linear Inequalities and Related Problems, eds. H. Kuhn and A.W. Tucker, Priceton University Press, Princeton, pp. 275-284.

GAME THEORY AND RELATED TOPICS
O. Moeschlin, D. Pallaschke (eds.)

COMPLEMENTARITY PROBLEMS AND VON NEUMANN EQUILIBRIA

H. Meister
Department of Mathematics
University of Hagen

The equilibria of a special von Neumann model are characterized as the solutions of a complementarity problem which can be solved by means of methods from graph theory and combinatorics without using any fixed point theorem, if certain nondegeneracy conditions are fulfilled.

INTRODUCTION

A usual economic model consists of four nonnegative p×q-matrices A,B,F,G, which are due to the following meaning:

A	outputmatrix	B	costmatrix
F	inputmatrix	G	revenuematrix

The matrices A and F transform intensity vectors $x = (x_1,\ldots,x_p) \in \mathbb{R}_+^p$ into commodity bundles xA, xF, the matrices B and G transform price vectors $y = (y_1,\ldots,y_q) \in \mathbb{R}_+^q$ into revenue vectors By^T *) and costvectors Gy^T, respectively. The lines of the given matrices correspond with different production processes, the columns with certain manipulations. An equilibrium of such a von Neumann model is defined as an intensity vector $\bar{x} \in \mathbb{R}_+^p$ and a price vector $\bar{y} \in \mathbb{R}_+^q$ together with a factor $\alpha>0$ of technological growth and a factor $\beta>0$ of capital growth satisfying

---

*) T *means the transposition operator*

$$(N)\qquad \begin{aligned} &\bar{x}(A-\alpha F) \geqq 0 \\ &(\beta B-G)\bar{y}^T \geqq 0 \\ &\bar{x}(A-\alpha F)\bar{y}^T = \bar{x}\,(\beta B-G)\bar{y}^T = 0\,. \end{aligned}$$

The first inequality says that the bundle of goods needed at the beginning of a production period may not surmount the one produced before, the second that the production costs exceed the revenue of the previous period. The last two equations show that the whole worth of the produced commodities is equal to the worth of the required goods - or in other words: surplus is worthless - and the whole profit is equal to the investment capital - processes which work with loss, are not pushed forward. Of course, only the nontrivial solutions of (N) are of economic interest. So by searching for solutions $(\bar{x},\bar{y},\alpha,\beta)$ of (N) we may restrict ourselves to such ones with

$$\bar{x}\in S^p\ ,\ \bar{y}\in S^q\ ,$$

where

$$S^n := \{(x_1,\dots,x_n)\in \mathbf{R}^n_+ \mid \sum_{i=1}^{n} x_i = 1\}.$$

In [1] the existence of a solution of (N) is proved by using von Neumann's fixed point theorem where the model has to satisfy some additional postulations. The following proof of existence is constructive and demonstrates the close connection between the described von Neumann model and the theory of complementarity problems. For that purpose let

$$e^n := (\underbrace{1,\dots,1}_{n\text{ times}})^T$$

and

$$E := \left.\begin{pmatrix} 1\dots1 \\ \vdots\quad\ \vdots \\ 1\dots1 \end{pmatrix}\right\}\ p\text{ rows}\quad (q\text{ columns})$$

Instead of the system (N) we may solve

$$(N')\quad \begin{aligned} &((1-\lambda)A - \lambda F + E)^T \overline{x}^T - v^T - e^q = 0 \\ &((1-\mu)B - \mu G + E)\ \overline{y}^T - u^T - e^p = 0 \\ &\overline{x}e^p - 1 = 0 \\ &\overline{y}e^q - 1 = 0 \\ &\overline{x}u^T = \overline{y}\ v^T = 0 \\ &\overline{x},u \geqq 0\ ;\ \overline{y},v \geqq 0\ ;\ \lambda,\mu \in\ ]0,1[\,, \end{aligned}$$

because for every solution $(\overline{x},\overline{y},\alpha,\beta) \in S^p \times S^q \times\ ]0,\infty[^2$ of (N) a solution $(\overline{x},\overline{y}\,;u,v\,;\lambda,\mu)$ of (N') is given by

$$u := \frac{1}{1+\beta}\,\overline{y}(\beta B-G)^T\ ,\quad v := \frac{1}{1+\alpha}\,\overline{x}(A-\alpha F)$$

$$\lambda := \frac{\alpha}{1+\alpha}\ ,\quad \mu := \frac{1}{1+\beta}\ .$$

And for every solution $(\overline{x},\ \overline{y}\,;u,v\,;\alpha,\beta)$ of (N') $(\overline{x},\overline{y},\frac{\lambda}{1-\lambda},\frac{1-\mu}{\mu})$ is a solution of (N) in $S^p{\times}S^q{\times}]0,\infty[^2$. In order to be able to prove that (N') is solvable we have to list some properties of complementarity problems. Of course, (N') is a complementarity problem, but unfortunately no linear one for which the theory is very high developed, as it is well-known.

ONEPARAMETRIC COMPLEMENTARITY PROBLEMS

A complementarity problem is given by a continuously differentiable function $f\colon \mathbb{R}^k \times \mathbb{R}^k \times \mathbb{R}^m \longrightarrow \mathbb{R}^n$ (where $k,m,n \in \mathbb{N}$; but we also admit of the case that either $k=0$ or $m=0$, where the argument set of f takes the form $\mathbb{R}^m$ and $\mathbb{R}^k \times \mathbb{R}^k$, respectively). We try to find the set of solutions of

$$KP(f)\quad \begin{aligned} &f(z,z^*,w) = 0 \\ &z(z^*)^T = 0 \\ &z,z^* \in \mathbb{R}^k_+\ ,\ w \in [0,1]^m\ . \end{aligned}$$

For our purposes we are content with an investigation of a special class of complementarity problems.

Definition

The complementarity problem KP(f) is said to be oneparametric, if m+k-n=1 holds.

For the sake of simplicity we introduce some specifications.
For $x = (x_1,\dots,x_{2k+m}) \in \mathbb{R}_+^{2k} \times [0,1]^m$ let

$$\mathrm{spt}(x) := \{i \in \{1,\dots,2k\} \mid x_i > 0\} \cup \{i \in \{2k+1,\dots,2k+m\} \mid 0 < x_i < 1\}$$

be the support of the vector x. Furthermore let

$$L_{KP(f)} := \{x \in \mathbb{R}_+^{2k} \times [0,1]^m \mid x \text{ solves } KP(f)\}.$$

For every $S \subset \{1,\dots,2k+m\}$ let

$$H_S := \{x \in L_{KP(f)} \mid \mathrm{spt}(x) = S\},$$

and let $D_S f(x)$ be the matrix built by the columns of the functional matrix $Df(x)$ at a point $x \in \mathbb{R}^{2k+m}$ corresponding with the set S where the sequence of the columns is the original one.

Definition

The complementarity problem KP(f) is said to be nondegenerate, if

$$\mathrm{rg}\, D_{\mathrm{spt}(x)} f(x) = n \qquad (x \in L_{KP(f)})$$

holds (the rank of matrix A is denoted by rg A).

If KP(f) is a oneparametric nondegenerate complementarity problem, then

$$L_{KP(f)} = \bigcup_{|S|=n} H_S \cup \bigcup_{|S|=n+1} H_S .$$

Of course, every solution (z,z^*,w) of KP(f) has at least k components equal to zero. This holds in consequence of $z(z^*)^T = 0$; $z,z^* \geqq 0$. So we have

$$|\mathrm{spt}(z,z^*,w)| \leqq 2k+m-k = n+1.$$

On the other side the nondegeneracy of KP(f) includes

$$|\mathrm{spt}(z,z^*,w)| \geqq n.$$

Additionally let $L_{KP(f)}$ be compact. Then H_S is a finite set for $|S|=n$ and a union of finitely many smooth curves without ramification points for $|S|=n+1$ as it is well-known from the theory of implicite defined curves. Every point on the boundary of such a curve belongs to a set H_S with $|S|=n$. The following lemma shows that, on the other hand every point of $\bigcup_{|S|=n} H_S$ is the end of such a curve.

Lemma 1

If KP(f) is a oneparametric nondegenerate complementarity problem, then every point $(z,z^*,w) \in \bigcup_{|S|=n} H_S$ is the end of just one or just two curves in $\bigcup_{|S|=n+1} H_S$ belonging to different sets H_S according as w has a component out of $\{0,1\}$ or not.

Proof:

Let $(z,z^*,w) \in H_S$ with $|S|=n$. In consequence of $k+m-1=n$ and $z(z^*)^T=0$; $z,z^* \geq 0$ the vector (z,z^*) either has just k components equal to zero and w has just one component out of $\{0,1\}$ or (z,z^*) has just k+1 components equal to zero and w has no component out of $\{0,1\}$. In the first case let $w_1 \epsilon \{0,1\}$. Then, because of

$$H_{S\cup\{i\}} = \emptyset \qquad (i \in 5\ ,\ i \neq 2k+1)$$

(z,z^*,w) can only be the end of a curve included in $H_{S\cup\{2k+1\}}$. The theorem about implicite defined functions shows us together with the nondegeneracy of our problem the existence of exactly one curve of this kind. In the second case there is just one pair (z_1,z_1^*) of components with $z_1=z_1^*=0$. Because of

$$H_{S\cup\{i\}} = \emptyset \qquad (i \notin S\ ,\ i \neq 1\ ,\ i \neq k+1)$$

we have to respect only curves included in $H_{S\cup\{1\}}$ or in $H_{S\cup\{k+1\}}$. The theorem about implicite defined functions shows together with the nondegeneracy that there exist just two curves (one in $H_{S\cup\{1\}}$ and one in $H_{S\cup\{k+1\}}$) which have (z,z^*,w) as an end.⌋

In conformity with the terminology of graph theory, in consequence of Lemma 1 the one points of $\bigcup_{|S|=n} H_S$ which are an end point of just one curve will be denoted by <u>nodes of degree 1</u>, the others which are end points of two curves will be denoted by <u>nodes of degree 2</u>. Thus we are able to formulate the following theorem as it is well-known from graph theory.

Theorem 1

If KP(f) is a oneparametric nondegenerate complementarity problem with a compact set $L_{KP(f)}$ of solutions then every connected component of $L_{KP(f)}$ is a piecewise smooth rectifiable curve without ramification points which is either closed and contains only nodes of degree 2 or contains exactly two nodes of degree 1 which are the end points.

Let KP(f) be oneparametric and nondegenerate, and let $L_{KP(f)}$ be compact. Furthermore let $g:\mathbb{R}^k\times\mathbb{R}^k\times\mathbb{R}^m \longrightarrow \mathbb{R}^n$ be a continuously differentiable function, for which the complementarity problem

$$\text{KP}\begin{pmatrix}f\\g\end{pmatrix}\qquad \begin{aligned} f(z,z^*,w) &= 0\\ g(z,z^*,w) &= 0\\ z(z^*)^T &= 0\\ z,z^*\in\mathbb{R}^k_+\ ,\ & w\in[0,1]^m\end{aligned}$$

is nondegenerate, too. Then, we can state the following about the zeros of g in $L_{KP(f)}$:

Lemma 2

Every zero $\overline{x}$ of g in $L_{KP(f)}$ lies on a curve in $\bigcup_{|S|=n+1} H_S$. Thereby the function g changes its sign in $\overline{x}$ on this curve.

Proof:

Let $\overline{x}\in L_{KP(f)}$ and g $(\overline{x})=0$. Because of nondegeneracy of KP $\begin{pmatrix}f\\g\end{pmatrix}$ we have then $|\text{spt}(\overline{x})| = n+1$ and

$$\det D_{\text{spt}(\overline{x})}\begin{pmatrix}f\\g\end{pmatrix}(\overline{x}) \neq 0. \tag{1}$$

Let

$$t \longrightarrow x(t)$$

be a local parametrisation of the smooth curve in $H_{\text{spt}(\overline{x})}$, on which $\overline{x}$ lies where

$$x(0) = \overline{x} \text{ and } \frac{dx}{dt}(0) \neq 0.$$

Then we have the equalities

$$Df(\overline{x})\circ\left(\frac{dx}{dt}(0)\right)^T = \frac{df}{dt}(\overline{x}) = 0$$

and

$$Dg(\overline{x})\circ\left(\frac{dx}{dt}(0)\right)^T = \frac{dg}{dt}(\overline{x}) .$$

So in consequence of (1)

$$\frac{dg}{dt}(\overline{x}) = 0$$

must hold. ⌟

A CONSTRUCTIVE PROOF FOR THE EXISTENCE OF SOLUTIONS TO (N')

Let us introduce the following functions f,g,h.
For $(x,y\,;u,v\,;\lambda,\mu)\in\mathbb{R}^{p+q}\times\mathbb{R}^{p+q}\times\mathbb{R}^2$ let

$$f(x,y\,;u,v\,;\lambda,\mu) := \begin{pmatrix}((1-\lambda)A-\lambda F+E)^T x^T - v^T - e^q\\ ((1-\mu)B-\mu G+E)y^T - u^T - e^p\end{pmatrix}$$

$$g(x,y\,;u,v\,;\lambda,\mu) := xe^p - 1$$

$$h(x,y\,;u,v\,;\lambda,\mu) := ye^q - 1 .$$

For every function $\tilde{f}$ in the variables $(x,y\,;u,v\,;\eta)\in\mathbb{R}^{p+q}\times\mathbb{R}^{p+q}\times\mathbb{R}^s$ let

$$\mathrm{KP}^*(\tilde{f})\qquad \begin{array}{c}\tilde{f}(x,y\,;u,v\,;\eta) = 0\\ (x,y)(u,v)^T = 0\\ (x,y)\,,\,(u,v)\geq 0\;,\;\eta\in[0,1]^s .\end{array}$$

Let us suppose now that the problems $\mathrm{KP}^*(\tilde{f})$ are nondegenerate for
$\tilde{f} = f\,,\begin{pmatrix}f\\g\end{pmatrix}\,,\begin{pmatrix}f\\g\\h\end{pmatrix}$.

Furthermore let us fix

$$f(\cdot;\cdot,\mu) : (x,y;u.v;\lambda) \to f(x,y;u,v;\lambda,\mu)$$

$$f(\cdot;\lambda,\cdot) : (x,y;u,v;\mu) \to f(x,y;u,v;\lambda,\mu)$$

$$g(\cdot;\cdot,\mu) : (x,y;u,v;\lambda) \to g(x,y;u,v;\lambda,\mu) = xe^p-1 .$$

Thus a short consideration shows the nondegeneracy for the following complementarity problems, too:

$$KP^*(f(\cdot;\cdot,0)), KP^*(f(\cdot;\cdot,1)), KP^*(f(\cdot;0,\cdot)), KP^*(f(\cdot;1,\cdot)),$$

$$KP^* \begin{pmatrix} f \\ g \end{pmatrix} (\cdot;\cdot,0) \quad .$$

In consequence of the compactness of the set of solutions to $KP^*(f)$ all the appearing sets of solutions are compact.
Let us list some properties of the solutions to $KP^*(f)$.

Lemma 3

For every solution $(x,y\,;\,u,v\,;\,\lambda,\mu)$ to $KP^*(f)$ the following implications are valid:

$$\lambda=0 \implies xe^p<1 \quad , \quad \lambda=1 \implies xe^p>1$$

$$\mu=0 \implies ye^q<1 \quad , \quad \mu=1 \implies ye^q>1 \, .$$

Proof:

First of all we have

$$((1-\lambda)A - \lambda F + E)^T x^T - v^T - e^q = 0$$

$$((1-\mu)B - \mu G + E)y^T - u^T - e^p = 0$$

$$(x,y)(u,v)^T = 0$$

$$(x,y)\,,\,(u,v) \geqq 0 \;;\; \lambda,\mu \in [0,1].$$

We shall prove the first implication as an example. The remaining statements may be deduced in a quite similar way.
As it is easily to be seen $v>0$ cannot hold. Otherwise we should have $y=0$ and $u=-(e^p)^T$. This would be a contradiction to $u \geqq 0$. Thus there exists a row $((1-\lambda)A - \lambda F)_i^T$ of the matrix $((1-\lambda)A - \lambda F)^T$, for which

$$((1-\lambda)A - \lambda F)_i^T \, x^T + xe^p - 1 = 0$$

is satisfied. Hence,

(1) $$\lambda=0 \implies xe^p \leqq 1$$

may be derived from $x \geqq 0$, $A \geqq 0$. Let $(x,y\,;\,u,v\,;\,\mu)$ be a solution of $KP^*(f(\cdot;0,\cdot))$ with $g(\cdot;0,\cdot) = 0$.

Because of Lemma 2 there are points $(x',y';u',v';\mu')$ in the set of solutions to $KP^*(f(\cdot;0,\cdot))$ which satisfy $x'e^p-1 = g(x',y';u',v';0,\mu') > 0$. This is incompatible with (1). Thus we get the stronger implication

$$\lambda=0 \implies xe^p<1$$

for every solution $(x,y;u,v;\lambda,\mu)$ to $KP^*(f)$. ⌋

Now we are occupied with the set of solutions to the oneparametric nondegenerate complementarity problem $KP^*(f(\cdot;\cdot,0))$. Of course, as this set is compact the statement of theorem 1 turns out to be true. Moreover we have

<u>Lemma 4</u>

(1) Every node of degree 1 in the set of solutions to $KP^*(f(\cdot;\cdot,0))$ is a solution $(x,y;u,v;\lambda)$ with $\lambda\in\{0,1\}$ and vice versa.

(2) The number of solutions $(x,y;u,v;0)$ to $KP^*(f(\cdot;\cdot,0))$ is odd.

(3) The number of curves in the set of solutions to $KP^*(f(\cdot;\cdot,0))$ which have a solution $(\bar{x},\bar{y};\bar{u},\bar{v};0)$ as one end and a solution $(\bar{\bar{x}},\bar{\bar{y}};\bar{\bar{u}},\bar{\bar{v}};1)$ as the other end is odd.

(4) The number of solutions to $KP^*\begin{pmatrix}f\\g\end{pmatrix}(\cdot;\cdot,0)$ is odd.

Proof:

(1) This statement may be derived directly from lemma 1.

(2) A point $(x,y;u,v;0)$ is a solution to $KP^*(f(\cdot;\cdot,0))$, if it satisfies the relations

$$(A+E)^T x^T - v^T - e^q = 0$$

$$(B+E)\, y^T - u^T - e^p = 0$$

$$(x,y)(u,v)^T = 0$$

$$(x,y)\,,\,(u,v) \geq 0$$

and vice versa. The solutions of this system are in a one to one correspondence with the equilibria of the bimatrixgame $\Gamma(A,B)$ (compare with [2]) which is nondegenerate in consequence of the nondegeneracy of $KP^*(f)$. As such a game has an odd number of equilibria (compare with [3]), we have proofed the second statement, too.

(3) This may be derived from (1) and (2) together with theorem 1.

(4) Because of lemma 1 $g(\cdot;\cdot,0)$ changes its sign at every zero on a curve in the set of solutions to $KP^*(f(\cdot;\cdot,0))$. Lemma 3 shows then that the number of zeros of $g(\cdot;\cdot,0)$ on a curve in the set of solutions to $KP^*(f(\cdot;\cdot,0))$ which connects a solution $(\bar{x},\bar{y};\bar{u},\bar{v};0)$ with a solution $(\bar{\bar{x}},\bar{\bar{y}};\bar{\bar{u}},\bar{\bar{v}};1)$ must be odd; on every other curve it must be even. Hence, by a consideration of (3) the number of zeros of $g(\cdot;\cdot,0)$ on the set of solutions to $KP^*(f(\cdot;\cdot,0))$ is odd. This leads to statement (4). ⌟

By these means we get similar results about the set of solutions to the oneparametric nondegenerate complementarity problem $KP^*\begin{pmatrix} f \\ g \end{pmatrix}$, for which the statement of theorem 1 turns out to be true again.

Lemma 5

(1) Every node of degree 1 in the set of solutions to $KP^*\begin{pmatrix} f \\ g \end{pmatrix}$ is a solution $(x,y;u,v;\lambda,\mu)$ with $\mu\in\{0,1\}$ and vice versa.

(2) The number of solutions $(x,y;u,v;\lambda,0)$ to $KP^*\begin{pmatrix} f \\ g \end{pmatrix}$ is odd.

(3) The number of curves in the set of solutions to $KP^*\begin{pmatrix} f \\ g \end{pmatrix}$ which have a solution $(\bar{x},\bar{y};\bar{u},\bar{v};\bar{\lambda},0)$ as one end and a solution $(\bar{\bar{x}},\bar{\bar{y}};\bar{\bar{u}},\bar{\bar{v}};\bar{\bar{\lambda}},1)$ as the other end, is odd.

(4) The number of solutions to $KP^*\begin{pmatrix} f \\ g \\ h \end{pmatrix}$ is odd.

Proof:

(1) Because of lemma 3 every solution $(x,y;u,v;\lambda,\mu)$ to $KP^*\begin{pmatrix} f \\ g \end{pmatrix}$ satisfies

$$(\lambda,\mu)\in\,]0,1[\times[0,1].$$

So the first statement may be derived from lemma 1.

(2) As every solution $(x,y;u,v;\lambda,0)$ to $KP^*\begin{pmatrix} f \\ g \end{pmatrix}$ corresponds with a solution $(x,y;u,v;\lambda)$ to $KP^*\begin{pmatrix} f \\ g \end{pmatrix}(\cdot;\cdot,0)$, statement (2) is a consequence of lemma 4 (4).

(3) This may be derived from (1) and (2) together with theorem 1.

(4) The proof of this statement runs quite similarly to the one of lemma 4 (4). ⌟

Lemma 5 (4) leads to the result that there exist solutions of (N') in form of a statement about the parity of the number of solutions . As all our statements are essentially based on auxiliary means out of graph theory and combinatorics, the following theorem is proved in a constructive way.

Theorem 2

The number of solutions to (N') is odd with respect to the given nondegeneracy postulations.

Proof:

Because of lemma 3 every solution $(x,y\,;u,v\,;\lambda,\mu)$ of $KP^*\begin{pmatrix}f\\g\\h\end{pmatrix}$ belongs to $\mathbb{R}^{p+q}\times\mathbb{R}^{p+q}\times\,]0,1[^2$. By these means the set of solutions of (N') is equal to the one of $KP^*\begin{pmatrix}f\\g\\h\end{pmatrix}$. Lemma 5 (4) leads to the statement of the theorem.$\lrcorner$

References

[1] J. Łoš: Extended von Neumann models and game theory; Computing Equilibria: How and Why (North-Holland, 1976); 141-157.

[2] M. Bastian: Lineare Komplementaritätsprobleme im Operations Research und in der Wirtschaftstheorie; Mathematical Systems in Economics 25, 62-67.

[3] C.E. Lemke, J.T. Howson jr.: Equilibrium points of bimatrix games; J.Soc.Indust. Appl.Math., Vol.12, No.2, 1964; 413-423.

GAME THEORY AND RELATED TOPICS
O. Moeschlin, D. Pallaschke (eds.)

NOTES ON OPTIMALITY AND FEASIBILITY OF INFORMATIONALLY DECENTRALIZED ALLOCATION MECHANISMS *

Andrew Postlewaite

Department of Economics
University of Illinois at Urbana-Champaign

David Schmeidler

Department of Economics
Tel-Aviv University

I. INTRODUCTION TO ALLOCATION MECHANISMS

In 1973 Gerard Debreu reviewed and summarized four major developments in general equilibrium theory [D.2]. The developments to which Debreu referred were the relation of the core and competitive equilibria, existence of equilibria in more general settings (i.e., measure theoretic models), computation of competitive equilibria, and topological properties of competitive equilibria derived by differential topological methods. Looking now, we would add another major line of research which differs from the above in that it alters the basic parameters of the model. Previously the conceptual framework consisted of a description of an economy in terms of agents characteristics (preferences and endowments and in the case of production, the technological possibilities).

* *The research was partially supported by the National Science Foundation of the United States of America, grant number - NSF SOC 77-27403*

The analysis focused on the correspondence between the economies so described and their competitive allocations. However it was never specified how these allocations arose. There was no specification of how the trades were made. Using the language of Hurwicz [H.2], the analysis was of the competitive "performance correspondence" without specifying what the mechanism is which implements this performance correspondence. The first efforts in this attempt to specify how outcomes arose was to cast the problem in a non-cooperative game theoretic framework. A mechanism was to be a precise specification of the set of strategies (or signals) which were available to an agent and the outcome (or allocation) which would arise from any simultaneous choice of strategies by the agents. We will call the function relating outcomes to joint strategies a strategic outcome function.

As an example, Hurwicz considered the set of classical pure exchange economies and let an agent's strategy be the announcement of a (classical) utility function . The outcome would be (assuming uniqueness) the competitive allocation. A problem arises, though, in that for many economies some agents would find that if they announced a utility function other than their true utility function, the outcome would be preferable to the one which would have arisen had they revealed correctly. In game theoretic terms, correct revelation is not a dominant strategy equilibrium; in Hurwicz's terms, this mechanism is not incentive compatible. But what of other mechanism? Perhaps there is some mechanism which has utility functions as strategies for the agents and is incentive compatible.It is trivial to design such a mechanism - simply let the outcome be no trade regardless of the utility functions announced. But here the correspondence between the economies and the outcomes thus realized (i.e., the performance function) is undesirable from an economic point of view in that the outcomes will not in general be efficient.

The question of interest then is whether or not, when economically interesting restrictions are placed on a performance correspondence, there exists an incentive compatible mechanism which will realize (or implement) such a performance correspondence. Hurwicz [H.2] showed that no Pareto efficient and individually rational performance function can be implemented with an incentive compatible mechanism.

This might seem to leave an economic planner in a quandary. If he designs a strategic outcome function which picks Pareto efficient and individually rational outcomes when people reveal their preferences correctly, then in some economies some agents will find it in their own best interest to misrepresent their preferences. The outlook is not so dismal for the planner however. He can ask what the outcome would be even if agents misrepresented their preferences. The initial concern was that if agents misrepresented their preferences, this would destroy any efficiency properties that the planner might have designed, but this may not necessarily so. It is at least conceivable that the combination of agents' strategic behavior leads to "nice" outcomes. If agents are playing strategically in their announcement of preferences, we are lead to consider as the outcome of the game not dominant strategy equilibria, but rather Nash equilibria. At the same time the planner might ask why the strategies of an agent should be announcement of preferences. Why must the form of the strategies be even related to agents' characteristics? Possibly by choosing some quite abstract set of available strategies for agents, the planner might be able to avoid some of the problems Hurwicz pointed to. So we leave the realm of strategies in which an agent necessarily has a correct or "truthful" strategy. Instead the agents are faced with some arbitrary sets of strategies of which they are to choose whichever they wish, and an outcome will be selected depending upon these strategies. This has been done for models with public goods. In Schmeidler [Sc.1] there is an attempt to introduce voting to a model of an Arrow-Debreu economy with public goods. Schmeidler proposed a mechanism whereby people are taxed proportionately according to the value of their private goods and the tax revenue is used to finance public goods according to each taxpayer specifications.This individual "earmarking" is aggregated to determine the quantity of each public good to be produced. The valuation of the private goods is made at the equilibrium prices. An equilibrium price and Nash equilibrium of individual decisions as to how to allocate the taxes is shown to exist. However, the Nash equilibria of this mechanism are not Pareto efficient.

Groves and Ledyard [GL] suggested a different outcome function. They considered an economy with both public and private goods and designed a mechanism, i. e., strategy sets for the agents and a strategic outcome function such that the Nash equilibria are always efficient. One might think that the work of Groves and Ledyard solves the planner's problem of designing a mechanism to implement a desirable performance function. There are several problems however. First, they do not allow full strategic behavior on the part of agents. Rather, they assume that the agents take as given the prices of both public and private goods (but not their taxes).

While this may be appropriate for addressing the free rider problem, as they do, it leaves unanswered the question of what will happen when agents take into account their effect on these prices. More importantly however, there may not exist an equilibrium in their model. There are in fact large families of economies in which equilibria fail to exist.

II. SOCIAL CHOICE MODELS

A parallel development has occured in the field of social choice. In [A.1] Arrow introduced the modern approach to social choice theory. In our language he asked whether or not there existed performance correspondences which satisfied a priori desirable characteristics such as Pareto efficiency, independence of irrelevant alternatives, and non-dictatorship. Whereas in general equilibrium theory the existence of "desirable" performance correspondences (e. g., competitive equilibria) was a cornerstone of the early work, Arrow obtained an impossibility theorem. Although Arrow was aware of the possibility of strategic behavior on the part of agents, he specifically avoids consideration of it. The very structure of the problem, i. e., considering performance correspondences which are maps from preference profiles into outcomes, preempts the consideration of agents' behavior. In [Fa] Farquharson altered the basic framework of social choice to allow strategic behavior. He considered a system of sequential majority voting in subsets of the alternatives to determine a final outcome. Every voter now takes into account how his voting interacts with the votes cast by other agents to determine this final outcome.Farquharson suggested various notions of stability which would define an equilibrium. In this way he has presented a model with well-defined strategies sets available to the agents, a strategic outcome function, and an equilibrium concept.

The next decisive step in the line of research was by Gibbard [G] and Satterthwaite [Sa]. Gibbard generalized the structure of Farquharson, removing the restrictions on the specific form of the strategy sets and allowing arbitrary strategic outcome functions. Within this framework he showed that the only outcome functions which have dominant strategy equilibria for all profiles of preferences are dictatorial (under an assumption that the minimum number of outcomes is three). In other words there are no non-dictatorial mechanisms which have dominant strategy equilibria for all profiles of preferences. In the social choice framework this theorem is equivalent to the following: If the strategy sets are taken to be the set of preferences then there exists no non-dictatorial strategic outcome function (with at least three outcomes in the range) for which the truth is a dominant strategy equilibrium for all profiles of preferences. The same result was obtained independently by Satterthwaite. This result is analogous to the result of Hurwicz on economic mechanisms. Hurwicz has an assumption of individual rationality (non-coerciveness) which essentially played the role of non-dictatorship in the Gibbard-Satterthwaite results.

One of the most important conclusions for us arising from the work of Gibbard and Satterthwaite is that within the framework of strategic outcome functions, the insistence on dominant strategy equilibria rules out all reasonable performance functions. Later work by others (e. g. Pattanaik [Pa] and Sengupta [Se] points out that one can weaken somewhat the concept of dominant strategy equilibria and one still obtains impossibility results.

These impossibility theorems led Hurwicz and Schmeidler [HSc] to the Nash solution concept for the strategic outcome function. Precisely, they asked that for any profile of preferences, there exist a Nash equilibrium and that all Nash equilibria be Pareto efficient. If one is considering dominant strategy equilibria, Pareto efficiency relative to the codomaine of the outcome function is a straightforward consequence. Of course, when changing to the solution concept of Nash equilibria one no longer has Pareto efficiency automatically, hence this restriction was added as a desirable criterion for the performance function. Non-dictatorship, or the stronger property, symmetry across people, would also be a desirable characteristic. In this framework existence of a class of mechanisms, i. e. description of strategy sets and strategic outcome functions, which had these desired characteristics was proved when there are at least three agents. Maskin [Ma] independently attained some of these results and extended others.

III. NASH EQUILIBRIA AND ALLOCATION MECHANISMS

In attempting to apply these results to the other problem of designing economic mechanisms, one confronts several problems. The framework in social choice in general considers only a finite number of outcomes, whereas the economic problem we are trying to analyse generally has an infinite number of outcomes, e. g. allocations. While this problem is not particularly difficult to surmount, there is a more basic difficulty. In the economic model we have the agents' endowments as a parameter of the economy in addition to the preferences. One would like to have the performance function be individually rational (or non-coercive) with respect to these endowments. There is no natural way to embed this concept into the social choice framework without destroying the existence results. Further since the endowments are a parameter of the economic model, the set of feasible outcomes in this economic model are <u>not</u> independent of the agents' characteristics. In the social choice model the set of feasible alternatives is the same regardless of the agent's characteristics. Nevertheless these results provide us with valuable insights into the design of economic mechanisms with desirable features.

Within this framework there are a number of economic models in addition to those mentioned before which can be analyzed using this concept of economic mechanisms. Unless otherwise mentioned, these models deal only with pure exchange private goods economies. Shubik [Sh] introduced a market clearing rule which Shapley and Shubik [SS] used in a general equilibrium model in strategic outcome function form. In this model one commodity is used as money. The strategies of agents consist of bids of the commodity money to purchase other goods, and offers to sell quantities of these other goods. Using the market clearing rule Shubik proposed together with a bankruptcy rule, a Nash equilibrium is shown to exist. Regardless of the agent's strategies the outcome is feasible. This is accomplished by making the strategies available to an agent depend on his initial endowment. The Nash outcomes in this model are individually rational. This is guaranteed since an agent has a strategy the possibility of not participating in the market. While the Nash outcomes are in general not Pareto efficient, in a modified version of this model [PoS] it was shown by Postlewaite and Schmeidler that the Nash outcome is asymptotically efficient. That is, if there are sufficiently many agents, none of them "too large" then the percentage of resources of the economy which is wasted due to the inefficiency is small.

Wilson [W] constructed an example of mechanism in which the strategy sets are feasible net trades (i.e., the strategy sets depend on initial endowments) for all but one agent. This central agent has only one strategy which is determined by his true preferences. The non-central agents propose feasible net trades and the central agent accepts the utility maximal (with respect to his preferences) subset of these trades. The strategic outcome function is such that the outcome is feasible for all choices of strategies and the Nash outcome is in the core, a fortiori it is efficient and individually rational. All the examples considered so far, considered only strategy sets which did not depend on agents' preferences. To the extent that the available strategies differed for different agents, the dependence was on endowments alone. If the preferences of agents are known or observable by others, we need only a computational scheme to achieve a particular performance function. We want to deal only with mechanisms which preserve informational decentralization with respect to preferences. That is we do not allow any dependence of strategy sets or strategic outcome correspondences on preferences, which we take to be unobservable. If the central agent in Wilson's example is allowed to pick as a strategy any strategy appropriate for some preference relation, the efficiency of the

Nash allocations disappears.

A mechanism discovered by Schmeidler [Sc.2] has the strategy sets the same for all agents and therefore independent of their characteristics, both preferences and endowments. A strategy for an agent consists of a pair, a price and compatible net trade, that is a net trade with value 0 at this price. The agents who have announced the same price trade, and to the extent that their aggregate net trade is not 0 , they are rationed proportionately. Given the strategies chosen by the other agents, an agent has a strategy which will give him as an outcome his Walrasian demand for any price announced by other agents. The Nash outcomes of this game (when there are at least three agents) are precisely the Walrasian allocations. This mechanism thus implements the competitive performance correspondence when Nash equilibrium is the solution concept. The price paid for this achievement is non-feasibility for some (non-equilibrium) strategy choices. That is, for some non-equilibrium strategy profiles, the net trades some agents are to carry out are not feasible given their initial endowments.

Hurwicz [H.4] also constructed a mechanism with this property that the Nash outcomes are precisely the Walrasian outcomes and strategy sets identical for all agents. The strategic outcome function is different from that used by Schmeidler however. Here an agent trades at the average price announced by others, and the price he announces affects other agents only. More importantly he also constructed a similar mechanism with these properties for economies with public goods. Here the Nash outcomes coincide with Lindahl equilibria. Both examples however have the same characteristic as Schmeidler's: non-equilibrium strategies may lead to non-feasible outcomes.

It is worthwhile to add a comment at this point. The non-feasibility of these mechanisms is not an oddity which can be rectified by a simple ad hoc change in the mechanism, such as making a rule which states that no trade will take place in the event of non-feasibility. This type of change fundamentally alters this mechanism and destroys either the existence or optimality of its Nash equilibria.

More generally, Hurwics, Maskin and Postlewaite [HMP] show that the Walrasian performance function cannot be implemented by a feasible outcome function. A proof of this result can be found in Appendix C.

In a general vein Hurwicz [H.5] has explored the relationship between individually rational and Pareta efficient performance correspondences implemented by Nash equilibria of strategic outcome functions and the Walrasian (competitive) performance function (Lindahl performance correspondence in the case of public goods). He showed that if the Nash equilibrium correspondence is upper semi-continuous, then for any economy the Walrasian allocations must be in the performance correspondence i.e., Walrasian allocations must be Nash allocations. Hurwicz also shows that with somewhat more restrictive assumptions, all Nash allocations must be Walrasian. To get this result Hurwicz assumes that for any strategies chosen by the other agents, the set of outcomes available to the particular agent is convex (assuming free disposal). The upper semi-continuity required for the first half of the theorem can be justified on the grounds that small changes in the parameters of an economy should cause an equilibrium to change slightly. The convexity assumption used for the other direction seem less compelling. Thus while in general for mechanisms which have individually rational, Pareto efficient Nash equilibria we expect that Walrasian allocations will be Nash, Nash allocations may not be Walrasian if the convexity assumption fails.

IV. FEASIBILITY OF ALLOCATION MECHANISMS

We are interested in the design of an economic mechanism which yields feasible allocations for any joint set of strategies chosen. This is of particular importance since as was mentioned in the introduction, it is impossible to design a strategic outcome function for which dominant strategy equilibria will be Pareto efficient and individually rational. Thus our attention will be focused on Nash equilibria instead. But here an agent's optimal strategy will depend upon the strategies chosen by other agents. Non-equilibrium strategy choices would seem more likely, then, when Nash equilibrium is our solution concept.

It is quite clear that if feasibility is to be obtained, then either agents' available strategies or the stregic outcome function will have to depend on agents' endowments. If both were independent of endowments, then for some set of agents' strategy choices which result in a non-zero net trade for some agent, we could replace that agent with another whose initial endowment was so small as to make the net trade infeasible. Thus the only hope we have is to introduce dependence of some sort on initial endowments. We will focus on dependence of the strategy sets on initial endowments maintaining an assumption that the strategic outcome function depends on the initial endowments only through the strategy set dependence. This seems most consistent with the notion of decentralization.

We will still ask that the strategy sets do <u>not</u> depend on preferences which we take to be unobservable. In the spirit of decentralization, we will also ask that the strategy set of an agent depend only on his <u>own</u> endowment, not the endowment of others. Thus we have a mapping from endowments into an abstract space which associates with any initial endowment the permissible strategies for any agent with that endowment in any economy.

This dependence of strategy sets on this part of the agent's characteristics is reasonable in that the endowments are at least potentially observable. An audit of agent can objectively determine whether or not he has a stated endowment as opposed to the general impossibility of determining his preferences.

One question which remains however is to what extent an agent's endowment can be precisely known. A demand to exhibit endowment would prevent an agent from overstating endowment. But whether or not an agent can successfully <u>hide</u> or withhold endowment is a more difficult problem. If agents have complete property rights over their own endowment, a further possibility available to agents is the destruction or elimination of some part of this endowment.

If only a ceiling can be put on an agent's endowment, another avenue of strategic behavior is opened for an agent. By "announcing" an endowment less than his actual endowment, he changes the set of strategies available to him. This could ultimately leave him in a better position than if he reported his endowment correctly.

In [Po], Postlewaite considered mechanisms in which the strategy sets were announcements of endowments; preferences were assumed to be known. It was shown that for any mechanism which yields Pareto efficient, individually rational outcomes, correct announcement <u>cannot</u> be a dominant strategy equilibrium. This is analogous to Hurwicz's result with preferences as the strategies.

As we have seen above, we must give up some of the informational decentralization if we are to design a feasible mechanism whose Nash equilibria are Pareto-optimal. Hurwicz, Maskin and Postlewaite [HMP] introduced such a mechanism whose Nash equilibria are individually rational as well. More specifically, the Nash equilibria of their mechanism coincide with the constrained Walrasian equilibria (see Appendix D). This last term means that each consumes maximizes his utility over the budget set constrained to feasible bundles. The information that must be known by the mechanism is vector of endowments.

There is another variant of the mechanism in which the endowments are not known. Here the individuals as part of their strategy state their endowments. It is assumed that it can be verified that they don't claim to have more endowment than they actually have, but many state that they have less. Again in this variant the mechanism is feasible and its Nash equilibria coincide with the constrained Walrasian equilibria. A similar informational constraint leads to the feasibility of the mechanism in [PaS., PoS, SS].

V. INFORMATIONAL REQUIREMENTS FOR COMPUTING NASH EQUILIBRIA

In our introduction we presented the development of the literature on mechanisms as being necessary to describe how allocations were to arise through the interactions of agents' actions or choices of strategies. As we stated, the first efforts in this area utilized the dominant strategy equilibrium concept. After the impossibility results of Hurwicz, attention was shifted to the concept of Nash equilibrium. It was with this notion of solution that the positive results of Groves and Ledyard, Hurwicz and Schmeidler, Schmeidler, Wilson, Maskin, and others were derived.

There is an essential difference between the two solution concepts however. If there is a dominant strategy equilibrium, each agent has a best response independent of the other agents' choices of strategies. Regardless of what strategies they choose, he can uniformly pick this strategy. He does not need any information of what strategies they choose to calculate his optimal behavior. Viewed in this light, it is quite plausible that if there is a dominant strategy equilibrium, the agents in an economy will arrive at it, at least if they know the strategic outcome function.

The use of Nash equilibrium presents the agents with a much more complex problem. Now an agent trying to choose his "best" strategy finds that in general this best strategy will change as the other agents change their strategies. Thus in addition to knowing the strategic outcome function, an agent may also need to know the strategies of all other agents in order to determine his optimal strategy.

This presents something of a problem. In the previous section we stated that one very desirable characteristic of a mechanism is that it be feasible. Regardless of the strategies chosen by agents, we would like there to be a feasible outcome. But now suppose that the agents in an economy choose some non-Nash profile of strategies resulting in an outcome. Will the agents in fact realize that this is a non-equilibrium position? Clearly they will if they know the messages or strategies of all other agents. But in some environments, particularly in large economies, this is an heroic assumption. If an agent is supposed to know the messages of all other agents, he obviously has to be in possession of "arbitrarily large" amounts of information if we consider economies with arbitrarily many agents.

This leads to a question of the amount or size of the information an agent needs to determine his optimal strategy in response to those of other agents. Ideally, an agent might see only the outcome to him from a joint strategy choice and be able to determine whether or not his choice is in fact optimal and if not, what his optimal strategy is. Failing this one would hope that at least an aggregate "summary" of other agents' messages or strategies would suffice. An aggregate summarization would be a function from the strategy spaces of the other agents into a euclidean space of the same dimension as the individual strategy space. Examples of summarizations would be summations of the agents strategies, averages of them, etc. Then as the number of agents grows, the "size" of the information needed by an agent would not change as we consider larger economies. Is there then a mechanism which yields individually rational and Pareto efficient Nash outcomes along with an aggregate summarization procedure which guarantees agents' will know their optimal strategy given the aggregate summarization? A somewhat weaker criterion would be that there is a finite dimensional space which would contain enough information for agents to be capable of knowing their optimal response, where the dimension of the space is independent of the number of agents in the economy. The mechanism introduced by Hurwicz [H.4] whose Nash equilibria coincide with Walrasian equilibria satisfies the latter weaker condition. The dimension of the space which contains the summarized information is four times the number of commodities.

As another example we consider the form of the model of Shapley and Shubik [SS] introduced in [PaS]and analysed also in [PoS]. Here the strategy sets of the agents are contained in a euclidean space of dimension 2ℓ where ℓ is the number of commodities. Given a joint choice of strategies, an outcome (allocation) arises. From one agent's share of the allocation, it is not generally possible for an agent to determine whether or not his strategy is optimal. Suppose though, that he is given aggregate information in the form of the sum of the other agents strategies. He then knows the combined quantities of the offers to buy and sell all goods by the other agents. This information is sufficient for the agent to calculate his optimal response to the other agents' strategies. We note that the size (dimension) of this information he needs depends only on the number of commodities, not on the number of agents. Also note that the strategic outcomefunction of Hurwicz, Maskin and Postlewaite [HMP] for the case that the endowments are known (appendix D) also has a summary function. That is if a person knows a vector of dimension four plus twice the number of commodities he is able to compute his best response or arbitrarily good responses in the case that there is no best response (a situation that may happen with this outcome function).

VI. INFORMATIONAL REQUIREMENTS FOR OUTCOME FUNCTIONS

The question of informational efficiency deals with the amount of information which must be transferred within the system. An obvious aspect of this question, not treated in the previous section, is the size or dimension of the strategy (or message) space. This question has been treated by Hurwicz [H.3] and Mount and

Reiter [MR]. The question here is what is the minimal "size" of information agents must send so as to be able to effect individually rational and Pareto efficient allocations. Mount and Reiter present general framework in which questions of informational efficiency can be investigated. More specifically, in their framework each agent sends a message in some space M. When their messages are consistant (in a well specified sense) an outcome arises from the messages. They present a system of messages consisting essentially of prices and net trade allocations which results in the competitive performance function, and show that no other system of messages can use a message space which is of smaller dimension and still accomplish this. That is, prices and proposed trades are essentially the most efficient messages which can effect competitive equilibria. Hurwicz [H.3] with a somewhat less general framework obtained similar results, though the assumptions used were not identical. Hurwicz also showed that informational requirements could not be reduced by considering different Pareto efficient individually rational performance correspondences than the competitive one.

The results of Mount and Reiter and Hurwicz are not directly applicable to the model of strategic outcome functions. Strategic behavior on the part of agents is not considered in their work on information. Rather, they asked what was the smallest message space which would contain enough information to affect "good" allocations <u>if</u> agents followed some prescribed procedure for chosing messages. The fact is that these messages are in general <u>not</u> dominant strategies if agents consider "manipulating".

If we turn to strategic outcome functions, the use of prices and net trades, as messages (strategies) still suffices to generate Pareto optimal and individually rational Nash equilibria. For instance this is the case in Schmeidler's outcome function in [Sc.2]. We note that in [PoS] the mechanism is feasible asymptotically efficient as the number of agents gets large, and has message spaces of dimension twice the number of commodities. The mechanism in [HMP] is feasible, achieves Pareto optimal and individually rational Nash equilibria and has strategy spaces of dimension twice the number of commodities.

MATHEMATICAL APPENDICES

APPENDIX A: SOCIAL CHOICE

Let T and A be finite non-empty sets and for each $t \in T$, let S_t be a non-empty set also. Ω is the set of transitive, reflexive and total binary relations on A. Denote $\underline{S} \equiv X_{t\in T} S_t$. A function f: $\underline{S} \to A$ is called a strategic outcome function (SOF). In the case that $S_t=\Omega$ for all $t\in T$, i.e., $f:\Omega^T \to A$, f is called a social choice function (SCF). We refer to elements of T as persons, A as alternatives, S_t as strategies, Ω as preferences. An element $\underline{P} = (P_t)_{t\Omega T} \in \Omega^T$ is called a profile of preferences, where an element $\underline{s} = (s_t)_{t\in T} \in \underline{S}$ is called a (strategy) selection. Given a SOF f and $\underline{P} \in \Omega^T$, a selection $\underline{s}^*$ is called a dominant strategy equilibrium (DE) if for all $h\in T$ and all $s\in S_h$ and $\underline{s}\in\underline{S}$: $f(\underline{s}|h,s^*_h)P_h f(\underline{s})$ where $(\underline{s}|h,s^*_h)= (r_t)_{t\in T}$ with $r_t=s_t$ for all $t\neq h$ and $r_h=s^*_h$. An SOF f is said to be straightforward if it has a DE for every $\underline{P}\in\Omega^T$. A social choice function is said to be non-manipulable if for every $\underline{P}\in\Omega^T$, $\underline{P}$ is a DE. An SOF f is said to be dictatorial if there is an $h\in T$ (called the dictator) such that for all $\underline{s}$, $\underline{s}'\in\underline{S}$, $s_h = s'_h \Rightarrow f(\underline{s}) = f(\underline{s}')$.

<u>Theorem</u>: (Gibbard-Satterthwaite): If the image of a straightforward SOF f contains at least three alternatives, then f is dictatorial.

An SCF f is said to be vetoproof if for all $\underline{P}\in\Omega^T$ and for all $x\in A$: $\#\{t\in T \mid$ for all $y\in A\ xP_t y\} \geq \# T-1 \Rightarrow f(\underline{P})=x$. An SCF f is said to be monotonic if for all $\underline{P}, \underline{P}'\in\Omega^T$ and $x\in A$ if $f(\underline{P})=x$ and for all $y\in A$ and for all $t\in T, xP_t y = xP'_t y$, then $f(\underline{P}')=x$. Given an SOF f and $\underline{P}\in\Omega^T$, a selection $\underline{s}^*\in\underline{S}$ is said to be a Nash equilibrium (NE) if for all $h\in T$ and all $s\in S_h$: $f(\underline{s}^*)P_h f(s^*|h,s)$.

<u>Theorem</u>: (Maskin): For any vetoproof and monotonic SCF f, there exist a SOF f' such that for all $\underline{P}\in\Omega^T$, $f(\underline{P}) = \{ f'(\underline{s}) \mid \underline{s}$ is a NE for $\underline{P}\}$. (In this case we say that f' implements f.)

APPENDIX B: ECONOMIC ENVIRONMENTS

For this appendix we will modify somewhat the definitions of the previous appendix. Here the set of alternatives $A \equiv \{\underline{\chi}=(\chi_t)_{t\in T}\in(R^L)^T \mid \Sigma_{t\in T}\chi_t=o\}$ where R^L is the Euclidean space whose coordinates are indexed by elements of the finite non-empty set L (the commodity set). In this section we refer to elements of T also as traders or agents and elements of R^L as nettrades, where χ_t is the net trade of agent t. The definition of strategic outcome functions, Nash equlilibria, dominant strategy equilibria, and straightforwardness carry over from the previous appendix; however, here we are not interested in all profiles. We are interested primarily in modelling Arrow-Debreu pure exchange economies. When an agent compares two alternatives, he considers only his net trades. Furthermore, this agent's preferences over his net trades should satisfy the standard assumptions of convexity continuity and monotonicity. More formally let us denote by R^L_+ the non-negative octant of R^L and by R the real numbers and agree that inequalities in R^L hold coordinatewise. Set $\theta \equiv \{P\in \Omega \mid \exists o<w\in R^L_+$ and a continuous, quasi-concave, strictly monotonic function $u: R^L_+ \to R$ such that for all $\chi, y\in A$: $\chi P y$ iff $[y+w\in R^L_t$ or $u(\chi+w)\geqq u(y+w)$ when $\chi+w$ and $y+w\in R^L_+]\}$. The analogue of a social choice function (correspondence) of the previous appendix is a prescription correspondence $g:\underline{\theta} \twoheadrightarrow A$ where $\underline{\theta} = \theta^T$. Two prescription correspondences which are commonly used are the Walrasian correspondence (WE) and PEIR correspondence (Pareto efficient and individually rational). Given $\underline{P}\in\underline{\theta}$, .$WE(\underline{P}) = \{\underline{\chi}=(\chi_t)_{t\in T}\in A \mid \exists o\neq p\in R^L_+$ such that for all $t\in T$, $\chi\in R^L$: $p\chi_t=pw_t$ and $p\chi\leq pw_t \Rightarrow \chi_t P_t x\}$ where w_t is from the definition of θ above. Given $\underline{P}\in\underline{\theta}$, $PEIR(\underline{P}) \equiv \{\underline{\chi}=(\chi_t)_{t\in T} \in A \mid$ for all $t\in T$ $\chi_t P_t w_t\} \cap \{\underline{\chi} = (\chi_t)_{t\in T} \in A \mid$ for all $\underline{y} \in A$: [for all $t\in T$ $y_t P_t \chi_t$] $\Rightarrow$ [for all $t\in T$ $\chi_t P_t y_t$]$\}$. Given a SOF $f:\underline{S} \to A$ its Nash performence correspondence NEf is the correspondence from $\underline{\theta}$ to A that assignes to each $\underline{P}$ in $\underline{\theta}$ the set $NEf(\underline{P}) \equiv \{f(\underline{s})\in A \mid \underline{s}$ is a NE for $\underline{P}$ (and f)$\}$ in A.

Given a SOF f and a $\underline{P}$ in $\underline{\theta}$ a selection $\underline{s}^*$ in $\underline{S}$ is said to be a strong equilibrium, SE, (for f and $\underline{P}$) if for any subset C of T and any selection $\underline{s}$ in $\underline{S}$: ([for all t in C, f $(\underline{r})$ $P_t f(\underline{s}^*)$] $\Rightarrow$ for all t in C, $f(\underline{s}^*) P_t f(\underline{r})$], where $r_t = s_t$ for t in C and $r_t = s^*_t$, for t in $T\setminus C$).

<u>Theorem:</u> (Schmeidler) 1. If $\#T>2$ then there is a SOF f such that $WE(\cdot)=NEf(\cdot)$ on $\underline{\theta}$. 2. If $\#T>2$ than there is a SOF f such that for $\underline{P}$ in $\underline{\theta}$ induced by a differentiable and strictly quasiconcave(utility) function: $WE(\underline{P}) = NE_f(\underline{P}) = \{\underline{s}\in\underline{S} \mid \underline{s}$ is a SE for $\underline{P}$ and f$\}$.

<u>Theorem:</u> (Hurwicz): Let a SOF f be given such that on $\underline{\theta}$, $NEf(\cdot)\subset PEIR(\cdot)$ and $NEf(\cdot)$ is upper semicontinuous. Then on $\underline{\theta}$, $WE(\cdot)\subset NEf(\cdot)$.

IF S_t = S for all t in T and a SOF $f:\underline{S} \to A$ is symmetric across persons in T then it is said to be totally (informationally) decentralized. This is the case for the SOFs of this appendix. However in order to get feasibility one has to give up some of the informational decentralization. Indeed the domain of the SOF of the last appendix includes information on the initial endowments.

Formally we have a function $f:\underline{S} \times (R^L_{++})^T \to A$ and a partition of $\underline{\theta}$ to sets $\{\underline{\theta}_{\underline{w}} \mid \underline{w}\in (R^L_{++})^T\}$ so that $\underline{P} = (P_t)_{t\in T}\in\underline{\theta}_{\underline{w}}$ iff for all t in T, P_t is induced by w_t. All previous definitions apply under the convention that wherever $\underline{P}\in\underline{\theta}_{\underline{w}}$ we use the SOF f restricted to $\underline{S} \times \{\underline{w}\}$.

APPENDIX C: IMPOSSIBLILITY OF A WALRASIAN AND FEASIBLE MECHANISM

Proposition: No mechanism which has a strategic outcome function which is independent of preferences and yields feasible outcomes for all strategy profiles can have Nash equilibrium allocations coincident with Walrasian allocations for all economies.

Proof: Suppose there is a mechanism which has a strategic outcome function independent of preferences and for all economies has equivalence of the set of Nash equilibrium allocations and Walrasian allocations. We will construct two economies, each with three agents and two commodities, showing that general feasibility is impossible. In the first economy, e, agents 1 and 2 have the same utility function $U^1(x,y) = U^2(x,y) = y - e^{-x}$ and identical endowments $w^1 = w^2 = (1,3)$. Agent 3 has utility function $U^3(x,y)=xy$ and endowment $(2,6)$. It is straightforward to show that price $P=(1,1)$ is competitive resulting in the Walrasian allocation $x^1 = x^2 = (0,4)$ and $x^3 = (4,4)$. By assumption this must be a Nash allocation, i.e., there exists a message triple (m_1,m_2,m_3) which is a Nash equilibrium which results in the Walrasian allocation above.

The second economy, $\bar{e}$, will be identical to the original in all respects except for the utility function of the third agent. We let $\bar{U}^3(x,y) = x^{1+\varepsilon}y$. The marginal rate of substitution for agent 3 at the point (4,4) is now $1+\varepsilon$, which is different from the price ratio. Hence the Walrasian allocation for the original economy e is not Walrasian in $\bar{e}$. Since we assumed that the set of Nash equilibrium allocations is equivalent to the set of Walrasian allocations, (m_1,m_2,m_3) is not a Nash equilibrium in $\bar{e}$. It is clear that if either agent 1 had a message m_1^* such that (m_1^*,m_2,m_3) resulted in an outcome which was strictly preferred to (0,4), this would contradict the fact that (m_1,m_2,m_3) is a Nash equilibrium in e, and similarly for agent 2. Thus if (m_1,m_2,m_3) is not a Nash equilibrium in $\bar{e}$, agent 3 must have a message m_3^* with the outcome resulting from (m_1,m_2,m_3^*) yielding strictly higher utility than (4,4) according to the utility function $\bar{U}^3$. In Figure 1 the indifference curves for both U^3 and $\bar{U}^3$ through the point (4,4) are shown and represented II and $\overline{II}$ respectively. It is straightforward to verify that $\overline{II}$ lies above II to the left of the vertical line through (4,4) and below it to the right.

The outcome to agent 3 from the message triple (m_1,m_2,m_3^*) must lie above $\overline{II}$. If this outcome were on or to the left of the vertical line through (4,4) it would be above II as well as $\overline{II}$. Since this contradicts the fact that (m_1,m_2,m_3) is a Nash equilibrium in e, the outcome to 3 must lie to the right of the vertical line through (4,4). But this means he must receive more than 4 units of good x, while the total quantity is only 4. Thus if (m_1,m_2,m_3) is not a Nash equilibrium in $\bar{e}$, there must be non-feasible outcomes for some strategy profiles.

There are three agents in this example because the examples of Hurwicz and Schmeidler which have equivalence of Walrasian allocations and Nash allocations work for economies in which there are at least three agents. It is straightforward to modify the example so as to provide the same effect with either two agents or more than three. The specific utility functions for agent three are not critical; the only requirement is that to the left of the vertical line the new indifference curve should not be below the original indifference curve and that the previously Walrasian allocation not be Walrasian with the new indifference curve. The essential feature is that the Walrasian equilibrium give all of one commodity to this agent. Any preferences for the other agents which permit a Walrasian equilibrium with this feature could have been incorporated instead of those used.

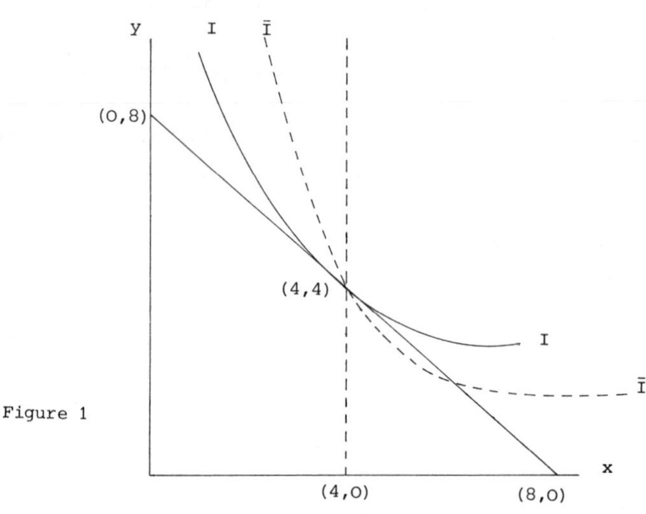

Figure 1

APPENDIX D: FEASIBLE MECHANISM GUARANTEEING CONSTRAINED WALRAS EQUILIBRIA

T is the finite set of agents $\#T > 2$

$L = 1, \ldots, \ell$ is the set of commodities

P^{ℓ} = ℓ dimensional price simplex

$S_t = \{(p,x) \in P^{\ell} \times R^{\ell} \mid 0 \leq x,\ p \cdot x = p \cdot w_t\}$

The outcome function is defined as follows:

1. If $\exists$ i,j,k$\in$T such that p_i, p_j, p_k are distinct

$$h_t = \frac{\|x_t\|}{\Sigma \|x_{t'}\|} w \quad \forall t \qquad \text{where } w = \Sigma w_t$$

2. If $\exists$ only two prices p und p' announced and at least two people announce each p and p'

$$h_t = w_t \; \forall_t$$

3. If $\exists$ $\bar{p}$ such that $p_t = \bar{p}$ for all t (3.1) and $\Sigma_{t\in T} x_t \neq w$ then $h_t = w_t$
(3.2) and $\Sigma_{t\in T} x_t = w$ then $h_t = x_t \; \forall t$

4. If there exist $\bar{p}$ and m s.t. $p_m \neq \bar{p}$, $p_t = \bar{p}$
for all $t \neq m$ then, $h_m = (\bar{p}\cdot w_m / \bar{p}\cdot x_m) x_m$
$h_t = (1/\!\!/ \, T-1)(w-h_m) \quad t \neq m$

<u>Definition</u>: An allocation $(\tilde{y}_t)_{t\in T}$ and a price p are a <u>constrained Walrasian equilibrium</u> if

i) $\forall t$, $p\cdot\tilde{y}_t = p\cdot w_t$
ii) $\forall t$, $\tilde{y}_t \gtrsim_t y \; \forall \; y \leq w$ such that $p\cdot y \leq p\cdot w_t$
iii) $\Sigma_{t\in T} \tilde{y}_t = w$

<u>Theorem</u>: (Hurwicz, Maskin and Postlewaite): The set of Nash equilibrium allocations (N.E.) for h coincides with the set of constrained Walrasian equilibrium allocations (C.W.E.).

REFERENCES:

[A.1] Arrow, K., Social Choice and Individual Values, Yale University Press, New Haven, Connecticutt, 1951.

[D.1] Debréu, G., <u>Theory of Value</u>. New York: John Wiley, 1959, Chapter 7.

[D.2] Debreu, G., "Four Aspects of General Equilibrium Theory", Proceedings of the International Congress of Mathematicians, Vancouver, 1974.

[Fa] Farquharson, R., <u>Theory of Voting</u>, Yale University Press, New Haven, 1969.

[G] Gibbard, A., "Manipulation of Voting Schemes: A General Result," <u>Econometrica 41</u> (1973) 587-602

[Gl] Groves, T. and J. Ledyard, "Optimal Allocation of Public Goods: A Solution to "Free Rider" Problem," <u>Econometrica</u> 45 (1977), 783-810.

[H.1] Hurwicz, L., "On the Interaction Between Information and Incentives in Organizations," <u>Communication and Control</u>, (1977), Ed., Klaus Klippendorf, Gordon and Breach, New York.

[H.2] Hurwicz, L., "On Informationally Decentralized Systems," Ch. 14, <u>Decisions and Organization</u>, C.B. McGurie and Roy Radner (Eds), Amsterdam (1972).

[H.3] Hurwicz, L., "On the Dimensional Requirements of Informationally Decentralized Pareto-Satisfactory Processes," (Manuscript), presented at the Conference Seminar on Decentralization, Northwestern University, Evanston, Illinois, February, 1972.

[H.4] Hurwicz,L., "Outcome Functions Yielding Walrasian and Lindahl Allocations at Nash Equilibrium Points," mimeo, November, 1976.

[H.5] Hurwicz, L., "On Allocations Attainable Through Nash Equilibria," mimeo, April 1977.

[HMP] Hurwicz, L., E. Maskin and A. Postlewaite, "Constrained Walrasian Equilibria,"

[HSC] Hurwicz, L. and D. Schmeidler, "Construction of Outcome Functions Guaranteeing Existence and Pareto Optimality of Nash Equilibria," mimeo, 1976.

[HS] Hurwicz, L. and L. Shapiro, "Incentive Structures Maximizing Residual Gain Under Incomplete Information," Discussion Paper No. 77-83, Center for Economic Research, University of Minnosota, Minneapolis.

[Ma] Maskin, E., "On Game Forms with Efficient Nash Equilibria," mimeo, Presented at Standford University, July 1977.

[MR] Mount, K. and S. Reiter, "The Informational Size of Message Spaces," Journal of Economic Theory 8 (1974), 161-192

[PA] Pattanaik, P., "Counter-Threats and Strategic Manipulation under Voting Schemes," Review of Economic Studies, 43 (1976), 11-18.

[PaS] Pazner, E. and D. Schmeidler, "Existence of Non-Walrasian Nash-Equilibria,"

[Po] Postlewaite, A. "Manipulation via Endowments," to appear in Review of Economic Studies.

[PoS] Postlewaite, A. and D. Schmeidler, "Approximate Efficiency of Non-Walrasian Nash Equilibria," to appear in Econometrica.

[Sa] Satterthwaite, M., "Strategy-Proofness andArrow's Conditions: Existence and Correspondence Theorems for Voting Procedures and Social Welfare Functions," Journal of Economic Theory, 10 (1975), 187-217

[Sc1] Schmeidler, D., "An Individually Rational Procedure for Planning the Provition of Public Goods, "Paper presented to NSD-SSRC conference on "Individual Rationality: Preference Revelation and Computation Cost in Models of Economic Equilibrium, "University of Massachusetts, Amherst, (July 1973), mimeo.

[Sc2] Schmeidler, D., "A Remark on Microeconomic Models of an Economy and a Game-Theoretic Interpretation of Walras Equilibria," mimeo, Minneapolis, March 1976. To appear in Econometrica.

[Se] Sengupta, M., "On a Difficulty in the Analysis of Strategic Voting," to appear in Econometrica.

[SS] Shapley, L. and M. Shubik, "Trade Using One Commodity as a Means of Payment," Discussion Paper R-1851-NSF, April 1976, Rand Corporation, Santa Monica, Ca.

[Sh] Shubik, M., "A Theory of Money and Financial Institutions: Commodity Money, Oligopoly Credit and Bankruptcy in a General Equilibrium Model," Western Economic Journal, 11 (1973), 24-38.

[W] Wilson, R., "A Competitive Model of Exchange," mimeo (1976).

GAME THEORY AND RELATED TOPICS
O. Moeschlin, D. Pallaschke (eds.)

SHAPLEY'S VALUE AND FAIR SOLUTIONS OF LOCATION CONFLICTS

W.F. Richter
Institute of Statistics
and Mathematical Economics
University of Karlsruhe, F.R.G.

This paper deals with fair solutions of conflicts which arise when locating some public project that everybody values positively. Such location conflicts deserve interest not only by their concrete meaning. In the figurative sense they play an important part in the theory of collective choice and public goods. They allow some representation which shows strong similarities to cooperative games.

Shapley's value is adapted to define a notion of fair locations. An existence theorem is supplied. These fair locations are then studied within a replica-model. They turn out to converge to the centre of gravity for most natural specializations.

INTRODUCTION

Consider n agents, $\Omega = \{1,..,n\}$, that take an interest in the location that is chosen for a public utility like a park. What is a fair or just location if the project is positively valued by everybody concerned? The question of financing is assumed to be settled in advance, thus not mattering in the sequel.

A. Ostmann (1978) will have been the first who most rigorously analyzed the distributive implications of locational choices. Whereas this paper concentrates on projects that everybody values positively Ostmann restricts attention to those projects like nuclear plants that everybody prefers to locate as far away as possible.

LOCATION CONFLICTS

We assume a *planning area* Q, a non-empty, convex, and closed subset of the *location space* $\mathbb{R}^m$, the Euclidean space of dimension m. Q defines the set of locations that are feasible for any coalition $S \subseteq \Omega$.

Individual $i \in \Omega$ is supposed to value alternative locations $x = (x_1,..,x_m) \in \mathbb{R}^m$ according to a convex and continuous *distance* or *(dis-)utility function* $d_i : \mathbb{R}^m \to \mathbb{R}$. The most natural example is certainly given by the Euclidean distance $d_i(x) = |x-z^i|$ with $x,z^i \in \mathbb{R}^m$. Individual $i \in \Omega$ is supposed to prefer x to y if $d_i(x) < d_i(y)$ which may explain why we call d_i disutility function.

We assume that $\{x \in Q \mid d_i(x) \leq u\}$ is bounded in $\mathbb{R}^m$ for all $u \in \mathbb{R}$, $i \in \Omega$.

$Y^i := \{y \in Q \mid d_i(y) = \min_{x \in Q} d_i(x)\}$ is called i^{th} *area of satiation*. The conditions imply $Y^i \neq \emptyset$. To rule out trivial conflicts we finally require $\bigcap_{i \in \Omega} Y^i = \emptyset$.

If all noted conditions are met we call $\Sigma = (Q,d)$ with $d = (d_1,..,d_n)$ *(n-person-) location conflict*. The set of all such Σ is denoted by $\not{\Sigma}$. Σ is said to be *Euclidean* if $d_i(x) = |x-z^i|$ $(z^i \in \mathbb{R}^m,\ i \in \Omega)$ and *strictly convex* if d_i is strictly convex for all $i \in \Omega$. $A \subseteq \mathbb{R}^n$ is called *convex from below* if for all u^0, $u^1 \in A$ some $v \in A$ exists s.t. $2v \leq u^0 + u^1$. We write $u \leq 0$ iff $u_i \leq 0$ $(i \in \Omega)$ and $u < 0$ iff $u \leq 0$, $u \neq 0$. Thus $u \in \mathbb{R}^n_+$ means $u \geq 0$.

REPRESENTATION BY CORRESPONDENCES

We shall represent location conflicts by correspondences defining sets of feasible payoff vectors. Let $\underline{\underline{P}}(\Omega)$ denote the power set of Ω. If $V : \underline{\underline{P}}(\Omega) \to \mathbb{R}^n$ defines a correspondence we implicitely assume $\emptyset \neq V(S) \subseteq \mathbb{R}^n_S := \{u \in \mathbb{R}^n \mid u_i = 0 \text{ for } i \notin S\}$

Definition: $V : \underline{\underline{P}}(\Omega) \to \mathbb{R}^n$ is called *(n-person-) location correspondence*, if $0 \notin V(\Omega)$ and $V(S) \subseteq \mathbb{R}^n_+$ is convex from below $(S \subseteq \Omega)$. Let $\mathbb{V}$ be the set of all such V.

The definition implies $V(\emptyset) = \{0\}$ for $V \in \mathbb{V}$. The projection of $\mathbb{R}^n$ on $\mathbb{R}^n_S$ will be denoted by proj_S.

There is a natural way of embedding $\not{\Sigma}$ into $\mathbb{V}$ according to

$$\Sigma \mapsto V_\Sigma\ ,\ V_\Sigma(\Omega) := \{d(x)-\underline{u} \mid x \in Q\}\ ,\ V_\Sigma(S) := \text{proj}_S V_\Sigma(\Omega)\ ,$$

$\{\underline{u}_i\} := d_i(Y^i)$. $\underline{u}$ is called *bliss point* and constitutes the analogon to the threat point of classical game theory. $0 \notin V_\Sigma(\Omega)$ follows from $\cap Y^i = \emptyset$. Let $\mathbb{V}_\Sigma \subseteq \mathbb{V}$ be the image of $\not{\Sigma}$.

Our notation is chosen such that the formal similarities between location correspondences and cooperative games (without side-payments) are stressed. This should not obscure the fact that there are important notional differences. Location conflicts naturally define some bliss but no threat point. Whereas the bliss point is not feasible for Ω the threat point is generally feasible for the grand coalition of cooperative games. Finally we speak of coalitions though cooperation makes no sense in its true game-theoretic meaning.

LOCATION CONCEPTS

We have to introduce some further symbols. Fix $V : \underline{\underline{P}}(\Omega) \to \mathbb{R}^n$, $S \subseteq \Omega$, and $\lambda \in \mathbb{R}^n_+$.

$$U(\lambda;S) := U^V(\lambda;S) := \{\bar{u} \in V(S) \mid \bar{\lambda}\cdot\bar{u} = \min \bar{\lambda}\cdot V(S)\}$$

$$U(\lambda;S) := \begin{cases} & \text{if } \operatorname{proj}_S(\lambda) \neq 0 \text{ or } S = \emptyset, \\ \bigcup_{\bar{\lambda}:\, \operatorname{proj}_S(\bar{\lambda})>0} U(\bar{\lambda};S) & \text{else;} \end{cases}$$

$$U(\lambda) := U(\lambda;\Omega),$$

$$P(S) := P^V(S) := \{\bar{u} \in V(S) \mid u_i < \bar{u}_i \ (i \in S) \text{ implies } u \notin V(S)\} .$$

$u \in P(S)$ is called *Pareto payoff for S*. Obviously $U(\lambda;S) \subseteq P(S)$ for $\lambda \geq 0$. The non-trivial equality stated by the following lemma is a consequence of a theorem by König (1978, p. 8).

<u>Lemma 1</u>: $P^V(S) = \bigcup_{\lambda \geq 0} U^V(\lambda;S)$ if $V(S)$ is convex from below.

From |5| we quote

<u>Lemma 2</u>: Let $V \in \mathbb{V}_\Sigma$, $P^V =: P$, $S \subseteq T$.

a) $P(S)$ is compact and non-empty;

b) $P(S) \subseteq \operatorname{proj}_S P(T)$;

c) $\operatorname{proj}_S P(T) \subseteq P(S) + \operatorname{proj}_S \mathbb{R}^n_+$.

A standard result which may be deduced from |1| (p. 30) is given by

<u>Lemma 3</u>: $V : \underline{\underline{P}}(\Omega) \to \mathbb{R}^n$. Let $P^V(S)$ be compact and non-empty. Then the same holds for $U^V(\lambda;S) = U(\lambda;S)$ $(\lambda \geq 0)$. $U(\cdot;S)$ is upper hemi-continuous and the function

$$v(\cdot;S) : \lambda \mapsto \inf \lambda \cdot V(S) = \inf \lambda \cdot P(S)$$

is continuous.

The determination of fair locations in Σ will be understood as the problem of choosing fair payoffs out of the feasibility set $V_\Sigma(\Omega)$. Which outcomes should be selected? A minimal requirement of collective choice is Pareto efficiency:

<u>Definition</u>: $\mathbb{V}_0 \subseteq \mathbb{V}$. A correspondence $\Psi : \mathbb{V}_0 \to \mathbb{R}^n$ is called *location concept for* $\mathbb{V}_0$ if $\Psi(V) \subseteq P^V(\Omega)$ $(V \in \mathbb{V}_0)$.

To demand Pareto efficiency, only, leaves a wide scope for defining Ψ. We are not going to discuss alternative axioms that a fair location concept might or should meet. An axiomatic characterization of location concepts for

$$\mathbb{V}_\Omega := \{V \in \mathbb{V} \mid V(S) = \{0\} \ (S \neq \Omega)\}$$

can be found in |5|. The main result is that under fairly reasonable assumptions any location concept for $\mathbb{V}_\Omega^c := \mathbb{V}_\Omega \cap \mathbb{V}^c$ with $\mathbb{V}^c := \{V \in \mathbb{V} \mid V(S)$ closed and convex$\}$ is induced by norm-minimization, i.e.

$$\forall\ \Psi\ \exists\ \|\cdot\| \text{ in } \mathbf{R}^n \text{ s.t. } \Psi(V) = \Psi^{\|\cdot\|}(V) \text{ for all } V \in \mathbb{V}_\Omega^c$$

$$\text{where } \bar{u} \in \Psi^{\|\cdot\|}(V) \text{ iff } \bar{u} = \min_{u \in V(\Omega)} \|u\| .$$

Natural interest deserves $\Psi^p := \Psi^{\|\cdot\|_p}$ where $\|\cdot\|_p$ denotes the p-norm of $\mathbf{R}^n$. A direct application of Ψ^p to $\mathcal{Z}$ is not possible because of $\mathbb{V}_\Sigma \not\subseteq \mathbb{V}_\Omega$ and $\mathbb{V}_\Sigma \not\subseteq \mathbb{V}^c$.

The first difficulty, $\mathbb{V}_\Sigma \not\subseteq \mathbb{V}_\Omega$, is easily overcome by applying the natural projection $j : \mathbb{V} \to \mathbb{V}_\Omega$ with $j(V)(\Omega) := V(\Omega)$ and $j(V)(S) := \{0\}$ for $S \neq \Omega$. Embedding $\mathcal{Z}$ into $\mathbb{V}_\Omega$ via j means to restrict attention to the grand coalition, alone.

As to $\mathbb{V}_\Sigma \not\subseteq \mathbb{V}^c$ suppose Ψ is location concept for $\mathbb{V}^c$. We extend the domaine of Ψ by resorting to an axiom of independence of irrelevant alternatives. Denote by $\overline{coA}$ the closed hull of coA, the convex hull of A. The notation $\overline{coV}$ implies pointwise application of $\overline{co\cdot}$ to $V \in \mathbb{V}$.

$$\mathbb{V}^\Psi := \{V \in \mathbb{V} \mid \Psi(\overline{coV}) \cap V(\Omega) \neq \emptyset\}$$

$$\Psi(V) := \Psi(\overline{coV}) \cap V(\Omega) \quad (V \in \mathbb{V}^\Psi) .$$

Clearly, Ψ is location concept for $\mathbb{V}^\Psi$ if it is for $\mathbb{V}^c$. Because of $j(\mathbb{V}_\Sigma) \subseteq \mathbb{V}^{\Psi^p}$ (cf. |5|) we may put

$$\Psi^p : \mathbb{V}_\Sigma \to \mathbf{R}^n , \quad V_\Sigma \mapsto \Psi^p \cdot j(V_\Sigma) .$$

Ψ^p allows for nice interpretations if $p = 1,2,\infty$ and if restricted to Euclidean $\Sigma \in \mathcal{Z}$. The corresponding solutions in the planning space $\mathbf{R}^m$ then turn out to generalize

Fermat points for $p = 1$,
the centre of gravity for $p = 2$,
the circumcentre for $p = \infty$.

Fermat points minimize $\Sigma|x-z^i|$ for $x \in \mathbf{R}^m$.

Whenever $V(\Omega)$ is convex from below we obtain $u \in \Psi^2(V)$ iff $\lambda \in \mathbf{R}_+^n$ exists s.t.

$$u \in U(\lambda) \quad \text{and} \quad u = \lambda. \tag{1}$$

This formula (1) reminds of Nash's bargaining solution which can be characterized by $u_i\lambda_i = \text{const}$ $(i \in \Omega)$.

THE SHAPLEY* LOCATION CONCEPT

The main purpose of this paper is to define location concepts Ψ that explicitely take into regard the well-being of coalitions S different from Ω. Formally we would like to extend the domaine of Ψ beyond the scope of $\mathbb{V}_\Omega$. The example we follow is given by Shapley's value for cooperative games without side-payments.

Given $V \in \mathbb{V}$ and some weights vector $\lambda \in \mathbb{R}^n_+$ the "value of the coalition S" is defined as

$$v(\lambda;S) \;:=\; \inf \lambda \cdot V(S) \;.$$

Because of $V(\emptyset) = \{0\}$ $v(\lambda;\emptyset) = 0$. Hence $v(\lambda;\cdot)$ defines a cooperative game in characteristic function form. For such $v(\lambda;\cdot)$ we formally calculate Shapley's value:

$$\phi_\bullet(\lambda) \;=\; \phi^V_\bullet(\lambda) : \Omega \to \mathbb{R} \;,$$

$$\phi_i(\lambda) \;:=\; \frac{1}{n!} \sum_{\pi \in \Pi} \left(v(\lambda;S^\pi_{\pi(i)}) - v(\lambda;S^\pi_{\pi(i)}\backslash i)\right)$$

with Π := {permutations of Ω}, $S\backslash i \;:=\; S\backslash\{i\}$, and

$$S^\pi_{\pi(i)} \;:=\; \{j \in \Omega \mid \pi(j) \leq \pi(i)\} \;.$$

Remark 1: Let $\lambda \in \mathbb{R}^n_+$, $V \in \mathbb{V}$.

a) $\sum_{i \in \Omega} \phi_i(\lambda) \;=\; v(\lambda;\Omega)$;

b) $V \in \mathbb{V}_\Sigma$, $u(\lambda;S) \in U(\lambda;S)$, $i \in S \subseteq \Omega$ imply

$$v(\lambda;S) \;\geq\; v(\lambda;S\backslash i) + \lambda_i u_i(\lambda;S) \;.$$

a) follows by definition and b) by referring to lemma 2a) and c).

The way Shapley defines his value for games without side-payments suggests to call $u \in V(\Omega)$ Shapley payoff if some $\lambda \in \mathbb{R}^n_+$ exists such that

$$\lambda_i u_i \;=\; \phi_i(\lambda) \qquad (i \in \Omega) \;. \tag{2}$$

Assume $\mathbb{V}_0 \subseteq \mathbb{V}$ is such that for all $V \in \mathbb{V}_0$ there exists a solution of (2). Taking these solutions u depending on V we can define a Shapley location concept $\Phi : \mathbb{V}_0 \to \mathbb{R}^n$. Unfortunately Φ performs quite unsatisfactorily as has been shown in |5|. There are good reasons to reject such Shapley location concept. Instead, we use

$$\lambda_i^2 \;=\; \phi_i(\lambda) \qquad (i \in \Omega) \tag{3}$$

as a defining relation for a new location concept:

Fix $\mathbb{V}^* \subseteq \mathbb{V}$ such that for all $V \in \mathbb{V}^*$

$$\Phi^*(V) := \bigcup_{\lambda \in \Lambda^V} U(\lambda) \quad \text{with}$$

$$\Lambda^V := \{\lambda \in \mathbb{R}^n_+ \mid \lambda_i^2 = \phi_i(\lambda) > 0 \quad (i \in \Omega)\}$$

is non-empty.

Definition: $\Phi^* : \mathbb{V}^* \to \mathbb{R}^n$ is called *Shapley* location concept for* $\mathbb{V}^*$.

We say that the Shapley* payoff $\bar{u}$ *is supported by* $\bar{\lambda}$ if $\bar{u} \in U(\bar{\lambda})$, $\bar{\lambda} >> 0$ and (3) applies for $\bar{\lambda}$. The condition $\lambda >> 0$ (i.e. $\lambda_i > 0 \ \forall i$) rules out trivial solutions.

To give a rough idea of (3) consider first the definition of Ψ^2. This location concept is characterized by (1), namely the condition $u = \lambda$. Thus welfare weights λ are determined by quantities u which in the context of location conflicts $\Sigma \in \mathbb{Z}$ represent distances, objectively measurable quantities. $\lambda_i^{-1} \phi_i(\lambda)$ may be interpreted as a socially adjusted distance. The Shapley* location concept makes $\lambda_i^{-1} \phi_i(\lambda)$ and not the objective distance u_i to determine the welfare weight λ_i.

Note that $V \in \mathbb{V}_\Omega$ implies $\phi_i(\lambda) = \frac{1}{n} v(\lambda;\Omega)$ and hence a solution $\bar{\lambda}$ of (3) holds $\bar{\lambda}_i = \text{const}$ $(i \in \Omega)$. We thus obtain

Remark 2: $\Phi^* = \Psi^1$ if restricted to $\mathbb{V}_\Omega \cap \mathbb{V}^*$.

Theorem 1: $\mathbb{V}_\Sigma \subseteq \mathbb{V}^*$.

Proof: Fix some $V \in \mathbb{V}_\Sigma$. We shall prove that $\Lambda^V \neq \emptyset$. $\Phi^*(V) \neq \emptyset$ then follows by lemmata 2 and 3.

For $r \in \mathbb{N}$ put

$$\Lambda^r := \{\lambda \in \mathbb{R}^n \mid r^{-1} \leq \lambda_i \leq r, \ i \in \Omega\}$$

$$f_i^r(\lambda) := \min(r, \max(r^{-1}, \lambda_i^{-1} \phi_i(\lambda))).$$

$f^r : \Lambda^r \to \Lambda^r$ then is continuous. By Brower's fixed-point-theorem $\lambda^r \in \Lambda^r$ exists with $f^r(\lambda^r) = \lambda^r$. We show that for sufficiently large $r \in \mathbb{N}$ λ^r lies in the interior of Λ^r. Hence $\lambda^r \in \Lambda^V$.

Case 1: Suppose we have for some subsequence $\mathbb{N}^0 \subseteq \mathbb{N}$ and some $j \in \Omega$ $\lambda_j^r = r \to \infty$ $(r \in \mathbb{N}^0)$. Then $r^2 \leq \phi_j(\lambda^r)$. As $P^V(S)$ is compact we can find $c \geq 0$ such that $u_i \leq c$ for all $u \in U(\lambda;S)$, $S \subseteq \Omega$, $\lambda > 0$, $i \in \Omega$. A contradiction then follows from:

$$r^2 \leq \phi_j(\lambda^r) \leq \frac{1}{n!} \sum_{\pi \in \Pi} v(\lambda^r; S^\pi_{\pi(j)}) \leq \frac{1}{n!} \sum_{\pi \in \Pi} c \sum_{i \in S^\pi_{\pi(j)}} \lambda_i^r$$

$$\leq \text{const } r.$$

Case 2: Suppose $\lambda_j^r = r^{-1} \to 0 \quad (j \in \Omega,\ r \in \mathbb{N}^0)$. Hence $\lambda_i^r \geq \lambda_j^r$ and

$$r^{-1} \geq (\lambda_j^r)^{-1}\, \phi_j(\lambda^r) \to 0 \quad (r \in \mathbb{N}^0) \ . \tag{4}$$

As $0 \in V(\emptyset)\backslash V(\Omega)$ we may fix $\bar{S} \subseteq \Omega$ with $0 \in V(\bar{S}\backslash j)\ V(\bar{S})$. Put $\mu^r := (\sum_{i\in\Omega} \lambda_i^r)^{-1} \lambda^r$. For $r \in \mathbb{N}^1 \subseteq \mathbb{N}^0$ we have $\mu^r \to \bar{\mu} > 0$, $\mu_i^r \geq \mu_j^r$ $(i \in \Omega)$

$$(\lambda_j^r)^{-1}\, \phi_j(\lambda^r) = (\mu_j^r)^{-1}\, \phi_j(\mu^r) \geq (\mu_j^r)^{-1} \frac{1}{n!} (v(\mu^r;\bar{S}) - v(\mu^r;\bar{S}\backslash j))$$

$$= (\mu_j^r)^{-1} \frac{1}{n!}\, v(\mu^r;\bar{S})$$

where the inequality follows from remark 1b) and the last equality from $0 \in V(\bar{S}\backslash j)$. Let $u(\cdot;\bar{S})$ denote a selection of $U(\cdot;\bar{S})$. As $U(\cdot;\bar{S})$ is compact-valued and upper-hemi continuous $\mathbb{N}^2 \subseteq \mathbb{N}^1$ exists such that

$$u^r := u(\mu^r;\bar{S}) \to \bar{u} \in U(\bar{\mu};\bar{S}) \ , \neq 0 \quad (r \in \mathbb{N}^2) \ .$$

A contradiction to (4) then follows by considering

$$(\lambda_j^r)^{-1}\, \phi_j(\lambda^r) \geq \frac{1}{n!} \sum_{i\in\bar{S}} \frac{\mu_i^r}{\mu_i^r}\, u_i^r \geq \frac{1}{n!} \sum_{i\in\bar{S}} u_i^r$$

$$\to \frac{1}{n!} \sum_{i\in\bar{S}} \bar{u}_i > 0 \ .$$

q.e.d.

ASYMPTOTIC BEHAVIOUR

From the theory of cooperative games and markets the following scheme is well-known: (Cf. |4|)

$$\text{Nash} \xrightarrow[\text{extension}]{} \text{Shapley} \xrightarrow[\text{replication}]{} \text{Walras} \ .$$

In the remaining part of this paper we show that in the same sense the sequence

$$\psi^1 \xrightarrow[\text{extension}]{} \text{Shapley}^* \xrightarrow[\text{replication}]{} \psi^2$$

holds for a fairly general class of location conflicts.

The rth - replication $\Sigma^r = (Q^r,d^r)$ of $\Sigma^1 = \Sigma = (Q,d)$ is defined as follows:

$$\Omega^r := \Omega \times \{1,..,r\} \ni i\rho \ , \quad Q^r := Q \ ,$$

$$d_{i\rho} = d_{i\rho}^r := d_i \quad (i \in \Omega,\ 1 \leq \rho \leq r) \ .$$

We shall study $\Phi^*(V_{\Sigma^r})$ for $r \to \infty$. Because of inherent symmetry we prefer an adapted notation:

Fix $r \in \mathbb{N}$. $\mathbb{N}_0 := \mathbb{N} \cup \{0\}$. $\text{prof}(S) := s \in \mathbb{N}_0^n$ $(S \subseteq \Omega^r)$ with $s_i := |\{j\rho \in S \mid j=i\}|$.

For $\lambda \in \mathbb{R}^n_+$, $\text{prof}(S) = s$, $\lambda\times s := (\lambda_i s_i)_{i\in\Omega} \in \mathbb{R}^n$ let

$$\tilde{v}(\lambda;S) = \tilde{v}^r(\lambda;S) = \tilde{v}(\lambda;s) := \inf_{x\in Q} (\lambda\times s)\cdot(d(x) - \underline{u}) ;$$

$$\underline{u}_i := \min d_i(Q) ;$$

$$\Pi = \Pi^r = \{\text{permutations } \pi \text{ of } \Omega^r\} ;$$

$S^{\pi}_{\pi(i\rho)} := \{j\rho' \in \Omega^r \mid \pi(j\rho') \leq \pi(i\rho)\}$ where $\leq$ denotes some total ordering of Ω^r ;

$$\tilde{\phi}^r_i(\lambda) = \tilde{\phi}^r_{i\rho}(\lambda) := \frac{1}{(rn)!} \sum_{\pi\in\Pi} (\tilde{v}(\lambda;S^{\pi}_{\pi(i\rho)}) - \tilde{v}(\lambda;S^{\pi}_{\pi(i\rho)}\setminus i\rho)) ;$$

$\tilde{U}(\lambda;s) = \tilde{U}^r(\lambda;s) \subseteq \mathbb{R}^n$ where $\tilde{U}(\lambda;0) := \{0\}$,

$\tilde{U}(\lambda;s) := \{\bar{u} = d(x)-\underline{u} \in \mathbb{R}^n_+ \mid x \in Q,\ (\lambda\times s)\cdot\bar{u} = \tilde{v}(\lambda;s)\}$ if $\lambda\times s \neq 0$,

$\tilde{U}(\lambda;s) := \bigcup_{\substack{\bar{\lambda}>0,\\ \bar{\lambda}\times s\neq 0}} \tilde{U}(\bar{\lambda};s)$ if $\lambda\times s = 0,\ s \neq 0$,

$\tilde{U}(\lambda) := \tilde{U}(\lambda;e)$ where $e = (1,..,1) \in \mathbb{R}^n$.

Note that for $0 < t \in \mathbb{R}$ $\tilde{U}(\lambda;s) = \tilde{U}(\lambda;ts) = \tilde{U}(t\lambda;s)$.

Definition: $\Sigma \in \mathcal{Z}$. Let $\Phi^*(V_{\Sigma^r}) := \Phi^{*,\text{sym}}(V_{\Sigma^r}) \subseteq \mathbb{R}^n_+$ be the set of all $u^r \in \tilde{U}(\lambda^r)$ $(\lambda^r \in \mathbb{R}^n_+)$ such that $(\lambda^r_i)^2 = \tilde{\phi}^r_i(\lambda^r) > 0$ for $i \in \Omega$. We then say that such λ^r *supports* u^r .

$$\Phi^{*a}(V_{\Sigma}) := \limsup_{r\to\infty} \Phi^*(V_{\Sigma^r})$$

$$= \{\bar{u} \in \mathbb{R}^n \mid \exists\, \mathbb{N}^0 \subseteq \mathbb{N},\ u^r \in \Phi^*(V_{\Sigma^r}) : u^r \xrightarrow[r\in\mathbb{N}^0]{} \bar{u}\}.$$

Φ^{*a} is called *asymptotic Shapley* location concept for* $\mathbb{V}_{\Sigma}$.

Note that Φ^{*a} is well-defined, which means that Φ^{*a} is location concept for $\mathbb{V}_{\Sigma}$: $\Phi^{*a}(V) \subseteq P^V(\Omega)$ for $V \in \mathbb{V}_{\Sigma}$ follows by compactness of $P^V(\Omega)$ (lemma 2); Non-emptiness of $\Phi^{*a}(V_{\Sigma})$ is a consequence of $\Phi^*(V_{\Sigma^r}) \neq \emptyset$. The latter can be shown to hold just in the same way as theorem 1 has been proved.

We shall now compute Φ^{*a} for selected $\Sigma \in \mathcal{Z}$. For that purpose we need some further symbols: Fix $r \in \mathbb{N}$, $i \in \Omega$, $\sigma \in \{0,..,rn-1\}$ and put

$$A^{\sigma} = A^{\sigma,r,i} := \{s \in \mathbb{N}^n_0 \mid \sum_{j\in\Omega} s_j = \sigma ,\ 0 \leq s_j \leq r-\delta_{ij}\ (j \in \Omega)\} ,$$

δ_{ij} : Kronecker symbol,

$$P^{\sigma}(s) = P^{\sigma,r,i}(\{s\}) := \binom{rn-1}{\sigma} \prod_{j\in\Omega} \binom{r-\delta_{ij}}{s_j} .$$

$P^{\sigma}(\cdot)$ defines a probability measure on $(A^{\sigma}, \underline{P}(A^{\sigma}))$. From |6|, p.87, we may quote

Lemma 4: For all $\delta \in (0,1)$ there is some σ_0 with $\sigma_0^{-1} < \delta$ such that $rn > \sigma \geq \sigma_0$ then always implies

$$P^{\sigma}(\{\|\frac{s}{\sigma} - \frac{e}{n}\|_{\infty} > \delta\}) < \delta.$$

Let $e^i \in \mathbb{R}^n$ be the ith unit vector. By direct computation which makes use of remark 1b) we obtain

Remark 3: $\tilde{\phi}_i^r(\lambda) \geq \frac{\lambda_i}{rn} \sum_{\sigma=o}^{rn-1} \sum_{s \in A^{\sigma}} P^{\sigma}(s)\, u_i(\lambda;s+e^i)$ where $u(\lambda;\cdot)$ denotes some selection of $\tilde{U}(\lambda;\cdot)$.

$\tilde{U}(\cdot)$ is upper-hemi continuous for $\underset{\sim}{V} \in \mathbb{V}_{\Sigma}$. Let the mapping $C : \mathbb{R}_+^n \to \underline{P}(V(\Omega))$ denote the "continuous fraction" of $\tilde{U}(\cdot)$:

$$C(\lambda) := \{u \in \tilde{U}(\lambda) \mid \varepsilon>0 \quad \delta>0 \text{ s.t. } |\lambda'-\lambda| < \delta\ ,\ \lambda' \geq 0 \text{ implies the existence of } u' \in \tilde{U}(\lambda') \text{ with } |u'-u| < \varepsilon\}\ .$$

Theorem 2: Let $u^r \in \Phi^*(V_{\Sigma^r})$ be supported by λ^r, $\mu^r := \lambda^r/\|\lambda^r\|_1 \to \bar{\mu}$, $u^r \to \bar{u}$ $(r \in \mathbb{N}^1 \subseteq \mathbb{N})$ and $\bar{u} \in C(\bar{\mu})$. Then $\bar{u} \in \Psi^2(V_{\Sigma})$. Furthermore $\lambda^r \to \bar{\lambda}$ $(r \in \mathbb{N}^2 \subseteq \mathbb{N}^1)$ implies $\bar{u} = \bar{\lambda} \neq 0$.

Proof: Fix $i \in \Omega$, $\varepsilon > 0$. As $\bar{u} \in C(\bar{\mu})$ there exists $\delta > 0$ such that $\|\frac{s}{\sigma} - \frac{e}{n}\|_{\infty} < \delta$, $r \geq r_{\delta}$ implies the existence of $u(\mu^r;s+e^i) \in \tilde{U}(\mu^r;s+e^i)$ with $|u(\mu^r;s+e^i) - \bar{u}| < \varepsilon$. By remark 3 and lemma 4 we conclude:

$$\|\lambda^r\|_1^{-1}\, \tilde{\phi}_i^r(\lambda^r) = \tilde{\phi}_i^r(\mu^r) \geq \frac{\mu_i^r}{rn} \sum_{\sigma=0}^{rn-1} \sum_{\substack{s \in A^{\sigma} \\ \|\frac{s}{\sigma} - \frac{e}{n}\|_{\infty} \leq \delta}} P^{\sigma}(s)\, u_i(\mu^r;s+e^i)$$

$$\geq (\bar{u}_i - \varepsilon) \frac{\mu_i^r}{rn} \sum_{\sigma=\sigma_0}^{rn-1} \sum_{s \in A^{\sigma}} P^{\sigma}(s) \cdots$$

$$\geq (\bar{u}_i - \varepsilon)\,(1 - \delta)\, \frac{\mu_i^r}{rn} (rn - \sigma_0)\ .$$

The supporting property of λ^r implies

$$\lambda_i^r \geq (\bar{u}_i - \varepsilon)\,(1 - \delta)\,(1 - \frac{\sigma_0}{rn})\ .$$

Suppose $\lambda^r \to \bar{\lambda}$ $(r \in \mathbb{N}^2 \subseteq \mathbb{N})$. Then $\bar{\lambda} \geq \bar{u}$. Because of

$$\sum_{i \in \Omega} (\lambda_i^r)^2 = \sum_{i \in \Omega} \tilde{\phi}_i^r(\lambda^r) = \tilde{v}(\lambda^r;\Omega) = \lambda^r \cdot u^r \tag{5}$$

$\{\lambda^r \mid r \in \mathbb{N}\}$ is bounded such that a convergent subsequence exists. (5) also implies $\bar{\lambda} = \bar{u}$. As $0 \notin V_\Sigma(\Omega)$ $\bar{\lambda} \neq 0$. $\bar{u} \in \tilde{U}(\bar{\lambda}) = \tilde{U}(\bar{\mu})$. Hence

$|\bar{u}| = \min\{|u| \mid u \in V_\Sigma(\Omega)\}$ and $\bar{u} \in \Psi^2(V_\Sigma)$. q.e.d.

Corollary: $\Phi^{*a}(V_\Sigma) = \Psi^2(V_\Sigma)$ for strictly convex $\Sigma \in \mathcal{Z}$.

Proof: $\tilde{U}(\cdot)$ is point-valued and hence $C = \tilde{U}$. As $\Psi^2(V_\Sigma)$ defines a singleton $\bar{u} \in V_\Sigma(\Omega)$ the assertion follows by the above theorem.

Unfortunately Euclidean location conflicts are not strictly convex. Thus $C(\lambda) \neq \tilde{U}(\lambda)$ holds in general. Consider e.g. $\Sigma = (Q,d)$ with $m = 1$, $\Omega = \{1,2\}$, $d_i(x) = |x-z^i|$. One easily computes $\Phi^{*a}(V_\Sigma) = P^{V_\Sigma}(\Omega)$.

Close inspection of theorem 2 reveals

Lemma 5: Let u^r be supported by λ^r, $\lambda^r \to \bar{\lambda}$, $u^r \to \bar{u}$ and $\mu^r := \lambda^r / \|\lambda^r\|_1 \to \bar{\mu}$. If $\bar{\lambda}_j = 0$ for some $j \in \Omega$ then $\bar{u} \in \tilde{U}(\bar{\mu})$ exists with $\bar{u}_j = 0$.

Theorem 3: Let $\Sigma = (Q,d)$ be Euclidean, $n \geq 3$, $d_i(x) = |x-z^i|$, $z^i \in Q$ $(i \in \Omega)$. If $z^i \notin co\{z^j, z^k\}$ for all $i \neq j,k$ then

$$\Phi^{*a}(V_\Sigma) = \Psi^2(V_\Sigma) .$$

Proof: Assume that $u^r \in \Phi^*(V_{\Sigma^r})$ is supported by λ^r, $\lambda^r/\|\lambda^r\|_1 =: \mu^r \to \bar{\mu}$, $u^r \to \tilde{u}$ $(r \in \mathbb{N}^1)$. The assertion follows from theorem 2 if we can show that $\tilde{U}(\bar{\mu})$ is point-valued.

Suppose $\tilde{U}(\bar{\mu})$ is multi-valued, say

$$\bar{\mu}\cdot d(x^o) = \bar{\mu}\cdot d(x^1) = \min \bar{\mu}\cdot d(Q)$$

and $d_j(x^o) \neq d_j(x^1)$. Hence $\bar{\mu}_j = 0$ must hold. As $\{\lambda^r \mid r \in \mathbb{N}\}$ is bounded $\lambda^r \to \bar{\lambda}$ $(r \in \mathbb{N}^2 \subseteq \mathbb{N}^1)$ with $\bar{\lambda}_j = 0$. By the above lemma 5 $\bar{u} \in U(\bar{\mu})$ exists with $\bar{u}_j = 0$. Hence $\bar{u} = d(z^j)$. Let $\bar{\mu}_i > 0$. If $\bar{\mu}_k = 0$ for all $k \neq i$ then $z^i = z^j$ which contradicts the assumptions. Thus there are at least two indices $i \neq k$ with $\bar{\mu}_i, \bar{\mu}_k > 0$. But then $t \in \mathbb{R}$ exists such that $z^j = tz^i + (1-t)z^k$ which is again contradictory. q.e.d.

This paper is part of the author's thesis submitted for the certification of habilitation at Karlsruhe University. Results were presented at the European Meeting of the Econometric Society held at Geneva in September 1978 and at the seminar on location theory held at Bielefeld in October 1978.
I would like to thank A. Ostmann and J. Rosenmüller for valuable support and constructive criticism.

References

|1| W. Hildenbrand: Core and equilibria of a large economy (Princeton, 1974).

|2| H. König: Neue Methoden und Resultate aus Funktionalanalysis und konvexer Analysis, Operations Research Verfahren XXVIII (1978) 6-16.

|3| A. Mas-Colell: Competitive and value allocations of large exchange economies, J. Econ. Theory 14 (1977) 419-438.

|4| A. Ostmann: Fair play und Standortparadigma, Thesis (Universität Karlsruhe, 1978).

|5| W. Richter: A game-theoretic approach to location-allocation conflicts (forthcoming).

|6| J. Rosenmüller: Kooperative Spiele und Märkte (Berlin, 1971).

|7| L.S. Shapley: Utility comparison and the theory of games, La décision, CNRS (Paris, 1969).

LIST OF PARTICIPANTS

ALBERS, W., Prof.,	Institut f. Mathematische Wirtschaftsforschung, Universität Bielefeld, D-4800 Bielefeld, Fed.-Rep. of Germany.
ARMBRUSTER, W., Dr.,	Institut für Angewandte Mathematik, Universität Heidelberg, D-6900 Heidelberg, Fed.-Rep. of Germany.
BÖGE, W., Prof.,	Institut für Angewandte Mathematik, Universität Heidelberg, D-6900 Heidelberg, Fed.-Rep. of Germany.
BOL, G., PD Dr.,	Institut für Statistik und Mathematische Wirtschaftstheorie, Universität Karlsruhe, D-7500 Karlsruhe, Fed.-Rep. of Germany.
BRUYNEEL, G., Dr.,	Seminarie voor Wiskundige Beleidstechnieken en Beheersinformatica, Sint-Pietersnieuwstraat 41, Rijksuniversiteit Gent, B-9000 Gent, Belgium.
DIERKER, E., Prof.,	Institut für Gesellschafts- und Wirtschaftswissenschaften, Wirtschaftstheoretische Abteilung, Universität Bonn, D-5300 Bonn, Fed.-Rep. of Germany.
DOBERKAT, E.-E.,Dr.,	Fachbereich Mathematik, Fernuniversität Hagen, D-5800 Hagen, Fed.-Rep. of Germany.
DORE, M.H.I., Prof.,	Department of Economics and Political Science, University of Saskatchewan, Saskatoon, S7N OWO, Canada.
EBERL,W.,jun.,PD Dr.,	Abteilung für Statistik, Universität Dortmund, D-4600 Dortmund, Fed.-Rep. of Germany.
EGEA, M., Dr.,	Département de Mathématiques, Université Claude - Bernard, Lyon I, F-69621 Villeurbanne, France.
EICHHORN, W., Prof.,	Institut für Wirtschaftstheorie und Operations Research, Universität Karlsruhe, D-7500 Karlsruhe, Fed.-Rep. of Germany.
EISELE, Th., Dr.,	Institut für Angewandte Mathematik, Universität Heidelberg, D-6900 Heidelberg, Fed.-Rep. of Germany.
FAN, K., Prof.,	Department of Mathematics, University of California at Santa Barbara, Santa Barbara, CA. 93106, USA.
FENSKE, Chr., Prof.,	Mathematisches Institut, Universität Giessen, D-6300 Giessen, Fed.-Rep. of Germany.
FILUS, Lidia, Dr.,	Szkoła Główna Planowania i Statystyki, Zakład Matematyki, 02-521 Warsaw, Poland.
FUCHSSTEINER,B.,Prof.,	Fachbereich Mathematik, Gesamthochschule Paderborn, D-4790 Paderborn, Fed.-Rep. of Germany.

GREENBERG, J., Prof., Center for Study of Public Choice, Virginia Polytechnic Institute and State University, Blacksburg, VA. 24061, USA.

HENN, R., Prof., Institut für Statistik und Mathematische Wirtschaftstheorie, Universität Karlsruhe, D-7500 Karlsruhe, Fed.-Rep. of Germany.

HÖPFINGER, E.,PD Dr., Institut für Wirtschaftstheorie und Operations Research, Universität Karlsruhe, D-7500 Karlsruhe, Fed.-Rep. of Germany.

JACOBS, K., Prof., Mathematisches Institut, Universität Erlangen - Nürnberg, D-8520 Erlangen, Fed.-Rep. of Germany.

JAEGER, A., Prof., Institut für Unternehmensführung und Unternehmensforschung, Ruhr-Universität, D-4630 Bochum, Fed.-Rep. of Germany.

JANSEN, M.J.M.,Prof., Department of Mathematics, Catholic University Toernooiveld Nijmegen, The Netherlands.

KLEIN, Chr., Dr., Gesellschaft für Mathematik und Datenverarbeitung, Schloß Birlinghoven, D-5205 St. Augustin 1, Fed.-Rep. of Germany.

KRAUSE, U., Prof., Fachbereich Mathematik, Universität Bremen, D-2800 Bremen, Fed.-Rep. of Germany.

LORENTZ, R., Dr., Gesellschaft für Mathematik und Datenverarbeitung, Schloß Birlinghoven, D-5205 St. Augustin 1, Fed.-Rep. of Germany.

ŁOŚ, J., Prof., Institute of Computer Science, Polish Academy of Sciences, P.O. Box 22, 00-901 Warsaw, Poland.

ŁOŚ, Maria, Dr., Institute of Computer Science, Polish Academy of Sciences, P.O. Box 22, 00-901 Warsaw, Poland.

MASCHLER, M., Prof., Institute of Mathematics, The Hebrew University of Jerusalem, Jerusalem, Israel.

MEISTER, H.,Dipl.-Math., Fachbereich Mathematik, Fernuniversität Hagen, D-5800 Hagen, Fed.-Rep. of Germany.

MERTENS, J.F., Prof., Center for Operations Research & Econometrics, - CORE - Université Catholique de Louvain, B-1348 Louvain-La-Neuve, Belgium.

MOESCHLIN, O., Prof., Fachbereich Mathematik, Fernuniversität Hagen, D-5800 Hagen Fed.-Rep. of Germany.

MYERSON, R.B., Prof., Graduate School of Management, Nathaniel Leverone Hall, Northwestern University, Evanston, IL. 60201, USA.

NEYMAN, A., Prof., Department of Mathematics, University of California at Berkeley, Berkeley, CA. 94720, USA.

OSTMANN, A., Dr., Institut f. Mathematische Wirtschaftsforschung, Universität Bielefeld, D-4800 Bielefeld, Fed.-Rep. of Germany.

PALLASCHKE, D., Prof., Institut für Angewandte Mathematik, Universität Bonn, D-5300 Bonn, Fed.-Rep. of Germany.

PELEG, B., Prof., Institute of Mathematics, The Hebrew University of Jerusalem, Jerusalem, Israel.

PLACHKY, D., Prof., Institut für Mathematische Statistik, Universität Münster, D-4400 Münster, Fed.-Rep. of Germany.

PRZEWORSKA-ROLEWICZ, Danuta, Prof., Institute of Mathematics, Polish Academy of Sciences, 00-950 Warsaw, Poland.

PUPPE, D., Prof., Mathematisches Institut, Universität Heidelberg, D-6900 Heidelberg, Fed.-Rep. of Germany.

RICHTER, W., Dr., Institut für Statistik und Mathematische Wirtschaftstheorie, Universität Karlsruhe, D-7500 Karlsruhe, Fed.-Rep. of Germany.

RIEDER, U., PD Dr., Institut für Mathematische Statistik, Universität Karlsruhe D-7500 Karlsruhe, Fed.-Rep. of Germany.

ROBIN, M., Prof., Institut de Recherche d'Informatique et d'Automatique, - IRIA - F-78150 Le Chesnay, France.

ROLEWICZ, S., Prof., Institute of Mathematics, Polish Academy of Sciences, 00-950 Warsaw, Poland.

ROTHBLUM, U., Prof., Yale School of Organization and Management, New Haven, CT. 06520, USA.

SCHMEIDLER, D., Prof., Faculty of Social Sciences, Department of Economics, Tel-Aviv University, Ramat-Aviv, Tel-Aviv, Israel.

SCHUBERT, H., Prof., Mathematisches Institut, Universität Düsseldorf, D-4000 Düsseldorf, Fed.-Rep. of Germany.

SPERNER, E., Prof., Fachbereich Mathematik, Mathematisches Seminar, Universität Hamburg, D-2000 Hamburg 13, Fed.-Rep. of Germany.

TAUMAN, Y., Prof., Center for Operations Research & Econometrics, - CORE - Université Catholique de Louvain, B-1348 Louvain-La-Neuve, Belgium.

TROTHA, H.,von, Dr., Gesellschaft für Mathematik und Datenverarbeitung, Schloß Birlinghoven, D-5205 St. Augustin 1, Fed.-Rep. of Germany.

TYS, S.H., Prof., Department of Mathematics, Catholic University Toernooiveld Nijmegen, The Netherlands.

VAHRENKAMP, R.,PD Dr., Abteilung für Angewandte Systemanalyse, Kernforschungszentrum Karlsruhe, Universität Karlsruhe, D-7500 Karlsruhe, Fed.-Rep. of Germany.

VAN DER WAL, J.,Prof., Department of Mathematics, Eindhoven University of Technology, Eindhoven, The Netherlands.

WEBER, R.J., Prof.,	Department of Economics, Cowles Foundation for Research in Economics, Yale University, New Haven, CT. 06520, USA.
WESSELS, J., Prof.,	Department of Mathematics, Eindhoven University of Technology, Eindhoven, The Netherlands.
WIECZOREK, A., Prof.,	Institute of Computer Science, Polish Academy of Sciences, P.O. Box 22, 00-901 Warsaw, Poland.
WINKELS, H.M., Dr.,	Institut für Unternehmensführung und Unternehmensforschung, Ruhr-Univeristät, D-4630 Bochum, Fed.-Rep. of Germany.
YOUNG, H.P., Dr.,	International Institute for Applied Systems Analysis, - IIASA - Schloss Laxenburg, A-2361 Laxenburg, Austria.
ZAMIR, Sh., Prof.,	The Faculty of Social Sciences, Department of Statistics, The Hebrew University of Jerusalem, Jerusalem, Israel.

AUTHOR INDEX